Crusading and the
Crusader States

Recovering the Past

Series Editors: Edward Acton and Eric Evans

'Recovering the Past' aims to present both students and the general reader with authoritative interpretations of major historical topics. Drawing on the results of recent research by the author and others, the volumes both explain the importance of the themes they address and explore how historians have come to radically different conclusions about key issues in past politics, society and cultures. The lifeblood of history courses around controversy and reinterpretation; these accessibly written volumes help to explain why. Readers can use them for readable accounts and for the stimulus of engagement with lively debate.

Crusading and the Crusader States
Andrew Jotischky

Rome and Her Empire
David Shotter

The British Empire
Philippa Levine

The Age of the French Revolution
David Andress

The Third Reich
Dick Geary

The Rise and Fall of Communism
Stephen White

The Impact of the Industrial Revolution
Peter Gatrell

Imperial Russia
Roger Bartlett

Crusading and the Crusader States

Andrew Jotischky

Routledge
Taylor & Francis Group

LONDON AND NEW YORK

First published 2004 by Pearson Education Limited

Published 2013 by Routledge
2 Park Square, Milton Park, Abingdon, Oxon OX14 4RN
711 Third Avenue, New York, NY 10017, USA

Routledge is an imprint of the Taylor & Francis Group, an informa business

Copyright © 2004, Taylor & Francis.

The right of Andrew Jotischky to be identified as author of this work has been
asserted by him in accordance with the Copyright, Designs and Patents Act 1988.

ISBN 13: 978-0-582-41851-6 (pbk)

British Library Cataloguing in Publication Data
A CIP catalogue record for this book can be obtained from the British Library

Library of Congress Cataloging-in-Publication Data
Jotischky, Andrew, 1965–
    Crusading and the crusader states / Andrew Jotischky.— 1st ed.
        p. cm. — (Recovering the past)
    Includes bibliographical references and index.
    ISBN 0–582–41851–8 (alk. paper)
    1. Crusades.   2. Middle Ages—History.   3. Church history—Middle Ages,
600–1500.   I. Title.   II. Series.

D157.J68 2004
909.07—dc22

2004048428

Set by 35 in 10/13.5pt Sabon

Contents

List of maps and genealogical tables

Royal house of Jerusalem and Cyprus

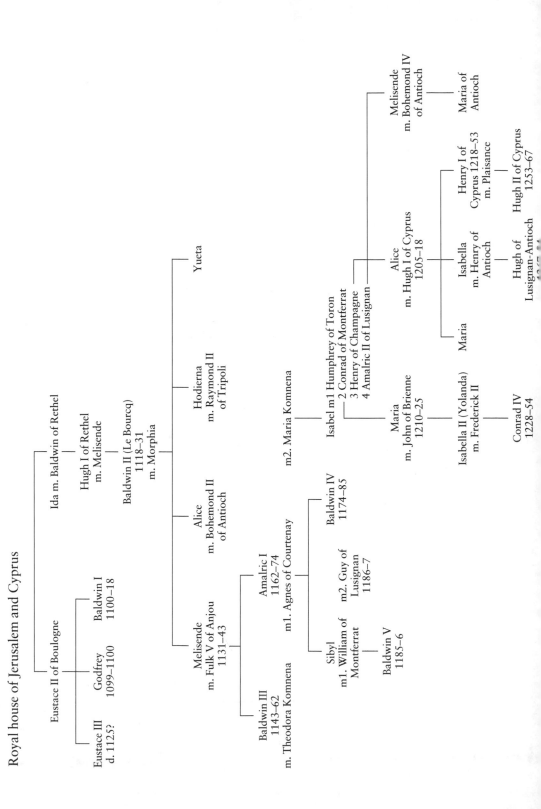

Chronology of main events

1169	Franco-Byzantine invasion of Egypt stalled by Shirkuh
1171	Saladin suppresses Fatimid caliphate
1174	Saladin usurps Nur ad-Din's heir in Damascus
	Accession of Baldwin IV in Jerusalem
1180	Sybil marries Guy of Lusignan
1186	Sybil and Guy mount *coup* for throne of Jerusalem
1187	July: Saladin defeats Franks at Hattin
	October: Jerusalem falls to Saladin
1189	Guy of Lusignan begins siege of Acre
1191	May: Richard I conquers Cyprus
	June: Acre falls to crusading army
	September: Richard I defeats Saladin at Arsuf
1192	Richard secures coastline, returns to West
1193	Death of Saladin
1202	Fourth Crusade sacks Zara
1203	Fourth Crusade installs Alexios IV on Byzantine throne
1204	Fourth Crusade sacks Constantinople
1208	Albigensian Crusade launched against Cathar heretics
1212	Castilian–Aragonese victory over Muslims at Las Navas de Tolosa
1215	Fourth Lateran Council
1217	Fifth Crusade launched
1219	Fifth Crusade captures Damietta
1221	Fifth Crusade defeated
1225	Emperor Frederick II marries heiress to Jerusalem, Yolanda
1229	Frederick II's peaceful recovery of Jerusalem
1239	Theobald of Champagne and Richard of Cornwall secure Jerusalem
1244	August: Jerusalem falls to Kharazmian Turks
	October: Franks defeated at La Forbie
1245	Pope Innocent IV launches crusade against Frederick II
1249	Louis IX of France captures Damietta
1250	Louis IX's army defeated at Mansurah; Louis captured
1258	Baghdad sacked by Mongols
	Abbasid khalifate ends
1260	Mamluks defeat Mongols at 'Ain Jalud
1265	Baibars captures Caesarea
1268	Antioch falls to Mamluks
1270	Louis IX dies at Tunis on crusade
1274	New crusading plans discussed at Second Council of Lyons

Preface

When I was approached a few years ago to write this book for the series 'Recovering the Past', my initial reaction was to wonder whether another general introductory book on the Crusades was really necessary. Crusading as a subject of study has always attracted a wide range of writing from the popular as well as the academic end of the spectrum. Scores of introductory books on the Crusades are already available, many of them reproducing the same information. Could anyone add anything to a story already so familiar? Could anyone retell the events as vividly as Runciman or Grousset, or survey the field with the magisterial gaze of Mayer or Richard? More worryingly, was it possible to write a single-volume study of such an expanding and complex field of scholarly study as the Crusades has now become without misleading readers by faults of omission or flaws in balance? These fears are still real enough, but I am grateful, for all that, to have been given the opportunity by Edward Acton and Eric Evans, the **Series** Editors, to try my hand at this book. Writing it, though every bit as difficult as I had feared, has been more fun than I had expected. It has also been more helpful than I could have imagined for my own understanding of the Crusades to have had to explain them within the parameters of a book designed for readers new to the subject. These, however, are selfish reasons for writing a book, and I hope that the tale told here will also prove to be of some value to others, for the following reasons. First, because no matter how familiar the tale, if a series of historical events is important enough in the development of human societies, it can stand any number of retellings. The essence of historical writing is that it is, after all, a personal reflection on the meaning of events and evidence, and for this reason no two historians will ever write quite the same book on the Crusades. Second, my purpose is not to write a narrative history of the Crusades, but to provide an analysis of the overall picture that is at the same time concise and as comprehensive as possible. For this reason, readers who are expecting detailed reconstructions of battles, political intrigues or stirring acts of heroism may be disappointed. I hope they will feel recompensed, however, by the relation of individual events and personalities to the broader currents within the history of crusading. I have tried to put myself in the position of an intelligent reader who comes unprepared to the subject and who would like to understand, as succinctly as possible, how crusading as a practice and theory evolved; how crusades were planned and executed; why they were considered to be so essential a part of medieval society; how crusading produced a distinctive western society in the Near East and how that society sustained itself; and, not least, why crusading continued to exercise an appeal to European society despite repeated military failures. A third reason for attempting this book is that the subject has never been so crucial to our understanding of the future, as well as the past, of western society. When I first started to study the Crusades as an undergraduate, more than twenty years ago, it was regarded by my contemporaries as a signally antiquarian pursuit. Nobody I have spoken to in the last few years shares this belief. The idea religiously justified violence, so far from being, as Gibbon assumed, consigned to a pre-Enlightenment past, is a fact of contemporary, no less than it was of medieval, society. An earlier generation of crusader historians could look back at medieval holy war, both Christian and Islamic, with the distant gaze of rational scepticism. At the time of writing, conflicts justified, if not explained by religious differences are taking place on almost every continent. Many of those engaged in such conflicts use the language of holy war and crusading, whether accurately or not, to appeal to a moral justification for violence. In perception if not in historical fact, there is a direct correlation between the Crusades and current conflicts, especially but by no means exclusively in the regions in which crusading ideals were first applied. Solutions to such conflicts can only come from understanding both of the historical realities and of the emotive responses generated by partial knowledge of those realities.

The help and encouragement of several people have been invaluable in writing this book. I would like to thank Eric Evans for suggesting it to me, my editor, Heather McCallum, for friendly but firm encouragement, and Julie Knight for her efficiency in seeing it through the Press. I would also like to thank Jonathan Phillips for suggestions and improvements to the text. Thanks are also due to my colleagues in the History Department at Lancaster University, for reading drafts and for lightening administrative burdens during the final stages of the book. Support in other ways has come from my family: from my parents, from my wife Caroline, and from my children, who have become alarmingly familiar with Tancreds, Baldwins and Innocents. Finally, I would like to thank all the students in my Crusades courses at Manchester, Oxford and Lancaster since 1992, from whom I have learnt so much.

Andrew Jotischky
July 2004

Publisher acknowledgements

The publishers are grateful to the following for their permission to reproduce copyright material:

Plate 13 © CORBIS; Plate 14 © Michael Nicholson/CORBIS; Plate 15 and 16 Akg-images/British Library; Plate 17 © A. Griffiths Belt/CORBIS

CHAPTER 1

Problems in crusading historiography

Crusading as a subject has expanded in recent years to include new fields of enquiry. In this chapter we will examine how crusading historiography has changed to include new geographical areas and new definitions, and focus on two fundamental issues in current writing: why people went on crusades, and what forms the western settlement in the Near East took.

Crusading is a perennially attractive subject for students of History. The difficulty posed by understanding the mindset of medieval popes, bishops and knights is tempered by the vivid colour of many of the surviving sources, and the illusion of closeness in time lent by the relevance of much of the subject matter to the contemporary world. Since a large number of general studies of the Crusades already exist, it may be worth laying before the reader what this book sets out to do, and what it does not propose to attempt. First, this is a book designed with undergraduate historians in mind. I have therefore tried to approach the subject from the point of view of the student by posing – and trying as far as possible to answer – the kinds of questions that my own students over the past twelve years of teaching the subject at university have themselves asked and debated in seminars. This means that I have given priority to analysis and interpretation over the narrative of events. Some degree of narrative is of course inevitable, if only to provide a context for analysis, but the story has been told so many times and in so many books that it has not seemed worthwhile to tell it again. There are, therefore, no set-piece descriptions of individual crusades, let alone battles. The best scholarly surveys of the Crusades – those by H.E. Mayer (1972, new edn 1988) and Jean Richard (1999) – are still invaluable for the detail that they provide to superimpose on the sketch in the following pages. Despite this set of priorities, however,

I hope that this book will also appeal to the general reader, who perhaps comes to the Crusades already knowing something of the story and its chief personalities, but would like an accessible yet sharply focused discussion of how the Crusades came about, what they attempted to achieve, and how they shaped the history of the East Mediterranean in the Middle Ages.

The study of the Crusades has expanded so rapidly in recent years that the parameters of the subject are much wider than they were a generation ago. This has made for some difficult choices in judging the scope of this book. One consequence of the growth of interest in the Crusades among historians has been that they are now almost open-ended. A convenient, if in some respects misleading, starting date is provided by the appeal that launched the First Crusade in 1095. For closure, however, historians must follow their own preferences. In doing so, they expose one of the fault-lines in recent crusading historiography. In determining when crusading ended, one defines what a crusade was, and what must remain outside that definition.

Although the fall of the last western possession on the mainland of the Crusader States in 1291 is still accepted by many as a sensible end-point (e.g. Richard 1999), this is in many ways an artifice. Crusades continued to be launched along exactly the same lines of organisation and with the same aspirations for another fifty years or so after 1291, and even after the emphasis shifted from the recovery of the Holy Land to the more generalised struggle against the advancing Ottomans, crusading ideals as promoted and expressed by the European nobility continued to be recognisable as part of the same tradition that had inspired the generation of 1095. Moreover, the last of the Crusader States, the kingdom of Cyprus, did not fall until the second half of the sixteenth century, and one can go further still: even the seventeenth-century wars against the Turks have been included in crusading history. My ending point in this book is the 1330s, the decade that saw the last large-scale expedition launched from the West with the specific single aim of recovering the Holy Land. I recognise, however, that there is much to be said about crusading and crusade mentalités in the later Middle Ages that for reasons of length can find no place in this book.

Geographical parameters are also more problematic than they might at first appear. Everyone knows that the crusades were wars of the Cross against the Crescent, but only comparatively recently have historians from outside Spain included the Iberian peninsula in their horizons when considering crusading. Equally, the crusades against pagans in the Baltic region have largely been understood within the context of the western colonisation of north-eastern Europe rather than of crusading. There are good reasons for emphasising the particularities of the northern crusades and the Spanish *Reconquista*, for in both cases these were wars whose inception cannot be explained solely by the development of crusading ideals. Some historians are still reluctant to see these wars as occupying the same place in the mindset of participants or contemporary commentators. Nevertheless, I have included discussion of both in this book because the course of warfare and Christian settlement in Iberia and the Baltic was shaped by largely the same aspirations and ideologies as in the East Mediterranean. For similar reasons I have also found space for brief accounts of crusades against heretics within Western Europe and against Christian enemies of the papacy. Beyond the discussion in the Introduction, however, I have not attempted to broach the question of what constituted a crusade.

The selection of what to include and what to omit largely represents trends in crusading historiography. One benefit of this book will, I hope, be to guide readers to the most recent treatments of different aspects of the subject. For this reason the references in the text are mostly to secondary literature that provides systematic discussion of topics in greater depth than this book can allow. Because I have in mind a readership whose proficiency is largely limited to English, I have restricted myself as far as possible to references to works in English. This may perhaps be the moment to stress that many substantial and important contributions to crusading scholarship have also been made by scholars writing in French, German, Italian, Spanish and other languages.

What does it mean to study the crusades?

There is scarcely any need in the present day to justify writing a book about the Crusades. A subject that a generation ago was seen by many professional historians as peripheral to their concerns with the political and social development of Europe has in the past fifteen years seen an explosion of interest and new research. In part, this is because the subject itself has changed; even more, perhaps, because the world has changed. The durability of conflicts inspired by religious ideology and fuelled by a growing sense of religious and racial hatred no longer appears foreign to present-day concerns or political discourse. In an age of increasing extremism in political and religious values – as much in the western world as in the regions traditionally associated with crusading – the actions of crusaders and their opponents can appear not as the exotic expressions of a more violent world in the distant past, but as a phase in a cycle of conflict. It would be understandable to see the crusades of the twelfth to the fourteenth centuries as echoes of present-day conflicts. As the fourteenth-century Arab historian Ibn Khaldun put it, 'the past resembles the present more than one drop of water resembles another'. Such a view, however, would also be misleading. The Crusades came about for specific reasons and in a specific set of circumstances. Although it may be true that the East Mediterranean still suffers from the legacy of the wars initiated by the western Church and knighthood in 1095, the circumstances in which those conflicts began reflect developments in European and West Asian societies during the eleventh century. There is no historical inevitability about war between Cross and Crescent, or between the values of Christian and Islamic cultures.

This may be one good reason for studying the Crusades today. Another is that crusading historiography has become increasingly sophisticated and wide-ranging. This means that the student who wants to know how medieval society worked on a number of different levels can do so through a study of crusading. Learning about the Crusades entails the study of the structures of western society as it developed from the eleventh to the fourteenth centuries. Crusading naturally evolved over this period. The expeditions of a thirteenth-century prince such as Louis IX of France or Edward I of England differed markedly from the First Crusade of 1096–9 in organisation, method, and to some extent even in aspiration. It could scarcely be otherwise, for every facet of crusade planning and execution was affected by advances in governmental procedure, in the increasingly complex organisation of the Church, in military technology and in changing ideologies that governed Christian conceptions of Islam. But this was not one-way traffic. The continuing aspiration in the West to recover Jerusalem itself brought about many of the changes in medieval society that constitutional historians have hailed as advances. An example is the introduction of the first income tax in Europe in 1183 to finance crusading. In ecclesiastical structures too, aspirations concerned with the Holy Land were pervasive. The wide-ranging reform of the Church envisaged by the Fourth Lateran Council of 1215 was articulated by the papacy by the need to make crusading more efficient; thus, fundamental changes to the ways in which bishops ran their dioceses all over Christendom were brought about in order to provide more systematic preaching of the Cross.

Crusading, then, teaches us about how and why medieval society evolved in the ways that it did. But crusading was above all a religious activity, governed by the expectation of a spiritual reward. This expectation, loosely articulated by Pope Urban II in 1095, was by c.1200 to be defined as a legalistic exchange between the participant and the Church, as the mediator of God's will, and articulated in a legally binding vow. The circumstances and experiences of crusading itself brought about this change. Crusaders, their dependants and those to whom they owed services, wanted to know what they could expect in the event of death, injury or the loss of property as a result of crusading. This is not to say that the Crusades by themselves brought about the growth of a more systematic theology of penance or a more complex canon law, but rather that crusading was one of the experiences that made possible such evolutions in the workings of the Church.

Historians of the Crusades have always been concerned with the apparent paradox of holy war. To understand what Urban II meant in 1095 when he preached an armed pilgrimage for the recovery of the Holy Sepulchre in Jerusalem, we must approach the Crusade from long range, by tracing the evolution of a theory of penitential violence through the social and political turbulence accompanying the papal reform movement of the eleventh century. Urban II's message would have been impossible without the development of a papal office with a clearly defined ideology. This is an eleventh-century story. Equally, however, Urban II's preaching only took on significance in the ears of his listeners, and in the ways in which they retold what they had heard. This means that we must understand what it was about Urban's appeal that was so resonant to late eleventh-century society. Crusading, therefore, takes us backwards into the workings of knightly families: how they learned and expressed their piety, how ideals of devotion and social responsibility became fused with the habits of warfare.

The western settlement in the Levant came about because of a particular combination of circumstances in the East as well as in the West. One of the most significant developments in crusading historiography in recent years has been the progress made in understanding the physical and human environment created by the settlement. This has also extended conceptual boundaries. We know far more about the ways in which western settlers lived in their new lands than we did ten years ago, and this knowledge has helped both to complement and to challenge older assumptions about the mentalité of the western knighthood. Disciplinary boundaries have also been crossed, bringing into the study of crusading new genres and sources. One field in which this has been most notable is art history. Study of the visual culture of the Crusader States has posed old questions in new ways, by examining the cross-fertilisation of ideas and taste among artists and patrons in the East Mediterranean. This in turn has reinvigorated an old debate about the 'colonialism' of the western settlement. As yet, crusade historians have been wary of exploring the possibilities of postcolonial theory in order to reshape the debate, but recent studies of interaction of settlers with indigenous peoples have suggested new approaches.

Crusading studies today are genuinely interdisciplinary. The contributions of literary scholars, art historians, archaeologists, canon lawyers, liturgists, theologians and codicologists have all helped to sharpen awareness of the central role that crusading played in the lives of Europeans from the end of the eleventh century to the middle of the fourteenth. This book cannot include more than a selection of this new work, but I hope that it will at least convey how important crusading as an ideal and a set of practices was to the development of Europe. In so doing, I trust it will also demonstrate how complex crusades were for contemporaries, and why answers to even the most apparently simple of questions – what *were* the Crusades? – are so difficult to reach today.

Three historiographical themes

The following brief discussions are designed to raise a few of the conceptual questions that have dominated crusading studies. I make no attempt to provide systematic answers, but simply to make readers aware of the distinct lines of argument, the related questions raised by each and the overall significance of each theme to the broader subject of crusading

What was a crusade?

The first difficulty in determining what medieval people – whether literate or illiterate – meant by 'the crusades' is that there was no single word or phrase to describe either the ideal or the practice. For one historian, 'crusading' is in itself a misleading term before the reign of Innocent III (1198–1216): 'There was holy war in the twelfth century which attracted spiritual benefits of various sorts. For some of these wars, randomly except in the case of those to the Holy Land, warriors adopted the cross' (Tyerman 1998: 19). Two premises underlie this view: first, that there was no systematic process according to which the Church or the knighthood classified holy wars or the spiritual rewards accruing from them; and second, that in the absence of any such system, the expeditions to the Holy Land still occupied a different place from any others. Few could argue with this as an account of the situation in 1095–9, but it becomes a controversial argument when pushed as far as the end of the twelfth century.

This view, if pushed to its logical conclusion, ends by resisting all definition, at least up to a point when it can be established that medieval society had, through canon law, constructed such a definition for itself. There is much to commend such an approach, at least in so far as it forces us to confront the fact that the word 'crusade' is not known in any source until almost a hundred years after 1095. The danger, however, is that it can lead to a kind of relativism that either includes any and all holy wars as crusades on the grounds that there was no litmus test by which contemporaries could distinguish one from another, or that resists the use of the word altogether. The argument over how to define crusades is in fact analogous to the controversy over universals in the twelfth-century cathedral schools. The nominalist approach resists defining the crusade closely because the word did not exist: if people had agreed on what constituted a crusade, they would have invented a word for it. In contrast, the realist position insists that the idea must have existed independent of language, and that despite the apparently random use of words such as 'journey' or 'pilgrimage' in the sources, it is possible to determine what for contemporaries was or was not a crusade. Between these two poles – which I have deliberately caricatured – are many different shades of opinion. My aim in this section is to explore some of the implications of different approaches.

The First Crusade has usually been seen as the blueprint for a successive series of wars following more or less the same pattern: the preaching of a holy war authorised by the pope and accompanied by the offer of spiritual rewards, followed by a military expedition undertaken by the nobility, sometimes including royal princes, against territory held by the Turks or any other enemy considered equally threatening to the stability of Christendom. But although the First Crusade was certainly the first such war, it has been argued that it does not necessarily follow that it was designed to inaugurate a movement, or that in fact it even established a pattern for future holy wars (Tyerman 1998: 8–29). The events of 1095–9 – or, as some historians prefer, 1095–1101 – can be seen as simply a peculiar combination of circumstance and opportunity that was not intended to be repeated. Moreover, far from initiating new ideas about holy war, the First Crusade simply reflected ideas that had been around since Pope Gregory VII (1073–85), if not before. In so far as the exercise was repeated in 1146–9, it was because no new or original thinking had been done about what a crusade was; indeed, Eugenius III's self-consciously modelled his bull of 1145/6 on Urban II's preaching in 1095–6. Many historians would probably agree with the broad outlines of this argument, though there is a case for seeing actions such as Bohemond's assault on the

Byzantine Empire in 1107, the siege of Zaragoza in 1119 and the attack on Damascus in 1129 as partaking of at least some of the elements mentioned above. In any case, it is one thing to argue that the First Crusade was essentially a 'traditional' exercise that did not provide a launching pad for new ideas about holy war, and quite another to conclude that it can have had no profound impact on knightly sensibilities. The First Crusade may have been a 'traditional' war in so far as the terms on which warriors participated were already recognisable to them, but it created new political conditions in the East Mediterranean that in turn offered new opportunities and expectations for the knighthood.

Some historians anxious to find some necessary condition by which a holy war can be recognised as a crusade have fixed on the assumption of the cross by warriors as the defining feature of a crusade (Richard 1999: 260). By the end of the twelfth century the rite of taking the cross was accompanied by a vow binding in ecclesiastical law. This would appear to widen the definition so as to allow expeditions that owe nothing to papal preaching to count as crusades, because sometimes kings – Henry VI and Frederick II of Germany and St Louis of France are examples – took the cross on their own authority, and had their actions retrospectively confirmed by the pope. Moreover, a bewildering array of personal expeditions and armed pilgrimages by members of the nobility followed the first crusaders throughout the twelfth century, among them the king of Norway in 1111/12 and the count of Flanders in 1177. They were not preached as holy wars by the papacy, though the Anglo-Norman knights who went to the defence of Outremer in 1186 were responding to an appeal from the patriarch of Jerusalem. We know very little about the state of mind of these knights, or whether they thought that what they were doing was the equivalent of crusading. They may have taken vows, but in few cases before the 1180s do we know anything about them.

Even this wider definition, however, still requires crusaders to have responded to an initiative offered by the Church, because the crusading vow was made to a cleric and formed a contract with God. Even if crusaders could take the cross at any time, without having to wait for a specific preaching campaign, the terms of crusading were still set by the Church. Two problems arise from such a definition. First, definitions of what constituted 'taking the cross' remained fluid even after Innocent III's pontificate. Second, can we be confident that it was the Church's model of crusading that informed lay mentalities? Did a knight – Conrad of Montferrat, for example, who arrived in Outremer in 1187 just in time to undertake the defence of Tyre – consider that he had earned spiritual privileges even if he had not been granted an indulgence as part of the contract represented by taking the cross? (And, we might ask, if he did not, why would he have bothered?) Fundamentally, how one answers the question 'what was a crusade?' will depend on the kind of weight one gives to the different forces shaping medieval society. Of course crusades were regarded by contemporaries as religious acts, but does this mean that any war that could be rationalized as contributing to the defence of Christendom was likely to be regarded as a crusade, regardless of whether it was subject to the same spiritual rewards as the expeditions to the East? In other words, did the consensus about what was meant by crusading come from the participants themselves, or from those who articulated, refined and shaped the developing ideal that lay behind the practice?

The most recent authoritative survey of the historiography divides cru sader historians into four categories: 'generalists', 'traditionalists', 'plural-ists' and 'popularists' (Constable 2001). Of these, the last category largely represents historians of an earlier generation, notably Alphandéry and Dupront (1954–9), but one current historian, Flori (1999) has followed the same premises in his work, arguing that the First Crusade was an outpouring of the kind of lay piety shaped by reform tendencies in the eleventh century. This approach, which has been marginalised in recent years, may yet enjoy a revival as new understanding of the lay impulses behind the reform movement (Moore 2000; Cushing 2004) are absorbed by crusader historians.

The three other categories are more prominent in current historio-graphy. The 'generalist' approach was pioneered by Carl Erdmann, whose work *The Origin of the Idea of the Crusade* (1977 [1935]) sought to place the First Crusade within a long standing tradition of holy war and ecclesi-astical politics, and thus to argue against its particular distinctiveness. Among current theorists of crusading, many, though not all, of Tyerman's arguments fall into this 'generalist' tradition. Another category, the 'tradi-tionalists', best exemplified by Mayer and Richard, sees crusading to the Holy Land as having been higher in status than in any other theatre, or even – in Mayer's case – discounts most other forms of crusading altogether. Despite his generalist credentials, Tyerman has also subscribed to the view that the expeditions to the east were the benchmark of crusading. For 'tra-ditionalists', what was really important about crusading was the objective.

In direct contrast to the 'traditionalist' position is the 'pluralist' view, which sees all forms of crusading as having been equal in the eyes of contemporaries, and the defining characteristic of the crusade as being not the direction or target but the papal offer of a spiritual reward. A manifesto for the 'pluralist' position has argued vigorously against the idea of a 'league table' of crusades (Housley 1992: 3–4) in which Jerusalem occupied the top place. The implications of 'pluralism' can be seen in a variety of recent works (e.g. Riley-Smith 1977/2002, 1987, 1995; Siberry 1985; Housley 1986, 1992; Maier 1996; Phillips 1996); indeed, one consequence of the hold that the 'pluralist' approach has enjoyed on crusading studies can be seen in the widening of crusading historiography to include regions other than the East Mediterranean and enemies of Christendom other than the Turks.

Such categorisation of the historiography doubtless has its uses, but it can also be misleading. Very few crusade historians (with the possible exception of Mayer) are comfortably one thing or another; there are elements of 'pluralism' in Tyerman's arguments, just as there are 'traditionalist' concerns in Housley's. Although I have tried in this book to follow no particular agenda, the result will probably seem moderately 'traditionalist'. Most of my discussion concerns crusading to the east and the Crusader States established there. Partly this reflects my own interests as an historian, and partly the constraints on space. Although other theatres of holy war find a place here, I have, with regrets, omitted Cyprus after the thirteenth century.

Undergraduate students whom I have taught often argue in seminars and essays that the *Reconquista* in Spain, the Albigensian Crusade, the Baltic Crusades, and the wars against 'heretics' and political opponents of the papacy in Italy were 'diversions' from the 'original' ideal of the crusade, and even that they devalued that ideal. They are right to argue that the occurrence of such wars was not caused simply by what happened at Clermont in November 1095: the *Reconquista* was already under way, and the later wars against the Baltic pagans would have happened anyway, while the idea of penitential war against 'heretical' enemies of the papacy was formulated twenty years before Clermont. But one can turn the question on its head to argue that the crusades for the recovery of the Holy Sepulchre were a diversion in a tradition of penitential violence that was already in place in the 1070s and that continued alongside the crusades to the East throughout the Middle Ages. As the crusades to the East engendered a distinctive vocabulary, so that vocabulary came to be applied to other holy wars; but the use of the vocabulary did not create them. Holy wars were normative to medieval society; it was the consistent application of the idea to the Holy Land that was anomalous.

A further consideration is that what contemporaries thought the function of crusading was changed over time. There is little attempt to theorise in the contemporary writing about the First Crusade, although certain ideas of about what constitutes a just war are present in the chronicle accounts of Pope Urban II's preaching. By the mid-thirteenth century, however, sophisticated views about the nature and rights of Christian society had emerged. The decrees of the Fourth Lateran Council (1215) provide a blueprint, but the canon law commentaries of Innocent IV (1243–54) and his contemporaries, notably Hostiensis, articulated a more analytical approach to the crusade. Innocent III, Gregory IX (1227–41) and Innocent IV, the lawyer-popes who dominated the office in the first half of the thirteenth century, all shared the fundamental view that the pope, as the head of Christian society, should rightfully enjoy sovereignty not only over that society, but over non-Christian societies too. This theory of sovereignty could be applied to the language of holy war, with the result that the implications of holy war could be defined in sharper, but also in broader ways. If one uses the terminology of the indulgence as a guide, one can argue that Innocent III still appeared to uphold a legal distinction between holy war for the recovery of Jerusalem and for other just wars. It is more difficult to see such a distinction in the crusading bulls of Innocent IV. One reason for this is that the defence of the Holy Land had become a separate issue from the wider problem of relations between Christian and Islamic societies. The deeply reflective thinking about the nature of crusading that is evident in the memoranda produced for the Second Council of Lyons (1274) shows how far the problem has shifted, from the recovery of territory that ought by rights to belong to Christendom to the more general threat posed by the existence of Islam, a religion that denied what seemed to Christians to be self-evidently true. This is not to say that the recovery of what had been lost in the Holy Land did not still preoccupy popes and crusade planners, nor that such an objective still had spiritual dimensions. But the greater complexity in which the question of Islam was framed demanded a wider set of responses, of which holy war was only one. Once the fundamental question was understood in terms of sovereignty, the defence of the Holy Land could not be so easily distinguished from – for example – the defence of Sicily from the Hohenstaufen, or the Latin Empire of Constantinople from the Byzantines.

The motivation of crusaders

A related question to what constituted a crusade is why people took the cross, and what they expected to result from this action. For the most part, the question has been framed in reference to the First Crusade, and in these terms has provided enduring controversy among historians. The view that has prevailed in recent years is that the crusade was so popular because it met the spiritual aspirations of the landed classes who provided military leadership and whose lives were often shaped by military experience. The chief proponents of this view, particularly Riley-Smith (1986, 1997) and Bull (1993), have rejected an older thesis advanced by Georges Duby, and to some extent still followed by medieval historians who consider the Crusades in the broader context of European expansion (Bartlett 1993). Duby's (1971) study of landed families in the Maconnais region of Burgundy led him to the conclusion that the crusade functioned as a solution to a sociological problem, namely the preponderance of younger sons of noble families who were effectively dispossessed by the feudal custom of passing on lands to the eldest son. The landless sons could either enter the religious life or attach themselves as household knights to a great lord and hope eventually to win patronage from him. The crusade offered economic advancement and the opportunity for increased social and political status for younger sons of landed families, in the form of new lands and wealth to be won in the East (Duby 1977: 120). This model, however, has not stood up to the rigorous examination to which it has been subjected in the last generation of crusading studies. One problem is that it is based on a regional study with little application to the situation in the West as a whole. The disinheriting of younger sons may have been a problem in regions where feudal custom prevailed, but in most of Germany and in France south of the Loire, division of territories among a wider kinship group was still customary on the eve of the first crusade, and such areas recruited well for the crusade (Bull 1993).

The assumption that the knighthood was attracted by the prospect of wealth and lands is resilient, perhaps in part because human nature suggests that it is so plausible. It is certainly difficult absolutely to disprove, because although there is evidence of families or individuals whose positions were advanced as a direct consequence of the crusade, it is hard to see what evidence could have survived that this was their motivation in taking the cross. Knights who went on the crusade and survived found their feats celebrated in verse and chronicle, and family myths grew up around crusading ancestors. Whatever a knight's motivation might have been, once it became clear that the crusade was being articulated in terms of self-sacrifice, courage and piety, he was unlikely to protest that he had in fact only taken the cross for material gain. This is not to say that profit, where it could be gained in the form of loot, for example, was not welcomed. In a much-quoted passage from the anonymous *Gesta Francorum*, the crusaders before the battle of Dorylaeum (1097) are urged to 'Stand fast together, trusting in Christ and in the victory of the holy cross; for today, if God is willing, you will all gain great booty' (Hill 1962: 19–20). In a similar vein, Martin, Cistercian abbot of Pairis, told prospective crusaders in 1200 that 'the land for which you are headed is by far richer and more fertile than this land, and it is easily possible that likewise many from your ranks will acquire a greater prosperity even in material goods than they will have remembered enjoying back here' (Andrea 1997: 71). Knights probably saw little contradiction between spiritual and material motivations; why should they not be rewarded in this life for the service of Christ?

Some of the contemporary chroniclers of the crusade, such as Robert of Rheims and Ekkehard of Aura, attempted to rationalise Urban's message by making him explain the endemic violence in the West in terms of land hunger, and appeal to knights to seek new lands outside Europe. This explanation is attractive if we see crusading as part of a broader colonisation movement of which the Spanish *Reconquista*, the Norman settlement in south Italy and Sicily, and German expansion east of the Elbe also form part. Crusading to the East may, from our perspective, seem to fit into this pattern. But quite apart from the reductionism implicit in such a model, it is difficult to see how the western knighthood, which maintained its livelihood from agricultural surpluses for which arable land was needed, could have been seduced by the prospect of settling a land with much poorer agricultural resources. Easier lands could be had, without fighting for them, through the internal colonisation of marginal lands in the West, which was a feature of eleventh- and twelfth-century Europe. The most recent research indicates, moreover, that very few surviving crusaders did in fact settle in the newly conquered territories (Riley-Smith 1997: 19). An assumption that securing new lands was a strong motivating force is perhaps suggested by the disproportionate evidence provided by the careers of some of the crusaders who became prominent in the East, such as Tancred and Baldwin of Edessa. But unless we argue that large numbers of crusaders had intended to settle but were so disillusioned by what they found in the east that they turned back for home to try their lot as landless knights again, it looks as though the inducement of new lands in the east was not a powerful motivation. It could be argued, of course, that crusaders were easily seduced by the prospect of settlement because they knew nothing of the reality of conditions in the east – an argument bolstered by the contention of some historians that genuine confusion existed in the minds of ill-educated crusaders between the heavenly and the earthly

Jerusalem (Alphandéry, Mayer). But it is dangerous to assume widespread ignorance or lack of mental grasp where it cannot be proven, and accurate knowledge about conditions in the east was available at least among the top echelons of eleventh-century society, from pilgrimage accounts, monasteries, like Monte Cassino, that had connections with the Holy Land, and papal envoys to Constantinople.

Plenty of crusaders, in any case, were already landowners in the West. The most celebrated example of a lord disposing of his property to be able to fulfil his crusading vow is Robert of Normandy's vifgage of his duchy to his brother, William II. It is clear from the terms of the contract, however, that Robert intended to return; the difficulties he encountered in regaining his inheritance from Henry I after his return (he was defeated in 1106 and spent the rest of his life in prison) are testament to the strong desire to participate in the crusade. Robert's spectacular attempts to raise money, and the collapse of his ambitions in the West as a result of taking the cross, only form on a larger scale an example of what must have been the experience of many knights. Other great lords who may seem in hindsight to have been motivated by the desire to establish new lordships for themselves are more difficult to assess. Raymond of Toulouse was giving up a wealthy county to settle in the East, but Godfrey of Bouillon's position was less secure. Both may have been genuinely inspired by a religious conviction that their lives would be spiritually enriched by permanent service to God in the Holy Land. The desire to settle new lands in the East can be interpreted in more than one way, and need not imply that a crusader's motivation was materialistic rather than spiritual. The archetypal 'empire-builder', Bohemond, probably did not intend to claim Antioch for himself until after it had become clear to him that the Byzantines were not prepared to take sufficient risks for its return. Opportunist he may have been, but this does not make him oblivious to pious conviction.

It is easier to be sceptical about 'new settlement' arguments than to prove that crusaders' motivations were entirely spiritual. One problem is that such evidence as has been uncovered about individual motivation accounts for only about 10 per cent of participants on the First Crusade. Another is the nature of the evidence itself. One recent development in the study of early crusading has been the use of charters to shed light on what departing crusaders thought about their actions. Although this method had already been suggested by Constable (1982), it has gained special prominence in the works of Riley-Smith (1986, 1997) and Bull (1993), which base their arguments for the spirituality of crusaders on the language of charters drawn up usually between crusaders and religious houses. The functions of such documents are varied: sometimes they are simple financial agreements reflecting the crusader's need to liquidise assets in order to fund his expedition; others reflect wider-ranging negotiations between the crusader and the religious house as rival landowners, and seek to bring about peace and security in the locality for the duration of the crusade. Although the financial details evidenced by such documents confirms the crushing expenses incurred by crusaders – and thereby provides ammunition against the argument that crusaders took the cross for economic enrichment – it is the language of the charters that has been most productively exploited by historians in search of spiritual motivation. The preambles to these charters often use phrases or formulae that seem to indicate a particular kind of devotion to Jerusalem or the Holy Sepulchre. Moreover, the recognition by knights that making peace with neighbouring landowners with whom they had been at war was a religious obligation suggests that the crusade was undertaken by many as a penance for sins confessed.

As even those who have made most use of such evidence are aware, however, charters cannot be taken at face value (Riley-Smith 1986: 38–9). Charters of all kinds were emblematic documents in medieval society. In an age of mass illiteracy, they had a kind of mystique as objects that went far beyond the information they contained. But the information was also vital, both as a legal guarantee of the transaction being described, and as confirmation that whatever was being granted was in the first place within the grantor's power. This explains why the language of the preambles was so formulaic and high-flown. But it was the language not of the crusaders themselves but of the monks who wrote them; often it employed references to the Bible or the Church Fathers that would have been unfamiliar to laypeople. Even the form in which most charters survive emphasises the clerical origin of the documents – the surviving charters used by crusade historians usually come from the cartularies, or charter collections, of religious houses, and are thus copies retained or made by those institutions rather than being the originals in the possession of the crusader's family. This is not to say that charters have nothing to tell us about the motivation of crusaders, but simply that they should be understood as a normative form of the expression of power, in which conventional formulae had an important function. This is true of all charters, not simply those of crusaders.

The whole question of whether crusaders' motivation reflected spiritual or 'material' concerns is premised in the first place on an assumption that the crusader exercised a free choice of whether or not to take the cross.

This certainly seems to have been Urban II's intention in 1095–6, but in practice it is unlikely that most crusaders made such a choice entirely free of other considerations. Urban's recruitment method was to approach the heads of noble houses who, if they took the cross, recruited from among the warriors who were dependent on them. As one crusade historian has recently expressed it, all members of the knightly class were part of a *mouvance*, which might best be translated as 'patronage network' (France 1996: 8). Decisions about whether or not to take the cross, in common with most other choices, were made with reference to the *mouvance*. In theory, taking the cross remained voluntary, but it was natural for large numbers of feudal dependents to accompany their lords on crusade. Peer or kinship pressure was also a factor. Joinville, by his own account, took the cross with other lords, among them his cousin, the count of Saarbruck (Shaw 1963: 191). Joinville's memoir suggests that taking the cross was perceived as voluntary but at the same time worked through existing relationships, and similar conclusions have been drawn from the data assessed for the Fifth Crusade (Powell 1986: 67–89). Political imperatives might also play a role, as they apparently did in recruiting by the Lord Edward in 1270.

Why most individual crusaders took the cross will never be known. The best that the sources can do is to suggest what the prevailing impulses in society were; but the historian's interpretation of a given source will always depend on understanding the purpose for which it was written and its intended audience. Of course it is always possible to say that because charters, chronicles and even crusaders' letters were written by clerics, they will reflect the pietistic and spiritualised language of the profession, rather than the aspirations of lay participants. As I hope will become clearer, however, a clear separation between the two did not exist at the time of the First Crusade. If we reject these sources altogether, we are left either with unsustainable sociological models or with assumptions about knightly behaviour based on the thirst for adventure or a general liking for warfare. Many knights doubtless did enjoy war, but this can scarcely be an adequate explanation for the sacrifices they made and the hardships they endured in order to practise it in such uncongenial conditions.

Colonialism vs integration

One early strand of crusading studies arose from an interest in the physical presence of the western settlement in the Levant. This approach was dominated by French or Francophone historians of the late nineteenth and first half of the twentieth centuries, in particular Rey (1866), Madelin (1916, 1918) and Grousset (1934–6). The tone of these works – though not completely their content – is characterised by the title of Madelin's essay of 1916 'Frankish Syria', expanded into his book *French Expansion from Syria to the Rhine* of 1918, and echoed in Grousset's formulation of 'a Franco-Syrian nation' (1934: 287). Another historian in this tradition, Dodu, argued that Frankish settlement in the Crusader States was benevolent to the indigenous peoples and that the Franks assimilated culturally with them (1914: 52–3, 75). The thrust of such writing was certainly, as critics were to point out, polemical. Madelin, R.C. Smail argued, wrote from a conviction of 'the ability of Frenchmen to rule other peoples for their own good', and his work was designed to celebrate the French colonial achievement (1956: 41). It is no coincidence, of course, that much of this writing coincided with a period of intensified nationalism during and following the First World War. These historians' analysis of the Frankish – for which read French – 'genius for colonisation' identified as particular benefits the capacity to absorb cultural models, to blend racial and religious elements without causing rancour, and to impose just rule on colonists and colonised alike (Smail 1956: 42–3).

These assumptions were severely criticised in the 1950s by Smail, Joshua Prawer and to some extent by Claude Cahen. The model these historians suggested in place of the 'assimilated society' of the earlier French school has largely prevailed until very recently. This model, scorning the Frankish taste for the luxuries of the Arabic-speaking world as inadequate basis for a genuinely assimilated society, gives priority instead to social and legal institutions. It is less significant, according to this argument, that the Franks ate sugar, employed Arabic-speaking doctors and wore eastern clothing, than that they erected a legal system in which the basic distinction was between Franks and non-Franks, or that they found no role for indigenous peoples in government. As Prawer observed, the only official function of the indigenous Muslims was to mourn at the funerals of Frankish kings. In fact, so far from being assimilationist, Frankish rule in the Crusader States showed the worst traits of colonialism: it was economically exploitative, segregationalist, and designed to protect the interests of a small ruling group.

The context in which the 'segregation' model was initially promoted was a study of military institutions in the Crusader States. As Cahen observed, the small number of the conquerors made their dispersal throughout the Levant too dangerous, and they thus grouped themselves in the fortified towns of the coastline, leaving the interior to the indigenous peasantry (1940: 327). This deduction also informed the study of crusader castle building (Smail 1956: 204–5), which was thought to have been primarily an attempt to dominate a dangerous hinterland from fixed defensive points. The general picture was one of a small but militarily powerful aristocracy with a defensive mindset, which banded together within a narrow area for collective security. Smail characterised the Crusader States as government by force; that force being necessitated by the justifiable fear of revolt on the part of the indigenous people, and represented by the field army and the castles.

Another approach was signalled by Prawer's study of legal and social institutions. Analysis of the complex legal tradition of the kingdom of Jerusalem – so far as it can be reconstructed for the twelfth century – indicates that the crusaders imported largely existing laws from the West for their own self-government. However, the indigenous peoples were largely excluded from Frankish justice, except when Frankish and indigenous claims clashed in the courts, when the rulers' laws prevailed. Similarly, the social structure of the Kingdom reflected the separation of Frank and non-Frank. The few names of Arab origin in governmental records and charters are exceptions to the rule that the landowners were exclusively Frankish. The religious settlement showed the same tendency towards segregation. Although there was no attempt to extirpate Islam, its growth was discouraged through a policy of prohibiting the building of mosques. Muslims and Jews were forbidden from living in the towns, which were Frankish preserves. The indigenous Christians, mostly Greek Orthodox, were disadvantaged by the policy of replacing their hierarchy with Latin bishops (Prawer 1985: 59–116).

Some important assumptions underlay these conclusions. One is that because the Franks had for their own safety to settle in restricted areas, and chose the fortified towns, the nature of the Frankish settlement was chiefly urban. From this assumption follow further deductions about the nature of feudalism and feudal services in the Crusader States. Another is that the indigenous population invariably opposed the western settlement. Challenges to both these assumptions have recently been made by Ellenblum (1998: 3–37), in a study of Frankish rural settlement in the kingdom of Jerusalem. These challenges, even if accepted, do not necessarily undermine scepticism of 'the benevolent colonialism' model, but taken alongside recent research in the fields of religious and cultural history, they cause the 'segregationalist' picture to become fragmented. Indeed, our understanding of colonialism and the ways in which colonial regimes behaved is more complex than was the case when Smail and

Prawer were writing. If, as Smail argued, the Francophone 'assimilationist' school represented by Madelin, Dodu and Grousset was writing to reinforce a positive vision of French colonialism, he himself was writing in a period of disillusionment accompanying British colonial withdrawal. It must have been tempting to conclude that the Crusader States, characterised by 'the imposition of a numerically small military aristocracy over the mass of the native population', in which '[T]his ruling class exploited the subject peoples economically by means of social arrangements which they found in existence, and which were akin to those they had known in Europe' (Smail 1956: 62–3) had shared the fate of all colonial powers in the modern age.

Similarly, Ellenblum suggests, Prawer's approach to the segregation observed in Crusader laws and social structures was 'influenced . . . by the manner in which Zionism interpreted the relation between the European immigrants and the local Muslim population' (1998: 10). The reference to Zionism encapsulates the political complexities inherent in approaching the subject of European settlement imposed by force on an inhabited landscape, even at a distance of several centuries. It should be noted, however, that early medieval governments imposed over different peoples had traditionally allowed each ethnic group to use its own laws in governing them selves. As long as they rendered the services due to the government, they were left a degree of autonomy. This was the principle adopted by the Frankish rulers in the Crusader States.

Any formulation of the problem, however, that uses the vocabulary of colonialism, or that refers generally to a European ruling class, is itself guilty of implicit assumptions. The Frankish ruling class did not consist of a single racially and culturally distinct body that imported customs from a colonial motherland of 'Europe'. Unlike in most nineteenth–twentieth century colonial situations, the crusade settlers did not come from what they recognised as a single motherland. Although we tend to follow some of the contemporary sources in speaking of 'Franks', we should be aware that this designation represents a process in which a 'Frankish' identity came into being (Murray 1995: 59–73). The settlers comprised northern French, Flemings, Lotharingians, Burgundians, Normans, Provençals, Italians and Germans, speaking a number of different vernaculars, and following different social and legal customs. Although it is certainly true that this disparate ruling class exploited the indigenous people economically, it did not do so – in contrast to most modern colonial powers – for the benefit of, or at the initiative of, a mother-country, but rather for themselves. Nor was this settlement intended (with the exception of the mercantile quarters in the ports) to channel the resources of the colonised state back 'home' for the enrichment of the colonists' countries of origin.

The assumption that the Franks settled almost exclusively in the towns of the conquered territories because of the need for collective security is, Ellenblum claims, erroneous. In a study of crusader castle-building, he argues that for most of the twelfth century, fortified places were designed and functioned as administrative units as part of a pattern of rural settlement, rather than as a means to oppress a recalcitrant peasantry (1996: 517–52). Moreover, his research into rural settlement – using methodologies borrowed from archaeology and historical geography – demonstrates a strong Frankish presence in the interior of the kingdom, in manor houses and villages as well as castles (1998, *passim*). The evidence for an overwhelmingly urban Frankish presence was stressed in previous accounts because of the assumption of the need for collective security; but this was itself predicated on an assumption that the indigenous people were hostile to the new settlers. This assumption, Ellenblum argues, blurs distinctions between the indigenous communities, particularly among the Muslims. It is easy to forget that the territories conquered by the crusaders in 1098–9 had only a couple of generations earlier been conquered by a Turkish military aristocracy from central Asia, which ruled over the indigenous Arabic-speaking population in exactly the manner attributed by Smail to the Franks. If collective security were needed by the Franks, should it not also have been by the Turks? It is natural to suppose that the Turkish ruling class might have attracted the loyalty of at least the Muslim indigenous population. Some current opinions, however, suggest that the indigenous population was evenly divided between Muslims and Christians, while even the Muslims may not have been Islamicised until the beginning of the eleventh century (Morony 1990: 135–50; Gil 1992: 170–2).

A further assumption is that the indigenous Christians were just as likely to have identified with the Turks as with the Franks (Smail 1956: 63). This is certainly the impression given by some episodes in the source material; equally, others suggest the contrary. We must be careful to distinguish between different periods in the Frankish settlement. There were certainly complaints about the collaboration of indigenous Christians with the Turks after 1187, when most of them had come under Turkish rule again, but if we take the period 1099–1291 as a whole, there are remarkably few examples of collaboration or resistance on the part of any indigenous peoples to Frankish rule. It may be argued that this was because they were in no position to resist militarily; but then why was the collective security described by Smail and Cahen necessary? A further argument has been that the Frankish treatment of the Orthodox Church – to which most of the indigenous Christians belonged – drove them into passive opposition to the régime. Recent research, however, suggests that the Orthodox Church was more vibrant than has been thought: monasticism, for example, staged a revival under crusader rule (Jotischky 1995: 65–100; Pahlitzsch 2001).

In arguing against the 'assimilationist' model, Smail pointed out that he was using largely the same sources of evidence as the proponents of that model, but attaching different weight to them. He accepted that the Franks adopted local crafts, diet, clothing and building techniques, but did not think this as significant as did Rey, Grousset and Munro, for example. It would be surprising if the Franks had not made use of local resources, but the fact that they did does not amount to the creation of a 'Franco-Syrian nation' (Smail 1956: 43–4). The sticking-point here may be the word 'nation', rather than 'Franco-Syrian'. By the time of the Third Crusade, western Franks no longer thought of the aristocracy of the Crusader States as *culturally* similar, even though they might have been *racially* identical. The question of 'assimilation' or 'segregation' rather depends on what one chooses to privilege: cultural, political, legal or any other considerations. For although a Frankish taste for sugar, or silks, may only be evidence of pragmatism rather than assimilation, can one say the same of – for example – the Frankish aristocratic use of icons in private devotion? Art historians have convincingly demonstrated that a 'hybrid' style emerged in the visual culture of the twelfth and thirteenth centuries in the Crusader States; a style that catered directly to aristocratic tastes. Artistic patronage, however, has never been simply a matter of aesthetic taste: it is loaded with political and cultural signification. In employing artists trained in Byzantine styles and working in Byzantine iconographic traditions, in adapting Orthodox forms such as the icon to western religious devotions, royal and noble patrons were contributing towards the creation of a new aristocracy. This was noted by westerners as much in Joinville's admiration for the ostentatious splendour of John of Ibelin's household as in Jacques de Vitry's scorn for the 'orientalised' Franks of Acre (Stewart 1895: 67–8; Shaw 1963: 202–5). The epitome of this culture can be found in thirteenth- and fourteenth-century Cyprus, but the process of evolution had begun in twelfth-century Palestine. As one art historian has recently observed, what characterised the cultural taste of the Cypriot crusader aristocracy was a conscious eclecticism designed to display wealth and authority (Carr 1995: 239–75). Taking a lead from the princely houses of Jerusalem and Antioch, the Frankish aristocracy in the East Mediterranean married among the Armenian and Greek nobility, producing a genuinely indigenous aristocratic class whose tastes and choices in self-projection were determined by more than simply pragmatism.

The Frankish aristocracy of the East did not behave markedly differently from that of the West: both spent rather than saved, but those in the East could afford more glittering treasures, and married from the eastern nobility rather than, as the French or English nobility did, Castilians, Savoyards, Provençals or Scandinavians. The point is that the aristocracy throughout Christendom were evolving in this period into an international caste distinct from the people they ruled. At the highest level of society, therefore, from which most of our evidence comes, discussion of whether the Frankish settlers assimilated with or segregated themselves from the indigenous people, though productive in stimulating new research, may in the end be an attempt to answer the wrong question. For the middling and lower ranks among the Frankish settlers, such evidence as can be reconstructed from the archaeological and documentary record suggests that, whatever attitudes to the indigenous peoples may have been, westerners had little choice but to live in close proximity with them.

CHAPTER 2

The papacy, the knighthood and the eastern Mediterranean

This chapter examines the background to crusading in the west and the east. The main themes are:

- the impact of reforming ideals in the papacy in the eleventh century;
- the religious convictions of the western knighthood;
- the political fragmentation of the Islamic and Byzantine worlds.

The papacy on the eve of the First Crusade

Modern historians have tended to follow their medieval predecessors in seeing the Council of Clermont, in November 1095, as the starting point of the crusades. If any single episode is to be chosen, the final day of the council, on which Urban II publicly proclaimed the expedition to the east to an audience comprised of senior clerics and some of the nobility of the Auvergne, certainly has more claim than any other. But we cannot hope to understand Urban's dramatic words at Clermont, still less the events of the First Crusade, without first considering the context in which the pope delivered his rousing message of holy war. If we are to trace his own thinking on the subject of sacralised violence, we need first to examine how the nature of the papal office that he held and its ideological direction had changed during the eleventh century. We must also consider the nature of the audience, and ask why his preaching struck such a responsive note in western European society. Finally, we need to look beyond Clermont, indeed beyond Rome and the West, to explore the geographical context of the first crusade. The crusade was conceived and born in the West, but it was only successful because of a coincidence of political factors in the Islamic and Byzantine worlds, and its impact was felt throughout the eastern Mediterranean.

Urban II was far from being the first pope to proclaim a holy war. The idea of warfare sanctioned by God may not have been part of the mindset of the early Church, but it formed part of both the historic Jewish and adopted Hellenistic heritage of Christianity. Constantine, the first Christian emperor of the Roman world, attributed victory over his pagan enemies to God's intervention, and the defence of the empire against barbarian peoples who refused to convert to Christianity could be seen as wars fought on behalf of God's kingdom on earth. Throughout the early Middle Ages, lay rulers and popes alike took up arms in defence of Christian civilisation against non-Christian enemies who were thought to threaten an order that God had imposed on the world. Contemporaries regarded the wars of Christian kings against Muslims in the Iberian peninsula, against the pagan Vikings in Britain and Ireland, and against the Saxons in northern Germany, as necessary blood-letting in defence of Christianity. Sometimes, as in ninth- and tenth-century Italy, popes issued bulls in which attempts were made to articulate the notion and circum-stances of holy war (Gieysztor 1948: 3–27; Erdmann 1977: 113–16; Bull 1993: 64–6; Bachrach 2003). It was in the eleventh century, however, that a rational theory of holy war came to fruition in a circle of reforming clergy in the papal curia. In order to understand what prompted such thinking, and how Urban II could make use of it in the 1090s, we must first chart the changing nature of the papal office itself in this period.

A revolution occurred in the highest echelons of the Church in the mid-eleventh century. The papal office had become, by the 1040s, scarcely more than a localised bishopric with little influence outside the city of Rome itself. The honour in which the successors of St Peter were in theory still held had been dissipated by a succession of unsuitable and incom-petent popes. The problem lay only partly in the men who occupied the papal throne; far more in the methods by which popes were chosen. The office had come to be dominated by rival Roman noble families, who filled it with relatives and protégés as a means of manipulating political power. In 1046, however, Emperor Henry III marched an army to Rome, deposed two rival candidates for the throne and installed his own nominee. It was the beginning of a reform movement that was to have the most profound effect imaginable on the whole of western political society as well as on the internal organisation and pastoral mission of the Church itself. The reformers, who coalesced around the figure of Leo IX (1049–54), developed a programme based on three main issues: the abolition of simony and of clerical marriage, and what they termed the liberty of the Church. Although all three had a largely moral basis, they came to have a political effect. The 'liberty of the church', for example, meant in practical terms the freedom of the papacy to perpetuate its own reform programme through a new procedure for election to the throne of St Peter by the college of cardinals. This reform measure, implemented by Nicholas II in 1059, was essential for the momentum of reform. It meant that a pope could ensure the continuity of his ideals by appointing to the college of cardinals like-minded clergy, by whom – and usually from among whom – his successor would be chosen. Thus we see from the middle of the eleventh century a group characterised by an international flavour (Leo IX was from Alsace, Victor II a Swabian, Stephen IX from Lorraine, Nicholas II from Alsace or Burgundy, Urban II a Frenchman) and by monastic experience and ideals.

The political effect of this system was to deny influence to the Roman noble families who had for so long dominated the office. But it also led, in the long run, to the marginalising of the emperor who had begun the reform process. Eventually, during the pontificate of Gregory VII (1073–85), the reform programme led to outright war between the papacy and the German imperial claimant, Henry IV. Although the immediate cause of the conflict known (somewhat misleadingly) as the Investiture Contest was the issue of appointment to bishoprics, the course of the war itself shaped and refined the reform agenda. When Henry IV invaded Italy with the intention of deposing Gregory in favour of his own candidate, the pope needed to find an effective military response.

In his recourse to war, Gregory VII was following in the footsteps of Leo IX, who had raised and led an army – to defeat, as it happened – against the Normans of southern Italy in 1053. Another reforming predecessor, Alexander II (1061–73), sent the papal banner to accompany William of Normandy's invasion of England in 1066, and may have offered a form of indulgence to Spanish and French knights fighting the Muslims; he was certainly interested in the progress of the Reconquista (Bull 1993: 72–6).

The language of righteous violence pervades Gregory's correspondence with secular lords. He warned a rebellious vassal of the king of Dalmatia, for example, that he would 'unsheath the sword of St Peter against [your] presumption', and a German critic speaks of Gregory as inciting men to violence. In fact, violence at a localised level had accompanied the reform movement throughout Italy. Historians have begun to realise that the reformers were not simply intellectuals imposing high-minded ideals on the Church, but that much of the impetus for reform came from the 'grass roots' in urban communities, and that it was often enforced by crowd violence. Gregory himself had supported the Patarini, a radical group of agitators in Milan who demanded the reform of their clergy (Cowdrey 1968: 25–48).

The defence of the papacy and its reform programme against Henry IV, however, demanded a more sophisticated set of alliances than could be found on the streets of Italian cities. Alexander II, anticipating the need for military resources, had developed a system of recruitment to the papal cause by means of oaths taken by members of the nobility to defend the papacy. This was extended by Gregory VII, with the result that by his death in 1085 a network of the *fideles Sancti Petri* had been created throughout Italy, northern Spain and the French and imperial lands. The term itself does not translate easily, but Gregory seems to have under-stood a form of vassalage owed by such oath-takers to the pope (Erdmann 1977: 206–8; Robinson 1990: 307–9). Among the most prominent in the 1080s were William Tête-Hardi, count of Burgundy, Matilda, countess of Tuscany, Robert, count of Flanders, Welf IV, duke of Bavaria, Hugh of Lusignan, a Poitevin baron, and probably Raymond of St-Gilles, count of Toulouse. Some of these, notably Hugh and Raymond, were later to be crusaders, while the families of others preserved the tradition of support for the papacy (Riley-Smith 1997: 44–6). Being a *fidelis sancti Petri* did not obviate one's secular ties of patronage and vassalage; the count of Burgundy, for example, was an imperial vassal, and the count of Flanders the vassal of the king of France. But Gregory clearly expected that lay sup-porters of reform would privilege their vassalage to St Peter above any earthly ties.

Matilda, perhaps the most important lay supporter of Gregory VII, not only raised an army for the pope, but also recruited scholars to the cause. Three of these, Anselm of Lucca, Bonizo of Sutri and John of Mantua, became particularly interested in developing a theology of holy war to sup-port the activity in which their patron was engaging. Their ideas derived ultimately from St Augustine of Hippo (354–430), whose own thoughts on justifiable violence had arisen in the context of the treatment of heresy (Robinson 1973: 169–92; Riley-Smith 1980: 177–92). Bonizo and John of Mantua may have been cautious of going farther than the notion that death in a just war merited the status of martyrdom (Riley-Smith 1997: 47–9), but Anselm of Lucca, more radically, took up Gregory VII's own idea that warfare could itself be a penitential activity. Here again, Gregory was not an innovator; he simply took to its logical conclusions a premise that had been part of the reform programme for a generation. Leo IX had promised absolution of sins to the army he led in 1053, but this seems to have been interpreted by contemporaries either as a general absolution given by a priest acting as confessor, or as a dispensation of penances that might remain uncompleted if soldiers died in battle. Gregory VII, however, was criticised by one of his clerical opponents, Sigebert of Gembloux, for ordering Matilda to fight against Henry IV for the remission of her sins; the distinction here being that a particular case of warfare was singled out by the pope as so meritorious that it could absolve her of sins (Cowdrey 1998: 657).

It is important to understand how revolutionary this distinction was. It is true that the Church had as yet no firm penitential system; indeed, such a system evolved partly through the experience of the first crusade and its antecedents. Contemporaries did, however, see that something quite novel was being thought through by the papacy, and not all of them approved of it (Gilchrist 1985: 37–45). The traditional position had always been that sins of violence were punishable by the Church through penances assigned by the sinner's confessor in relation to the gravity of the sin committed. H.E.J. Cowdrey, in a recent examination of the shift in Gregory VII's thinking on the subject, contrasts the traditional penance assigned in 1073 to Peter Raymundi, son of the count of Barcelona, for the murder of his stepmother, with later penances (1997: 21–35). Peter faced 24 years of penance, which included a prohibition on bearing arms, save for self-defence, or – because of the geographical context – for war against the Muslims. The problem was that a great lord such as Peter could hardly engage in political life, or even hope to maintain his family and property, without engaging in warfare. If those who confessed to acts of violence and performed penances designed to satisfy those sins were removed from the arena of conflict, anarchy, rather than peace, would be the result. The solution to which Gregory VII seemed to turn was to distinguish between licit and illicit acts of violence. In 1075 he urged the laity of Chiusi to use force if necessary to expel a provost guilty of sexual sins, assuring them that they could thereby gain remission of their sins (Cowdrey 1993: 51–64). In his final encyclical, in 1084, Gregory called on the faithful laity to help defend the Roman Church by armed service as well as by spiritual support, as a way of gaining absolution of all their sins: '[He] divided those who bore arms into opposing camps – those who bore them from base motives in the corrupt service of this world, and those who bore them in the service of God and for the purposes of the Christian religion. Such service ennobled, and it raised their bearing of arms above the reproach of sinfulness' (Cowdrey 1997: 33). Where once all acts of violence had been judged inherently sinful and thus required penance, now certain prescribed acts of violence were not only permissible, but could actually form penances in themselves. The path to Clermont lay open.

The papal reform programme presumed the position of the pope as head of the Church. This was less obvious to contemporaries than it might seem, for until the mid-eleventh century few popes had even attempted to impose ecclesiastical policy or doctrinal interpretation on bishops outside Italy, and few bishops would in any case have listened. Gregory VII met resistance from many German bishops who found his interference in the affairs of their sees intolerable, and even a reform-minded bishop such as Lanfranc of Canterbury could not stomach papal intervention in England (Cowdrey 1998: 463). Gregory VII used the image of the Church as a body; a body has only a single head, and that head was logically the successor of St Peter, whom Jesus had entrusted with responsibility for the disciples. The Petrine commission became the basis for a theory of papal primacy not only over the Church in the sense of its 'professional members' – clergy, monks, nuns and so on – but in the sense of the whole of Christian society. All who were baptised were members of the Church, and thus subject to papal authority.

The doctrine of papal primacy was clearly central to Gregory's relations with secular powers; but it had an effect far wider than the Investiture Contest. For in claiming headship over all Christians, the popes undermined the traditional view of the Church's hierarchy held by eastern Christians. Greek Orthodox tradition maintained that the pope was merely one of the five patriarchs of the Church. The bishop of Rome was held in particular honour by virtue of occupying the see of St Peter, but Christian doctrine could only be determined in an ecumenical council at which all five patriarchates – Rome, Constantinople, Antioch, Alexandria and Jerusalem – were represented. Of these, Rome, Constantinople and, to a lesser degree Antioch, were considered senior, if only because the patriarchs of Jerusalem and Alexandria lived under the sufferance of Muslim governments. Relations between Rome and Constantinople fluctuated during the era of papal reform. Leo IX's attempt to heal a breach that had been opened as much by neglect as by design failed when, in 1054, his envoy to Constantinople, Cardinal Humbert, caused offence to the patriarch by excommunicating him publicly. Mutual recriminations followed, but although there were important theological issues over which the

Roman and Greek Orthodox traditions differed, these were not so insurmountable as to entail separation of the Churches. Indeed, even while
Orthodox patriarchs declined to restore the names of popes to the diptychs
(the formal gesture of friendship through prayer), the two Churches
remained in communion with each other. Technically, the events of 1054
therefore did not constitute a schism between the Churches.

Theological differences between the Churches arose through the separate and organic development of different customs in East and West in the
early Middle Ages – quite naturally, in an era when contact between them
was limited. One of the main points at issue concerned the Roman
Church's addition to the Creed, in the ninth century, of the phrase *filioque*
('and from the Son'), to show that the holy spirit proceeded from both
Father and Son, rather than, as in the earliest formulation of the Creed
upheld by the Orthodox, from God the Father alone. Another was the use
of unleavened bread (*azymes*) in the Eucharist in the Roman Church. Both
of these positions could be, and were, attacked and defended in theological
terms during the course of the Middle Ages, but they were essentially
differences of custom rather than fundamental differences of dogma. The
problem was that debates between Roman and Orthodox representatives
rarely took place in a context that was free of political overtones, and all
such debates seemed to founder on the much more fundamental question
of authority in the Church.

At the beginning of his pontificate, Gregory VII conceived the idea of
mounting an armed expedition to the East, in order to provide aid for the
Byzantine Empire in its struggle against the Seljuq Turks who were over-
running Asia Minor. Although the plan never materialised into action,
it provides an indication of the effect that the ideology of the papal reform
had on papal policy with regard to the Byzantines (Cowdrey 1982:
27–40). Gregory, however, became embroiled in Byzantine politics to
the extent of supporting the Norman attack on Byzantium in 1081, and
excommunicated Emperor Alexios I Komnenos (1081–1118) for over-
throwing his predecessors. Urban II (1088–99) chose rapprochement
rather than confrontation with the Orthodox world, and was invited to
submit a confession of faith to a local synod in Constantinople, which
could then decide whether his name should be restored to the diptychs.
Urban was adept enough to maintain good diplomatic relations with
the Orthodox Church, while at the same time declining to submit Roman
traditions to the scrutiny of the Orthodox senior clergy. The attempt at
the Council of Bari (1089) to accommodate the Orthodox rite of the
Greek-speaking population of southern Italy while insisting that the Greek

clergy recognise the papal primacy, demonstrates both his political skill and his view of the universal Church.

Urban had more room for manoeuvre than the Byzantines. Alexios Komnenos, the Byzantine emperor, realised that he needed to maintain good relations with the papacy in order to prevent a further attack from the Adriatic by the Normans of Italy. It is a measure of how far the papacy had progressed from the time of Leo IX that the popes were now not only the guarantors of political stability in the Mediterranean, but also a conduit by which the whole of the western military class might be reached. In March 1095 Alexios sent envoys to the pope at the council of Piacenza to ask him to recruit western knights to serve in his armies against the Seljuqs. It was this appeal that Urban relayed to his audience at Clermont eight months later. On one level, we need look no further in order to understand why Urban called for an armed expedition to the East at Clermont. But Urban transformed the emperor's message into something more complex and far-reaching; something designed to appeal to the piety of the knighthood. It is not enough, therefore, to approach Clermont only from Urban's point of view; we must also chart the patterns of religious belief and practice among his audience.

The western knighthood on the eve of the First Crusade

One of the chroniclers of the First Crusade, Guibert of Nogent, famously recorded Urban II's words at Clermont as offering 'a new way of attaining salvation'. To Guibert, the attraction of crusading to the secular warrior was the fluidity it established between two ways of life that had previously been regarded as incompatible – the knighthood and the religious life.

God has established holy wars in our day, so that the order of knights and their followers . . . can find a new way of attaining salvation. Now they need not abandon secular affairs completely by choosing the monastic life, or any other religious profession, as was once customary. Now they can to some degree win God's grace while pursuing their own way of life, with the freedoms and in the dress to which they are accustomed. (Guibert of Nogent, Gesta Dei per Francos I.)

Guibert was neither present at Clermont nor an eye-witness to the crusade, and his narrative relies heavily on other chroniclers. From his monastery in northern France, however, Guibert had the necessary distance to think about what the crusade meant to contemporaries. He had no doubt about the novelty of the enterprise. Guibert's perception that the crusade offered an alternative to the religious life was shared by other contemporaries. Ralph of Caen, analysing the motives of the crusader Tancred, the future ruler of Antioch, portrayed a man aware of the spiritual dangers of leading a violent life. Tancred was apparently considering giving up knighthood to follow a religious life, when he heard reports of Urban's preaching and took the cross (Ralph of Caen: 605–6). Recent analysis of charters has yielded anecdotal evidence of individuals such as the Limousin knight Brunet of Treuil, who changed his mind about entering the cloister at the last minute, persuaded someone else to take his place, and the abbey he was about to join to use the proceeds of the gift he had made them to buy equipment for his expedition (Riley-Smith 1997: 70). Why did such men see a war preached by the pope as an alternative to the religious life? And why in the first place should they have been attracted to a form of warfare that was articulated in terms of an act of piety?

Warfare was endemic in western Europe in the eleventh century. Wars consisted, however, mostly of small-scale localised conflicts, rather than larger affairs between sovereign powers (France 1999: 1–15). Even when a great lord such as the duke of Normandy or the count of Anjou went to war, the numbers involved on each side usually consisted of no more than a few thousand, and the strategies employed were simply those of petty disputes between neighbours, albeit on a larger scale. Warfare was an inescapable part of being a property owner, because ownership had to be constantly maintained and defended. Moreover, in an age when the public authority mediated through courts of law lay in the hands of the greatest landowners, recourse to law was not always a viable alternative to taking up arms. At all levels of society, armed conflict was simply one aspect of the political process. This did not necessarily mean that eleventh-century Europe was a period of anarchy. Disputes over ownership of land, and the economic and jurisdictional rights that accompanied such ownership, could be solved by a process of negotiation and consensus between various parties, sometimes but not invariably involving a court of law presided over by a greater landowner who claimed to embody public authority. But such solutions were often arrived at only after one side or other had used military violence. Violence was not a last resort, but an opening gambit in the political discourse.

The moral problems raised by these forms of social organisation were grave, and recognised as such by contemporaries. During the eleventh century, recognition of the moral dangers inherent in engaging in warfare probably became more widespread. We have already seen how the reform papacy tried to deal with the problem by permitting, even encouraging, certain acts of warfare. But how did the ideas in Gregory VII's circle filter through to the average knight in, say, Burgundy, Anjou or Swabia? Some of the great lords of the West were in direct contact with popes; thus, for example, Urban II was able to appeal personally to Raymond of St-Gilles and Fulk IV, count of Anjou. An examination of Gregory VII's register reveals numerous letters to named individuals. Beyond this circle of the highest aristocracy, however, any presumed link between Gregorian reform ideals and knightly piety hinges on an actual link between senior ecclesiastics close to Gregory and the local communities of monks and canons throughout Europe. It is worth remembering that among the papal reformers themselves, monks were a strong presence. Stephen IX and Victor III were both abbots of Monte Cassino, while Gregory VII and Urban II had been Cluniac monks. Others in the reforming circle, such as Peter Damiani, were monastic founders. Reforming ideals derived from monastic ideals, and through the monasteries, could therefore reach the landowning laity. As I suggested above, the papal reform movement was a manifestation of the desire for a more effective Church, which at the popular level in Italy took the form of riots and public protests against immoral or incompetent clergy. In France, England and Germany, lay piety tended to take the form of family investment in monasteries, or of practices of devotion such as pilgrimages. New monasteries and other religious communities were founded, and older ones endowed, at an increasing rate from about the middle of the eleventh century onward. Guibert de Nogent, himself a Benedictine monk, described how 'in manors and towns, cities and castles, even in the woods and fields, there suddenly emerged swarms of monks who spread out busily in all directions; those places where wild animals had made their homes and the caves where robbers had lurked were suddenly devoted to the name of God and to the veneration of the saints' (Guibert of Nogent I: 11).

Moreover, monasteries were not as isolated from the secular world as one might expect. Guibert of Nogent, for example, was aware that although the profession of arms was diametrically opposite to that of prayer, the people who chose the latter were usually drawn from the same families as the former. Monasticism in particular, and the religious life more generally (meaning the clergy attached to a cathedral, a collegiate foundation or community of canons, rather than the parish clergy), was an aristocratic profession in this period. Knights had brothers, sisters, cousins, uncles, nephews, nieces in monasteries and convents. Guibert himself came from a landed knightly family in northern France, and had entered the cloister as a child. He understood knights and knightly aspirations, and as an abbot with a responsibility for maintaining and maximising the landed endowment of his community, appreciated the close ties between landed families and their local religious communities. As Marcus Bull has observed, 'The important contacts were everywhere, played out thousands of times every year in chapterhouses, abbey churches, and cathedrals . . . as laymen . . . interacted with religious bodies' (1993: 115).

Families' investment in monasteries took different forms. Children might be placed in monasteries as oblates, with the expectation that they would grow up in and come to serve the cloister, while preserving an important biological and emotional bond between the family and the monastery. Adult entry to monasteries might occur at any time in an individual's life; Guibert's mother, for example, became a nun after her husband's death. Previous experience, even of warfare, was not unwelcome; thus, for example, Hildebert Grioard, abbot of Uzèrche, in the Limousin, in the early twelfth century, had been a household knight at Ségur (Bull 1993: 126–7). Becoming a monk or nun was the surest way of ensuring one's salvation; having a relative in a monastery guaranteed that the family would be remembered in the monastery's liturgical devotion, through prayers and masses offered by the community.

Gifts to religious houses, in the form of objects, land, property or the profits from it, were a normative part of the relationship between the landowning classes and the monasteries (Bouchard 1987: 171–246; Rosenwein 1989: 109–43; Bull 1993: 157–66). It is easy to think of this relationship as a form of spiritual bribery: in return for material prosperity, the monks prayed for the souls of the donors. But the theology and the sociological reality underlying such practices are more complex. Historians of the early Middle Ages have become accustomed to thinking of their period in terms borrowed from anthropological discourse, for example, as a 'gift-giving culture' (White 1988). Gifts should be thought of not as bribes, but as assurances of friendship, alliance and mutual cooperation. The monasteries performed a social function for lay society. It was generally accepted that only a life devoted to God in the cloister was an assurance of salvation, but clearly not everyone could be a monk, nun or canon regular. Those who took on such a life assumed a responsibility for the rest of society, for the work they did in the cloister – the maintenance of a constant round of liturgical devotions – was a surrogate penance for those outside (Southern 1970: 224–8). This theory was articulated by Peter the Venerable, abbot of Cluny, in the 1120s.

According to Peter, the donors could, by virtue of their gifts, share in the spiritual reward earned by the monks through their daily lives of prayer and fasting (Constable 1967: I, 84).

These normative practices were reinforced by penances imposed for sins. Laypeople could choose their confessors; for most, the parish priest had to suffice, but aristocratic penitents might use their own chaplains, or go to the abbot of a monastery with which they had a particular relationship. Friendships between monks and prominent landowners, based largely on pastoral letters of advice and injunction, were one way of directing the penances of the influential among the laity. Among the most common of penances was pilgrimage to the shrine of a saint, and in the eleventh century the most arduous, and thus the most meritorious pilgrimage, was that to the Holy Sepulchre in Jerusalem.

Pilgrimage to the Holy Land began in the fourth century with the construction of shrine churches in places considered holy because of the presence of Jesus and the Apostles. The most important of these from the start were the Church of the Holy Sepulchre in Jerusalem (which by 1095 comprised Calvary, the site of the crucifixion, as well as Christ's tomb) and the Church of the Nativity in Bethlehem. Pilgrimages in the later Roman Empire seem to have been largely devotional exercises, but in the early Middle Ages a penitential element becomes discernible both in the surviving accounts of pilgrims and in the references to pilgrims in other kinds of sources. The apparent increase in pilgrimage to the Holy Land in the eleventh century has been explained partly by the opening of Hungary as a safe route for pilgrims after *c.*1000, and partly by a few contemporary reports of apocalyptic expectations in *c.*1033. The most often quoted comes from Ralph Glaber, a Burgundian monk writing in the 1040s, who wrote that 'an innumerable multitude of people from the whole world, greater than anyone could have hoped to see, began to travel to the Sepulchre of the Saviour at Jerusalem' (France 1989: 198–9). Rhetorical exaggeration apart, it does seem clear that pilgrimage to Jerusalem became more extensive not only in terms of numbers but also of the social spread of pilgrims. Glaber's pilgrims included townspeople, and the massed pilgrimage organised and financed by Richard, duke of Normandy in 1026/7 attracted people from across the social scale. For the most part, however, it was the landowning classes who had the leisure and resources to initiate pilgrimages. Clergy as well as laypeople undertook pilgrimages as penances: Fulk III of Anjou, who went to the Holy Sepulchre three times in expiation of his sins, is perhaps the most famous secular pilgrim of the eleventh century (France 1989: 60–1, 212–15), but the lay lords of whom we know are outnumbered by senior clergy such as Richard, abbot of Saint-Vannes, Aldouin, bishop of Limoges, Hugh, bishop of Chalons, Liebert of Cambrai, Poppo of Stavelot, Odilo of Cluny and Thierry of Saint Evroul. Pilgrimage was one devotional activity common to both the lay and the spiritual nobility, and some historians have argued that the knighthood was taught to make penitential pilgrimages by the monks (Cowdrey 1973: 285–311).

The conceptual link between the crusade and the practice of pilgrimage lies not only in the obvious fact that both focused on the Holy Sepulchre, but also in the use made by contemporary chroniclers of the crusade of the imagery of pilgrimage. Scholars tracing the origins of the idea of the crusade have traditionally used pilgrimage as part of the framework on which crusading was built (Delaruelle 1944: 37–42; Alphandéry and Dupront 1954; Blake 1970: 11–31; Cowdrey 1973: 285–311). The other main planks in this framework have included the *Reconquista* in Spain and the Peace of God movement of the first half of the eleventh century. Both, however, have recently suffered serious blows. The historical orthodoxy in Spanish scholarship, in which the *Reconquista* was seen as a 'proto-crusade' (Menéndez Pidal 1956) that attracted and taught French knights who thereby became conditioned to the idea of holy war against the Muslims, has been rejected by Fletcher (1987: 31–47), who argued that religious ideology is only discernible in the *Reconquista* after the first crusade, and by Bull (1993: 70–114), who demonstrated that the French military contribution to the *Reconquista* was in any case negligible. The Peace of God movement, which began in Aquitaine around the end of the tenth century, was thought by an earlier generation of historians (Erdmann 1977: 57–94) to have contributed to a general mindset among the knighthood, in which knights had become used to the idea that warfare was morally wrong and that it ought to be restricted to certain categories of people and to certain times. Secular lords were encouraged to take oaths and join 'peace leagues' to maintain these restrictions, and to help punish those who contravened their oaths by engaging in violent acts outside the set terms. A connection with the first crusade is apparent in Urban II's recourse to a 'general peace' as part of his organisation of the expedition in 1095 (Cowdrey 1970: 42–6). More recent scholarship has been less sure of the value of the Peace movement to understanding the reception of the crusade message in 1095. Bull (1993: 21–69) argues that even in the areas where it was most noticeable, its effectiveness was limited, and that by the middle of the eleventh century it had in any case fizzled out. It is, he says, 'very difficult to imagine circumstances in which any coherent knightly ethos could have been central to the ideology and administration of the Peace programme' (1993: 57). In regions where power was more centralised, such as Normandy and Flanders, there was no real tradition of peace leagues, but the practical advantages of the idea were seized upon by the central authority (the duke or count) as a means of controlling his baronage.

It may be unarguable that the peace leagues were ineffective in the long run; indeed, as the bishop of Bourges discovered in 1038, enforcing and maintaining the peace oaths tended only to contribute to the cycle of violence (Head 1987: 513–29). But this need not mean that they had no effect on the changing character of the secular knighthood. Peace leagues in Aquitaine, like riots against corrupt bishops in Italy, were attempts by local communities of laity and clergy to redress what they considered to be injustices in the face of the impotence of public authority. Of course, not all knights in areas where peace leagues had been tried were convinced of their value; just as not all knights founded or endowed religious houses, and relatively few made the pilgrimage to the Holy Land. Some landowning families, however, can be identified as having participated in all these activities from one generation to another, just as crusading itself was to become in part a family tradition (Riley-Smith 1992: 101–8; 1997: 93–105). The veneration of particular saints or cults in certain families has been cited as evidence of a 'predisposition to crusading' (Riley-Smith 1997: 94). But it is in any case a mistake to assume that we can separate the landowning classes into 'good' and 'bad' knights; into one group that supported papal reform ideals, performed sincere penances for their sins, gave generously to local monasteries, made pilgrimages, upheld the peace and refrained from attacking their neighbours' property – and thus answered Urban's appeal to go on crusade – and another that rejected such idealism. Most probably fell into both categories.

Preparations for the crusade can be extraordinarily revealing of the remorse felt by knights for their sins of violence. The owners of the castle of Mezenc in Burgundy had robbed the peasants in villages belonging to the nearby abbey of St Chaffre du Monastier, but in a charter drawn up before their departure for Jerusalem, they promised to abandon their practices of extorting food and animals from the villages. The bishop of Le Puy absolved them on the strength of their commitment to the crusade (Riley-Smith 1997: 113–14). But, as has recently been pointed out, crusaders who returned home in 1099 did not behave markedly differently from before they had set out; whatever their state of idealism when they took the cross, it appears to have been transitory (Tyerman 1998: 11–12).

The case of Thomas of Marle is instructive. A landowner in the Ile de France, Thomas is best known to posterity from the pages of the biography of King Louis VI of France, where he appears as the worst kind of feudal villain, an exploiter of his own peasantry and a menace to civil society (Cusimano and Moorhead 1992: 106–9). Guibert de Nogent was more explicit in his description:

His thirst for blood was so unprecedented in recent times that people who are considered cruel in fact appear less so when they are butchering animals than Thomas did when killing humans . . . When he compelled his prisoners . . . to pay ransoms, he had them hung up by their testicles (sometimes doing this with his own hands), and when, as often happened, the weight was too great, their bodies ruptured and their entrails spilled out. (Benton 1984: 184–5)

Yet Thomas was also a crusader, whose prowess at the battle of Dorylaeum earned him an epithet in the contemporary *Chanson d'Antioche* as 'valiant . . . with a loyal heart'. Riley-Smith has suggested that Thomas' proclivity to violence came from a disturbed personality as a result of his upbringing; he certainly made the most of opportunities provided by the crusade to shed blood, taking part in the massacre of Jews in the Rhineland (Riley-Smith 1997: 156–7). Does this mean that he took the cross cynically because it offered such unparalleled opportunities for indulging in violence without the dangers of divine judgement? Or that he, presumably in common with many other crusaders, took the cross precisely because he knew that he had committed heinous sins? On the other hand, not all those whom Urban might have expected to take the cross on the strength of a 'predisposition for crusading', did so. Fulk IV of Anjou resisted an appeal made by Urban in person, despite the fact that he had already made the pilgrimage to Jerusalem.

The question of the motivation for taking the cross has been rephrased by some historians. The model once proposed, according to which the primary motivation must be understood in sociological and economic rather than spiritual terms, is now accepted by few. This model, based on the study of knightly families in the Maconnais region of Burgundy, suggested that the practice of inheritance of family land by the oldest son led to the disenfranchisement of younger sons by denying them a landed base, and thus made the prospect of military adventure in the East more appealing (Duby 1971: 283; 1977: 120). Some support for this model appears to be advanced by one version of Urban II's preaching at Clermont, in which the pope suggests that one reason for violence among knightly families is that the land to be shared out among them is insufficient. The problems with applying this model have been pointed out forcefully (Riley-Smith 1986: 44–7): it cannot be proved that younger sons were in fact predominant among crusaders; the response to the crusade was just as strong in regions where partible inheritance rather than primogeniture was still customary; it is inconceivable that landless younger sons would have been capable of raising the necessary money to embark on the expedition; the landless had plenty of safer opportunities to colonise new lands in the West. Such arguments, on both sides, presuppose that knights were in a position to exercise individual choices about whether to take the cross. In theory, the crusade vow was voluntary. But many knights – especially but not exclusively the landless – relied for their livelihoods on decisions made about war and peace by lords to whose patronage they were tied. Great lords, such as the crusade leaders themselves, had to raise armies, and the only means at their disposal was from among the traditional 'feudal' constituency of knights to whom they were bound in relationships based on marriage, conditions of land tenure, neighbourliness, or simple friendship.

Ultimately, an individual's motivation for taking the cross is unknowable. The whole question, indeed, can sometimes appear to be subject to a circular logic. Because we know that at Clermont Urban II appealed to an increasingly articulate piety among the knighthood, the normative devotional and penitential practices of eleventh-century aristocratic society can appear to us as so many stages on the road to Clermont. Pilgrimage and endowment of monasteries, however, were staple features of aristocratic life, and would have continued to be so even had the crusade never been preached. It is worth remembering that although the First Crusade was an event of great magnitude, only a fraction of the nobility chose to participate. Moreover, the First Crusade cannot be interpreted as the logical outcome of such devotional practices. Although much recent crusading historiography has focused on the social and spiritual conditions in which the crusade was launched, we must also consider the political context in the theatre of the war itself – the eastern Mediterranean.

The eastern Mediterranean on the eve of the crusade

According to a tradition recounted by Albert of Aachen and repeated by William of Tyre, which was to have a powerful hold on the later European imagination, the impetus for the First Crusade came from a wandering monk, Peter the Hermit. Peter had apparently suffered persecution at the hands of the Muslims while on a pilgrimage to Jerusalem, and had been entrusted with an appeal from the patriarch of the city to the pope to launch a military campaign so that pilgrims could travel safely (Blake and Morris 1985: 79–108; Morris 1997: 21–34). There is indeed some scattered evidence of Christian pilgrims being maltreated by Muslims while on pilgrimage. The Norman pilgrims of 1027 were pelted with stones, as was Ulrich of Briesgau in c.1040, and the German pilgrims of 1064, though a group of several hundred, were attacked near Ramla by brigands (Runciman 1958: 76; Favreau-Lilie 1995: 321–41). Pilgrimage was never completely safe: even in the relatively stable conditions imposed by the Mamluk sultanate in the fourteenth century, pilgrims were sometimes attacked by hostile locals (Jotischky 2004), and in the early years after the crusader conquest, armed escorts were necessary for the journey to the Jordan. But in almost all such cases, violence was spontaneous, and committed by local groups over whom the political authorities had little control, rather than being part of an anti-Christian policy.

Such dangers were more prevalent, naturally, at a time when centralised political authority was weak. It would not be surprising, therefore, to find some germ of truth behind the legend of Peter the Hermit. For the Islamic Near East experienced profound political change in the eleventh century, as a result of which existing ethnic and religious tensions had become severely exacerbated by the time of the first crusade. Although it may be true that the crusade could only have been as attractive as it proved to the western knighthood because of developments in the West, the conditions that made it feasible from a practical standpoint were the fragmentation suffered by the Seljuq Empire in the 1090s.

The Seljuqs, a tribe of Ghuzz Turks from central Asia, conquered Iran, Iraq and the Near East between the 1040s and 1060s. Recent converts to Islam, they adopted Sunni traditions and ideology. For much of the second half of the eleventh century, Syria and Palestine were battlegrounds between the Seljuqs and the Fatimid rulers of Egypt, who were Shi'ite Muslims. In theory, the Islamic world was divided in allegiance to two khalifates: the Abbasid dynasty in Baghdad, recognised by Sunnis, and the Fatimid dynasty in Cairo, recognised by Shi'ites. In fact both khalifates were largely honorific offices by this period: the Abbasid khalifate had ceased to exercise any real political influence over affairs in the Islamic world in the tenth century, and the power behind the Fatimid khalif was effectively the vizier. Although the majority of Muslims in Syria and Palestine were Sunnis, there were significant pockets of Shi'ite adherence, particularly in Damascus and Aleppo. The separation within the Dar al-Islam had originally been a political matter, concerning the dynastic succession after the death of the Prophet Muhammad, but by the eleventh century each tradition considered the other as heretical, and suppression of Shi'ism was regarded by the Seljuqs as a religious duty. This focus on the internal homogeneity of the Islamic lands resulted in neglect of the world beyond Islam, which was to prove devastating in the 1090s.

The Seljuqs had assimilated quickly into Muslim society. The original Turkish tribes which had formed the basis of Seljuq military power in the mid-eleventh century were supplemented by regular armies recruited from enslaved peoples (*mamluks*). The sultans realised that military power by itself was not enough: the prosperity of the Islamic Near East rested on agriculture and urban commerce. The cities, in which the majority of the population was Arab, continued to provide the intellectual, economic and cultural leadership of the Seljuq Empire. The term 'empire' is in any case misleading, for the Seljuqs never ruled their territories as a single centralised state, but rather as a collection of provinces, no two of which shared the same ethnic patterns. Moreover, Seljuqs tended to fragment authority across ruling families, with the consequence that provinces or cities did not always act in concert with each other. It was largely the personality of Sultan Malikshah (1055–92), rather than any structural unity, that gave the impression of centralised rule over the Near East in the generation before the First Crusade.

In truth, there was little reason for Muslims in the Near East to pay much attention to events outside the Dar al-Islam. The age of Islamic territorial expansion was long past, and save for in the Iberian peninsula, contact with Europeans was minimal. Apart from trade, Europe had little to offer an Islamic world that had for centuries enjoyed a higher level of technological and intellectual development. Muslims, like the Byzantines, tended to regard western Europeans as culturally backward and interested chiefly in warfare. Despite political uncertainties, moreover, late eleventh-century Syria and Palestine were economically prosperous. Although the economy of the Near East was based to a greater degree on urban commerce than was the case in much of the West, agricultural production was more important than has sometimes been appreciated. The Holy Land may not have been flowing with milk and honey, but it provided substantial crops of sugar cane, figs, olives and grapes. Moreover, being far from the centres of power in the Islamic world, it remained relatively peaceful until the advent of the crusaders. Within the Islamic world, it enjoyed a certain cultural eclecticism conferred by its geographical location as a crossroads between East and West. A pilgrim from al-Andalus, Ibn al-'Arabi, found scholars from Iran and Baghdad as well as from his own native Spain studying at the al-Aqsa mosque, and remarked on the *madrasas* (religious schools) of both Hanafite and Shafi'ite traditions (Hillenbrand 1999: 49)

As we have seen, the immediate event that precipitated Urban II's appeal in November 1095 was an embassy sent by the Byzantine emperor requesting military aid for the recovery of Asia Minor. How did this situation arise in the first place? Sporadic raiding during the 1060s prompted Romanus IV Diogenes to lead an army to confront the Seljuqs, but he was routed by Alp Arslan at Manzikert, in eastern Anatolia, in 1071, and subsequently the whole of Asia Minor was lost. For the first time since the eighth century, Muslim armies had reached the Bosphorus. Yet from a position of apparent strength, the disintegration of the Seljuq protectorate began quite suddenly in 1092, with the death of the vizier Nizam al-Mulk, who had in practice been the real ruler of the empire, followed closely by that of the sultan Malikshah. By coincidence, in the same year the Fatimid khalif al-Mustansir also died. One Islamic historian has recently compared the year 1092 in its effect on Islam to the fall of the Iron Curtain in Europe in 1989: 'familiar political entities gave way to disorientation and disunity' (Hillenbrand 1999: 33).

The resulting civil war between Malikshah's sons, Barkaru and Muhammad, which depleted the dynasty's resources, demonstrated that the real centre of gravity in the Dar al-Islam lay in Iran and Iraq (Holt 1986: 10–15). Consequently Syria and Palestine presented opportunities for ambitious and decisive leaders to construct new political entities. The crusaders were to find that many of the towns on the littoral were virtually independent of higher authority; governors and emirs were free to negotiate such terms as seemed most advantageous to them. The collapse of the Seljuq Empire empowered autonomous Arab emirs, such as the rulers of Shaizar, in western Syria, to follow their own policies. To such local Arab dynasties, the crusaders were scarcely more alien than the central Asian Seljuqs. Moreover, all the lands under Seljuq and Fatimid control contained large populations of Christians, mostly Greek Orthodox in Syria and Palestine and Copts in Egypt. Although there is little indication that these indigenous Christians identified with the western co-religionists in the 1090s, the Armenians of Cilicia, who occupied a political penumbra in the Seljuq territories, were prepared to ally with the crusaders.

The Fatimid khalifate in Egypt had also been weakened by revolts from its armed forces – the consequence of a long-established Egyptian policy of using Turkish Mamluks, Sudanese and Berbers to serve the khalifate's military needs. From 1074 to 1094 Egypt was effectively ruled on behalf of the khalif by an Armenian mamluk, Badr al-Jamali, and his Armenian army. After his death, the legitimate heir to the khalifate, Nizar, was passed over by the new vizier al-Afdal in favour of a younger brother. Nizar's rebellion, coupled with outbreaks of plague from 1097 to 1100 fragmented Fatimid authority further. However, the enmity between Fatimids and Seljuqs enabled al-Afdal to capture Jerusalem itself in 1098, while the crusaders were besieging Antioch. It is an irony that the crusaders called to free the Holy Sepulchre from the Turks found that they were fighting an Egyptian, not a Seljuq, army in 1099.

There are indications in the Arabic sources that a sense of foreboding hung over the Islamic world during the 1090s. The political crisis appeared to confirm the story told by the stars – as the chronicler al-Azimi observed, when the Franks appeared, Saturn was in Virgo, a sure sign in the Arabic astrological tradition of misfortune and devastation (Hillenbrand 1999: 37). It is perhaps in this context that we should read the anecdote told in *Gesta Francorum* of the astrological prediction that the Christians would prevail over Kerboga at Antioch (Hill 1962: 53–5) – a story obviously told from a crusader perspective, but perhaps in knowledge gained at Antioch of low morale in the Near East. However far the western knighthood may have been conditioned to holy war against enemies of the Church in the late eleventh century, there could have been no better time to strike against the Islamic world than the 1090s. The question that immediately suggests itself is whether the Byzantine emperor Alexios Komnenos, from whom the original appeal for a military expedition originated, knew this.

If the Islamic world was experiencing fragmentation on the eve of the First Crusade, the Byzantine Empire had suffered prolonged crisis throughout the eleventh century (Ostogorsky 1969: 316–50; Treadgold 1997: 583–611). The scale of the difficulties faced by Alexios Komnenos can only be appreciated if we compare the state of the empire in 1095 with its situation seventy years earlier. When Basil II, known as 'Bulgar-slayer', died in 1025, the imperial frontiers stretched as far to the east as the mountains that separate Iran from Turkey. The Bulgarian kingdom had been annexed, the Byzantine hold on southern Italy appeared secure, and the Mediterranean had been cleared of pirates by a dominant Byzantine fleet. For 150 years, from the last quarter of the ninth century onward, Byzantine armies were virtually unchallenged.

The arrival of new forces on all frontiers during the mid-eleventh century, however, placed intolerable strains on the resources of the state. In the west, the Norman adventurers who began to serve as mercenaries in southern Italy in the 1030s – fighting sometimes for the Byzantine governor, sometimes for the Lombard dynasts – gradually became the dominant power. The remarkable Hauteville family, originally from Normandy, established itself in the Greek-speaking provinces of Apulia and Calabria in the mid-eleventh century. Led by the head of the family in Italy, Robert Guiscard, in 1071 the Normans demonstrated their mastery of the region by their capture of Bari from the Byzantine governor (Loud 2000: 133–7). In the north, the Pechenegs raided unchecked across the Danube frontier, despoiling the resources of Thrace. The diversion of Byzantine forces to meet these threats stripped the eastern frontier, encouraging the Seljuqs to raid at will. It was Emperor Romanos IV's attempt to settle the eastern frontier in 1071 that led to the disaster of Manzikert, in which an ethnically mixed Byzantine army was smashed by Alp Arslan. Asia Minor itself was lost as a consequence – not as the immediate result of the defeat, but in a way that reveals how internal political instability in the empire exacerbated the military failings. Romanos was deposed while in captivity in favour of Michael VII Dukas, and in the resulting civil war, he approached the Seljuqs as allies to restore him to his throne. When Romanos was captured by Dukas and blinded, Alp Arslan had an excuse to invade and annex Asia Minor. At the same time, Michael Dukas courted Guiscard in Italy as a potential ally against the Seljuqs, offering him titles, salaries and a marriage alliance between his son and the Norman's daughter in return.

The Byzantine policy of attempting to balance hostile neighbours through alliances that played one off against another had often worked in the past, but it was inherently dangerous (Harris 2003: 35–51). When Michael VII was himself overthrown by a rival faction in 1078, Guiscard played the same card as Alp Arslan had done seven years earlier, justifying his invasion of the empire from the west in the 1080s as the fulfilment of his alliance with Michael. But instability in the imperial hierarchy was probably a symptom rather than a cause of the decline of Byzantine authority in the Mediterranean world. Of the thirteen emperors ruling between 1025 and 1081, eight were in power for less than three years; only one lasted more than ten years. Five were deposed, one more was probably murdered, and another forced to abdicate. It is easy to dismiss the imperial candidates as incompetent – Constantine VIII was addicted to pleasure, Michael IV an epileptic, Michael V unbalanced, Michael VI too old to be effective (Charanis 1958: 193–4) – but this does not explain the malaise that allowed such figures to rise to the top. The imperial office was not invariably hereditary; in the Byzantine conception of power, it was the office itself, rather than one's family, that conferred authority. Perhaps surprisingly, however, Byzantines had little compunction in ridding themselves of emperors, with the result that palace coups generated from within the civil administration were frequent.

The fundamental dynamic of power in eleventh-century Byzantium was the tension between the interests of the civil service and the landed aristocracy. Michael Psellos, the philosopher who served many of the eleventh-century emperors, thought that those emperors who had come to power through palace service tended to neglect the army, in the hope of curbing the power of landed magnates who might otherwise threaten their own positions. The result was that the military resources that were needed to meet predators on all fronts was simply inadequate. The emperors themselves, however, were victims of social and economic developments in the provinces that they seemed unable to control. At the height of Byzantine military strength, the backbone of the army was the free peasantry of Thrace, Greece and especially Asia Minor. By the mid-eleventh century, however, the system of 'pronoia' according to which landholders rendered military service had ceased to function to the benefit of the imperial throne – partly because so many estates, both secular and ecclesiastical, had secured exemption from taxation and jurisdiction (Treadgold 1997: 680–3). Moreover, the tendency for great landowners to accumulate large estates at the expense of the smallholders meant that military resources coalesced into the hands of a few, who could then form sufficiently powerful cliques to challenge the emperor.

After 1071, the situation deteriorated markedly, as the emperors lost the ability to recruit from Asia Minor altogether. Byzantine armies had always included large numbers of foreign troops, but by the late eleventh century emperors had little choice but to rely on mercenaries. These troops, moreover, came from the very peoples who under different circumstances might be enemies of the empire; thus, for example, a chrysobull of 1088 mentions Pechenegs and Normans, as well as Norsemen, English, Rus and Germans under imperial command. In the twelfth century the English – 'the barbarians who bear the two-edged sword on their shoulders', as they are described in one contemporary document – were particularly prominent (Vasiliev 1937: 39–40). In the late eleventh century the Normans were the preferred mercenaries, though they had a tendency to be disloyal: three Norman captains, Hervé, Robert Crispin and Roussel of Bailleul, at different times rebelled against the emperor. In 1090, Alexios recruited 500 Flemish knights when the count of Flanders passed through Constantinople on his return from a pilgrimage to the Holy Land. Troops were recruited by direct contact with military leaders. Robert Guiscard, for example, was offered a Byzantine military title, and salaries for his men. The norm was for mercenaries to be associated with success in war in the form of conquered land, which would then be held through the arrangements customary under 'pronoia'.

Recent scholarship has tended to find more encouraging signs in the underlying condition of the empire. It is certainly an exaggeration to portray the Byzantine military capacity as moribund; after all, a Byzantine army accompanied the First Crusade as far as Antioch (France 1971: 131–47), and in the wake of the crusade, the Komnenoi were able to reconquer much of their territory in western Asia Minor. The economic decline, too, has been reassessed. One traditional indication of decline, debasement of the coinage, may in fact have been a deliberate policy in response to greater demand for coin (Hendy 1970: 47–8). Byzantine markets were able to provide the crusader armies with supplies, and foreign visitors admired the fertility and climate that allowed for two annual harvests in some regions. The Italian merchants who settled in Constantinople, numbering several thousand, were engaged not simply in long-distance luxury trade, but also in the export of grain, wine and meat from Byzantine farms.

The political situation that Alexios Komnenos faced in the 1090s was complex. He doubtless knew of the collapse of the Seljuq Empire far to the east, but even before the death of Malikshah, Asia Minor had been a battleground between Kilij Arslan, heir of the victor of Manzikert, and the rival Danishmend emirate. His scope for interference in Asia Minor, however, was constrained by a threat from Serbia in 1094 and from a Cuman invasion of Thrace in 1095. Alexios had already ruled for fourteen years, but the Cuman threat was probably part of a rebellion by Nikephoros Diogenes, son of the Romanos IV who had been deposed after 1071. Employing western knights to strike against the Seljuqs in 1095 was simply a continuation of the policy adopted in 1090, when the Flemish recruits had defended Adrianople. Alexios' appeal in 1095 was probably part of a well-established pattern of diplomacy on Alexios' part. Since 1089 he had been working towards rapprochement with Urban II, and an ecumenical council had been mooted at which fundamental differences could be ironed out. An account of the acquisition of important relics from the Holy Land by the monastery of Cormery, near Tours, in 1103, reveals how Alexios had been acting even before 1095 as an intermediary and broker in enabling prominent westerners to become acquainted with the situation in the East. Alexios seems to have placed himself at the centre of a network of dealings with the West, of which the relics from Jerusalem that found their way to Cormery, and a western force that secured the defences of Nicomedia, were just two visible results (Shepard 1996: 117–18).

The western chronicles of the First Crusade are apt to present the expedition as a bolt from the blue. But wherever we look, we see signs to suggest that nobody need have been taken by surprise. Jerusalem had become increasingly familiar to laity and clerics alike as a symbol of penitential devotion. The Seljuq hold on the holy city was weaker than ever before. A display of military might in the eastern Mediterranean, such as had been proposed by Gregory VII as early as 1074, would also bring the opportunity to reinforce the principle of papal sovereignty. The Byzantines welcomed western military aid for specific campaigns against the Seljuqs. Elements of the western knighthood had already demonstrated a willingness to be directed by papal ideals in bearing arms. But these factors cannot explain in isolation what happened between 1095 and 1099. In order to understand how the crusade was put in motion we need to examine in some detail the Council of Clermont and the impact of Urban II's preaching on western society.

CHAPTER 3

Crusade and settlement, 1095–*c*.1118

Pope Urban II launched the crusade in 1095. This chapter examines how he presented his message, how it was understood by the European knighthood, and how the course of the crusade was affected by these ideas. The establishment of the Crusader States are then examined in turn, with a focus on political and constitutional developments.

Preaching the crusade

On 27 November 1095, Urban II brought the proceedings at the Council of Clermont to a close with a dramatic public oration. Invited archbishops, bishops and abbots had attended the previous sessions of the council, but on the final day crowds of lesser clergy, and probably laypeople as well, gathered in a field outside the town to hear the pope. We do not know exactly what he said, but from the four surviving accounts of his preaching, as well as from incidental observations in other sources, there can be no doubt that this was the occasion chosen for the public launch of the expedition we know as the First Crusade. The content of Urban's preaching, while it can never be known in full, can be reconstructed sufficiently to provide us with the main themes of his vision of the crusade. The implications of what he said, however, are still a matter for debate.

It can even be argued that for many crusaders, Clermont itself was an event of only minor significance. The narrative account of the crusade closest to the events themselves, the *Gesta Francorum*, does not even mention Clermont, saying only 'When that time had already come, of which the lord Jesus warns his faithful people every day ... there was a great stirring of heart throughout the Frankish lands' (Hill 1962: 1). The *Gesta Francorum* was the work of a follower of Bohemond, the son of Robert

Guiscard. Bohemond had not been approached by Urban in person as a potential recruit and had apparently not even heard of the expedition until word reached him of the passage of an army through southern Italy, as he was besieging Amalfi (Hill 1962: 7). But when he came to revise the narrative of the *Gesta Francorum* for his own account, Robert of Rheims, who had been present at Clermont, left his readers in no doubt that the council marked the formal beginning of the expedition. Three other chroniclers of the crusade, Fulcher of Chartres, Baldric of Bourgeuil and Guibert of Nogent, also used the same strategy, and in their hands the occasion of Urban's preaching became an opportunity to flaunt their own rhetorical training. But neither Raymond of Aguilers, who like the author of the *Gesta Francorum* was a participant on the crusade, nor Albert of Aachen, saw Clermont as the natural starting-point for their accounts. The central figure of Raymond's account, Raymond of Saint-Gilles, count of Toulouse, had already been sounded out by Urban and decided to take the cross before Clermont, though he waited before making a public announcement until after the council. In Albert's chronicle, however, the omission of Clermont serves the broader narrative, in which the crusade was initiated not by Urban II but by Peter the Hermit. Albert was a subject of the German emperor, and the main protagonist of his chronicle, Godfrey of Bouillon, had fought against the pope for Henry IV. He was reluctant, therefore, to credit Urban with initiating the expedition, preferring to see it as a broad-based alliance of Christians against the Muslim threat to the eastern Church. (It is significant that Albert was also the chronicler who was least partisan in his writing about the Byzantine role.) It is worth noting that those chroniclers who did see Clermont as the event that initiated the crusade were those representing northern French interests.

For those chroniclers who wished to emphasise the papal role in the expedition of 1096–9, however, the Council of Clermont was an event of overriding importance. The explanation for this lies partly in the theatrical nature of such councils, and partly in the immediate context of Clermont itself. When he became pope in 1088, Urban II cut a rather unimpressive figure. Like his predecessor Victor III, his sphere of action was limited to southern Italy, where he could rely on Norman protection. The early years of his pontificate were characterised by efforts to restore the papal position in Italy. To this end he resumed diplomatic contact with Alexios Komnenos, and succeeded in securing both Byzantine recognition of his own status, and assurances that Alexios would not weaken this, or the Norman position in Italy, by supporting Henry IV. Although he seemed more amenable than Gregory VII in his dealings with the Orthodox, it is

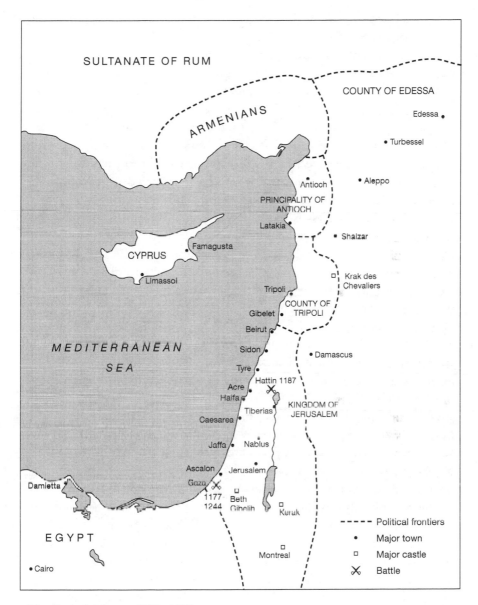

The Crusader States, 1099–1187

worth noting that he took care not to be drawn into theological debate, declining for example to send the required statement of faith to Constantinople, which the patriarch had stipulated as the necessary requirement before an ecumenical council could be held. Urban also managed to build up sufficient political weight through a network of alliances to return to

Rome in 1094. Once installed in the Lateran, he immediately called a council for spring 1095 at Piacenza.

The Council of Piacenza can appear the 'poor relation' of Clermont, but its significance in its own right should not be overlooked. Most of the business related to securing the momentum of the reform programme in Italy, now that imperial forces were in retreat. The restored prestige of the papacy was evidenced by the presence of representatives from Peter of Aragon, who became a papal vassal, and from Alexios Komnenos, who appealed to the pope to recruit a western army to fight for the emperor in Asia Minor. Our source for Piacenza, the often-overlooked chronicle of Bernold, says that the pope urged knights to take an oath to confirm their willingness to do so. This is surely an early indication of the strategy he would employ to recruit crusaders. Piacenza may have suggested to Urban the possibility and desirability of a military expedition to the east, but it was in the months between May and November 1095 that his thinking clarified, and even after Clermont he appears to have refined his vision still further (Cowdrey 1970: 177–88).

From Piacenza onward, the planning for the crusade went hand in hand with the reaffirmation of papal supremacy. At Piacenza Henry IV's estranged wife had testified against him; afterwards his heir Conrad became a papal vassal (Robinson 1999: 290–1). Accompanied by an enormous household that included two cardinals and an archbishop, Urban crossed the Alps into France. He progressed up the Rhone Valley into Burgundy, visited Cluny, where he dedicated the altar in the abbey church, then turned southwest into the Auvergne, reaching Clermont by the middle of November. The visit to Cluny was a powerful symbolic act: Urban himself was a Cluniac monk, and the dedication of the altar doubtless reminded observers of Cluny's direct dependence on the papal see, created at its foundation almost two centuries earlier. Urban's tour through France in 1095–6 was self-advertisement on a grand scale. After Clermont he continued west into the Limousin, then in the New Year north to Anjou and south again to Bordeaux and Montpellier before returning by way of Provence to Italy (Becker 1997: 127–40). No pope in living memory had made himself so visible; and wherever he stopped, Urban dedicated churches and altars. Besides Clermont, he held councils at Tours and Nîmes. He disciplined and corrected where necessary, visited secular lords sympathetic to reform, and collected in his train yet more bishops. At the great feasts of the Assumption and Christmas he wore his tiara and processed publicly. The regal theatricality of his progress and behaviour was surely part of the message he intended to convey (Becker 1988: 436–46).

Three main themes emerge from the different accounts of Urban's preaching at Clermont: the danger that the eastern Church faced at the hands of the barbaric Turks; the sacred quality of Jerusalem for Christians; and the rewards to be gained by those who took the cross. The first two of these are treated with obvious rhetorical licence by the chroniclers who paraphrased his words, but we can surmise the effect of heightened emotion that the pope intended to produce in his audience. Although at least two of the chroniclers, Fulcher of Chartres and Robert of Rheims, were probably present, no two reports lay exactly the same emphasis on these themes, and some do not report all three: Fulcher, for example, makes no mention of Jerusalem, referring only to the eastern Christians and to the Seljuq conquest of Asia Minor (Fink and Ryan 1969: 65–8). The most intractable problem for historians, therefore, has been to decide how important Jerusalem itself was to Urban. Following Erdmann, Mayer argued that Urban did not speak of Jerusalem at all at Clermont, on the grounds that so potent a symbol would distract the audience from the real message, which was the aid to be rendered to the Byzantine Empire, as Alexios' envoys had requested at Piacenza. Yet it is clear from other evidence that Urban did mention Jerusalem at later stages of his preaching in 1096 – for example, in letters to Bologna and Vallombrosa (Riley-Smith and Riley-Smith 1981: 38). Mayer concludes that Urban gradually altered his vision in 1096 as control of the expedition slipped out of his hands and as events came to be shaped instead by those who responded to his appeal ([1988]: 35–6). The impassioned speech about the sanctity of Jerusalem in Baldric of Bourgeuil's account can thus be explained by hindsight, for by the time the chronicle was written, the crusade was long over and its extraordinary military success must have made it appear as though Jerusalem had always been the original target.

Most historians, however, follow the more orthodox line that Urban all along intended Jerusalem to be the aim of the crusade (for recent examples see Flori 1997: 267–85; Riley-Smith 1997: 59–62; Richard 1999: 22–3). For one thing, it would have been difficult to interest western knights in such numbers in a war against the Turks on behalf of the eastern Christians of Asia Minor. Previous western recruits had been offered financial inducements to fight for the emperor, but at Clermont Urban spoke only of the responsibility to aid co-religionists, with vague references to profits that might be made. The examination of charters drawn up between departing crusaders and monastic houses with which they came to financial arrangements has revealed an emphasis not only on Jerusalem, but on the Holy Sepulchre itself as the specific goal of the crusade (Riley-Smith

1997: 62–66). This appears to confirm the impression conveyed by the language of the chronicles, namely that crusaders thought of the expedition in the same terms as a pilgrimage to Jerusalem (e.g. Fink and Ryan 1969: 76). Mayer has remarked that none of the accounts of Clermont make much use of the pilgrimage tradition in reporting Urban's appeal (1972 [1988]: 32) but in fact the language with which Guibert de Nogent describes Jerusalem is very close to that found in contemporary pilgrimage accounts (Levine 1997: 42–4). Moreover, the only surviving decree from Clermont that has a bearing on the crusade seems to make a specific connection between the two in proclaiming that 'for anyone who, out of devotion alone, and not for the sake of honour or wealth sets out to free the church of God in Jerusalem, that journey shall be considered as equal to all penance' (Somerville 1972: 74). The expedition was to be a penitential pilgrimage of the kind to which the western aristocracy had become accustomed; the difference was that in this case pilgrims were allowed to bear arms. But westerners did not make pilgrimages to Constantinople, or to Asia Minor; the association of the indulgence must have recalled Jerusalem in people's minds.

A further point is the supposition, made initially by Erdmann, that Urban was primarily interested in answering Alexios' appeal because it provided an opportunity to reinforce the issue of papal primacy with the backing of military might in the region. Eastern Christians 'liberated' by western arms would have been more likely to subject themselves to papal authority. This argument presumes that Urban had little interest in Jerusalem for its own sake. Yet, as Cowdrey has recently shown (1997: 65–83), Urban brought with him into France, to use in the altars he dedicated, relics from the Lateran collection. Many of these relics were of Christ himself (for example, his foreskin and his sandals), or of his Passion (beans from the Last Supper; the reed and sponge). 'No one who trod the Lateran palace', argues Cowdrey, 'could well fail to be mindful of the Holy Land and of Jerusalem' (1997: 70). Urban's predecessor as pope, Victor III, had been abbot of Monte Cassino at a time when there was considerable contact between the abbey and the Orthodox patriarchate of Jerusalem. Urban himself corresponded with Symeon II, the Orthodox patriarch, during the course of the crusade. Such contact, indeed, was a natural application of the theory of papal primacy over the whole Church. If Urban wished to extend recognition of papal authority to the patriarchates of Constantinople and Antioch, why not also over Jerusalem?

The fundamental distinction between the two arguments lies in whether we credit Urban with altering the terms of Alexios' appeal, and transforming the appeal for aid for the Byzantines in reconquering Asia Minor into the broader aim of liberating Jerusalem, or whether we regard this change in focus as the result either of Urban's own evolving ideas or of pressure from the recipients of Urban's message. The distinction might be regarded as academic, for after all, whatever Urban's intentions might have been, the route to Jerusalem led through Asia Minor, and the defeat of Kilij Arslan by the crusaders in 1097 made Byzantine recovery of much of their former territory possible. But it is in fact a critical point, not only for understanding how the crusade worked in practice, but for determining the nature and extent of the influence that the papacy had on the lay nobility, and how far such influence could be applied in terms of practical organisation.

A crucial element in the puzzle is the nature and meaning of the indulgence itself. No papal bull exists to confirm the decree of the council, although Eugenius III was to specify in his bull for the Second Crusade that he was repeating what Urban had enacted for the First. Moreover, the language of the decree, and of Urban's own letter to the Bolognese of 1096, differs slightly from that in his letter to the Flemish (Riley-Smith and Riley-Smith 1981: 38), in which he refers not to lifting penances imposed for sins but to remission of sins altogether. Mayer has argued that even the 'remission of sins' meant, in Urban's mind, simply an absolution from necessary penances that might have been imposed by a confessor; in other words, that the Clermont indulgence had an earthly but not a heavenly effect (1988: 32–3). The problem with making such distinctions is that it is far from clear that anyone in the 1090s thought in such precise terms. It was not until the systematic definitions of twelfth-century theologians such as Hugh of St Victor and canonists such as Huguccio that the distinctions between the authority of the Church to dispense equivalent penance accrued as a consequence of sin, and the authority to remit sin altogether, were articulated. If we assume that such a distinction was meaningful to Urban, then we are led back to the notion that what he himself preached at Clermont was taken out of his hands by those who responded to the appeal and interpreted according to what they wished to hear. Whether Urban intended to make such theological distinctions about the punitive consequences of sin in this world and the next, it must have been clear to his audiences that the war he was proposing fell into the same category as the penitential violence encouraged by Gregory VII. The main difference was the target – no longer the enemies of church reform, but of 'the Church of God' itself.

The means by which Urban's message was disseminated can be reconstructed only to a degree. He recruited the count of Toulouse and

Adhémar, bishop of Le Puy directly, and tried the same with Fulk IV of Anjou. Fulcher and Baldric both say that bishops were encouraged to preach the crusade, and we know that in the archdioceses of Lyons, Milan and Venice this was done (Riley-Smith 1997: 75). Presumably the many bishops who accompanied him throughout France had ample opportunity to learn his intentions, although it may also be the case that his original ideas were refined in such discussions with prelates. Most of those who took the cross, however, must have learned about the crusade – as Bohemond did – from hearsay and rumour, and it would not be surprising if the message had become distorted according to the political, theological and geographical grasp of the audiences. However they understood the precise nature of the expedition, there can have been little doubt about the general impression made by the pope. The most obvious point about Clermont and Urban's subsequent preaching is also the most fundamental. The pope was urging the knighthood of Europe to take up arms. Gregory VII had presumed that the loyalty owed by the German nobility to the Holy See counted above that owed to their king; Urban II took this still further at Clermont by assuming that all fighting men were subject to the authority of the pope in their use of arms. Whatever else it may have initiated, Clermont was the highest expression of papal reform ideals.

The First Crusade, 1096–1102

Although Urban devoted considerable efforts to publicising the crusade, he set limits on recruitment. He does not seem to have envisaged participation by the Normans of southern Italy or by German knights – though the recruitment of Godfrey of Bouillon, duke of Lorraine, an imperial vassal, must have been particularly pleasing. He wrote to the counts of Roussillon, Besalu, Cerdagne and Ampurias, whose lands lay close to Muslim territory in Spain, assuring them that their armed struggle against the Almoravids was the equivalent of the expedition to the east and discouraging them from setting out (Riley-Smith and Riley-Smith 1981: 40). In 1099, when new armies were recruited in the west to reinforce the first contingents, Paschal II renewed this prohibition. Although the sources can give the impression that representatives from all regions in the West set out in 1096, recruitment followed a regional pattern. Following the example of their great magnates, large numbers from Provence, Flanders, Normandy, the Ile de France, Lorraine, Swabia and the Rhineland set out for the East, but not until 1099, when the dukes of Aquitaine, Burgundy and Bavaria, the archbishop of Lyons and the count of Nevers took the cross, did equivalent recruitment from those regions, along with northern Italy, take an effect.

Urban also tried to limit participation to those who would be militarily useful. Women and non-combatants were dissuaded, and although several bishops and clergy took the cross, monks were prohibited from going. These prohibitions did not all have the desired effect, for many crusaders took their families with them, despite Urban's assurance of protection for those left behind. The abbot of Corfe tried to equip a boatload of his monks to join the expedition, but had to abandon the plan when the archbishop of Canterbury found out. The total number of crusaders will never be known, but the best recent estimates suggest that at the siege of Nicaea in 1097 there were 70,000 men and women, of whom perhaps a tenth were knights. By the time they reached Jerusalem in June 1099, the army had dwindled to about 12,000, with only 1,200–1,300 knights (France 1994: 122–42; see also Riley-Smith 2002: 13–28). The rate of human wastage due to death in battle, disease or starvation, and to desertion, was appalling, but the army that conquered Jerusalem was still, by the standards of the day, a sizeable one.

The great magnates each led their own contingents, and the extent of communication between them before they reached Constantinople seems to have been minimal. Each army chose a different route, though all went overland as far as possible. Much has been made in some histories of the crusade about the so-called 'People's Crusade', the contingents led by Peter the Hermit and Walter Sans-Avoir, and supposedly made up of the urban poor and peasants with no experience of fighting, many of whom did not even reach Asia Minor (e.g. Runciman 1954: 121–33; Payne 1984: 38–46). In fact several of these were noblemen, such as William 'the Carpenter', viscount of Melun, Emich, count of Leiningen and Hugh, count palatine of Tübingen (Riley-Smith 1986: 51). Walter himself was not as poor as his name suggests – he was in fact lord of Boissy-sans-Avoir (Riley-Smith 1997: 224). One reason why these crusaders have attracted negative attention is that in May 1096, Emich's contingent attacked the Jews of Speyer, Worms and Mainz; some crusaders then seem to have broken off from the march to attack Jews in the lower Rhine valley. It is difficult to explain how the anti-Turkish rhetoric of Urban's preaching came to be applied against the Jews, but it may have arisen from financial demands made of Jewish communities by crusaders, which then tragically exploded into violence, in which the local populations of the towns in question then joined (Chazan 1996: 3–106). The Church did not encourage violence against the Jews – in fact the bishop of Speyer protected those in his town – though neither did its general attitude towards them do much to prevent it. Envy of Jewish wealth has been suggested as a plausible reason for the attacks, but most crusaders had probably borrowed from or mortgaged to monastic houses rather than Jews, and in any case the extent of Jewish money lending in the 1090s can be exaggerated. It is worth noting that Emich of Leiningen's anti-Jewish activities seem to have been concerned with conversion rather than massacre, though deaths certainly resulted. A recent study has argued that the Jewish sources themselves exaggerated the extent of the attacks (Chazan 1996: 107–26), but in any case this was a sombre episode that indicates how little control there was over individual components of the crusading armies.

The armies arrived at Constantinople at different times between winter 1096 and spring 1097. The role of the Byzantines on the crusade was to become one of the major grievances of contemporary chroniclers, particularly those, such as the anonymous *Gesta Francorum* and Raymond d'Aguilers, who reflected the attitudes of their leaders. The Greek source most often used to represent the Byzantine viewpoint of the first crusade, Anna Komnena's *Alexiad*, pictures an emperor both overwhelmed by the unexpected scale of the western enterprise and unable to comprehend the mindset of the barbaric Franks (Sewter 1969: 308–9). There are good reasons, however, to disregard the tone, if not the content, of the *Alexiad*, which was written in the 1140s by a woman who was only a child at the time of the events being described, and who moreover was trying to eulogise her father, the emperor, in pursuit of a domestic political agenda (Shepard 1996: 108–9). Alexios had asked for western aid, and although he may have been surprised at the numbers who responded, the markets in and around Constantinople proved adequate to feed them. But the impression of an invading force conveyed to Anna Komnena is understandable given the appearance of different armies each with a distinct regional character and commanded by the leading magnate of that region. Moreover, if one of the leaders, Robert of Flanders, was the son of an old friend of Byzantium, another, Bohemond, had tried to conquer the Adriatic provinces of the empire with his father in the 1080s.

A hostile atmosphere had been created by skirmishes with imperial troops in the Balkans, mainly Pechenegs and Cumans, even before all the armies had reached Constantinople. The main problem, however, was to clarify the relationship between the crusaders whom Alexios thought he was taking under his banners, and the imperial throne. The oath of vassalage that he imposed on the leaders probably represented his own understanding of the kind of relations that existed among the western nobility. In eleventh-century France, for example, such oaths might be taken to cover specific situations or military enterprises. But the diversity of practice regarding concepts of vassalage is reflected in the refusal by some leaders, notably Raymond of Saint-Gilles and – initially, at least – Godfrey of Bouillon, to take the oath, while most of the northern barons – the counts of Blois, Vermandois and Flanders and the duke of Normandy – made no such difficulties. Bohemond, whom one might have expected to cause trouble, may even have asked to be taken directly into imperial service as *domestikos* of the East. Raymond's refusal was explained on religious grounds – in taking the cross he had taken service with God, and could not therefore swear allegiance for the same enterprise to any man. But he, or the northern leaders, could equally well have baulked on the grounds that they already held lordships from a king in the West, and could not owe vassalage to more than one lord. Such considerations had already, earlier in the decade, troubled some of Robert of Normandy's barons who owed allegiance to Robert for lands in Normandy and to William II for lands in England, and it is curious that they do not seem to have worried Robert at Constantinople. The difference may be that in 1096 the crusaders were taking a personal oath that had no application to territory; they swore, indeed, to restore territory conquered from the Turks to Alexios. One further question is raised by the oath: what lands did Alexios expect to have restored to him? Asia Minor and Antioch certainly; but beyond that, Syria and Palestine had once formed part of the Roman Empire, and the same logic could be applied to them as to the lands lost more recently to the Seljuqs. Alexios' grandson, later in the twelfth century, seems to have pursued such an argument.

The western chronicles, with the notable exception of Albert of Aachen, give the impression that the Byzantines themselves played little role in the military enterprise once the crusaders had crossed to Asia Minor. One chronicle, indeed, claims that Alexios cheated the crusaders out of their rightful spoils by accepting the surrender of Nicaea in June 1097, when all the hard work had been done by the westerners. The sense of common purpose with which they became imbued once the fighting had begun, and particularly after the hard-won battle of Dorylaeum (1 July 1097), in which Kilij Arslan was decisively beaten, must have helped to bond the crusaders together, but also to make them more suspicious of outsiders. Fulcher of Chartres found this cooperation among different peoples remarkable (Fink and Ryan 1969: 88). The fact that a Byzantine contingent under Tatikios accompanied the crusaders as far as Antioch, and took part in the first phase of the siege, is unappreciated by most chroniclers (France 1971: 131–47 and 1994: 243).

There is no doubt that the nine-month siege of Antioch (October 1097–June 1098) marked the critical phase of the expedition, and helped to shape its subsequent character. After the relatively easy passage through Asia Minor, in which the towns of Cilicia were taken, this was the first serious threat to the crusade. The privations of the siege were terrible. Northern Syria has cold winters, and even the capture of the port of St Symeon could not assure an adequate food supply. Fighting men and horses were lost to starvation and the inevitable disease caused by poor hygiene. Anna Komnena's reflection on Bohemond, though in a different context, is apt here: '[This] has taught me how hard it is to check all barbarians once they have set their hearts on something: there is nothing, however objectionable, which they will not bear when they have made up their minds once and for all to undergo self-inflicted suffering' (Sewter 1969: 367).

Rivalry among the leaders threatened the enterprise. Two of the younger commanders, Baldwin, younger brother of Godfrey of Bouillon, and Tancred, a relative of Bohemond, clashed over possession of Cilician towns (Asbridge 2000: 16–24), and Baldwin went off in 1097 to the north Syrian city of Edessa at the invitation of its Armenian governor, and took no part in the siege of Antioch. Bohemond and Raymond of Toulouse seem to have distrusted each other's motives. It is not clear at what point it became obvious that Bohemond had no intention of honouring the oath to return Antioch to Alexios, but it was probably even before the city fell. Bohemond was in a strong position, for he had been elected overall commander to coordinate resistance to Ridwan, the emir of Aleppo who belatedly sent an army to raise the siege, and his military skill secured a victory at the Lake Battle outside Antioch in February 1098 (France 1994: 245–51). By this time, Tatikios had left the siege, and when the crusaders finally entered the city after Bohemond had suborned a captain in the garrison, there was no line of communication between them and the Byzantines.

This was to prove a crucial factor in the aftermath of the siege. The crusaders found themselves besieged in turn within the city they had just captured, by a relieving army sent by Kerboga, atabeg of Mosul. Matters were so desperate that Stephen of Blois, the army's quartermaster, who had withdrawn before the entry into the city to the port of St Simeon, gave the expedition up as lost and retreated towards Constantinople (Brundage 1960: 380–95). He met Alexios at Philomelium on his way to join the crusade, but his news dissuaded the emperor from going any further. Not surprisingly, after the crusaders had managed, against the odds – and again under Bohemond's command – to drive Kerboga away (June 1098), they considered that Alexios had broken faith with them. By abandoning them to their fate, Alexios had lost Antioch to the crusaders.

It was not until six months after securing Antioch that the crusade continued. The reason for the delay was largely that nobody knew quite what to do next. Adhémar of Le Puy, whose authority alone of all the leaders was unquestioned, died of typhoid in the summer. The emperor had not shown up to claim his prize; so the crusaders wrote to Urban II, inviting him to take possession of the city. Bohemond alone seems to have had a solution to the dilemma – he would keep Antioch as a personal fief. It is remarkable that during the months of delay, the armies did not disperse; their leaders occupied them instead with securing the region that was to become the principality of Antioch (Asbridge 2000: 42–6). In fact, the insistence of the rank and file on marching south, and their willingness to accept Raymond's sole command in January 1099, is the strongest indica tion that for most crusaders, Jerusalem was the natural destination of the expedition.

Some of the towns along the route south, such as Tripoli, were glad to submit to the crusaders; others, like 'Arqah, had to be stormed. But on the whole resistance was light, for the defeat of Kerboga in the north and the Fatimid capture of Jerusalem while the crusaders were at Antioch had left Palestine under no single authority. The crusaders reached the holy city on 7 June, and took it on 15 July. The relatively short siege may appear something of an anticlimax after the travails at Antioch, but success was far from certain. Besides the casualties lost in the two years since crossing to Asia, the army had lost Bohemond, who had stayed behind in Antioch, and Baldwin, who had insinuated himself into mastery of Edessa. The remaining leaders had divided more or less into two factions, with Raymond and his Provençals on one side opposed by the northern French, Normans and Flemish. Jealous rivalry between the leaders increased as the crusade proceeded, but how much this owed to mutual antipathy between different cultures is unclear. Ralph of Caen, Tancred's biographer, declared that Normans and Provençals were as different from each other as ducks and chickens (Ralph of Caen 676), but recent research has suggested that the chroniclers came to see the crusade as an enterprise that bonded participants together, with the result that a new concept of a 'Frankish people' evolved as a consequence of the expedition itself (Murray 1995: 59–73; Bull 1997: 195–211). The siege of Jerusalem, partly because of mutual distrust among the leaders, went badly until two Genoese galleys arrived at Jaffa to supply the necessary wood for building siege engines (France 1994: 334–7). When the crusaders were finally ready to assault the walls, it was Godfrey of Bouillon, encamped on the north side of the city, who made the first breakthrough. He found the defenders were better equipped than the crusaders, and the fighting for control of the walls was desperate.

The ferocity of the battle on 15 July may explain in part the bloodletting that accompanied victory. No two accounts of the massacre are identical, but it seems that a large number of Muslims who had taken refuge in the al-Aqsa mosque were killed. Modern historians cannot agree on the scale of the massacre, but understandably, those writing from an eastern perspective have stressed the shock the event caused throughout the Islamic world (Kedar 1997: 288–9; Hillenbrand 1999: 63–6). On the other hand, western historians have argued that the contemporary descriptions are inconsistent and misleading, and that the treatment of the inhabitants, though brutal, fell within the norms observed in western warfare (France 1994: 355–6). At any rate, Raymond d'Aguilers' notorious description of blood rising to the height of the horses' bridles is an image taken from the biblical Book of the Apocalypse rather than a narrative description. There is independent evidence from a Jewish source that many of the Jews of the city were taken to Ascalon to be ransomed, while some Muslims seem to have been enslaved and deported to the countryside, and others may have escaped to Damascus (Prawer 1988: 23–30).

The respite for the victorious crusaders was brief, for in August a large Fatimid army had to be repulsed near Ascalon. As at the Lake Battle and Kerboga's assault on Antioch, this engagement alone could have ended the enterprise, for in each of these cases the Muslims greatly outnumbered the crusaders. One reason for the success of the crusade in attaining its target was superior military skill. The crusaders, having been surprised by Kilij Arslan at Dorylaeum, showed themselves to be adaptable in counteracting the eastern tactics that relied on drawing the charge of the heavily armed westerners until the knights' momentum was spent and they became separated from each other or could be surrounded. At close quarters, the superior weight of western horses and armour would invariably prevail. At Ascalon they benefited from surprise; in both battles at Antioch, Bohemond's resourcefulness in drawing up the army's dispositions was critical.

A further, less tangible element in the crusade's success was the confidence imbued by piety. Unless we are to discredit all the evidence provided by the chronicles, it seems clear that the expedition had a peculiarly religious character. Of course, armies had since the early Middle Ages regularly travelled and fought with clergy who blessed weapons, said Mass and heard confessions before battle (Bachrach 2003: 78–94), but on the First Crusade, the large number of bishops and other clergy, and the probable presence of unarmed pilgrims travelling with the armies for safety, ensured a constant provision of services and prayers. These may have increased in intensity as the expedition progressed (McGinn 1978: 33–72). The circumstances of the crusade certainly gave rise to unique religious experiences, the most celebrated – or notorious – being the 'holy lance' of Antioch. The discovery of the piece of metal claimed to be the lance used to pierce Christ's side at the crucifixion followed a vision experienced by a Provençal priest, Peter Bartholomew. Many crusaders, Adhémar among them, doubted the veracity of the relic, considering it part of a Provençal strategy to wrest the leadership away from Bohemond, but its usefulness as a sign of God's continuing support for the expedition was incontestable (Morris 1984: 33–45). The lance may not have won the battle against Kerboga, but if some crusaders fought more purposefully because they thought that it would, the strategy was from their point of view worthwhile.

A vision related by another priest, Peter Desiderius, persuaded the crusade leaders to hold a penitential procession around the walls of Jerusalem in June 1099. This was of course a highly symbolic act, consciously recalling the Israelites' procession around the walls of Jericho. This was one indication that the crusaders saw themselves as the 'new Israelites' replaying the events of the Old Testament; certainly the chroniclers, such as Raymond d'Aguilers, who called the Muslim enemy 'Amalekites', 'Canaanites', and so on, seem to have thought along these lines. By this stage of the expedition – three years after having left home, having suffered disease, hunger, extremes of temperature, and the loss of horses, friends and family – it must have been a suggestion hard for most crusaders to resist. It is all the more striking in that by this time, the clerical leadership of the crusade had declined through the deaths of Adhémar and many of the senior clergy and the appointment of others as bishops in newly conquered territories. Nevertheless, the penitential character of the expedition had already been stressed during the siege of Antioch, when harsh penalties were meted out to a couple caught in adultery, and prostitutes were expelled from the camp.

The disunity of the Islamic Near East has often been cited as a reason for the crusade's success. It is true that the crusaders profited from divisions between Kilij Arslan and Danishmend in Asia Minor, though in Syria they faced an army made up of forces from the supposedly rival Muslim leaders of Mosul, Homs and Damascus. To some extent this is in any case a red herring, for either Ridwan or the Fatimid army at Ascalon in August 1099 could by themselves have defeated the crusaders. The lack of co-ordinated resistance between June 1098 and June 1099 was the crusaders' reward for the defeat of Kerboga.

Political settlement in Jerusalem, 1099–c.1118

The intention of the crusade had been to recover territory, but no provision had been made for retaining, much less governing it. Even before the capture of Jerusalem, the crusaders had clearly begun to think in terms of some kind of ecclesiastical supervision, as appointments to bishoprics that were probably vacant, such as Albara in north Syria, and Lydda in Palestine, demonstrate (Hamilton 1980: 10, 11). A body of opinion in the army in July 1099 considered that Jerusalem itself should form an ecclesiastical lordship, as was common in many western cities, but in the event the claims of the secular leaders prevailed. The choice fell between the only two who had declared an intention to remain in Jerusalem, Godfrey and Raymond. The circumstances of Godfrey's election in July 1099 and the meaning of the title have been minutely studied (Riley-Smith 1979: 83–6; Murray 1990: 163–78). Although Raymond of Saint-Gilles justifiably regarded himself as the leader of the crusade from autumn 1098 onward, he was isolated among the leaders. The realisation that he commanded insufficient support to defend Jerusalem alone probably lay behind his pious refusal of a crown on the grounds that only Christ could wear a crown in Jerusalem. The title taken by Godfrey may have been a clever response to a stratagem on Raymond's part to dissuade others from assuming rulership but, as Murray has shown, it was an office already familiar to the duke of Lorraine (2000: 74–7).

The crucial factor underwriting Godfrey's authority, in any case, was the contingent of troops from Lorraine who remained with him in Jerusalem. The army of Godfrey and his brothers Eustace and Baldwin was composed overwhelmingly of vassals of the Ardennes–Bouillon dynasty (Murray 1992: 301–29). The willingness of sufficient numbers to remain behind in Godfrey's service assured the continuation of the dynasty after Godfrey's death in August 1100. By holding the Tower of David in the name of Baldwin until he could arrive in Jerusalem in November 1100, the Lorrainers foiled an attempt by Daimbert, the patriarch of Jerusalem, to seize power for himself. Daimbert, who had been associated with the crusade since its inception (he was among the bishops who accompanied

Urban II in France in 1095–6), was sent to the East by Urban with a Pisan fleet, but by the time he arrived in December 1099 the issue of rulership had been settled, and he had to content himself with what looks like a compromise – the complicity of Godfrey in deposing Arnulf of Choques from the patriarchate in his favour (Hamilton 1980: 16). Even before Godfrey's death, however, Daimbert had succeeded in extracting temporal concessions from him on behalf of the patriarchate, presumably in return for the use of his fleet. As in July 1099, however, Daimbert was away from Jerusalem at the critical moment – on this occasion besieging Haifa – and the opportunity to secure mastery of the city was lost.

Baldwin I could not afford to share his brother's scruples about accepting a crown. The royal title and, more important, the rite of anointing that accompanied coronation, were necessary to confirm Baldwin's position as a ruler chosen by God rather than by his vassals. Besides Daimbert, Baldwin also faced the recalcitrance of Tancred, who had attempted to seize Haifa in order to form an autonomous lordship based in the Galilee but with access to the sea. Once the initial succession problem had been resolved, most of Baldwin's reign was devoted to the extension of the new kingdom. At the time of his coronation, the kingdom consisted only of the city and its immediate environment, including Bethlehem, and a corridor leading to Jaffa and including Ramla. Baldwin's apparent dependence on the Pisans, Genoese and Venetians – and the consequent commercial privileges he assigned to them – is explained by the need to capture the port towns. Caesarea fell in 1101, Acre in 1104 and Sidon (thanks to a Norwegian fleet this time) in 1110, but not until the capture of Tyre in 1124 was the coastline secure, and Ascalon remained in Egyptian hands until 1153. The acquisition of new lands and properties was crucial to the state's survival, because in order to be able to defend it against external attack, the king needed a body of knights under arms; such a body, however, could only be maintained if he had the means to reward them with land, property or money. Defence, moreover, was a constant priority: although Baldwin defeated an Egyptian army in 1101, he was in turn badly beaten a year later, and further Egyptian attacks had to be sustained in 1105–6, 1107 and 1111–12, while in 1113 the dangerous Mawdud of Mosul invaded the kingdom with Damascene support. The interior of the kingdom remained insecure: in 1106–7, the visiting Russian abbot Daniel required a military escort to travel north from Jerusalem to the Galilee through Samaria.

All this had to be accomplished from a very small base of manpower. Proportions are easier to calculate than exact figures. Out of the 659 definite or probable first crusaders identified by Riley-Smith (1997:19), only 104 are known to have stayed in the East. Not all of these settled in Jerusalem, and some were clerics. Murray (2000: 98–9) reports figures of 250–300 knights at the first battle of Ramla (September 1101), but a year later only 200 fought, many of whom had recently arrived from the West. By 1105 Baldwin was able to put 500 knights into the field, and in 1110, 600. The increase in the numbers of knights as the reign progressed was partly due to new settlement. One of the biggest gaps in our knowledge of the early kingdom is the nature and extent of settlement in the East outside the major expeditions of 1096–1101, but one indication is Riley-Smith's calculation that of the 697 settlers from before 1131 whose circumstances are known, only 122 had taken the cross in 1096–1101. Some of Baldwin's manpower came from temporary help offered by knightly pilgrims, such as a group of English knights in 1106, or King Sigurd of Norway and his followers in 1110, but there must also have been a steady stream of new settlers from the West.

One characteristic of the new kingdom was the regional composition of its baronage. Besides the Lorrainer vassals of Godfrey and his brothers, many of whom seem to have died before 1100, a significant number of knights from Flanders, Picardy and the Ile de France formed part of the baronage of the new kingdom. The prominence of certain extended families is also noteworthy. The Montlhéry family, related by marriage to the Ardennes–Bouillon ruling dynasty, established an influential network thanks to the participation of fourteen members of the clan in the first crusade (Riley-Smith 1997: 171). By c.1115 Montlhérys held fiefs from the Euphrates to the Jordan, and they also numbered among them senior clerics such as Gilduin, abbot of Notre-Dame de Josaphat. The Courtenay dynasty that was to rule Edessa until the mid-twelfth century and subsequently to become one of the major baronial families of the kingdom of Jerusalem was descended from a Montlhéry daughter. William of Tyre, indeed, was later to remark on the extraordinary progenitive powers of the four daughters of Guy I and Hodierna of Montlhéry. Although well connected, the Montlhérys were not in origin a family of great importance outside a small area in the Ile de France. The first generation of settlement in the East provided them with opportunities to acquire power on a new scale.

That such families were so characteristic a feature of the new settlement also reveals in part why the crown was able, despite a shortage of manpower and vulnerability to Egyptian attack, to build up its authority in the new kingdom with comparative ease. The only great magnates to remain behind in the East were those who established themselves at the head of new political entities – Godfrey and his brother Baldwin, Bohemond in Antioch and Raymond of Saint-Gilles, after 1101, in Tripoli. With the exception of Tancred, who in any case soon removed himself from the scene, most of the fief-holders in the new kingdom had been directly dependent on or related to the ruling dynasty before 1099. Moreover, most can be classed as belonging to the lower ranks of the landowning classes – if we take the Montlhérys as an example, families whose influence had been limited to defined regions where they might have been castellans or hereditary lords of a castle. The kingdom of Jerusalem thus had, in its first generation at least, no 'higher baronage' corresponding to that in the West.

The process by which new land was distributed also contributed to this situation. Godfrey and Baldwin created few lordships beyond Haifa and Hebron, though after 1101 Baldwin was able to take Tiberias into his gift. In part this reflects the slow pace of expansion, but even new conquests, such as Arsuf (1101) and Acre became part of the royal domain. Baldwin seems to have been anxious to create a strong landed basis for royal power. Nevertheless, he had to provide for his knights the resources that would enable them to remain in the kingdom. One way in which the kingdom of Jerusalem is usually thought by historians to have differed from western feudal states is in its potential for allotting pecuniary as well as landed resources. Baldwin made use of revenues from various royal properties to establish 'money fiefs' (fiefs *en bezant*); for example, in 1107 Gerard the Chamberlain was given income from the revenues of Jaffa, a royal property, while Robert of Apulia was given the farm of tribute from the local Muslim population in Arsuf (Murray 2000: 113). Such tributes were a valuable source of cash – in 1106 and 1107 Baldwin accepted payments from the still unconquered cities of Sidon and Tyre in return for breaking off his sieges – that could be used in lieu of landed property to create nominal fiefs. Plunder from battle against the Egyptians or from raiding beyond the kingdom's borders was also a means of supplementing resources – according to Albert of Aachen, Baldwin's campaign southeast of the Jordan into the wadi Musa in 1112 was undertaken to gather booty from caravans. Because the kingdom was a more urban society than that in the West, greater potential existed for granting out as fiefs property in towns that could be let or sold, or from which commercial profits could be realised (Prawer 1980: 154). The royal demesne also included villages, with their agricultural resources, fortresses and fiscal rights such as the collection of the head-tax imposed on non-Franks.

The circumstances of the young kingdom's survival also contributed to the growing authority of the crown *vis-à-vis* the barons. The king was almost constantly at war, either for defence or to extend territory, and this naturally resulted in a high turnover of fief holders from death in battle or imprisonment. The lack of continuity among baronial families in the first generation after settlement has been extensively analysed (Tibble 1989: 5–65). For example, five unrelated knights held the lordship of Hebron between 1100 and 1135, while Galilee was held successively by four un-related lords before 1120. Not all of these changes in tenure are known for certain to have been caused by the death of the fief holder, but it is striking that in none of these instances did an heir succeed to the lordship. This may have been because the fief holder had not yet produced an heir – particularly in the early years after 1099, the number of eligible Frankish women for marriage must have been rather small. The rapid turnover of lords in two of the major fiefs in the kingdom is particularly remarkable when we consider that during this general period the baronial classes in the West were striving to make hereditary succession to lands and offices an accepted principle. Whether by design or force of circumstance, Baldwin I was able to achieve the reversion of fiefs to the crown, and thus to reserve the right of patronage to new lords.

The general picture of emerging feudal relations in the kingdom of Jerusalem thus suggests domination of the baronage by the crown. However, political events do not invariably complement this picture. In particular, the question of succession to the throne provided opportunities for the barons to determine policy. Thus, in 1116–17, pressure from the secular barons and the patriarch forced Baldwin I to repudiate his wife, Adelaide of Sicily. The marriage to Adelaide, contracted in 1112–13, had never been popular, because it guaranteed the succession to Adelaide's son from her first marriage, Roger II of Sicily, in the event that it produced no heirs. Such an arrangement ignored the claims of Baldwin's surviving brother, Eustace, count of Boulogne, who as a crusader in 1096–9 was better known, and probably more popular than Roger (Murray 2000: 15–16). When Baldwin fell ill in 1116, the barons were able to act through the instrument of a council presided over by the patriarch.

In 1118, when Baldwin died on campaign, the barons once again acted decisively to determine the succession. According to Albert of Aachen, Baldwin had designated Eustace as his heir, with his more distant relative, Baldwin of Le Bourcq, who had succeeded him as count of Edessa in 1100, as reserve choice in the event that Eustace declined the offer. William of Tyre, describing the events sixty years later, invoked the idea of a strict hereditary principle according to which Eustace should have succeeded. But a group of nobles, led by the former patriarch Arnulf of Choques and

Joscelin of Courtenay – who doubtless hoped to take over in Edessa – argued Baldwin's case on the grounds of pragmatism. Eustace, who had been count of Boulogne since 1088, was almost past campaigning age and the kingdom would be leaderless until he could come out to the East. Baldwin I died on 2 April, and Baldwin of Le Bourcq was crowned only days later. The speed of events makes this look very like a *coup d'état* on behalf of Baldwin – who by coincidence happened to be in Jerusalem at the time of Baldwin I's death. A war of succession might have ensued, because Eustace did in fact set out to claim his crown, but turned back when he heard of Baldwin II's coronation. If this was a *coup* engineered by Baldwin and a section of the baronage, who formed the rival factions? Murray (1994: 60–85; 2000: 120–3) surmises the existence of a 'legitimist party' who supported Eustace's claim. On the other side, Baldwin II, son of the count of Rethel, was related to the Montlhéry family, and it is certainly suggestive that the new king granted the important fief of Tiberias to William of Bures, a knightly vassal of the Montlhérys back in the West, and created a new fief, Jaffa, out of the royal demesne, for Hugh II of Le Puiset, another Montlhéry descendant. It certainly looks as though a closely related group of barons advanced each other's interests in order to gain influence in the new kingdom. In the process, they signalled the beginning of a more powerful baronial class.

At Baldwin I's death in 1118 the kingdom of Jerusalem, though still vulnerable, was a viable political territory. All but two of the ports as far north as Beirut were in royal hands. Capable use of small numbers of knights had secured the Egyptian frontier and extended Frankish influence east of the Jordan. A document of 1104 reveals that Baldwin even seems to have entertained the idea of conquering Egypt. Fiefs had been established through a resourceful combination of grants, while at the same time a royal demesne had been built up. Baldwin had encouraged settlement from the West and brought Christians from across the Jordan to settle in the kingdom. Much of this had been accomplished through the continued support of western arms, and without such aid it is difficult to imagine that Baldwin II would have had a kingdom to inherit. The extent to which western support could be channelled to constructive uses by the kings of Jerusalem would be crucial in the coming years.

Settlement in the north, 1098–1119

Occupation of towns in Cilicia and northern Syria probably began as a strategy to secure the First Crusade's passage into the vicinity of Antioch and to defend the army's flanks during the siege of the city. After the capture of Antioch, the leaders grabbed what they could in rather piecemeal fashion during the long delay before marching south to Jerusalem. The formation of the principality of Antioch was therefore the work of the First Crusade itself, rather than of Bohemond. Antioch's geographical situation, close to regional centres of Muslim population such as Aleppo, Homs and Shaizar, and between equally warlike rivals for power – the Danishmendids, the Seljuqs of Rum, the atabegs of north Syria and the Armenians of Cilicia – made it more vulnerable than Jerusalem. In addition to this, the Byzantine emperors never renounced their claim to a city they had controlled until 1085.

Bohemond himself had little impact on the state he founded. Captured by the Danishmend emir in 1100, he was only released (in an episode which became the subject of a romantic interlude in Orderic Vitalis' chronicle) in 1103, and returned to the West to raise an army for a new crusade in 1105. His invasion of the Byzantine Empire in fulfilment of this resulted in defeat in 1108, and he died in Apulia in 1112. His magnificent tomb in Canosa, Apulia, bears witness to the eclectic cultural ambience in which his life was played out. Tancred, who acted as regent for Bohemond and his son from 1100–3 and 1105–12, was responsible for the real work of construction and consolidation. It was he who recovered the Cilician towns of Tarsus, Mamistra and Adana from the Byzantines and captured Latakia. The Franks of north Syria, however, were undone by their own ambition, for the joint attempt to take Harran by forces under the leadership of Tancred, Bohemond, Baldwin of le Bourcq and Joscelin of Courtenay, led to a crushing defeat in 1104. As a result of this the Cilician towns shrugged off Antiochene rulership in favour of Byzantine; but more seriously, Tancred's control of the Jebel as-Summaq and Ruj valley, which acted as a natural defence against Aleppo, was ended. At least seven strongholds were lost to Ridwan.

It was the defeat at Harran that prompted Bohemond to return to the West to launch the crusade of 1106. In some ways this expedition was a precursor of the more notorious crusade of 1202–4, which also ended with an attack against the Byzantines rather than the Turks. What eventually transpired should not obscure the fact that the crusade was preached by papal commission throughout the West, and that contemporaries regarded it as an expedition to reinforce the conquests of 1097–9. Bohemond's insistence on besieging Durazzo seems to have been articulated as an act of vengeance against an ally who had proved to be faithless (Richard 1999: 129), but there was probably a conscious policy behind it. Bohemond reasoned that in order to secure his north Syrian possessions, he would have to defeat the Byzantines as well as Ridwan, and the blow may as well be struck in one Byzantine territory as in another. We should also note that while Bohemond was tying down imperial forces in the West, Tancred was able to retake Cilicia and Latakia.

The terms of the treaty of Devol (1108) have been interpreted in different ways. Lilie argued that the treaty, as recorded by Anna Komnena (Sewter 1969: 424–34) represented a complex network of clauses designed to restrict Bohemond's effectiveness in Syria, and, moreover, that it demonstrated Alexios' understanding of western feudal law (Lilie 1993: 77–8). In contrast, Asbridge has recently countered that the treaty derived from Greek as well as western precedents, and that Alexios wished to regard Antioch as falling under the umbrella of *pronoia* arrangements (Asbridge 2000: 95–8). The fundamental point is that Alexios, recognising the impossibility of driving Bohemond out of Antioch, tried to absorb him into the structure of Byzantine rule. Bohemond was to retain Antioch until his death, with the title of *dux*; his lands were to include St Simeon and the coast, the towns of Baghras and Artah and the Latin possessions in the Jebel as-Summaq. Latakia and Cilicia, however, were to revert to direct Byzantine rule. Lilie's conclusion that Bohemond did well out of the treaty (1993: 79), is hardly borne out by this attempt to restrict Antioch to its borders after Harran, for, as Asbridge points out, much of what the emperor granted to Bohemond (including Aleppo itself) was still in Muslim hands (2000: 98).

The treaty in any case remained academic, for Tancred, who continued to exercise the regency until his death in 1112, did not consider himself to be bound by it. By the time that Roger of Salerno succeeded him, the principality had been extended in all directions to encompass Cilicia to the northwest, Latakia, Jabala and Apamea to the south, the Jebel as-Summaq and, to the northeast, the strategically important town of Marash. Short of conquering the Muslim states of Shaizar or Aleppo, there was little room for further expansion. Consolidation, however, came to an abrupt end with the disastrous defeat known as the Field of Blood (1119). The defeat was all the more crushing for having followed a period after Ridwan's death (1114) in which Aleppo had more or less become an Antiochene tributary. Surrounded by Latin fortresses, the Aleppans abandoned a policy of appeasement and alliance, and appealed to Il-Ghazi, ruler of Mardin. Il-Ghazi probably had no thought of destroying Antiochene power, but intended only to restore to Aleppo strongholds that lay between the two cities (Asbridge 2000: 75). In the event, however, Roger allowed his army to be trapped near al-Atharib; on the Field of Blood, he himself was killed in a massacre from which only 150 Frankish knights escaped. Antioch was saved by the intervention of Baldwin II of Jerusalem, who between 1119 and 1123 restored the strongholds beyond the Orontes. Although its territorial integrity was secured, the principality would for much of the rest of the twelfth century rely periodically on aid from Jerusalem in order to retain its frontiers from Muslim threat and its independence from Byzantine claims to suzerainty.

In military and demographic terms, Antioch was the most important of the Frankish states in the north. The first to be established, however, was Edessa. Baldwin I, having been invited by the Armenian population to drive the Turks from the area between the Taurus mountains and the Euphrates, managed to profit from quarrels among the Armenians to seize control of Edessa early in 1098. Enough crusaders remained with him to establish a county, but the westerners were always a small minority in a population of Armenians, Syrian Orthodox Christians and Turks. Despite the departure of Baldwin I for Jerusalem in 1100, and the capture of the second count, Baldwin of Le Bourcq, and Joscelin of Courtenay in the attempt to capture the nearby city of Harran in 1104, Edessa survived the first generation of crusader settlement in vigorous health. Alexios Komnenos regarded it as part of Byzantine territory – in 1097 it was ruled by a Greek-appointed Armenian, Thoros – and may have granted it as a dukedom to Bohemond at Devol (Asbridge 2000: 98), but Antioch and Edessa practised mutual cooperation in military affairs. Bohemond's capture in 1100 came about as he was answering an appeal from Baldwin, and Franks from Antioch and Edessa fought in alliance at Harran. Close relations developed while Tancred acted as regent for Edessa during Baldwin's captivity. Antioch was prepared to stretch its own resources in order to protect Latin rule in Edessa, but whether this was out of a sense of confraternity between frontier states, or because Bohemond and Tancred cast envious eyes on the rich natural resources of its northern neighbour (WT 1986: 504–6) must remain conjectural. In fact, Baldwin of Le Bourcq had already done homage to Baldwin I in 1100, and although Tancred seems to have wanted Edessa to acknowledge Antiochene suzerainty, this was never enforced. In 1109–10 Tancred's attempt to claim Antiochene suzerainty over Edessa was rebuked by Baldwin. In 1108, on Baldwin's release from captivity, Tancred allied with Ridwan in order to lock the count out, but after a battle near Tell Bashir between Tancred and his Muslim ally, and Baldwin and the atabeg of Mosul on the other, order was restored. The most significant aspect of this unedifying scrap was the willingness of Latin and Turkish rulers to cooperate. After the 'righteous violence' against the Muslims on the crusade, the Franks were becoming sucked into the political complexities of the Near East. Edessa's vulnerability to Turkish attack was graphically demonstrated in 1110–11 when Mawdud, atabeg of Mosul, joined with Tughtegin of Damascus to ravage the county. The lands to the east of the Euphrates, which were hardest hit, never really recovered.

The last of the Latin states to be established as a result of the crusade was Tripoli. This was something of a consolation prize for Raymond of Saint-Gilles, who had been disappointed at Jerusalem. His leadership of the crusaders of 1101 doubtless owed a good deal to the expectation that he could still re-establish himself in the East. In the event, however, his humiliation was completed by being forced to give an undertaking to Tancred, then regent of Antioch, not to make any conquests on the coast south of Antioch. On Tancred's part, this was probably not so much dislike of Raymond as the fear that the count would insist on returning any conquests to the Byzantines. In the event, however, Raymond forswore his oath and by 1105, when he died, he had taken Tortosa and Gibelet and begun the siege of Tripoli. This was not completed until 1109, when his son, Bertrand, arrived. Betrand's arrival enabled the kingdom of Jerusalem to act, for the first but not the last time, as arbiter between the competing claims of the Latin rulers. For Bertrand was opposed by William Jordan, Raymond's cousin, who had maintained the siege of Tripoli since 1105, and who had called upon Tancred for help. Baldwin I took the opportunity to settle the grievances of Baldwin of Le Bourcq against Tancred regarding Edessa at the same time; outflanked, Tancred and William Jordan had to give way. Tancred gained something from the settlement on behalf of Antioch, however, for William Jordan was allowed to hold 'Arqah as a vassal of Antioch. Baldwin I gained far more for Jerusalem, however. Bertrand was given Tripoli and its environs, as a vassal of Jerusalem. In the event, even Antioch's suzerain status over the northern part of the embryonic state was curtailed by the sudden death of William Jordan and the reversion of 'Arqah to Bertrand. The deaths of Tancred and Bertrand in 1112 engendered closer relations between Tripoli and Antioch. Pons, Bertrand's son, married Tancred's widow Cecilia, and the county of Tripoli achieved a quiet stability.

It has been argued that the council of Tripoli in 1109–10 demonstrates the existence of a kind of confraternity among the Latin rulers in the East (Asbridge 2000: 117). It is difficult to know what the terms of such a confraternity might have been, beyond the recognition of the king of

Jerusalem as the senior member. Tancred was forced to accept terms in a council because he realised that he could not stand against the combined forces of Jerusalem and Edessa, together with Bertrand's Provençals. But Baldwin had no authority in feudal custom to summon Tancred, and according to Albert of Aachen, he resorted to a formula on behalf of the 'universal church of Jerusalem' in order to do so. The council demonstrated that, ten years after the crusade, the Latin rulers were using any method available to them to maintain and expand conquered territory, much as they and their ancestors had done in the West. They regarded conquests as heritable land to be ruled according to feudal custom as far as possible and passed on to heirs. Mutual protection was important because the loss of any of the lands to the Turks – or the Byzantines – would threaten all; but where given, it did not necessarily imply dependence. When the agreement was put to the test by Mawdud's invasion of Edessa in 1110, the Armenian chronicler Matthew of Edessa and the Arab historian Ibn al-Qalanisi believed that Tancred had to be begged to join Baldwin I and the other Latin rulers (Asbridge 2000: 119–20); and in the event, Tancred took little part in the campaign to fortify Edessa. There was little evidence, in truth, of the spirit of Frankish unity that had so impressed Fulcher of Chartres during the crusade.

CHAPTER 4

Politics and war in the Crusader States, 1118–87

The political development of the Crusader States in the twelfth century was marked by various factors. War against the neighbouring Muslim states was endemic, but the lack of political unity in Islam in the first half of the twelfth century allowed the Franks to establish strong kingship in Jerusalem. The main themes of this chapter are:

- how the Franks dealt with problems of dynastic succession;
- how extra military resources were found through the establishment of the Military Orders;
- how relations between the Franks in the East and in the West affected the security of the Crusader States.

Political developments in Outremer, 1118–44

The death of a king without a clearly recognised heir in the Middle Ages could not fail to threaten the stability of his kingdom. The succession of Baldwin II was effectively a *coup d'état* that polarised baronial opinion in the kingdom into a 'legitimist' faction that supported the Bouillon dynastic claim and a 'pragmatic' group that preferred to elect the closest relative already in the East (Murray 1994: 60–85; Riley-Smith 1997: 173–4). This dynamic was to repeat itself several times during the twelfth century; indeed, the relationship between the settler nobility, the ruling dynasty and their kin in the West is one of the most enduring ways of understanding the problems experienced by the kingdom of Jerusalem in this period (Smail 1969: 1–20; Phillips 1996: *passim*).

The succession of Baldwin II brought to the ascendancy a new kinship group, that of the Rethel–Courtenay–Montlhérys. Baldwin's patronage of those who had brought him to power – Joscelin of Tiberias, for example, who was a Courtenay, became count of Edessa, while Hugh of Le Puiset, a Montlhéry, was made count of Jaffa – effectively replaced a Lotharingian élite with a northern French one. Murray has suggested plausibly that members of this kin grouping in the West, whose semi-independence from the throne in the Capetian royal demesne had been circumscribed by the vigorous policies of Louis VI (1107–37), naturally gravitated to Outremer, where they might receive large rewards in a new frontier territory. The underlying tensions that this caused within the baronial class were revealed when Baldwin II was captured in north Syria in 1123. During the sixteen months of his captivity, an approach was made, presumably by the 'legitimist' barons in Jerusalem, to Charles, count of Flanders, to depose Baldwin in his absence and rule in his stead (Murray 1994: 69–75; 2000, 124–46). Eustace of Boulogne was probably considered too old (and he may in any case already have become a monk by 1123), which left Charles as the active head of the Ardennes–Bouillon dynasty. In the event Charles' refusal of the offer and Baldwin II's return to the kingdom in 1125 led to further royal patronage of 'new men', such as the Brisebarres, doubtless in an attempt to surround the throne with trusted supporters to counteract the influence of the 'legitimists'.

The insecurity of the new dynasty was underlined by the continuing problem of succession. Baldwin's four children were all daughters, which meant that a husband had to be sought for the eldest, Melisende. The decision to find a husband in the West to rule as Melisende's consort was natural. Whoever was chosen would be responsible for the military leadership of the kingdom; in the 1120s the Armenians of Cilicia were militarily not powerful enough, and the Byzantines too much of a threat in the north, to be considered – even if such a choice had been acceptable to the Latin settlers. But because he had married an Armenian princess while count of Edessa, Baldwin had no western in-laws who might supply an eligible bachelor; moreover, the danger from the dispossessed Boulogne dynasty had not vanished, for Eustace's daughter Matilda had married Stephen, count of Blois, who was himself a descendant of William the Conqueror, and who might relish the opportunity to press a claim to a throne. Baldwin's policies after his release centred on aggression, and Damascus had been identified from 1125 onward as a target. The search for a husband in 1127 can be linked to the attempt to recruit western knights for an assault on Damascus, which some historians identify as a 'crusade' (Phillips 1996: 19–43). This explains in large measure the choice of Fulk V, count of Anjou, as Melisende's consort. Already in middle age, Fulk had valuable military resources and, as ruler of a powerful territory in north-western France that had proved able to defy both Normandy and the Capetians, considerable experience of warfare. Moreover, from his first marriage Fulk already had an heir for Anjou, and was thus free to take up new responsibilities. Above all, perhaps, Fulk was familiar to the barons of Outremer from an armed pilgrimage undertaken in 1120, and may already at that stage have been an active supporter of the Templars (Barber 1994a: 11; Phillips 1996: 30).

The import of a western consort was a measure to which the kingdom of Jerusalem would be driven again in the 1170s, 1180s–90s and 1210s. In such cases, the military advantages of an alliance with a western dynasty were accompanied by an equivalent threat to the stability of royal/baronial relations posed by the arrival of the consort's entourage of household knights, for whom positions and offices had to be found to compensate those left behind in the West. The effect of Fulk's patronage after his suc-cession with Melisende in 1131 has been studied in depth by Mayer, who concluded that although Fulk did surround himself with 'new men', these were not invariably Angevins, and indeed that he retained some of Baldwin II's favourites, such as Walter Brisebarre (1988: 1–25). The overall effect of Fulk's arrival, however, can be assessed in the revolt of Hugh II of Le Puiset, count of Jaffa, who was one of Baldwin II's protégés. William of Tyre represented the revolt of 1134 as the result of a love affair between the count and Melisende, but this looks like the deliberate romanticising of a political crisis. The fact that Hugh's revolt failed for lack of sufficient support among the barons should not disguise its significance. Hugh was acting on behalf of the interests of Melisende, whom Fulk had a tendency to sideline early in his reign, and of their heir, Baldwin, born in 1129. Indeed, it was to guard against such an eventuality that on his deathbed Baldwin II named the two-year-old as joint heir with Melisende and Fulk. Another layer of interpretation is provided by Mayer's analysis of Hugh's background and adherents, for although Hugh's family came from the Ile de France, he had grown up in Apulia and came to represent the Normans in the kingdom, many of whom were dispossessed by Fulk in favour of Angevins (Mayer 1989: 23–5). Hugh was thus identifying himself with the cause of nobles whose hold on crown offices and castles had been threatened by newcomers. Such an analysis makes perfect sense in the wider European context, where for example the last decades of Norman rule in England were characterised by struggles between rivals for crown patronage.

The reign of Baldwin II also highlighted growing differences between the interests of the barons of Jerusalem and those of the northern states. Whereas the threat from Egypt had receded in the first decade of crusader occupation, the Seljuq rulers of north Syria posed constant dangers to Antioch and Edessa. The defeat at the Field of Blood left Antioch vulnerable, and the death of Prince Roger in the battle made a regency necessary until the legitimate heir, Bohemond II, came of age. During the first half of his reign, therefore, Baldwin II acted as military protector of Antioch, much to the disgust of Jerusalem barons who complained of his neglect of the kingdom. Indeed, Baldwin's (ultimately unsuccessful) strategy against Damascus, pursued from 1125–9, may have derived from his inclination to see Antioch's defence as integral to the security of Jerusalem. The marriage of his daughter Alice to Bohemond II cemented an apparently growing bond. All this changed abruptly, however, when Bohemond II was killed in 1130. The lengths to which Alice was prepared to go to retain sole authority – which included an appeal for assistance to Zengi, atabeg of Mosul – forced Baldwin to march north again to discipline his daughter. When Baldwin himself died a year later, however, Alice was able to count on support from Pons, count of Tripoli and Joscelin count of Edessa, and the new régime in Jerusalem appeared isolated.

Antioch now faced a similar dilemma to that faced by Jerusalem in the 1120s: the need to find a husband as consort for a female heir. In 1134, when Fulk intervened, Constance, the daughter of Alice and Bohemond II, was only six, and could not be married for another six years (Phillips 1996: 50). A betrothal could, however, be arranged once she was seven, and in the circumstances it made sense to initiate proceedings early. Alice favoured a Byzantine marriage, a proposal that might be politically difficult but that would ensure imperial protection for the principality, and some Antiochenes were pragmatic enough to support what would amount to an acknowledgement of Byzantine claims (Asbridge 2003: 29–47). But in the event, Fulk was asked to suggest a suitable husband, and chose Raymond of Poitiers, younger son of William IX of Aquitaine. So young was Constance, and so ambitious for power was Alice, that according to William of Tyre, Alice was duped into accepting Raymond for her daughter's hand by being led to assume that he was proposing to her! (WT 1986: 658–9). In fact the machinations over the marriage were probably more prosaic and revolved around Alice's desire to retain control over her heiress (Phillips 1996: 60–1, 65–6). Raymond was a revealing choice by the man who had himself been approached only a few years earlier to fulfil the same role for Jerusalem for which he now proposed Raymond. It was also a risky one, for although Raymond's pedigree was impeccable, and although he had crusading connections through his father, he was young (probably in his mid-twenties, though perhaps younger), and can hardly have had the same military or political experience as Fulk (Phillips 1996: 53–7). This was crucial at a time when Antioch was in considerable peril after Zengi's seizure of Aleppo in 1128. One might suspect Fulk of selecting someone over whom he thought he could exercise sufficient influence to reimpose a protectorate over Antioch, but if this was ever his intention, he was too preoccupied with the revolt of Hugh of Jaffa and its aftermath to intervene in the north after 1134.

The decision to import a husband from the West was even more politically charged in Antioch than it had been in Jerusalem, for it amounted to a repudiation of the Byzantine suzerainty implied by the treaty of Devol. John Komnenos (1118–43) invaded Cilicia in 1137, replacing Latin bishops with Greek ones, and began to besiege Antioch, forcing Raymond to submit. The Byzantine threat prompted Pope Innocent II to ponder military sanctions against the emperor. Runciman, writing from a markedly pro-Byzantine stance, observed that John's invasion only proved Alice right (1952: 213). The terms to which Raymond was forced to agree went beyond Devol, for he was now recognised as an imperial vassal for Antioch only until Aleppo and Shaizar were conquered, after which he would rule the new conquests and Antioch revert to direct imperial rule. Byzantine ambitions, however, outpaced their military capacities, and when John died in Syria in 1143, he was still trying to enforce his suzerainty over Antioch; the prospect of extending Frankish or Byzantine territory beyond the Amanus range looked remote.

The period between the two military disasters of the Field of Blood (1119) and the fall of Edessa (1144) saw the Latin states undergo different fortunes. Despite problems with succession, the kings of Jerusalem were able to benefit from quiescent frontiers to develop a solid fiscal and territorial basis for royal power. The striking feature of Hugh of Jaffa's revolt is that it failed to command support among a baronage that did not enjoy the territorial resources to resist the crown by military means. Meanwhile the boundaries of the kingdom were expanded by cumulative acts of expansion, such as the seizure of Banyas in 1140. Antioch, in contrast, had suffered from the revival of Byzantine power from one side and the threat posed by Zengi from the other, while enduring its own succession crisis. Like Jerusalem, however, Antioch was able to attract significant interest and support from the West.

The smaller states proved more vulnerable. Damascus, scared by Frankish aggression in 1129 and by Zengi's policies in the 1130s, saw an outlet for its own aggrandizement to the west, and in 1137 the county of Tripoli suffered a serious attack that deprived it of much of its eastern territory; in spring Count Pons was killed, and only a few months later his successor Raymond II was captured by Zengi. Tripoli's survival as an independent state was in jeopardy, and could only be assured through the protection of the military orders. To the north, Edessa, though territorially large and wealthy in natural resources, suffered from its exposed position. The city of Edessa itself was separated from the more protected fiefs in the county by the Euphrates, which meant that when the count was in his preferred stronghold of Turbessel (Tell Bashir), he could not easily respond to threats from the East. Ultimately, the counts simply had too few western forces at their disposal, and Edessa, primarily a mercantile centre, was defended largely by mercenaries. Joscelin II, who succeeded his crusading father, Joscelin of Tiberias, in 1131, has been blamed for succumbing too easily to Seljuq pressure (Nicholson 1958: 432, 446), but the real damage came from the weakening of Antioch and the inability of Joscelin II and Raymond of Poitiers to recreate the vigorous, if stormy, partnership between Count Baldwin and Tancred in the early period of Frankish settlement.

The fall of Edessa was almost accidental. Zengi's real ambition was the capture of Damascus, in order to secure total dominance in Syria, but from 1140 onward he had been preoccupied with enforcing his rule over the Artuqids to the north of his base at Mosul. When Joscelin, perhaps unwisely, involved himself in an alliance with the Artuqids, Zengi seized the opportunity, in December 1144, to encircle and capture Edessa. The aftermath emphasised the problems that the Franks of Edessa had always faced, for Zengi massacred the western defenders but spared the much larger indigenous Christian population, who in turn viewed Muslim rule with equanimity. The county was now cut into two, and when Joscelin's attempt to retake his capital in 1146 was repulsed with further serious losses, its continued existence as an independent territory was hopeless.

The loss of the first of the crusader states to have been established represented the most serious setback yet faced by the settlers. As well as being a blow to Frankish morale, it affirmed the danger of allowing the fragmented Seljuq lordships of north Syria to become concentrated in the hands of a single ambitious ruler. As we shall see, it also raised the question of how much the West was prepared to expend in the defence of the conquests of 1098–9.

The Military Orders

The most important institutional development in this period was the creation of the Military Orders. The most powerful of these, the Templars and the Hospitallers, were to amass such territorial and financial resources as to become virtually independent sovereign powers in the region. These resources, coupled with a spiritual role that was itself bolstered by wideranging privileges granted by the papacy, turned them into international conglomerations the like of which had never previously been known in Europe or the Near East.

The origins of the Military Orders did not presage such future greatness. The Hospital of St John, founded by merchants from Amalfi in the last quarter of the eleventh century, catered for pilgrims who needed accommodation or medical attention while in Jerusalem. The Hospital was originally part of the monastery near the Holy Sepulchre known as St Mary of the Latins, also an Amalfitan foundation, but by 1113 it had been recognised by the pope as a separate Order within the Church. The legend of the heroic suffering of the Latin administrator of the Hospital, Gerard, during the siege of 1099, at the hands of the Egyptians, made the institution popular with donors from the beginning of the Latin settlement. Royal patronage was noticeable as early as 1101, when Baldwin I gave the Hospital a tenth of the spoils of his victory at Ramla, and in 1112 the patriarch exempted it from paying tithes within his jurisdiction (Riley-Smith 1967: 32–43). Thus, from the start, valuable ecclesiastical exemptions went hand-in-hand with aristocratic bequests.

The Templars were a recognisable institution from 1120 onward, but originated from an indeterminate group of crusaders who had remained in Jerusalem in 1099 under a personal vow to the patriarch of Jerusalem and who formed a kind of military household in the Holy Sepulchre (Luttrell 1996: 193–202). From 1120 their purpose became much more sharply defined by their leader, Hugh of Payens, who declared the intent to provide military defence for the roads to pilgrimage sites, such as the river Jordan, that lay in areas where banditry was still rife, and for pilgrims travelling on them (Barber 1994a: 9–10). After papal recognition of the order at the Council of Troyes (1129), and particularly as a result of the articulation of their spiritual aims by St Bernard of Clairvaux, the Templars grew rapidly from the 1130s and attracted widespread recruitment and bequests. Like the Hospitallers, they benefited from generous papal privileges in the early stages of their development.

The status of the membership of both orders in their early stages reflects broader spiritual currents in Europe. The particular attachment to pilgrimage expressed in both orders' constitutions placed them firmly within a spirituality centred on the experience of personal devotion to the Holy Land. But the popularity of the Hospital can also be seen in the context of the growth in foundations dedicated to practical piety, and associated in particular with the Augustinian Rule, in the early twelfth century (Luttrell 1997: 40–7; Nicholson 2001: 1–3). The earliest Hospitallers were laymen – and women: the Amalfitans also founded a hospice for women pilgrims – who followed what was described in an Amalfitan chronicle as 'an almost religious way of life', which probably means that they had taken vows but were not full religious. Such groups of men and women were, around 1100, an increasingly common part of the European landscape. When the canons of the Holy Sepulchre became Augustinian regulars in 1114, the Hospital moved in that direction, and the rule written by the Grand Master Raymond of Puy (1120–1158/60) shows a direct dependence on the Augustinian rule (Riley-Smith 1967: 46–52). The Hospitaller rule shows an interest in care for the poor as well as for the sick, which may be part of the legacy of the original dependence on the Benedictine monastery of St Mary of the Latins in Jerusalem.

The Templars, unlike the Hospitallers, were a direct product of the crusades – not only because protection of the holy places was an avowed aim of the First Crusade, but also in as much as the notion of a knight dedicated to a religious purpose while remaining an active warrior was fully comprehensible after 1095. The Templar rule approved in 1129 required the brothers to take vows and to live in a community as monks, having first repudiated all ties of property and family. Although the European baronage seems to have taken up the Order enthusiastically, as testified by important bequests to Hugh in 1129–20, criticism of the ideal continued to be voiced throughout the twelfth century. John of Salisbury, for example, attacked the idea that the same body could spill blood and administer the sacraments (Riley-Smith 1967: 385–6; Barber 1994a: 41–2). The Gregorian reform movement of the eleventh century had sought to emphasise distinctions between the priestly and the secular, particularly in the lives of those entrusted with the sacraments, and Urban II had banned participation on the First Crusade by monks. It was the intervention of Bernard of Clairvaux, at the request of Hugh of Payens, and his coherent articulation of the Templar ideal in the 1130s, that ensured the order's popularity. In deliberately encouraging a blurring of the distinction between knight and monk, Bernard was pursuing a strategy that he had already used to recruit knights to the Cistercians. The core of the Cistercian reform programme was the transformation of secular institutions and ways of living into spiritual ones. For Bernard, there was no anomaly, for in his eyes the *only* acceptable way to be a knight was to dedicate one's sword to the service of God. Thus the Templar was 'a man of peace, even when drawn up for battle' (Greenia 1977: 140). It is not surprising that this idea should have proved as popular with the knightly class as the indulgence offered to crusaders in 1095 had been, for not only was the reward for this service eternal life, but this was the only sure way in which salvation could be guaranteed to military men.

A few early high-profile recruitments to the Templars helped not only to spread the order's popularity, but also to lay a basis for its wealth in landed property. Thierry I, count of Champagne, was an early donor to the order, while in England King Stephen (1135–54) granted royal manors, such as Cowley in Oxfordshire, but the most spectacular example of royal patronage came from Alfonso I of Aragon (1104–34), who in 1131 willed his kingdom jointly to the Temple, the Hospital and the Holy Sepulchre (Barber 1994a: 24–7). Magnates who had to alienate property on becoming a Templar might will part of it to the order; for others, bequests to the order were seen in the same light as those to a monastery – and, in addition, as contributing in a practical way to the defence of the Holy Land. The astonishing rapidity with which the Templars accrued wealth is attested by Louis VII's admission in 1148 that only loans from the order had enabled him to sustain his crusade.

Landed wealth in the West enabled the orders to channel resources in manpower to the East, but the orders' ecclesiastical privileges rendered them virtually free of any local episcopal jurisdiction. By 1154, the Hospitallers and Templars had each been granted a series of papal bulls that, taken together, allowed them to possess tithes from their lands, to elect their own Master, to build churches in newly colonised regions and to remain exempt from episcopal excommunications or interdicts. If the orders were popular with the lay knighthood, the clergy reacted strongly against such privileges. William of Tyre was scandalised by the behaviour of Hospitallers who, in the course of their dispute with the canons of the Holy Sepulchre in 1154, shot arrows into the canons' cloister (Riley-Smith 1967: 398–9), and in 1179 he was a member of the delegation of bishops from Outremer who brought complaints against the orders to the Third Lateran Council. Disputes dragged on into the thirteenth century: in 1228

James of Vitry, bishop of Acre, complained that the Hospitallers were even carrying off corpses to bury them in their own cemeteries, the point being that they could thereby channel burial fees away from parish churches to their own.

Whereas the Templars played from the start an aggressively military role, for example taking part in the siege of Damascus in 1129, the Hospitallers came only gradually to the profession of arms. Until the Second Crusade, however, both orders were primarily engaged in defence by garrisoning castles. The first indication of such a role by the Hospital was in 1136, when King Fulk gave the order the castle of Beth Gibelin, which lay roughly halfway between Bethlehem and Ascalon and which was thus a critical element in the defence of the southern frontier of the kingdom of Jerusalem. In 1144 Raymond II of Tripoli conceded much of his eastern frontier, including the great castle of Krak des Chevaliers, to the Hospital. Raymond, forced into this measure by his inability to defend his border, gave the land in freehold, which meant that the Hospitallers exercised rights as feudal lords in the region. Between 1160 and 1210 similar grants were made in parts of Antioch, Egypt and Armenia. In some of these cases, the Hospitallers were encouraged not only to garrison castles but also to establish colonists – hence the relevance of the papal privilege to build new churches. Lay pressure from outside the order encouraged Raymond of Puy (1128–58/60) to develop a military role and until the middle of the century the Hospitallers may even have relied largely on mercenaries. The first mention of military duties occurs in the Constitutions of 1182, where fighting is conceived as an extension of alms-giving and care of the poor. Raymond of Puy made an explicit connection between entry into the order and crusading: 'Whoever enters our fraternity will thereby be secure in God's mercy, just as if he himself had fought in Jerusalem' (Riley-Smith 1967: 58). The adoption of an aggressive role was clearly a matter of wider debate, for in 1178/80 Pope Alexander III forbad the Hospitallers to bear arms except in the specific circumstances of defending the True Cross or in besieging Muslim cities. The tumultuous events of the 1180s, however, and the need of the kings of Jerusalem to put huge armies into the field against Saladin, confirmed the trend towards the creation of a Hospitaller field army. By 1187, the order could contribute some 300 knights to the royal army, which more or less matched Templar military resources. When we consider that the total military strength of the kingdom from feudal sources amounted to about 670 knights (Edbury 1997: 124–6), the Military Orders' 600 is not only impressive but highly significant for the direction of royal policy.

The Templars' reputation as a military force was sealed by the Second Crusade. Everard des Barres, the order's Master in France, kept Louis VII not only afloat financially, but alive – when Louis' army was being mauled by the Turks in Anatolia, it was Everard who organised a fighting retreat that ensured that the king could withdraw and the crusade continue (Berry 1948: 124–5). Although the Templars' reputation suffered from their involvement in the failed siege of Damascus in 1148 and from some rather ambivalent behaviour at the siege of Ascalon in 1153 (WT 1986: 797–9), the kings of Jerusalem appreciated their capacity to provide a standing army of mounted knights and foot soldiers. This was particularly important during the 1160s, when King Amalric I (1163–74) depended on both Military Orders for his ambitious invasion of Egypt. Indeed, the reluctance of the Hospitallers to participate in the proposed campaign in 1177, after they had invested so disastrously in the failed expedition of 1169, demonstrated how crucial a part of the kingdom's military resources the orders had become.

In addition to extra manpower, the Military Orders contributed an organisation and discipline born of dedication to the defence of the Holy Land. Bernard of Clairvaux's *In Praise of the New Knighthood* (c.1130) stresses the Templars' seriousness of purpose, in contrast to the dice-playing buffoonery of secular knights (Greenia 1977: 127–41). Steadfastness in battle was ensured with the rule that those who fled the enemy in the field were expelled from the house to which they belonged (Upton-Ward 1992: 148). Contemporaries regarded Templar fanaticism in battle as fearsome, and this seems to be borne out by the recklessness of their Master Gerard of Ridefort in May 1187 (Barber 1994a: 180–1). Whether his brave but disastrous charge against impossible odds at the battle of Cresson served the best interests of the kingdom is another matter. The orders' value to the kingdom was appreciated by Saladin, who paid them the dubious honour of executing his Hospitaller and Templar prisoners at Hattin – even to the extent of first buying them back from their captors. According to 'Imad ad-Din, Saladin regarded the Military Orders as defilers of the holy places, which was, in a fashion, a backhanded compliment to Bernard's vision of the Templars as purifiers of the same sites (Gabrieli 1984: 138).

The success of the Hospitallers and Templars prompted imitation, but this was largely a matter of converting existing religious houses into military institutions (Forey 1995: 186). After the Third Crusade the German hospital at Acre became the Teutonic Order of St Mary, which came to play a significant role in the defence and political development of the

Crusader States in the thirteenth century, while the house of regular canons at Acre became the Order of St Thomas. Already by the 1140s there was a military order specifically for leper knights, the Order of St Lazarus (Barber 1994b: 439–56), while in Spain the Order of Calatrava was founded in the 1150s and the Order of Mountjoy in 1175 (Forey 1971: 250–66).

Although their accumulation of territorial wealth and ecclesiastical privileges provoked the jealousy of many, the Military Orders came to be seen as a solution to one of the gravest problems confronted by the Crusader States. As the grants of castles to the orders demonstrated, the lay magnates were simply unable to afford the cost of defending the frontiers. The settlers needed constant financial and military support from the West, and the orders offered the most reliable and efficient means of channelling western pious sentiment for the holy places into coin, muscle and iron: the tools of war.

The Second Crusade, 1145–50

Although Edessa fell on Christmas Eve 1144, there is no evidence that this was widely known in the West until late in 1145. The timing of Eugenius III's crusading bull *Quantum praedecessores* (1 December) and its relationship with Louis VII's announcement of his intention to make an armed pilgrimage at his Christmas court in 1145 has been much discussed (Berry 1958: 503–4; Grabois 1985: 94–104; Cole 1991: 40–52; Ferzoco 1992: 91–9; Rowe 1992: 79–89; Phillips 1996: 75–82; Richard 1999: 155–7). The likeliest scenario is that Louis had quite independently made a pilgrimage vow, perhaps in fulfilment of one made by his brother before his death; certainly Eugenius' bull was aimed at Louis and the French knighthood. For this reason it has even been suggested that the bull was a response to Louis' declared intent (Ferzoco 1992: 91–9). It seems clear, however, that the preaching of the crusade was a direct response to the crisis in the East, and that the initiative came initially from the East rather than from the papacy (Phillips 1996: 78–82).

In hindsight it is curious that nothing appears to have been done before the arrival in Italy of a legation led by Hugh, bishop of Jabala, in November 1145. News must have reached the West long before this, for the journey could be accomplished in a matter of weeks, and even allowing for inclement winter weather, by late spring at the latest the pope should have been aware of the disaster that had befallen Edessa. An appeal for military aid might logically have been sent in the first instance to lay magnates, particularly noble dynasties, such as those of Aquitaine and Anjou, who were closely related to the ruling families in the East. But recent research has revealed increasingly close ties between the Latin ecclesiastical hierarchy in the East and the papacy from *c.*1137 onward (Hiestand 2001: 33–5), and it seems inconceivable that the violent death of the archbishop of Edessa and the destruction of Latin churches, both of which are mentioned in *Quantum praedecessores*, would not have been known to the pope for almost a year after the event.

The reason for the long delay before the preaching of a crusade to recover Edessa may simply have been that such a course of action was not obvious to the pope. There is a tendency to regard the crusade of 1096–1101 as a kind of blueprint for what became a normative feature of papal policy. In 1145–6, however, Eugenius III seems to have been aware that he was doing something that no pope had done since 1095. His bull began with a reminder of Urban II's appeal and of the glorious response from the western knighthood. Although popes had licensed preachers to raise troops from the West since then – Pascal II in 1106 and Honorius II in 1127, for example – *Quantum praedecessores* marks the first attempt to re-create the conditions of the First Crusade.

There are important differences, however, between 1095 and 1145. For a start, the response to Eugenius' appeal was so sluggish that the bull was reissued at Easter 1146. This may have been because, unlike Urban II, Eugenius made no attempt to preach the crusade in person, but delegated this task to others. The role of the most famous of these preachers has led one historian to characterise this as 'Bernard of Clairvaux's crusade' (Hiestand 2001: 37). The character of the crusade was shaped, therefore, more by Bernard's vision of what it ought to be than by papal policy framed solely in response to the fall of Edessa. Although *Quantum praedecessores* leaves no doubt that the recovery of Edessa (or Rohais, as it was known in Latin) was the priority, Bernard's preaching seems to have relegated this to a secondary place. He emphasised instead the sinfulness of Christians, which had permitted such a disaster to occur, and presented the crusade as an opportunity for the knighthood to demonstrate its love for God and thereby to redeem itself. This was not a completely new departure, of course, for the chroniclers' reports of Urban II's preaching make the same link between sinfulness, political events and salvation, and Bernard had said something similar in his manifesto for the Templars, a decade or so earlier: 'If God has allowed (Jerusalem) to be attacked so often, this has been in order to give brave men an opportunity to show courage and win immortality' (Greenia 1977: 144). The second version of

Quantum praedecessores, moved by the same concern, attempted to regulate knightly behaviour and even clothing on the crusade in order to affirm the penitential quality of the enterprise. Consciously developing Urban's Clermont address, Eugenius offered a less ambiguous indulgence for the remission of sins for participants. The characteristic feature of Bernard's preaching in 1146 was its interior focus on the state of mind of the individual crusader. There is no doubt that it was successful; when he preached before the king and his court at Vézelay at Easter 1146, large numbers took the cross, whereas at Christmas the response to Louis VII's personal vow from among his barons had been lukewarm. When he preached to Conrad III's court in Germany, Bernard met with a similarly enthusiastic response – so enthusiastic that the intention may already have been present before he preached (Phillips 2001: 27).

It would be a mistake, however, to see the Second Crusade entirely in terms of Bernard's 'interior focus', or to assume that it was purely as a result of his preaching that the aim of the crusade became diffused. For one thing, Bernard was not the only influential preacher of the crusade. A Cistercian monk, Rudolf – usually labelled as a 'renegade' – preached an anti-Semitic message in the Rhineland that had tragic consequences for some Jewish communities; although Bernard's own preaching activity in the Rhineland is usually seen as an attempt to counteract Rudolf's inflammatory work, the precise link between his preaching and collective murders is still unclear (Richard 1999: 157–8), and at least one contemporary source mentioned Rudolf as a perfectly respectable crusade preacher in connection with Bernard (Edgington 2001: 57–8).

Another consideration is that Bernard's public preaching to Louis and his barons at Easter and to the German king Conrad III and his barons at Christmas 1146 did not constitute the limit of his activity on behalf of the crusade. In March 1147, Bernard, presumably with papal consent, authorised barons from eastern Germany who wished to take the cross to transfer their vows to a campaign against the pagan Slavs who constituted their northeastern frontier. This amounted to a recognition of the German territorial expansion east of the Elbe as a holy war for the purposes of conversion (Christiansen 1980: 50–9). Even the king of Denmark was encouraged to participate in this war (Jensen 2001: 164–79). At the same time, and by the same logic, Bernard seems to have encouraged crusaders to assist campaigns in the Iberian peninsula, notably the conquest of Lisbon by Afonso Henriques, king of Portugal, in 1147, and of Tortosa, in Catalonia, in 1148. These campaigns, which for long were regarded as side-shows to the main action in the East, have recently been the subject of considerable discussion, and a convincing argument has been made for see-ing them as part of an overall vision of the Second Crusade (Livermore 1990: 1–16; Phillips 1997: 485–97, following Constable 1953: 213–79; but see Forey 1994: 165–75 for an opposing view). Recruitment for the crusade in the Low Countries, England and the Rhineland focused on an assault on Lisbon en route to the East. The campaign for the recovery of Edessa, led by Louis VII and Conrad III, was thus only one thrust in a three-cornered assault on enemies of Christ on different frontiers of Christendom.

This conception of crusade strategy might suggest that a rather sophis-ticated theory of holy war was already in place in the papal curia; alterna-tively, it might argue that the idea of holy war and the indulgence premised on it was so popular in Christendom as a whole that lobbying by rulers on behalf of their own regional interests left the papacy no alternative but to extend the same privileges as had originally been considered appropriate for the East to other areas. One difficulty was that, serious though the loss of Edessa was, it did not even indirectly threaten the security of Jerusalem; in 1146, when Zengi was murdered, even Antioch seemed safe. If remis-sion of sins could be granted for a campaign in northern Syria, why not for other frontier areas? As has been pointed out by historians who take a wide view of crusading in the twelfth century, indulgences had been offered for war in Spain in the 1120s and even in Italy in 1135 (Housley 1992: 36).

Discussion of the Second Crusade is inevitably dominated by knowl-edge that it failed in its full objective. One of the crucial questions raised by the failure is how clear those objectives were from the beginning. If Louis was indeed committed to a personal pilgrimage to Jerusalem even before *Quantum praedecessores* (Grabois 1985: 95–7, 98–9), is it surpris-ing that he was deflected from the direction proposed by the papal bull? The contemporary evidence (including a letter from Conrad himself to Wibald, abbot of Stavelot) indicates that both Louis and Conrad thought they were going to campaign in north Syria (Phillips 1996: 84). Although traditionally Conrad has been seen as a reluctant participant, who was per-suaded by Bernard's rhetoric against his own and the pope's better judge-ment, a convincing argument has recently been made that his participation was as integral as that of Louis VII, and consistent with prevailing views of the imperial role (Phillips 2001: 15–31; but Hiestand 2001: 32–53, for the opposing view). Whatever their original goal might have been, events in the East changed their minds, for in 1146 Joscelin II's attempt to recapture Edessa resulted in its complete destruction, and probably made any hope of retaining the rest of the county impossible.

Another factor in the change of strategy was the role of the Byzantine Empire. The French came to regard Manuel Komnenos as an enemy of the crusade, and blamed their mauling at the hands of the Seljuqs in Asia Minor on Manuel's refusal to supply them with reliable guides. Even before the French reached Asia Minor, a contingent in the army proposed that they attack Constantinople instead. Our main source for the French campaign, Odo of Deuil, was anti-Byzantine, but the German chronicler Otto of Freising (like Odo a participant) shows awareness of the wider political situation. Manuel was not only related to Conrad III by marriage, but sought an active role in Mediterranean politics in support of Byzantine opposition to the Norman kingdom of Sicily – an opposition shared, of course, by the Germans. However, the crusade could not fail to embarrass Manuel, partly because he had recently (1146) concluded a treaty with the sultan of Iconium but, more important, because any crusading campaign in north Syria might jeopardise his own claims to suzerainty over Antioch and what was left of Edessa. (It is worth recalling here the agreement made between Prince Raymond and John Komnenos in 1137 for the handover of Antioch in the event that Aleppo and Shaizar were conquered.) Conversely, this explains Louis' change of strategy in 1148. Louis may not have been fully aware of the eastern situation before he arrived in Antioch in 1148, but once there, his reluctance to campaign in the north, as Raymond urged, is understandable, for any conquests made would be subject to Byzantine claims, and the king of France could not appear to be acting as the vassal of the Byzantine emperor. This is a more plausible explanation than either the insistence that he had to fulfil his pilgrimage vow before engaging in battle, or the romantic view that he left Antioch hurriedly because his wife, Eleanor, was having an affair with Prince Raymond (Runciman 1952: 279). Conrad's opinion no longer counted: he had been forced by a crushing defeat in Asia Minor to retire to Constantinople and would only rejoin the crusade in April 1148, once Louis had moved south to Acre.

At a grand council held near Acre in June 1148, Louis, Conrad and their magnates met the young king of Jerusalem Baldwin III and his mother Melisende, together with the prelates and barons of the kingdom, to decide how the crusading army would be deployed. This was a snub to Raymond, who wanted Louis to help him conquer Aleppo and Shaizar, so as to invalidate the 1137 Byzantine agreement. Although some barons advised an attack on Ascalon, the last Egyptian stronghold on the Mediterranean littoral, in the event the crusaders decided to besiege Damascus. Historians have usually been critical of this policy, for Damascus had traditionally been an ally of the kingdom of Jerusalem against Zengi, whose ambition the Damascenes feared (e.g. Runciman 1952: 281). With the benefit of hindsight, it is easy to criticise the decision to turn a Muslim 'neutral zone' into an active enemy, particularly since by 1154 Zengi's son Nur ad-Din had in any case achieved what his father had failed to do. Yet at the time there was every chance of success, and Damascus was a rich prize that would have extended the kingdom's territory and resources immeasurably. Baldwin III, who was already engaged in a struggle to assert his authority and to end his mother's regency, had every reason to aim at such an appenage. Moreover, although some of the barons, such as Welf VI of Bavaria, left the crusade, Louis' army was still large enough to expect success.

Damascus was surrounded by orchards, in which the crusader army camped towards the end of July 1148. Initial skirmishes, in which Conrad III distinguished himself, went well for the crusaders, but Unur had appealed to local provincial governors for reinforcements, and once these arrived they mounted a counter-attack that cleared the immediate danger to the city's walls. The orchards now proved excellent cover for lightly-armed Muslim raiders who forced the crusaders to move their camp to open ground. Unfortunately the new site, facing the city's eastern wall, gave little chance for an assault, and the counter-attacks from Damascus continued. After less than a week, during which arguments had already begun over the future of the city that was proving so difficult to besiege, the barons of the kingdom of Jerusalem withdrew their support and the crusaders, fearing that Nur ad-Din would arrive from Aleppo to throw his weight behind Damascus, were forced to withdraw. The feebleness of the siege (24–29 July) can be blamed on bad luck and poor tactics, but the strategy itself was not unreasonable. Damascus had been a target of crusader ambition since 1126, and its capture would have dented the Zengid dynastic expansion in southern Syria (Hoch 1992: 119–28 and 1996: 359–69).

The consequences of the Second Crusade's failure to accomplish any territorial gain in the East now appear rather complex. William of Tyre, writing thirty or so years later, asserted that 'from that day onward the situation of the Latins in the East became worse' (WT 1986: 770–1). For William, the crusade was a huge blow to the morale of the kingdom, and the failure to agree in the conduct of the campaign led to distrust between the West and the settlers. Some western contemporaries, contrasting the fortunes of the 1147–9 expedition with that of 1096–9, concluded that the crusade was an ideal that could easily become tarnished (Constable 1953:

213–79). Modern historians have on the whole concurred with this gloomy view (Smail 1969: 1–20; Mayer 1988: 103; Prawer 1972: 25), though the reconsideration of the Spanish campaigns of 1147–8 has led to a more positive reassessment. The most serious allegation is that the choice of Damascus threw Unur into the arms of Nur ad-Din, thus ending Damascene neutrality and hastening the process by which the Zengids built up a unified power base in Syria. Such an analysis, however, is too simplistic. On the eve of the crusade, the Franks were already undermining 'Unur's authority by supporting a vassal who had rebelled against him (Richard 1999: 155). But Unur's goal was above all to maintain Damascene independence, and when Nur ad-Din marched on Damascus in 1150, the Damascenes once again turned to the Franks for support. The alliance of 1140–7 seemed still to be in place (Hoch 2001: 188–200). The cause of Nur ad-Din's eventual ascendancy over Damascus was more complex than the military balance of forces.

Bernard of Clairvaux felt the humiliation of failure deeply, but was soon attempting to redeem it with another crusade. In June 1149 Prince Raymond was killed in Nur ad-Din's crushing victory at Inab. Like il-Ghazi thirty years earlier, Nur was unable to seize Antioch, but the principality was devastated. Plans for a recovery expedition from the West reached an advanced stage of planning, to judge from the correspondence between Bernard, Abbot Suger of Saint-Denis, who knew the king's mind intimately, and Abbot Peter of Cluny. Eugenius III, however, was unnerved by previous failure, and the plans were never realised (Phillips 1996: 100–10). Louis VII also seems to have been unwilling to become involved – presumably his objections to campaigning in north Syria in 1148 had even more force now, and indeed he may have preferred the idea of joining Roger II of Sicily in an attack on Constantinople (Reuter 2001: 150–63).

What can explain the failure of the Second Crusade in the East? Blame for military failure was variously apportioned at the time by Conrad III to Baldwin III and the barons of the kingdom, for giving bad advice; to the Military Orders for accepting bribes from Damascus; and to Louis VII for alienating the settlers by promising Damascus to his own vassal, Thierry of Flanders. The decision to raise the siege after only a week because of the approach of Zengi's sons from Aleppo and Mosul was not supported by all the crusaders (Richard 1999: 166–7), and bad feeling prevailed among the Franks. The crusade had begun with clear objectives and high expectations, but bad luck, the lukewarm support of the Byzantines in Asia Minor and a series of poor military performances in 1147 and 1148 showed how much could go wrong on such a complex expedition – and, in retrospect, how remarkable the success of 1096–9 had been.

The zenith and nadir of the Latin East, 1153–87

The conquest of Ascalon in 1153 appeared to signal a new chapter of expansion in the Latin East, and helped to offset the disappointment of 1148. More immediately, it marked the end of a potentially damaging period for the kingdom of Jerusalem, during which Melisende's refusal to give up the regency – even after her son had come of age – erupted briefly into civil war (1150–2). It is striking in this regard to recall her younger sister Alice's similar stance after the death of Bohemond II – though Alice at least had the excuse that her heir was a six-year-old girl. Strife in Jerusalem was particularly dangerous to the Crusader States, because the situation in the north had again become critical. Baldwin III, like his father Fulk in 1131 and his grandfather Baldwin II in 1119, had to act as regent for Antioch following the disaster of Inab in 1149. As before, the priority was to find a new husband for Constance, for her four children were still minors. Although Nur ad-Din had symbolically bathed in the Mediterranean after his triumphant campaign in Antioch in 1149, he seems to have had no thought of annexing the principality, and subsequently concentrated his resources on Damascus instead. But a year after Inab, Joscelin II was captured, leaving both Edessa and Antioch effectively leaderless. In these circumstances Baldwin III advised Joscelin's widow to take the drastic step of voluntarily bringing to an end the existence of the county of Edessa by selling what remained of it to Manuel Komnenos.

As has been pointed out, it was ironic that the Greeks, who were seen by many as the cause of the Second Crusade's failure, were now able to seize control of the same lands that the crusade had hoped to save (Phillips 1996: 123). Not only was this *realpolitik* on Baldwin's part – for he could not remain in the north indefinitely – it also recognised the contribution that a strong Byzantium might play in the political balance in north Syria. Constance might have accelerated this process, had she accepted the hand of the emperor's brother-in-law in 1152, but in the event she chose a western baron, Reynald of Châtillon. Reynald, whose reputation has been rescued in recent years (Richard 1989: 409–18), was not the penniless knight he was once thought to have been (Runciman 1952: 345), but the scion of an old aristocratic family from the Loire who had come to the East on an armed pilgrimage and served with Baldwin III at Ascalon. Given that Constance's choice was entirely personal, and the marriage secret, this was probably a genuine love match – but it also made good political sense for Antioch. A western lord of good pedigree and military experience, yet without a potentially destabilising entourage, enabled Antioch to pursue a policy of seeking western support in the face of Byzantine ambitions.

In 1157 cooperation between Baldwin III and Reynald resulted in the acquisition of Harim, a stronghold that had been lost to Antioch since 1119. However, in the following year Baldwin was required to play a critical diplomatic role that brought Byzantine influence fully into the picture. Reynald, who had been asked by Manuel to discipline the Armenian prince Thoros of Cilicia, a Byzantine vassal, ended up by joining with Thoros in an attack on the imperial island of Cyprus. This provided Manuel with the provocation he needed to accomplish what his father John had been on the point of doing when he died in 1143: the formal and final subjection of Antioch to Byzantine suzerainty. Faced by a Byzantine invasion, and in the knowledge of Nur ad-Din's growing strength to the east, Reynald appealed to Baldwin III for help. But Baldwin had in 1158 married Manuel's niece Theodora, and wanted to pursue a pro-Byzantine policy. He met Manuel in Cilicia in 1159, hunted with him and cemented a personal friendship. At Easter, he persuaded Reynald to submit to Manuel. It was a diplomatic triumph for Baldwin, who had assured a three-cornered defence of Christian north Syria against Nur ad-Din.

The Byzantine alliance with the kingdom of Jerusalem was to be a central feature of Near Eastern politics for the next twenty years. Although soon after his accession Amalric I raised suspicions of Byzantine intentions in petitioning letters to Louis VII, he, like his brother Baldwin, married a Byzantine princess, and it was in his reign (1163–74) that the closest links were forged through joint military activity, cultural interaction and formal diplomacy. Military imperatives demanded the continuation of the alliance, despite the misgivings of some of the settler barons. In 1154 Nur ad-Din had finally secured Damascus, and during the 1160s his authority throughout the Islamic Near East increased with his espousal of *jihad*. Although Reynald of Antioch was captured in 1161, and was to spend the next fifteen years imprisoned in Aleppo, Nur ad-Din was somehow kept at bay, even after the young Bohemond III, Constance's son by Raymond of Poitiers, was himself captured along with Raymond III of Tripoli and many other leading figures at the battle of 'Artah in 1164. Nur ad-Din is said to have declined to attack Antioch itself out of fear of Byzantine military strength (Richard 1999: 180).

Instead, military attention turned to the kingdom of Jerusalem itself, and to the attempt to curb Nur ad-Din's raiding across the Jordan from

Damascene territory, and ultimately, from 1167 onward, to forestalling his ambitions in Egypt. Western military support, though occasionally decisive, as at Ascalon in 1153, or simply helpful, as in Thierry of Flanders' armed pilgrimage of 1157–8, was sporadic. There were also setbacks, notably the defeat at Jacob's Ford in 1157, when over three hundred knights were killed and Baldwin III barely escaped with his life. The Byzantine Empire was better placed to provide consistent military support, and in Manuel Komnenos the Crusader States had a neighbour who was genuinely Francophile by inclination – to the extent of marrying Maria of Antioch, sister of the young ruler Bohemond III, after his first wife, the Austrian Irene of Salzburg, had died.

Amalric I first intervened in Egypt in 1163, but had to leave in a hurry when news arrived of the fall of Harin and Banyas. He explained to Louis VII that a Frankish attack was necessary if Egypt were not to fall to Nur ad-Din. Because this in effect did eventually happen, it is difficult to know whether Amalric was being prescient or merely trying to rationalise a military adventure that promised easy rewards. Egypt was certainly an attractive target. Its Fatimid rulers had, since the reign of Baldwin I, abandoned serious hopes of recovering the Palestinian territory they had once held; the government was weakened by internal rivalries and the army largely made up of slaves and mercenaries. But its wealth, assured by its role as intermediary in the trade between Italian towns and the Far East, had continued to grow. Amalric was to exploit this by demanding tribute. William of Tyre remarked on his cupidity but, as we have seen, the kings of Jerusalem had from the start relied on wealth in cash in order to maintain their position in the feudal hierarchy.

When Amalric invaded Egypt again in 1167, it was once again at the instigation of the vizier, Shawar. In the absence of government by the Khalifs, the Fatimids' viziers exercised full authority, but Shawar had needed the support of one of Nur ad-Din's generals, Shirkuh, to wrest control from a rival, and when he found the price of Shirkuh's help too high, he turned to Amalric. When Shirkuh seemed to be on the point of establishing a territory of his own in Egypt, Amalric invaded and, after several months of campaigning, forced Shirkuh to withdraw and secured for himself an annual tribute from Shawar. The success of the campaign persuaded Amalric that a full-scale conquest was possible, especially if Manuel Komnenos could be persuaded to help. Charters drawn up in 1168 in favour of the Hospitallers, the Pisans and some secular barons for fiefs in Egypt demonstrate the seriousness of the project, though the role of the Byzantines as joint conquerors remains uncertain (Lilie 1993: 189–93).

The eventual failure of the expedition at the beginning of 1169 proved to be disastrous for the kingdom of Jerusalem. Shawar was murdered and his place as vizier taken by Shirkuh; the final death blow to Fatimid Egypt was dealt by Shirkuh's nephew Saladin, who massacred the caliph's personal guard of Sudanese and Armenians and, in 1171, suppressed the Khalifate itself and declared himself sultan (Lyons and Jackson 1982: 47–57). Saladin had first emerged from obscurity as the defender of Alexandria in 1167, but it was only after Shirkuh's death that the ability and ambition that was to make him such a legendary figure became clear. Almost the only hope not shattered by this setback was the Byzantine alliance, which was formally ratified when Amalric paid a state visit to Constantinople in 1171. William of Tyre, who was part of the royal entourage, was impressed by everything he saw, but although he was pro-Byzantine in his sentiments, he stopped short of declaring that the king actually became Manuel Komnenos' vassal. The question of the relative status of king and emperor has divided historians, and some modern commentators have chosen to support the Byzantine perception, reported by John Kinnamos, that Amalric recognised Manuel as his overlord (Lilie 1993: 206–9; Phillips 1996: 211–12; Hamilton 2000: 31), though I prefer to maintain the distinction between protection and feudal suzerainty (La Monte 1932: 253–64). Amalric probably granted Manuel some recognition as protector of Greek Orthodoxy and of the Orthodox people in the kingdom of Jerusalem, but it seems unlikely that he went further.

When Amalric died in 1174, the danger facing the kingdom of Jerusalem had never been more acute. Amalric had foreseen, but been unable to prevent, the encirclement of the kingdom by Nur ad-Din and governments that he controlled. In 1171 Nur had finally added his father's base, Mosul, to his rule of Aleppo and Damascus, and when he died – coincidentally in the same year as Amalric – the Seljuq sultan's influence had been limited to Iran, and his own authority recognised in Syria and Egypt by the Khalif in Baghdad. Despite plans and appeals for military expeditions from the West throughout Amalric's reign, nothing had materialised.

Most unfortunate of all, Amalric's heir, Baldwin IV, was not only a boy of thirteen, but also a boy already afflicted with a mysterious ailment that looked dangerously like leprosy. William of Tyre's fears were to be confirmed a few years into Baldwin's reign, which René Grousset eloquently called 'an agony on horseback' (Grousset 1935, vol 2: 610). The years between Baldwin's accession and the fall of the kingdom to Saladin in 1187 are the most colourful, the most dramatic and the most often described in the historiography of the Crusader States. Baldwin himself displayed remarkable courage and forbearance in the face of his disease. William of Tyre, who was the boy's personal tutor, describes in a moving passage of his chronicle how he first became aware that the active and intelligent heir to the throne suffered, unsuspecting, from a condition that removed all sensation from his nerve endings (WT 1986: 961–2). At the time of his father's death, however, it was not certain that Baldwin's condition would develop into full lepromatous leprosy. Given that he would need a regent in any case until he attained his majority, the risk of crowning him was probably less than that of trying to find a better-qualified candidate (Hamilton 2000: 38–43). His sister, Sybil, who was two years his elder, had been brought up in a convent and had no experience of statecraft; in any case, the search for a suitable husband as consort would take time, and even if Baldwin should turn out to be unable to produce heirs, an undisputed succession would buy sufficient time for this to take place. On his accession Amalric I had been forced by baronial pressure to repudiate his wife, Agnes of Courtenay, who was the mother of Baldwin and Sybil, and his subsequent remarriage to Maria Komnena had produced another daughter, Isabel; but the girl was only two years old in 1174, and although the circumstances of the divorce from Agnes were murky, the pope had legitimised the children of this first marriage (Mayer 1992: 121 35, Hamilton 2000: 33–4).

The office of regent was at first filled by Miles of Plancy, Amalric's seneschal, perhaps unofficially. Either his unpopularity with the barons, or his failure to take advantage of Nur ad-Din's death by intervening in Egypt, led to his assassination in October 1174; but Miles' position had already been challenged by Raymond III of Tripoli. Raymond had only recently been released from captivity, and was comparatively unknown in the kingdom. His demand for the regency, based on his claim of being *plus dreit heir aparant* – the closest male heir to the king – was a constitutional novelty, but seems nevertheless to have been supported by the baronage as a whole (Riley-Smith 1973a: 101–4). Raymond's claim was based on the king's paternal side, for his mother, Hodierna, was a daughter of Baldwin II. It is striking, especially given what later transpired, that the claim was accepted by the king's maternal relations, the Courtenays – even though Bohemond III of Antioch, who was descended from another of Baldwin II's daughters, Alice, could make the same claim (Hamilton 2000: 89–94).

The most pressing matter to be resolved during Raymond's regency was the marriage of Sibyl, for by 1176 Baldwin's condition had worsened and it must have been apparent that he would be unable to father children. Once Sibyl had married and borne children, those children could eventually be Baldwin's heirs, but because lepers had a much shorter life-span than the average, it was also likely that the children's father would be required to act as military leader of the kingdom until they reached their majority. One interpretation of the choice of William, marquis of Montferrat, as Sybil's husband, is that this signalled a reversal of the pro-Byzantine policy of the previous two reigns. William was both a kinsman and a vassal of the western emperor Frederick Barbarossa, who moreover was at this period pursuing an anti-Byzantine policy of his own; Raymond of Tripoli, who was himself hostile to the Byzantines, may therefore have been trying to counterbalance a Byzantine 'protectorate' over the kingdom of Jerusalem with that of the western emperor (Hamilton 2000: 100–2). However, although both Baldwin III and Amalric had married Byzantine princesses, there was no precedent for a Byzantine male consort, whereas in seeking a husband from the West the barons were simply following the traditional policy of the Crusader States. Moreover, by the time the marriage took place (November 1176), the situation had changed. Barbarossa's defeat in Italy at the hands of a papal–Lombard force, to which Manuel Komnenos was allied, had lessened any influence he might have brought to bear, and an embassy to Constantinople led by Reynald of Châtillon that same winter renewed the alliance of 1171.

Baldwin IV's reign has in the past been understood in terms of factional rivalry between two 'parties' among the baronage, identified first among historians of the modern era by Marshall Baldwin (1936: *passim* and 1958: 590–621) and followed by Runciman (1952: 405). According to this view, which became the accepted orthodoxy, factionalism centred on rivalry between the 'court party' led by the Courtenay relatives of the king and their adherents, whose main characteristic was that they were recent arrivals in the East, and an 'old guard' among the baronage, led by Raymond of Tripoli and the Ibelin family, who represented the original Frankish settler families. More recent studies have demonstrated the flaws in such a view. For one thing, the Courtenays themselves had been in the East for as long as Raymond's family, while the Ibelins had come to prom-inence only in the reign of Fulk. Besides, the policies to which apparent rivals adhered cut across such 'factions' (Edbury 1993: 173–89). As has recently been shown, events in Baldwin's reign did eventually become a battleground between the king's maternal and paternal relations (Hamilton 2000: 154), but there is no real evidence for this before 1180. Indeed, up to that point, attention focused on the attempt to bring about a permanent western alliance, in the form first of William of Montferrat, then Philip, count of Flanders (1177–8). Both men found themselves faced with a renewal of the Byzantine alliance of the 1160s, with presumably the understanding that the allies would try once again to conquer Egypt. By 1176–7, the need was even more pressing than it had been ten years earlier, for Saladin had by now gained control of Damascus as well as Egypt, and threatened to be a more powerful enemy than Nur ad-Din.

Whatever William of Montferrat thought of the Byzantine alliance, his unexpected death after a short illness in the summer of 1177 left the kingdom highly vulnerable in the immediate term and the question of succession still unresolved. Thus, when Philip, count of Flanders arrived only weeks later to fulfil an armed pilgrimage, he was offered not only command of the army of Jerusalem in the proposed invasion of Egypt, but also the regency. From the kingdom's point of view, Philip was an ideal choice: he was wealthy and well regarded, he was related to the ruling house (his mother Sibyl had been the daughter of Fulk of Anjou by his first marriage) and his family had an impeccable crusading tradition – his father Thierry had been to the East four times, and his mother had retired to the convent at Bethany. But the offer left Philip in a quandary. As Hamilton has pointed out (2000: 123–4), the quarrels that marred his stay in the East arose over the status of Egypt. If it were to be treated as a new kingdom, then it might be worth Philip's while to accept Baldwin IV's offer. But the barons presumably saw Egypt, as they had done in the 1160s, as a new province of the kingdom of Jerusalem, and Philip was unwilling to risk his county of Flanders against an uncertain position as deputy to an ailing king – let alone to conquer territory that might be held under Byzantine suzerainty. When Philip, as a relative, demanded the right to advise on a new marriage for Sibyl, and suggested one of his own barons, Robert of Béthune, as a suitable husband, this was too much for some of the settler barons to stomach. In particular Baldwin of Ibelin hoped to marry Sibyl himself, and the chronicle of Ernoul, which represents the Ibelin family's perspective, suggests that a group of barons in the kingdom began to emerge that wanted to see Sibyl's hand – and with it the regency – pass to a native of the Crusader States.

It is easy to see how resentments and difficulties arose over Philip's role, but in retrospect the failure to coordinate the various forces – Flemish, Jerusalemite and Byzantine – in an assault on Egypt meant that a genuine opportunity to break Saladin had been missed. William of Tyre, whose chronicle from this point assumes a partisan approach, blamed Philip, but the anti-Byzantine Raymond of Tripoli, Bohemond III of Antioch and the Templars were all equally reluctant to commit themselves to the campaign. In the event, the bulk of the kingdom's forces campaigned inconclusively in the north, leaving Saladin free to invade from the south. The heroism of Baldwin IV and his small army, which included Reynald of Châtillon, the Ibelins and Joscelin III of Courtenay, the king's uncle, ensured an unexpected victory at Mont Gisard, near Gaza, in November 1177. When one considers that Baldwin, who had entirely lost the use of one arm, could ride a horse only by gripping with his knees, so as to leave his good arm free to hold a weapon, his personal courage is all the more remarkable.

The policy of seeking a husband for Sibyl from the West continued, and in 1179 it looked as though Hugh III, duke of Burgundy, was ready to accept the role. But 1180 was to prove a decisive year for two reasons: first, the death of Louis VII meant that Hugh changed his mind, fearing a succession crisis in France; and second, Sibyl chose as her own husband Guy of Lusignan. Guy has suffered from the fact that we rely so heavily for this period on William of Tyre and Ernoul, neither of whom had any time for him. He has been much pilloried by historians (e.g. Runciman 1952: 424) but more recently his reputation has begun to recover (Smail 1982: 159–76). The choice of Guy may have been as romantic as Runciman thought, but it reveals the canniness of Baldwin IV as well. In 1180, the succession question was forced by the threat of a *coup d'état* by Raymond III and Bohemond III; whether or not such an event was likely, Baldwin clearly believed that it was. William of Tyre says that the intention was to put Sibyl, whose heir Baldwin V had been born in 1177, on the throne. By marrying her off, Baldwin could ensure that such a measure would become pointless, because it would promote her husband to the regency and deprive Raymond the opportunity to exercise power through Sibyl. Guy was an unknown quantity as a ruler, but his Lusignan descent meant that he was connected to the Angevins, and thus offered similar hopes as William of Montferrat had done in 1176 – the protective mantle of a powerful western ruling dynasty.

A third critical event in 1180 was the death of Manuel Komnenos. Although the anti-western reaction to his reign in Constantinople did not set in until 1182, it must have been clear that his heir, who was a minor, would be unable to maintain the same investment in the kingdom of Jerusalem. Indeed, by 1185 the Byzantines had swung towards an alliance with Saladin, and the kingdom had no option but to look again to the West for protection.

Despite victory over Saladin at Le Forbelet in 1182, and a demonstration of the range of the kingdom's attacking power in Reynald's Red Sea raid in the same year (Hamilton 1978: 102–3), the situation remained critical. By 1183 the king's condition had become so bad that he could neither see nor use his limbs, and at the end of the year the young heir, Baldwin V, was crowned. The ceremony itself indicated that Baldwin IV now had misgivings about what Guy of Lusignan might do when he himself was dead – and in this sense recalls Baldwin II's naming of his two-year-old grandson as heir in 1131. Guy had incurred the king's displeasure by his conduct of the defensive campaign against Saladin's invasion of Galilee in summer 1183. It is apparent that Guy's marriage had polarised baronial opinion, and that Raymond of Tripoli's disappointment at being marginalised brought him sympathy from the Ibelins, Reynald of Sidon, William of Tyre and others who distrusted Baldwin IV's maternal relations. Some of the animosity on both sides seems to have been entirely personal – as, for example, the hatred of Raymond expressed by the new Master of the Templars, Gerard of Ridefort – and some the result of professional rivalry, for example William of Tyre's defamatory portrait of the patriarch of Jerusalem, Heraclius (Kedar 1982: 177–204). Resentments and distrust made it impossible for Guy to use the army with which he had been entrusted in the field, but although he had successfully outfaced Saladin, Baldwin IV nevertheless stripped him of the regency. It may be true that Guy had shown incompetence in his preparations (Hamilton 2000: 191), but the real problem lay in his inability to provide united leadership. Baldwin may even have tried to have Guy's marriage to Sibyl annulled, so serious were his misgivings; in response, Guy openly rebelled by refusing to attend a summons to the king's council, and in punishment was stripped of Jaffa, half of his fief.

Baldwin, in fear of civil war, was prepared to risk opening the succession question once again. Hamilton has pointed out that even if the marriage were annulled, any alternative candidate who agreed to marry Sibyl would do so in the knowledge that his regency would be a temporary affair, until Baldwin V came of age (2000: 196), though this argument may be weakened if, as has been suggested, it was already widely known that the young heir was so sickly that he was unlikely to reach adulthood (Riley-Smith 1973a: 104). In any case, when Baldwin IV finally died in 1185, the succession was a mess. On his death bed he appointed Raymond regent for Baldwin V, but Joscelin III, his great-uncle, as his personal guardian – perhaps in the hope that this might curtail Raymond's ambition to make himself king. Baldwin IV by now seems to have trusted nobody, for control of the royal castles was given not to the regent but to the Military Orders. Should Baldwin V die before coming of age, the whole question was to be thrown open to the pope, the western emperor and the kings of France and England to decide between the claims of Sibyl and Isabel. In other words, deadlock had been reached in the negotiations.

The boy king survived only a year before dying at the age of nine – in his way, as tragic a figure as his uncle. The events of the summer months of 1186 are most closely described by Ernoul, but because he was so openly hostile to Guy and Sibyl, his interpretation must be tempered by the reports of chroniclers admittedly less close to events, who report that Raymond was plotting to make himself king. The drama in which he was outmanoeuvred by Sibyl at her own coronation, in which she cleverly allowed everyone to believe that she would put Guy aside, before crowning him herself, has been told fully elsewhere (Runciman 1952: 446–9; as corrective, Kedar 1982: 196–7, Hamilton 2000: 217–21). Most of those hostile to Guy accepted the *fait accompli*, though Baldwin of Ramla, the head of the Ibelin family, left the kingdom, and when Saladin invaded in the spring of 1187, Raymond had not yet been reconciled to Guy, and had in fact made his own truce with Saladin in his capacity as count of Tripoli.

The prolonged succession crisis had weakened baronial unity, but the kingdom had survived such threats before – in 1150–2, for example, in 1134 and in 1123. Why did the kingdom crumble at the single decisive battle of Hattin in July 1187? One argument is that the pressure of Saladin's slow build-up of strength, culminating in the largest army ever assembled against the Franks, led inevitably to the kingdom's defeat. But Guy's army in 1187, if outnumbered, was also the largest that the kingdom had ever put into the field, and victory had been won before over Saladin from a less promising position. Saladin's coalition of forces, moreover, proved to be brittle when put to the test in 1190–2. A more persuasive argument is that the removal of Constantinople from the balance of power in the Near East since 1180 left Saladin free to dominate his Muslim enemies in the northeast at his leisure. For, although the Byzantine alliance had never delivered in military terms what it promised, it had prevented both Nur ad-Din and Saladin from following up partial victories, for fear of reprisals. Yet another course is to blame the West for failure to respond to the urgent appeal made by Heraclius' embassy in 1184. But although it is true that no crusade was launched until after Jerusalem had fallen, the history of military intervention in support of the Crusader States before 1187 does not allow us to assume that they guaranteed success in any case. The war between Angevins and Capetians did not help, but Henry II had sent large sums of money to the East, which was in fact used to pay for mercenaries in 1187. In fact, many crusaders did arrive in the Holy Land in 1186, but found that a four-year truce had been agreed with Saladin; one can under-stand the frustration of western knights who must have felt that they had been hoodwinked by appeals from the East (Phillips 1996: 263–4).

The internecine quarrels engendered by Guy and Sibyl's *coup* allowed Saladin time to complete his control of northern Syria by taking Mosul and Homs from the Zengids. The opportunity to attack him before he had cemented his hold on the Islamic Near East had passed. Nevertheless, the Franks do not appear to have been afraid of confrontation. Reynald of Châtillon's attack on a Muslim caravan in March 1187 has been defended (Hamilton 1978: 106–8), but it certainly gave Saladin the excuse to invade on the grounds that the truce had been broken. Early in May Gerard of Ridefort lost over a hundred Templar and Hospitaller knights in a fool-hardy attack on a large Muslim raiding party that weakened the kingdom's military capacity. Saladin seems to have indicated to Raymond, who had made his own truce with him in his role as count of Tripoli, that he was bound by family honour to retaliate for Reynald's attack on the caravan, because his sister had been travelling in it. However, rather than simply allowing Saladin to be seen to have fulfilled his debt of honour, the Franks sought a confrontation with his raiding party in the spring of 1187. Now Saladin could quite legitimately regard the truce as having been broken, and even Raymond, who had advised conciliation, realised that war was inevitable.

When Saladin launched a full-scale invasion in June, the stage was set for a repeat of the 1183 campaign. On this occasion, however, Saladin exploited baronial divisions by assaulting Raymond's town of Tiberias, where his wife and two sons were housed. As in 1183, Guy summoned the army of the Kingdom of Jerusalem to Sephoria. In the high court, Raymond advised the same policy that had produced a stalemate four years earlier – even at the expense of his own family. But this strategy, though sound from a military view, had in 1183 proved politically disas-trous for Guy, who had been dismissed from the regency by Baldwin IV for failure to engage Saladin. This – rather than the indecisiveness of character of which he is accused by Ernoul – is probably why he chose to follow the advice of Reynald of Châtillon and Gerard of Ridefort rather than that of Raymond, and risked all on an encounter with Saladin. In order to meet Saladin, however, Guy had to lead the kingdom's army across the water-less Galilee plain towards Tiberias in the heat of mid-summer. The success of such a strategy, in the heat of July, depended on a supply of water for horses and men, and in the end it was exhaustion, heat and thirst that defeated them. Saladin had split his forces, and throughout the march the

Franks were harassed by Turkish raiders, until finally, desperate for water for men and horses, they were surrounded on the shallow hill of Hattin, only a few miles from Tiberias (Kedar 1992: 190–207).

With the exception of a detachment of mounted knights led by Raymond, which broke through the Muslim ranks, the entire army was lost. The king himself and hundreds of knights were captured, and the True Cross, the kingdom's totemic standard, was never afterwards carried into battle by the Franks. Despite the size of Saladin'a army, there was nothing inevitable about a Muslim victory in 1187. Poor judgement, perhaps brought on by tensions among the barons, decided the outcome of the campaign, and thus of the kingdom. Guy was released after a year, but Saladin executed Reynald of Châtillon and all the Templar and Hospitaller prisoners. The kingdom lay open, and Jerusalem fell after brief resistance in October 1187. Tyre alone escaped, and it was from there that, two years after Hattin, the slow recovery of the kingdom was to begin.

CHAPTER 5

The Islamic reaction, 1097–1193

The Islamic world was slow to acknowledge the spiritual nature of crusading, and the first crusaders were regarded as just another invading force on an already fragmented map. Despite some military successes against the northern Crusader States, the political disunity of Islamic Syria meant that there was no concerted attempt to drive the Franks out. Under a succession of rulers from the 1130s–90s, however, there was a revival of the *jihad* ideal in Islam, which resulted in both political unification and a new religious spirit. This chapter examines the careers of Zengi, Nur ad-Din and Saladin as *jihad* leaders.

The first generation

It has become something of a commonplace among historians to explain the military success of the First Crusade by pointing to the lack of resistance from the Muslims. In fact the first crusaders faced stiff opposition: for example, from the Seljuqs in Asia Minor and north Syria in 1097–8 and 1101, and from the Egyptians in Jerusalem and at Ascalon and Ramla in 1099 and 1102. Any one of these encounters might by itself have ended the crusade. There is no doubt, however, that the Franks were an unexpected enemy, and that the novelty of an invasion from the West took Muslim rulers by surprise. Not unreasonably, given that the crusade was launched in response to a Byzantine appeal, some Arab sources confused the Franks with Byzantines. Most of the Arab historians who wrote about the First Crusade treated it as a territorial conquest – sometimes, as in the perceptive account of as-Sulami, followed later by Ibn al-Athir, with economic causation (Gabrieli 1984: 3–4).

Few Muslim contemporaries appreciated the religious roots of the western invasion of Syria and Palestine, and they were thus unable to form any concerted ideological response to the crusade. This does not mean, however, that the invasion did not inspire pious sentiments. Hillenbrand (1999: 69–71) has shown how the deep shock at the fall of Jerusalem was expressed in poetry, such as the lines by al-Abiwardi, a Baghdadi who learnt about the disaster from Syrian refugees: 'This is war, and the infidel's sword is naked in his hand, ready to be sheathed again in men's necks and skulls. This is war, and he who lies in the tomb at Medina seems to raise his voice and cry: "O sons of Hashim!" ' Laments such as this show a degree of religious sensibility, but it is directed inward, towards Islam, rather than outward, against the Franks as Christians. Thus an anonymous poet whose work has been preserved in a later history addresses his appeal to Muslims: 'Do you not owe an obligation to God and Islam, defending thereby young men and old? Respond to God! Woe to you! Respond!' Although the savage treatment of Muslim-held cities was condemned, the first crusade did not seem to give immediate rise to an anti-Christian response in the Dar al-Islam. Instead, Muslims were urged to examine their own commitment to Islam in order to protect fellow-religionists who had been attacked.

There is one exception to this general rule. Among the first generation of Muslim writers, the Damascene as-Sulami (1105) stands out for his broad understanding of the context of the crusade. He linked the invasion of the Near East to the Norman seizure of Sicily in the 1090s (as Ibn al-Athir, writing a century later, would also do), and argued that the crusade was a premeditated strike that revealed the extent of Frankish intelligence of the political disarray in the East since 1092. Moreover, as-Sulami realised that the goal of the crusade was symbolic as well as political, and that the conquest and tenure of Jerusalem was a religious imperative for the Franks. As Hillenbrand has pointed out (1999: 73), at the time that he was writing, most of the Levantine coast, including the important cities, were still held by Muslims, and his *Book of the Holy War* is thus as much an exhortation to resist an ongoing conquest as it is a lament for the fall of Jerusalem. As-Sulami knew that the Franks had persisted through luck as well as military prowess, and that they were far from invincible: their supplies were uncertain and the numbers of heavy cavalry limited.

The title of his work is revealing, for as-Sulami was perhaps the first Muslim writer of the Seljuq period to appeal to the *jihad*, an ideal that appeared to have lain dormant since the early expansion of Islam. The notion of war justified on religious grounds is sanctioned in the Qur'an, most notably in sura 9, in which those who pursue *jihad* will attain Paradise. The *jihad* ideal remained strong during the first two centuries of Islam, because it was necessary to rationalise the continued territorial expansion in pursuit of the aim demanded in the revelation to Muhammad: the submission of all people to the will of God. In its formulation by Islamic scholars and legists during this period, *jihad* seems to have been unique. Although in many respects Islam was open to outside influences, and although Christian ideas of holy war had already been formulated, there is no evidence that *jihad* owed anything to these ideas. Nevertheless, an important aspect of both Christian holy war and Islamic *jihad* was the theme of personal inner renewal. In the development of both religions, the relationship of the individual to society was critical; hence, although holy war and *jihad* were considered as means by which Christian or Islamic society could attain an idealised state of perfection, they were also incumbent as duties to be fulfilled by their followers. This did not, of course, mean that each Muslim was expected to be in a perpetual state of warfare against unbelievers. To be engaged in *jihad* was to be constantly striving for the subjection of the self to God's will; in other words, to strive to be a good Muslim.

This more abstract and interior understanding of *jihad*, besides being regarded by the earliest thinkers as more fundamental than its outward manifestation through war against unbelievers, also became a more pragmatic way of defining the concept. From the end of the eighth century onward, the impetus for further expansion slowed, and boundaries between the Dar al-Islam and the lands beyond became more or less fixed. As different legal schools of interpreting Islam developed, so a variety of theories of the requirements of *jihad* on individuals, rulers and society also emerged. The Hanafite school, for example, traditionally took a conservative line according to which peace with non-Muslims was a theoretical impossibility. On the other hand, the Hanabalite school argued for accommodation with unbelievers when it was in the interests of the Dar al-Islam. *Jihad*, in consequence, declined as a significant military ideal between the end of the eighth century and the end of the eleventh.

The argument that the *jihad* was non-existent during the generation after the first crusade, propounded most fully by Sivan (1968), has recently been criticised. Hillenbrand (1999: 104) argues instead that *jihad* ideals were always present among religious scholars, but that they had lacked military application for so long that it was difficult for the religious classes to infuse the ruling classes with enthusiasm for war waged for spiritual, rather than territorial or economic ends. That a sense of the moral imperative of war against the Franks did exist among Muslims is clear from an episode in 1111, when the Aleppan *qadi* Ibn al-Khashshab burst into a mosque in Baghdad, accompanied by a number of *sufis*, and smashed the *minbar* (pulpit) in protest against the inactivity of the sultan. But a successful *jihad* required, above all, the sacrifice of individual political aspirations to the greater aim of the spread of orthodox piety throughout the Dar al-Islam. Moreover according to religious orthodoxy, legitimacy to wage *jihad* rested solely with the khalif. The Abbasid khalifs had the spiritual authority, but lacked the political weight, to direct a *jihad*, and even if the sultan were responsive, he could do nothing without the co-operation of the various rulers of the mosaic of small territories that had emerged from the Seljuq Empire, who in their turn had no need of the *jihad* in order to satisfy their own political aspirations.

In fact, no single Muslim ruler in the Near East – not even the Seljuq sultan – enjoyed sufficient authority over his Muslim neighbours to make a unified stand against the Franks possible. The strongest power in the early twelfth century, Fatimid Egypt, formed an alliance with Damascus to annex the kingdom of Jerusalem in 1104–5, and tried to invade from Ascalon in 1105–6, 1111–12 and 1112–13, but after the fall of Sidon became increasingly ineffective. By then, the Fatimids may have decided that the Franks, who had shown some inclination but little capacity for an invasion of Egypt, could form a useful buffer state between themselves and the Seljuqs. As Shi'ites, the Fatimids in any case did not recognise the spiritual authority of the Abbasid khalif. Having taken advantage of the collapse of Seljuq central authority to seize Palestine in 1098, the Fatimids were unlikely to do anything that might lead to the return of the Seljuqs to that region.

Although the Seljuq rulers of Syria and Asia Minor posed more of a threat to the Frankish settlers than the Fatimids, this threat was defused by their inability, for the most part, to operate together. The one individual who might have accomplished a show of unity, the sultan Barkyaruq, was unwilling to leave Baghdad to go to the aid of the Syrians at the time of the First Crusade because he was fighting a war of succession with his brother. Once his brother, Muhammad, succeeded to the sultanate, he did send armies to Syria under the command of Mawdud, governor of Mosul, in 1110 and 1111–12, but nothing was accomplished. On the second occasion the ruler of Aleppo, Ridwan, refused to allow Mawdud to enter his city.

One reason for the failure of Muhammad's expeditions seems to have been that the local Seljuq rulers were singularly unenthusiastic about direct influence from Baghdad. Indeed, rulers of small territories such as Ridwan, Kerboga, Tughtegin of Damascus, or the autonomous Arab lords of Shaizar, seem to have preferred to allow the Franks to become another player on the complex chessboard of Syria and Palestine. Aleppo, Antioch, Edessa and the Danishmend emirate of eastern Asia Minor were all, when it suited them, prepared to form alliances that cut across the religious divide, rather than submit to the suzerainty of a greater power. When Mawdud returned to western Syria in 1113, he was murdered in the streets of Damascus, probably on the orders of the governor, Tughtegin. Two years later, the sultan found himself opposed by a mixed army composed of Franks from Jerusalem and Antioch and Muslims from Damascus and Aleppo. The Frankish presence was not, so far as most Muslim rulers were concerned, a religious issue but a political one, and they saw the *jihad* as a means by which the sultan could re-establish his authority in Syria at their expense.

The first signs that an alliance between the religious and military classes was emerging to confront the Franks occured in north Syria about twenty years after the seizure of Antioch. The *qadi* al-Khashshab of Aleppo seems to have influenced his own people with a degree of religious idealism in their struggle against Antioch, although the Aleppans' defender, il-Ghazi, despite his military success at the Field of Blood (1119) proved to be an inadequate role model for *jihad*. In 1124, however, the tomb of il-Ghazi's nephew Balaq was adorned with an inscription extolling his virtues in pursuing the *jihad* against the infidel. Balaq was described as the 'sword of those who fight the holy war, leader of the armies of the Muslims, vanquisher of the infidels and the polytheists' (the last being a standard insult against the Christians, for their belief in the Trinity) (Hillenbrand 1994: 60–9; 1999: 110–11). Balaq's greatest triumph came in the autumn of 1122, when he captured Joscelin of Edessa. With Antioch still leaderless after Roger's death, Baldwin II had little choice but to march north to his old territory, but he too was surprised and captured by the Euphrates. Balaq secured control of Aleppo by marrying Ridwan's daughter, and proceeded to reduce the fortresses still in Frankish hands.

Yet despite his mastery of the region, Balaq's downfall is indicative of the obstacles in the way of the *jihad*, for he was killed while besieging the fortress of a rebel vassal. Only weeks later, Tyre, the penultimate coastal city left in Muslim hands, surrendered to a Frankish/Venetian alliance. Balaq had been on the point of marching to raise the siege when he died. His successor in Aleppo, his cousin Timurtash, was uninterested in war against the Franks, and withdrew to Mardin, showing open contempt for the problems of western Syria. When Baldwin II (now released) tried to press home the advantage by besieging Aleppo, the delivery of the city came from Mosul, whose governor al-Bursuqi forced Baldwin to retreat.

And it was from Mosul, the chief city of Mesopotamia, that the first Muslim leader ruthless and able enough to combine the disparate forces of Syria was to emerge.

Zengi and Nur ad-Din: *jihad* revived

Although he was the first – and in some ways the most impressive – of the Muslim rulers who was to restore the *jihad* as a central feature of Near Eastern politics, Zengi remains a somewhat shadowy figure. His father, Aq Sunqur, was a close adviser to Sultan Malikshah, but in the succession crisis after 1092 his support for Barkyaruq cost him his life, and little is known of Zengi's formative years. By 1114–15, at the age of about thirty, he was in the employ of al-Bursuqi, the governor of Mosul. In 1122–3 he was entrusted with the governorship of Basra and Wasit, and in 1127 he returned to Mosul as governor. He is usually known, however, by the honorific title *atabeg* because, following Seljuq custom, the sultan appointed him as tutor to his sons. From Mosul, Zengi was well placed to extend his influence north, into Artaqid territory, as well as west into Syria. His early campaigns against these Turcomans won him a reputation for ferocity, but also provided a ready supply of troops.

Zengi was fortunate in that his arrival in Mosul coincided with the death of Tughtegin of Damascus (1128), and followed close on that of Balaq (1124). His seizure of Aleppo in 1128 would scarcely have been so easy had they still been alive. But Zengi's aspiration was Damascus, the key to Syria. Tughtegin's death weakened a state that had for a generation resisted both Frankish expansion and attempts by the sultan to extend his authority into Syria. Zengi, like Baldwin II, who tried to annex Damascus in 1129, was following the policy of all Near Eastern rulers, whether Muslim or Christian, in seeking to profit from the misfortune of his neighbours. Although Zengi is best known in western historiography as the conqueror of Edessa, and thus as the destroyer of the first of the crusader states, that victory was almost incidental to his strategy, which was the creation of a new axis of power centred on Mosul, Aleppo and Damascus. Throughout the 1130s, he cast an inescapable shadow over Damascus, but was unable either to capture the city or to persuade its ruler, Jamal al-Din and his successor Unur, to hand it over. Damascus fluctuated between supporting Zengi's campaigns in the north – as in 1138 when it provided troops for his defeat of an Antiochene–Byzantine army – and resisting his influence through alliance with the Kingdom of Jerusalem.

Zengi's union of Mosul with Aleppo had altered the balance of power in Syria by isolating Damascus. The policy of preserving Damascene independence was preserved, however, at the price of opening the door to the influence of the Ismaelite sect known as the Assassins. Tughtegin had been forced to give them Banyas as a reward for their support, and after his death they were able to continue what one historian has called their 'terrorist activities' under the shelter of Damascus (Gibb 1958: 455). The Assassins, who were staunch Shi'ites, were to be a thorn in the sides of all empire-builders in the region throughout the twelfth century.

One reason why the domination of Syria proved so difficult for Zengi – as it was to prove later in the century for Saladin – is that his attentions were frequently called elsewhere. For the two periods 1131-7 and 1140-3 he was preoccupied with power struggles in Iraq and operations against the Artaqids and Kurds to the north of Mosul. Yet it was the latter distraction that was to provide Zengi with his greatest triumph. When the young Artaqid prince Kara Arslan opened negotiations with Joscelin II of Edessa, Zengi threatened the county's eastern fortresses and, thanks to Joscelin's precipitate march east, found himself with an opportunity to attack and sack Edessa itself, in December 1144.

The conquest of Edessa sealed Zengi's reputation among both the Franks and Muslims. Because of the massacre of westerners, William of Tyre called him 'homicidal', transliterating his Turkish name into the Latin 'Sanguinus' ('bloody'). The inverse sentiment appears in the story told by Ibn al Athir, in which Zengi, in Paradise, tells another dead man that he has been pardoned all his wrongs for the single virtuous act of ejecting the Franks from Edessa (Hillenbrand 2001: 119). Ibn al-Athir, himself a Mosuli, was the historian of the Zengid clan, and therefore predisposed to elevate Zengi's reputation. Arab chroniclers writing from other perspectives, such as Saladin's partisan, al-Isfahani, painted a portrait of a tyrannical and cruel man who ruled by fear: 'He was like a leopard in character, like a lion in fury, not renouncing any severity, knowing no kindness . . . He was feared for his sudden assaults, shunned for his roughness, aggressive, insolent, [a bringer of] death to enemies and citizens' (cited in Hillenbrand 2001: 122). Hillenbrand has argued that 'even in the context of medieval Arab writers discussing the Turks the blood-curdling qualities of Zengi stand out as exceptional' (2001: 122). There can be no doubting his personal courage, his skills of generalship or his political ambition. He can be distinguished from earlier Muslim leaders who confronted the Franks by his quality of ruthlessness. But to what extent was Zengi a *jihad* leader; and does his career mark the emergence of a new kind of resistance to the Franks, a resistance stiffened by the ideological coherence offered by the *jihad* ideal?

Although the capture of Edessa in 1144 may have been incidental to the strategy he was pursuing in Syria, there is no doubt that it was beneficial in persuading other Muslims both of his ability to defeat the Franks and of the service that he rendered Islam in so doing. Thus Ibn al-Athir, writing in the early thirteenth century, commented that if Zengi had not made himself master of Syria, the Franks would have overrun it. Another factor in assuring Zengi's reputation was the situation in Damascus. For although Zengi can appear as an aggressor who tried to bully the Damascenes into handing him the city, he can also be seen as the protector of the Sunni majority against the double threat of the powerful Shi'ite Assassins within the city and the Franks who had already demonstrated their desire to conquer Damascus.

One respect in which Zengi differed from most of the warlords whom he succeeded was his use of eastern Seljuq institutions to bolster the infrastructure of his government. Thus Syria saw, under Zengi, the proliferation of the *madrasas* and *khanqas* that were already common in Iran. *Madrasas* were schools for the study of the Qur'an and Islamic law; *khanqas* were rather like the western monasteries, though with the important difference that the *sufis* for whom they were founded were not bound by vows of enclosure within the community. In sponsoring these institutions Zengi was cementing an alliance with the religious scholars. As one scholar has remarked (Irwin 1995: 229–30), Zengi's aim was to institute a 'moral rearmament' to complement his military campaigns. Together, rulers and the religious scholars would rout out corruption and negligence in the Dar al-Islam. As early as 1130, Zengi was calling upon Damascenes to co-operate in a *jihad*, but it soon became clear that Damascus itself was to be an initial target of this campaign. For Zengi, the Franks were merely part of the infidel threat to the Muslim community; in many respects, a far more dangerous threat was posed by the presence of the Shi'ite Assassins in Damascus. Zengi's *jihad* was targeted against moral and religious laxness within the Dar al-Islam, and against the heterodoxy promoted by the Shi'ites, as well as the infidel on its borders. Even al-Isfahani, no friend of the Zengids, called Zengi 'a pillar of the *jihad*', and this reputation seems to have been enhanced, rather than created, by his conquest of Edessa (Hillenbrand 2001: 124).

The significance of the alliance between religious scholars and military leaders should not be overlooked. The scholars, *sufis* and *qadis* trained in the schools founded by Zengi, who reached maturity under Nur ad-Din and Saladin, were members of the largely Arab urban intelligensia of Syria and Iraq. They were thus ethnically distinct from the Turkish (or, in the case of Saladin, Kurdish) Seljuqs who formed the military leadership of the Islamic Near East. Seljuq rule had been imposed on the Near East by conquest in the mid-eleventh century, and Seljuq rulers were still resented and distrusted by many noble Arab Syrian families, such as the Munqidh rulers of Shaizar, which lay within the spheres of influence of Aleppo and Antioch. The alliance cemented by Zengi laid the foundations for the kind of unity within the Islamic world that could eventually guarantee the revival of the *jihad* as a practical and realisable aim for the regeneration of Islam and for the restoration of its lost territories.

Zengi was assassinated by a slave in 1146, when he was at the height of his powers. It was left to his son, Nur ad-Din, to seize the prize for so long denied his father – Damascus. If the fall of Edessa marked the beginning of a *jihad*-inspired resurgence, the failure of the Franks to take Damascus on the Second Crusade assured its momentum. Although recent scholarship has explained the logic of the Frankish strategy, it remains the case that the Damascenes seem to have been genuinely shocked by the assault on them from a neighbour that had so recently been an ally. It may have been this shock, coupled with the martyrdom in the siege of Muslim clerics, that served to awaken a sense of *jihad* that had been absent among the pragmatic Damascenes up to that point. But just as decisive in promoting the *jihad* was Nur ad-Din's stunning victory at Inab in 1149, at which Raymond, prince of Antioch, was killed. The realisation that Nur was not only as formidable in the field as his father, but if anything more committed to an anti-Frankish *jihad*, coupled with the loss of the kingdom of Jerusalem as a viable ally against north Syrian expansion, decided Damascus' fate. In fact Nur ad-Din, having failed in 1149 to seize Mosul after the sudden death of his brother, had little choice but to concentrate his attentions on Syria. Unur, who had skilfully maintained Damascene independence, also died in 1149, and Nur ad-Din was able to demand troops from Damascus for a campaign to relieve Ascalon in 1150, pleading the cause of *jihad*. By this stage popular opinion in Damascus was with Nur ad-Din, no doubt in large part because he had brutally suppressed Assassin adherents in Aleppo and could be expected to show the same treatment to the sect that had become so feared in Damascus. In 1154, Nur ad-Din entered the city in triumph without having to shed blood, and Zengi's ambition of a Syria unified under a single rule had been accomplished.

The true significance of 1154 was revealed in Nur ad-Din's actions after taking control of Damascus. In addition to new *madrasas* and *khanqas*, he also established in 1163 a 'palace of justice' (*dar al-'adl*) where he himself sat as chief judge in cases brought by the citizenry. The seriousness with which Nur ad-Din took his judicial duties should be seen as part of the ideological facet of the *jihad* that was so characteristic a feature of his rule. In order to present himself as a fit leader of the Islamic community, Nur ad-Din realised that he had to fulfil the requirements of the just ruler (Rabbat 1995: 3–28). Religious foundations mushroomed in his reign; it has been estimated that of the forty-two *madrasas* built in Syria in this period, half were Nur ad-Din's own foundations (Hillenbrand 1999: 127). By building – literally – the facilities for Sunni Islamic education and training, he was creating a new generation of Sunni *qadis* and *imams*. Moreover, the religious classes had a direct role to play in his rulership. *Sufis* and legal scholars fought in his armies, while in Damascus public readings of religious tracts were held.

Although Arab chroniclers such as the Damascene Ibn al-Qalanisi present an image of the just ruler who from the start was inspired by zeal for the *jihad*, we should be wary of taking such assessments at face value. The Iranian Imad ad-Din al-Isfahani described him as 'the most chaste, pious, sagacious and virtuous of kings', and Ibn al-Athir saw in Nur ad-Din the restoration of the glorious age of the first khalifs. But Ibn al-Athir's history of Mosul is heavily biased towards the Zengid dynasty, while Imad ad-Din was in Nur ad-Din's service from 1166 onward. Following the chroniclers, Elisséeff, his modern biographer, has seen Nur ad-Din's career as guided by zeal for the *jihad*, in an apparently smooth progression from the imposition of internal justice, order and religious orthodoxy on Syria to the conquest of the infidel – internal *jihad*, in other words, followed by the outward extension of the ideal (Elisséeff 1967: II, 426). But, like Zengi, Nur ad-Din realised the practical value of the *jihad* ideal in building up his own territorial power. It is also possible to see the same pattern in quite a different light – as the exploitation of *jihad* ideals to accomplish mastery over the Islamic community, in order to provide him with the resources to extend his territorial ambitions to the west (Köhler 1991: 239, 277). It has been argued that Nur ad-Din experienced an inner conversion either as a result of illnesses in 1157 and 1159 (after which, in 1161, he undertook the pilgrimage to Mecca), or after his heavy defeat at the hands of the Franks in 1163. After this, according to Ibn al-'Adim, Nur self-consciously and publicly combined personal austerity with his prosecution of the *jihad* (Hillenbrand 1999: 134–5).

Nur ad-Din's most important innovation in the field of *jihad* was probably his adoption of Jerusalem as a central focus for the ideal. Although it had never been regarded as so important within Islam as Mecca and Medina, Jerusalem revived as a spiritual centre under Fatimid attentions in the eleventh century (Duri 1982: 355). It even became a substitute for pilgrims who could not reach Mecca. The spiritual significance of the city lay both in its role in the life of the Prophet, who had made his celebrated 'night journey' from Mecca to Jerusalem, and in the associations with the patriarchs of the Jewish Scriptures, who were also venerated in Islam. Religious poetry from the reign of Nur ad-Din complemented the increasing presence of Jerusalem in the works of religious scholars; for example, a poem by Ibn Munir spoke of fighting the crusaders 'until you see Jesus fleeing from Jerusalem' (Hillenbrand 1999: 150). Most famously, Nur ad-Din enshrined his aspiration for the conquest of Jerusalem in buildings, particularly in the *minbar* he commissioned to be installed in the al-Aqsa mosque once the crusaders had been evicted. Study of the *minbar* and its inscriptions has highlighted both how definitive a statement of intent it is and how unusually specific its message (Tabbaa 1986: 233–4). The *minbar* was constructed in 1168–9 and used in Aleppo until Saladin brought it to Jerusalem after the conquest in 1187.

The balance of power in north Syria established by Byzantine intervention in 1158–9, coupled with his failure to secure Mosul, and his defeat at Frankish hands in 1163, led Nur to intervene in Egyptian affairs. It is difficult to know whether this was a response to King Amalric's own involvement from 1163 onward, or whether Amalric wanted to forestall Seljuq intervention after the death of the Fatimid vizier in 1161. At any rate, in 1164 the new vizier, Shawar, appealed to Nur ad-Din to return him to office after being overthrown by a rival, Dirgham. When Nur ad-Din sent his Kurdish commander Shirkuh to reinstate Shawar, he hoped to gain immediate financial advantage in the form of one-third of Egyptian revenues, as well as a measure of control over Egyptian policy. But after Dirgham had been killed, Shawar reneged on his promise to Nur ad-Din and appealed to Amalric for protection. It was a dangerous game, but one that seemed to have paid off when the Franks forced Shirkuh to retreat, then themselves withdrew. Meanwhile, Nur ad-Din's crushing victory over the northern Franks at 'Artah (1164), at which the rulers of Antioch and Tripoli were captured, raised his prestige – and capacity to hire Turcoman troops – to even higher levels. By mid-1168 he had achieved mastery over all of Syria between Antioch and Mosul, and when his brother Qutb ad-Din died in 1170, Nur ad-Din lost no time in intervening decisively in the resulting succession dispute to instate his nephew Saif ad-Din as his vassal. Shirkuh led armies into Egypt in 1167 and early 1169, on the second occasion securing the defeat and death of Shawar and the final withdrawal of Amalric. This victory made Nur ad-Din the most powerful Muslim leader in the Near East, and indeed the most powerful since the death of Malikshah. Yet his final years were not to see the culmination of his aspir-ation to add Jerusalem to his conquests. Although he raided Tripoli and the Galilee and attacked Kerak, he stopped short of a full assault. When he died unexpectedly in 1174, little had been done to accomplish the *jihad* that poets, legists and religious scholars had supported in the public image that their work presented of Nur ad-Din. It was in many ways an unfinished career, yet one that has endured in the memory of Muslim communities who continue to regard Nur ad-Din with respect and admiration.

Saladin and the zenith of *jihad*

[H]e strengthened the cutting blades, gave drink to the terrible lances, and returned to his tents happy and content, received with welcome and gratitude, generous and appreciated . . . with glowing face, fragrant odour, radiant aspect, certain of victory and in firm possession of certainty . . . clearly drawing up his terms for recovering the debt owed to the Faith. (Gabrieli 1984: 126–7)

The description of Saladin before the battle of Hattin by his secretary 'Imad ad-Din conveys a sense of the manifest destiny he was about to seize. It was an image that Saladin himself was keen to present to fellow-Muslims, particularly those who had reason to suspect his ambition. Yet, although Hattin can sometimes appear to us as the crash of a tidal wave of *jihad*-inspired resistance to the crusade that had been building up since the 1130s, there was nothing inevitable about Saladin's rise to the top. Few, indeed, could have predicted, when Nur ad-Din was at the height of his powers, that he would be succeeded by this obscure Kurdish general. For Saladin conquered the kingdom of Jerusalem by first murdering, manoeuvring and bullying his way into Nur ad-Din's inheritance.

When Shirkuh died only three months after defeating Shawar in 1169, Saladin took over leadership of his uncle's mixed army of Turks and Kurds who constituted the occupying power in Egypt. Even after he had been appointed vizier, however, his authority was not unquestioned. As he explained to the khalif in Baghdad, his was not the only army in Egypt, for the Fatimid khalif, al-'Adid, had a large force of Armenians and Nubians loyal only to him (Lyons and Jackson 1982: 32). In order to impose Sunni authority over the Fatimid régime, Saladin first had to overcome the potential resistance offered by this force. He did so with a combination of economy and brutality that characterised his early career: having manufactured a plot against him, he murdered one of the most powerful courtiers, the eunuch Mu'tamin, then bottled up in the streets and squares of Cairo the Nubians and Armenians who had taken up arms in protest against his action, and burned them out. He still had to face the joint assault launched on Egypt in 1169 by the Franks and Byzantines, but the scale of his triumph in Cairo was such that al-'Adid, in recognition of his vizier's newly won authority, sent sufficient troops to defend Damietta against them.

By the time that Saladin's growing power in Egypt had attracted the jealous attention of Nur ad-Din, on whose behalf he held the vizicrate, he had already ensconced many of his relatives in positions of authority. His father Ayyub, his uncle and three brothers controlled the ports and were given *iqtas* (fiefs) from which troops could be raised. This delegation of authority among his immediate kinship was to be characteristic of Saladin's government as it expanded. It is tempting, indeed, to see the role that Saladin came to play in the Dar al-Islam as a means of creating sufficient resources to be able to reward the clan that kept him in power at the expense of his Muslim neighbours (Irwin 1995: 231–3). It is certainly significant that with Saladin the Islamic Near East saw the rise to power of a dynasty drawn from what had been an underclass, the Kurds. Much of the distrust Saladin was to face from Arab Syrians and Turks was based on racial hostility.

As an occupying Sunni ruler over a Fatimid state, Saladin had to tread carefully. Before the end of 1170 Nur ad-Din was urging him to replace Shi'ite worship with Sunni by establishing the Abbasid *khutba* in place of that pronounced for the Fatimid khalif, but Saladin was wary of the strong centre of Shi'ism focused on the person of the Fatimid khalif, al-'Adid. According to Ibn abi Tayy, a Syrian chronicler unsympathetic to Saladin, it was the vizier's reluctance to quash Shi'ism that led Nur to suspect his loyalty (Lyons and Jackson 1982: 44). In any case, Saladin was fortuitously spared having to decide, for al-'Adid, though still only twenty years old, fell gravely ill, and when it was already clear that he was dying, Saladin had the Abbasid *khutba* pronounced while parading his troops through the streets of Cairo. Although Saladin seems to have treated the khalif's children well, there was no question of their succeeding, and in September 1171 the Fatimid khalifate was abolished.

The death of Nur ad-Din and King Amalric I of Jerusalem in the same year (1174) removed two distinct threats to his power in Egypt, for Amalric had not given up hope of conquest, while Nur was on the point of curbing his lieutenant's authority by calling for a financial account of his government. By a stroke of fortune for Saladin, the succession to both was uncertain, for both heirs were minors, and Amalric's was already known to be sickly. In fact, at the time the death of Amalric, coming so soon after Nur ad-Din's, was less useful to Saladin than it appears now, for Baldwin IV's succession temporarily removed any danger to Syria from the Franks, and thus robbed Saladin of an excuse to take over the government of Damascus from Nur ad-Din's heir as-Salih on the grounds of protecting the Muslim community. For although Saladin's base in Egypt made him more powerful than any Syrian leader, Syria remained a better base from which to launch a *jihad*, and Saladin realised that it was only by making use of the *jihad* ideal that he would be able to attain dominance over his Muslim rivals. In the event the leading Damascene families argued among themselves, while Saif ad-Din, Nur's nephew, was too concerned with events in Mosul to show an interest in Syria, and it was a simple matter for Saladin to have himself invited to Syria to act as as-Salih's guardian.

Control over Damascus did not deliver the whole of Syria, still less confer the legitimacy within the Dar al-Islam that Saladin needed. Most of Saladin's career, indeed, was occupied in first trying to dislodge as-Salih from Aleppo, then in conquering Mosul from the Zengids. Interspersed with these campaigns were military activities directed against the Franks; for example, his attempted invasions of the Kingdom of Jerusalem in 1177 and 1183. After the glorious triumph of 1187, Saladin struggled to maintain his conquest against the fresh assault of the Third Crusade (1189–92), in the face of increasing reluctance by other Muslim leaders to recognise the authority he considered should have been guaranteed by his victories. Much the same question can be asked of Saladin as of Nur ad-Din: to what extent was the *jihad* an instrument of policy designed to confer legitimacy on his territorial ambitions?

In Saladin's case, such legitimacy was harder won, for Nur ad-Din was after all the son of Zengi, while Saladin blatantly disinherited the heirs to Fatimid Egypt, Damascus and Aleppo. In the case of Egypt, it could at least be said that he was acting on Nur ad-Din's instructions, and ultimately in the interests of the Abbasid khalifate, in order to spread orthodox Sunni Islam. Syria was a different matter, and here actions such as Saladin's marriage to Nur ad-Din's widow (which, cynical though it may appear, was standard medieval practice) had to be balanced by a constant stream of letters to the khalif designed to establish his position. Such letters contained a curious mixture of begging (for troops and money), boasting of victories, and demanding that he be confirmed as the leader of the *jihad* against the Franks. Clearly, Saladin's chancery was a vital part of his government, and one historian has attributed the initiative in his *jihad* propaganda to the religious intellectuals, especially al-Fadil and 'Imad ad-Din, who staffed it (Irwin 1995: 232). According to this argument, it was the men produced by the policies of Zengi and Nur ad-Din who were pushing the *jihad* to the top of the agenda.

Besides official communications must be placed sources such as the biography by Baha ad-Din Ibn Shaddad, *qadi* to Saladin's army from 1188 to 1193 (Richards 2001), or the highly rhetorical *Eloquent Exposition of the Conquest of Jerusalem* by 'Imad ad-Din, which was apparently recited in front of Saladin in 1192 (Hillenbrand 1999. 182). Because he naturally devoted most of his work to the period in which he knew Saladin most intimately, in other words after the conquest of the Latin Kingdom, it was impossible for Baha ad-Din not to present him as the ideal *jihad* warrior. But he also wanted to portray him as an ideal Muslim ruler, and to explain his success in *jihad* – in unspoken contrast to earlier warriors, such as il-Ghazi – as the natural consequence of his piety. Saladin embodied, for Baha ad-Din, the pillars of Islam; thus, in the biography, Saladin is made to renounce secular amusements and alcohol at the crucial moment of 1169. It is difficult to know how seriously such formulaic statements might be. One critic, al-Wahrani, painted a picture of licentiousness at Saladin's court in Egypt in 1177 that included homosexuality and drunkenness (Lyons and Jackson 1982: 118–19). Saladin's generosity – giving alms is one of the requirements of Islam – is a constant source of comment by Baha ad-Din. But the fact that when he died, Saladin had only 47 dirhems and a singe gold piece in his treasury (Richards 2001: 19; for other examples of generosity, 25–6) can be interpreted as financial negligence as well as generosity: as al-Fadil acknowledged, Saladin spent the fortune of Egypt to gain Syria and that of Syria to gain Mesopotamia, and that of Mesopotamia to conquer Palestine. But Saladin understood the importance of generosity as a weapon of policy, a point recognised by the equally perceptive William of Tyre, who acknowledged that Saladin used liberality as a means to exact loyalty (WT 1986: 1012–13). Ibn al-Athir, likewise, attributed the victory at 'Amid, near Mosul, in 1183 to Saladin's generosity, which he contrasted with the meanness to his own men shown by 'Amid's defender, Ibn Nisan. In this context, al-Fadil quoted a dictum of Khalif Harun al-Rashid: 'no money is wasted that receives a legacy of praise' (cited in Lyons and Jackson 1982: 368). Saladin knew, moreover, how to make financial exactions appear in the best light: when he abolished the unpopular service taxes that had been imposed by the Fatimid régime in Egypt, he replaced it with a compulsory *zakat*, or alms tax, which was not only sanctioned in Islamic law but required by Qur'anic injunction. Saladin was in fact perpetually short of money, and although Egypt appeared wealthy, the ostentatious splendour of the Fatimids may have disguised economic decline; by the 1190s its gold coinage had run out (Lyons and Jackson 1982: 49, 318). When he was appealing to the khalif for money during the siege of Acre in 1191, he declared that his personal resources amounted to only three estates, one in Egypt and two in Syria (Lyons and Jackson 1982: 324).

Baha ad-Din's testimony about the role of piety in the army is harder to doubt. Parts of the *hadith* – the customs of the Prophet that formed part of Islamic law – were recited to the army before battle. Like Nur ad-Din, Saladin attached religious scholars to his army. According to Baha ad-Din, he preferred to give battle on Fridays, when he could benefit from the prayers said on the Islamic holy day (Richards 2001: 72). After Hattin, the captured Hospitallers and Templars were given to *sufis* to be executed, to emphasise the religious duty of even non-combatant Muslims to wage *jihad*. (That *sufis* were unaccustomed to such duties doubtless prolonged the captives' agony.) The appeal to *jihad*, which became much more noticeable in Saladin's bureaucracy and in his policy after the submission of Aleppo in 1183, was necessary in order to amass an army of sufficient scale to conquer Jerusalem. The picture painted by the sources of a decisive religious awakening at the mid-point of his career looks rather formulaic, especially given that we encounter the same in contemporary accounts of Nur ad-Din. It is true that in 1175–6 and 1185 attempts were made on his life by the Shi'ite Assassins, which may have resulted in his devoting himself to the cause of Sunni orthodoxy more strongly. Like the later crusader Louis IX of France, Saladin is said to have dedicated himself to the recovery of Jerusalem in gratitude to God for delivery from serious illness. During this illness, which occurred in 1182, al-Fadil apparently encouraged Saladin to vow that he would hereafter devote himself to *jihad*, and eschew warfare against Muslims. After gaining Aleppo in 1183, Saladin could afford to take such a vow – although if he took one, he did not keep to it – for he had no more need to fight Muslims for control of Syria.

Modern scholars have tended to be sceptical about Saladin's religious motives. Ehrenkreutz (1972: 237) and Lyons and Jackson (1982: 240) argue that if he had died in 1185, when he was seriously ill, he would be remembered as nothing more than a self-aggrandising dynast in common with many of his predecessors, and Köhler (1991: 316) points to the many treaties he made with the Franks, and indeed, from 1183 onward, with the Byzantines. This scepticism has been less readily accepted by Hillenbrand (1999: 185–6), who argues that the public image presented by the ruler was more important than his personal piety. Even if we regard much of the chroniclers' account as stylised, we need not be unduly sceptical about Saladin's religious motives. It is perhaps easier for today's reader to take seriously sincere declarations of *jihad* than it was even a few years ago. Moreover, it is dangerous to dismiss as simply propaganda the personal piety of any medieval ruler: beliefs were nonetheless sincere for being subject to exploitation.

The attempted invasion of Jerusalem in 1177 may have been a piece of opportunism on Saladin's part, taking advantage of the evacuation of much of the kingdom's armed strength to the northern campaign. At this stage, the problem posed by the Franks was still largely their willingness to give military aid to Aleppo (as they had done in 1175) and thus to forestall Saladin's Syrian ambitions. In any case, the defeat at the hands of Baldwin IV at Gaza, though humiliating, was only a temporary setback, partly reversed by a victory near Banyas and the destruction of the Templar castle on the Jordan at Jacob's Ford, north of the Sea of Galilee, in the following year. Yet he could hardly claim to be the leader of the *jihad* on the basis of such sporadic affairs, especially while Kilij Arslan was winning a spectacular victory over the Byzantines in Anatolia (1178) In his letters to the khalif from this period 'the holy war is the dominant concept, but how it is to be pursued is left blurred' (Lyons and Jackson 1982: 156). In fact, he was in an impasse, for he could not pursue the *jihad* until he had the resources of Syria behind him, while on the other hand he had no claim to those resources without waging wars against Zengid Aleppo and Mosul that made his protestations of *jihad* sound hollow.

As in 1169 and 1174, it was the death of a rival – this time as-Salih – that gave Saladin his opportunity. His own claim to Aleppo was weak, but by 1183 he had been able to isolate it by gaining effective suzerainty over Mosul. It was as he returned to Syria that Saladin heard news of Reynald of Châtillon's Red Sea raid, which had already been dealt with by his brother al-'Adil in Egypt. Scholarly opinion is divided as to the significance of Reynald's campaign (Hamilton 1978: 106–8; Lyons and Jackson 1982: 185), but by exploiting rumours of the Frankish intent to raid Medina and steal the body of the Prophet, Saladin could at last convince the khalif that his leadership alone could protect the Dar al-Islam from the infidel. A letter to the khalif in which Saladin defends his resumption of war against the Mosulis also lays out his own strategy for the future: he had first come to Syria, he says, to fight the unbelievers, including the Assassins, but had been prevented from doing this by Mosuli interference and by their seizure of Aleppo. If he were given Mosul, then Jerusalem, Constantinople, Georgia, and even Spain would follow, 'until the word of God is supreme and the Abbasid caliphate has wiped the world clean, turning the churches into mosques' (cited in Lyons and Jackson 1982: 194).

Saladin finally began work on realising this dream with his invasion of the kingdom of Jerusalem in 1183, but Guy of Lusignan's caution fore-stalled him. Four years later, he had not only the further justification that Reynald of Châtillon had broken the truce, but the extra resources from his subjugation of Mosul behind him. By 1187, his own credibility was at stake, for having for so long demanded to be taken seriously as a *mujahid*, he now had to deliver Jerusalem to Islam. His numerical advantage over the Franks had been attained largely through the promise to recover Jerusalem, so he could not afford another inconclusive campaign. This is why he was prepared to divide his forces at Tiberias, and why he was so fortunate that Guy, following bad advice, took the bait and marched across the Galilee to meet him. Saladin's forces probably outnumbered Guy's by three to two, without counting reservists. It was, however, a loosely structured army, its size based on his having convinced and bullied other emirs across Syria and Mesopotamia that he would be victorious. As has been argued, 'profit and numbers were inextricably linked': once he started to lose, his allies, who were under no compulsion from the khalif to fight for him, would desert him (Lyons and Jackson 1982: 252–3, 286). The battle of Hattin itself was won by using the traditional tactics of outflanking and surrounding the enemy, but it was the ability to use the extra numbers to hold the ridge so that the Frankish cavalry could not break through that proved decisive. Nevertheless, Saladin had been handed his opportunity by Guy's determination to pursue him and to seek battle where there was, as yet, no need.

Saladin's victory at Hattin, though complete, posed new dilemmas. The strategic imperative was to seal the victory by taking the coastal towns and the strongholds in the interior of the kingdom, but the momentum and logic of *jihad* demanded that Jerusalem itself be taken. In the event, a single miscalculation – the length of time that it would take to besiege Tyre – was to cost Saladin much of the advantage he had won. For although Acre, the chief city, Haifa, Arsuf and Caesarea fell quickly, the defence of Tyre was rallied by Conrad of Montferrat, who arrived at Acre without realising that it had fallen, and turned back in the harbour as he was about to go ashore. Meanwhile Saladin was preoccupied with the assault on Jerusalem in September–October 1187. After brief resistance, Balian of Ibelin was able to persuade Saladin to offer terms to the defenders by threatening to kill all his Muslim prisoners and to destroy the Islamic holy places. The capture of Jerusalem, which was celebrated in contemporary poetry and religious literature (Hillenbrand 1999: 188–92), brought Saladin immense credit in the Dar al-Islam, but his failure to take Tyre in 1187 with the other Frankish towns gave the Franks a base from which to regroup. A further lapse, the release of Guy of Lusignan from captivity in 1188, also proved to be costly. Abjuring his oath not to take up arms against Saladin, Guy put together a tiny force from the rump of the barons who remained loyal to him. Denied entry to Tyre by Conrad, with nowhere to go, Guy encamped in front of Acre in 1189. Gradually his force attracted others, and provided a focus for the early recruits to the crusade launched by Gregory VIII, notably Henry of Champagne, until it grew into a full-scale blockade by land and sea.

Saladin's difficulties within the Dar al-Islam were far from ended by his successes. From the khalif's point of view, the continued existence of the kingdom of Jerusalem had provided a necessary distraction for Saladin. Indeed, according to 'Imad ad-Din, the khalif feared that quick victory over the Franks would free Saladin to realise his aspiration of an Islamic empire that would include Baghdad (Lyons and Jackson 1982: 280–1). The longer that Saladin struggled to relieve Acre, the harder it proved to keep his coalition together. Baha ad-Din reports the unravelling of the coalition through the desertion of some emirs, who clearly felt that they had won all they were likely to from Saladin (Richards 2001: 95–6, 133–5). Saladin's letters to the khalif had become increasingly desperate. 'Islam asks aid from you as a drowning man cries for help', he wrote, and it was with despair that he reported the dispersal of his army.

Moreover, although he could still expect reinforcements from al-'Adil in Egypt, Saladin knew that a crusade was on its way to Palestine. The unexpected death of Frederick Barbarossa in Anatolia provided some relief, but the arrival of the French and Angevin contingents under Philip Augustus and Richard I in summer 1191 proved decisive. The fall of Acre was Saladin's worst setback in the *jihad*. What was so damaging for Saladin was that, whereas he had lost battles against the Franks in the past, at Acre he had every advantage on his side and still failed. The fall of Acre was followed by the reverse at Arsuf in September, which, although it did not result in any territorial loss, assured Richard's reputation among the Franks while weakening Saladin's among the Muslims. During the next year, the two rulers fought themselves to a standstill: just as Richard did not have sufficient forces to take and keep Jerusalem, so Saladin was unable to retain intact his own conquests of 1187–9. Essentially Saladin fought a defensive campaign, spending six months inside Jerusalem and maintaining an armed front along the Jordan valley. During the protracted negotiations between the protagonists in 1192, Saladin's brother al-'Adil came to prominence, with the result that the treaty that both parties eventually accepted owed much to his view of how Muslim–Frankish relations might work along this frontier. Throughout the crusade, Saladin took advantage of the splits among the Franks so as to play Conrad of Montferrat and the native barons off against the Angevins. He also continued to receive Byzantine embassies, though it must have been obvious to him after the German army had forced its way through Byzantine lands in 1189–90 that his alliance with Constantinople had little practical value. The key to the negotiations was Jerusalem, for although both Saladin and Richard realised that its strategic value was minimal, neither could afford to give it up: for both sides, the city's ideological significance provided the basis on which they had in the first place been able to recruit armies.

After Richard's departure in September 1192, Saladin made plans for the pilgrimage to Mecca, but never carried them through. He wintered in Damascus, and died there in March 1193, having contracted a fever. His reputation among Muslims was assured by the recovery of Jerusalem, which was to remain in Muslim hands, save for a brief hiatus, for over seven hundred years. At his death, Egypt, Syria, Mosul and its environs and most of Palestine were under his control. In the Holy Land Jerusalem had been taken and there was little prospect of its loss to the Franks, who were restricted to a tiny coastal strip between Jaffa and Tyre, though large parts of Tripoli and Antioch remained unconquered. The *jihad* ideal, propagated by religious scholars, provided a coherent ideology for the future defence of his conquests. The outlook for crusading looked bleak.

CHAPTER 6

Crusader society

This chapter poses questions about the kind of society established by the Franks in the East. What were relations with the subject peoples – both Muslim and Christian – like? Were the native peoples allowed freedom of religious worship? What was their legal status, and did the legal framework really work in practice? Recent archaeological evidence is used to assess settlement patterns in the Holy Land, and the importance of the visual culture of the Crusader States is stressed through an examination of art and architecture. Finally, there is a brief explanation of the nature of the economy of the Crusader States: how was Crusader society fed, what were the main sources of income, and what did the crown do to stimulate trade?

Franks and natives: assimilation and marginalisation

The First Crusade was proclaimed by the papacy as a war to liberate the eastern Churches. Early encounters with the indigenous Christians of the East, however, were disillusioning for the Franks and must have been frightening for the 'liberated' peoples. The letter sent by the crusade leaders to Urban II announcing the conquest of Antioch made no distinction between Muslims and 'heretical' Christians (Fink and Ryan 1969: 111); indigenous Arabic-speaking Christians were among those slaughtered in Jerusalem in July 1099; and the Greek Orthodox monks who served the shrine of the Holy Sepulchre were expelled from the church along with the clergy of the separated eastern Churches. After the death of Adhémar of Le Puy, who had maintained contact with the Orthodox patriarch of Jerusalem, Symeon II, there was nobody in the crusading army with sufficient knowledge to guide the conquerors in the new relationships into which they entered with their subject peoples. Consequently, the first generation of the new settlement looks rather like a period of 'trial and error', in which the settlers gradually learnt how to deal with the Arabic-speaking peoples, both Christian and Muslim, who constituted the vast majority of the population.

In the 1120s, one of the settlers, Fulcher of Chartres, reflected that the capacity of the new state to assimilate peoples of different ethnic origins was in fact one of its virtues:

Consider and ponder how in our own days God has transformed the West into the East. We who were westerners have become easterners. Whoever was a Roman or a Frank has in this country become a Galilean or a Palestinian; people from Rheims or Chartres are now citizens of Tyre or Antioch. We have already forgotten our birth places, which are unknown to many of us, and no longer talked about. Some of us have already inherited property here. Some have married not only among their own people but Syrians, Armenians, or even converted Muslims . . . Words of different languages have become common to all nationalities, and a mutual faith unites people who know nothing of their ancestry. (Fink and Ryan 1969: 271–2)

Because Fulcher's motive in writing was in part to attract new settlement from the West, we cannot take his words at face value; nevertheless, they provide a glimpse of Frankish self-perception by a thoughtful and well-informed contemporary. The question of ethnic identity and its relationship to the state was one that interested Fulcher. Describing the journey of the first crusaders through Asia Minor, he remarked on the unity of purpose among the crusaders despite the diversity of language and identity. Just as the experience of common action forged a single 'Frankish' people on the first crusade, so the task of state-building created, in his eyes, a hybrid society of 'occidental-orientals' (Murray 1995: 59–73).

On one level, crusader society was no such thing. Even Fulcher was unable to envisage Muslims as forming part of the hybrid society so blessed by God; indeed, they were not permitted to enter the city of Jerusalem. Muslims were treated differently in civil and criminal law from Christians; however, non-Frankish Christians also suffered different treatment from Franks. Any society operates through the interplay of legal, social, political, cultural and religious forces, and in Outremer these forces were sometimes in a state of mutual tension. What the law said or implied about the place of one ethnic group or another was not invariably borne out by the evolving experience of state-building, or in the development of a visual culture. Most writing about crusader society has tended to coalesce on either side of a debate about colonialism, and to focus on the aptness of a colonial model for understanding the relations between Franks and natives. But we should also remember that Outremer was a frontier society, in which the subject people shared a common identity with populations on the other side of the political frontier. The conditions under which the subject peoples lived therefore reflected the latent threat that a Muslim conquest might be perceived by the native peoples as a war of liberation, in which they would themselves participate to overthrow Frankish rule.

Religious worship

Written crusader sources identify Muslims, Greeks, Syrians, Armenians, Jacobites, Georgians, Nestorians, Copts, Ethiopians, Maronites, Samaritans and Jews as distinctive religious groupings. Sometimes these communities could be identified with territorial units such as villages: in practice, most local rural communities comprised a homogeneous group with the same religious affiliation. All the towns, however, had mixed populations in which many of the different groups were represented, and here the interaction of different traditions was most noticeable. Pilgrimage accounts from the twelfth century remark on the diversity of religious traditions at the shrine churches, and attempts to categorise and understand these differences can be seen in treatises on the state of the Holy Land from the end of the twelfth century onward (Kedar 1998: 111–34).

Religious identity became an index of ethnic behaviour and customs for western observers like James of Vitry, bishop of Acre (1216–27). James overtly linked his critique of the religious customs and theology of Arabic-speaking Christians to negative racial observations: thus, for example, the 'Syrians' (Greek Orthodox), who were halfway between Greeks and Latins in religious customs, are described as 'vacillating', and 'duplicitous': 'they say one thing but think another'. Because they spoke Arabic and were culturally closer to Islam than western Christendom, James was suspicious of their loyalty both to their secular rulers and to their ecclesiastical superiors. While the Armenians and Georgians were respected because they ruled over sovereign territories of their own, and could thus enter into political relations with the Crusader States, the indigenous Greek and Syrian Orthodox were despised for their pusillanimity: 'everywhere they live, they pay tribute to someone . . . they are as useless as women in battle' (Stewart 1895: 67). Much of the vocabulary and ideas expressed by James derives from classical and earlier medieval sources, but the important point is that he chose to give racial overtones to religious customs.

Many of the customs of indigenous peoples – even of Christians – were unfamiliar, and therefore regarded with suspicion by the Frankish authorities, and perhaps by ordinary Franks also. James remarks, for example, on the practice of circumcision among the Syrian Orthodox, which they borrowed from Jewish and Muslim custom; on the lack of auricular confession and clerical celibacy; and on the Syrian Orthodox custom of branding the foreheads of children with the sign of the cross. However, in the twelfth century at least, there was no attempt to impose conformity of religious practice on the subject peoples. Indigenous Christians, Muslims and Jews were by and large permitted to observe their own traditions unmolested. Greek and Syrian Orthodox, Armenian, Georgian, Coptic and Ethiopian monasteries all flourished under Frankish rule. All Christian confessions were expected to continue their worship within their own churches, and to be guided by their own priests. James of Vitry's complaints about the continuation of Orthodox practices in his diocese indicate that the eastern Christians were accustomed to being left alone by their Latin bishops. James' arrival in the East, however, marks a change in policy by the papacy. In contrast with the attitude of *laissez-faire* that had prevailed throughout the twelfth century, James went into the churches of his Greek Orthodox flock, and even those of the Syrian Orthodox, who were not his responsibility, and hectored them about their errors in religious custom. Whether his preaching had any effect may be doubtful, but the fact that he tried to impose Roman conformity on them remains significant.

In respect of the religious hierarchy, however, the farther away in belief from Latin Christianity, the greater freedom of both ecclesiastical jurisdiction and religious custom applied in practice. Because the Greek Orthodox were technically members of the same Church as the Roman, they and their clergy were subject to the supervision of the same bishops as the Latins, and since no see could have two bishops, it was Latins who were appointed at the expense of Greek Orthodox. There were exceptional circumstances, determined by political events, when Greeks held office; for example, between 1165 and 1170, and during the commune of Antioch under Bohemond IV, when a Greek patriarch resided in Antioch, or, more obscurely, after 1159, when the Latin bishop of Laodicea, Gerard of Nazareth, was forced out of his see by a Greek (Jotischky 1997: 222–5; Hamilton 2001: 199–207). The Syrian Orthodox and Armenians, however, who were members of separate Churches, continued to elect their

own religious hierarchy throughout the crusader period without interference, and their church leaders even enjoyed the friendship of Latin
senior clergy.

The First Crusade was a war of liberation and conquest; it was not a
war for the extermination or conversion of Muslims. Thus, although the
Franks in the first ten years of the settlement did sometimes massacre
Muslim urban populations when they resisted conquest, there is little evidence of any policy of conversion. Some individual Muslims certainly did
convert: just after the conquest of Jerusalem, for example, the ruler of
Ramla was converted and fought with the crusaders against the Egyptians.
A few, such as the lord of Galilee, Walter Mahomet, even managed to
attain high status in Frankish baronial society. On the whole, however,
Frankish lords were reluctant to convert their Muslim peasantry, for
conversion to Christianity would end their servile status. The thirteenth-
century papacy, which took a more active interest in the ecclesiastical
affairs of the Crusader East than had been the case before 1187, found it
impossible to persuade landowners – even religious ones like the Military
Orders – to permit or undertake conversion on a large scale, although by
this time there was greater urgency to do so because of the growing fear
that the existence of Islam itself constituted a danger to Christendom
(Kedar 1984: 145–9), and despite a changing spiritual climate in the West
in which conversion and missionary activity played a prominent role.

Some mosques, particularly urban ones, were converted into churches –
most famously, the Qubbat as-Saqra in Jerusalem, which was reconsecrated as the Templum Domini, and the mosque in Caesarea, which
became the cathedral of St Peter. Sometimes, as in Ascalon after 1153, the
reconsecration of a mosque as a church represented a return to the situation before the Islamic conquest. But the Muslim traveller Ibn Jubair of
Granada, who visited the kingdom of Jerusalem for a few weeks in 1183,
reported functioning mosques in Tyre and public prayers in Acre, and it is
quite likely that had he visited other towns, he would have found a similar
situation (Kedar 1990: 138–9). In some cases, as for example in Sebastia,
where John the Baptist was buried, or the cave of Elijah on Mt Carmel,
Muslims were permitted to pray in a space consecrated for Christian worship. The most celebrated example of this occurred when the ruler of the
autonomous little emirate of Shaizar, Usama ibn Munqidh, was allowed to
pray in the Templar church inside what had once been the al-Aqsa mosque
in Jerusalem. In rural communities, there was probably little interruption
in Islamic worship. An episode in 1156 in the region southwest of Nablus,
when restrictions on Friday sermons were imposed by the Frankish lord on

his Muslim subjects, can be read as revealing religious intolerance; however, as Kedar observes (1990: 151–2), it also suggests that long-standing practices such as pilgrimages to Mecca, and visits by local jurists to Damascus to consult Qur'anic scholars, continued unmolested until they were discovered. What the Frankish lord objected to was not so much the religious content of the sermons as the drain they imposed on his resources by taking peasants away from their work. In the end, Frankish lords regarded the economic value of a largely passive and contented villeinage as more important than whether they practised Christianity or Islam.

Legal and social status

In law, the Franks effectively made no distinction on religious grounds between their subject peoples. The basic division of society was between Franks and non-Franks, not between Christians and non-Christians. The poll tax on *dhimmis* (non-Muslims who lived under Islamic rule), which had applied to all native Christians and Jews, was reversed so that a similar tax was now paid by indigenous peoples, whether Muslims, Jews or Christians. Because a lord could raise the normal rate arbitrarily, the tax must potentially have been punitive (Mayer 1978: 175–80). The laws of the kingdom of Jerusalem, which survive only from the thirteenth century, but which can be taken as an indication of earlier practice as well, are interested in comparative legal status primarily in a financial context. The financial compensation for the crime of assault, for example, regarded 'Syrians' – which included Christian and Muslim Arabs, as well as Jews – as both paying and receiving only half what was due to Franks. But the Cour de la Fonde, the court that had jurisdiction over all non-Franks living in towns in cases other than assault, showed clear prejudice both in its composition and in some assizes against Muslims (Kedar 1990, 164–5).

Fulcher of Chartres' enthusiastic encouragement of new settlers from the West also reveals something of the economic advantages of such divisions: 'those who had only a little money have innumerable bezants here; those who had not even an estate, here, through God's gift, already have a city' (Fink and Ryan 1969: 272). In effect, there was no Frankish peasant class in Outremer; nor was there any need for one, because the native peoples fulfilled the role by working the land, and thus of producing the agricultural needs of the population in conditions of servitude or of villeinage analogous to the system of manorial economy in the West. The Franks simply imposed their own feudal culture on to the existing relations of social and economic dependence that existed in pre-crusade Palestine and Syria. Since an agricultural economy quite similar to that in the West was already developing in the Near East by 1100, this made little effective difference to the vast majority of the subject peoples. Ibn Jubair famously remarked that the Muslim villeins he saw in Galilee seemed more prosperous and content than those living under Islamic rule outside the kingdom of Jerusalem (Broadhurst 1952: 317). What did make a difference in status and economic position was whether one lived in a town or in the countryside, for servitude in the sense of being tied to a particular piece of land applied only to the rural economy. In practice, however, this important distinction must have been relevant mostly to the indigenous Christians, for relatively few Muslims, except those in servitude in urban industries, such as glass-making, lived in towns. But indigenous Christians who lived in towns enjoyed higher legal status than those in the countryside, and this in turn enabled them to acquire wealth through commerce or industry.

The subject peoples enjoyed autonomy of varying degrees in self-government. Looking back to the foundation of the kingdom of Jerusalem, the thirteenth-century legal scholar and landowner John of Ibelin thought that the 'Syrians' (meaning the indigenous Christians) had asked Godfrey of Bouillon for the privilege of self-government through their own courts and following their own laws. There was little reason why Godfrey should not have granted this, for it was convenient for the Franks to use religion as a determining feature in local jurisdiction, given the homogeneity of most units of settlement. Besides, early medieval practice had always allowed different ethnic and indigenous groups to use their own laws, and the expansion of western settlement into Islamic territories in Sicily and Spain at the same time as the Near East saw comparable arrangements for internal self-government by religious minorities. Thus Muslim villages were allowed to appoint a headman (ra'is) to supervise the administration of Islamic law in the community, and to represent the village to its lord, and the same seems to have been true for indigenous Christian villages. Because rural communities tended to be homogeneous units, there can have been few occasions when cases arose that required judgement between Franks and non-Franks; in towns, however, where the populations were more mixed, such cases were heard in Frankish law.

Ecclesiastical law enforced the separation of Franks and Muslims. The decrees of the council of Nablus (1120), for example, enacted that sexual intercourse between Christians and Muslims was to be punished by castration for males; consenting women were to have their noses cut off.

Muslims were prohibited from wearing 'Frankish' clothing, in a reversal of the sumptuary laws to which Christians living under Islamic rule were subjected (Mayer 1982: 531–43). But the Nablus legislation may have portrayed an ideal rather than reality, and Fulcher of Chartres' description of an assimilated society, written in the following decade, may be closer to the actuality. The career of Usama ibn Munqidh testifies to the close relations that could develop between individual Muslims and Franks (Hitti 1929), and if Usama was an exceptional case because of his political importance as an independent frontier lord, there are also examples of Muslim and Arabic-speaking Christian doctors who were valued by the Frankish aristocracy for their skills. If Usama's horrifying description of Frankish medicine is accurate, the preference for eastern practices is hardly surprising.

The subject peoples of the kingdom of Jerusalem, whether Christian or Muslim, played no role in the public life of the state: they had no official representation in government. The situation seems to have been somewhat more favourable in Antioch where, as Usama indicates, Greeks could hold public office, and probably also in Edessa. In both Antioch and Edessa, eastern Christians – Greeks, Armenians and Syrians – constituted the majority of the population. But even where an eastern Christian community remained a powerless minority, it could not be entirely ignored, especially when it could find influential backing. Thus, for example, the Syrian Orthodox monastery of Mary Magdalene in Jerusalem complained in the 1130s about the seizure of some villages in its possession to the south of Jerusalem. Godfrey of Bouillon had originally granted this property to a Frankish knight after the First Crusade, but when the knight was captured by the Egyptians and disappeared from sight, the monastery was able to petition for its return from Baldwin I. When, to everyone's surprise, the knight returned as a hero from captivity in the 1130s, Fulk reconfirmed Godfrey's grant of 1099. But the monks found an ally in Queen Melisende who, herself half-Armenian, championed the interests of the native Christians. Fulk found himself caught between Frankish public opinion and the need to placate his wife, in whose name he ruled (Palmer 1992: 74–94). It is significant that the case occurred soon after the revolt of Hugh, count of Jaffa, which can be seen as an attempt to prevent Fulk from marginalising Melisende from government. Although the revolt failed, it must have sounded a warning for Fulk. Despite the lack of military or political authority with which James of Vitry was to taunt them, eastern Christians could on occasion use their influence with the royal dynasty to protect their interests.

Settlement patterns

Of the religious and ethnic groupings under crusader rule, some minorities were largely limited to a particular region – for example, the Maronites in the county of Tripoli and the northernmost parts of the kingdom of Jerusalem. Others, such as the Georgians, Copts and Ethiopians, comprised religious communities in or near Jerusalem. Armenians, besides a significant community in Jerusalem itself, could be found in all the major towns of Outremer, but were particularly numerous in Antioch and Edessa, and in the latter also formed a sizeable element in the countryside. The Jacobites, the name given by the Franks to members of the Syrian Orthodox Church, were likewise concentrated in the north, but a rural population was resettled by Baldwin I from east of the Jordan to villages to the south of Jerusalem, and they had a presence in the main towns. Jewish communities could be found both in the coastal towns – especially in Haifa – and comprising some villages in Galilee. The largest groups of subject peoples, the Arabic-speaking Greek Orthodox – 'Syrians' in most crusader sources – and the Muslims, were scattered throughout Outremer, with larger concentrations of Muslims in the kingdom of Jerusalem than in the northern states.

Recent research using methodologies borrowed from historical geography has suggested that the spatial distribution of Muslims and indigenous Christians in the kingdom of Jerusalem was more sharply delineated than has previously been supposed. The picture that emerges from place-name evidence, coupled with archaeological surveys of the distribution of churches, indicates that the Arabic-speaking Orthodox tended to live in villages located within defined areas of the kingdom (Ellenblum 1996: 213–76). From the archaeological record of surviving churches, it is possible to determine that Palestinian Christians in the Byzantine period lived largely around Jerusalem itself, and within an area of an arc sweeping from Jericho and the Jordan southwest as far as Hebron to the south, and westward to the coast around Ascalon, and including the areas around Lydda and Ramla. There were also Christian communities all the way up the coast, and further concentrations in Galilee and the far north between Acre and Tyre. Although not all the identifiable Byzantine churches survived into the crusader period, Ottoman census records from the sixteenth century confirm that most of these parishes remained Christian throughout the Middle Ages. The central areas of the kingdom, however – roughly corresponding to biblical Samaria – seem never to have been areas of predominantly Christian settlement. It was here, Ellenblum argues, that the Arab conquest of the seventh century filled a vacuum, after the destruction of the Samarian communities in the sixth century, and it was these regions that were therefore largely Muslim.

This model of spatial distribution argues for a much more even proportion of indigenous Christians to Muslims than most historians have previously allowed – perhaps even a 1:1 ratio. Contemporary observations, which are by their nature impressionistic, give conflicting views. According to the Frankish writer Ernoul, in the 1160s Muslims could be found in all the villages of the kingdom, and this seems to be confirmed by Ibn Jubair (Kedar 1990: 149). But Ernoul was reporting the words of the king of Armenia, a visitor to the kingdom, and Ibn Jubair only travelled in the Galilee and the environs of two northern towns. Moreover, another Spanish Muslim traveller, Ibn al-'Arabi, wrote that in 1093–5 the indigenous Christians occupied the whole of Palestine (Kedar 1990: 149), but he may have been speaking of the region of Jerusalem, which he visited. If we add to the mix the remark of 'Imad ad-Din that the villages around Nablus and Sidon and Beirut were largely Muslim, then the regional distribution model seems to be borne out. Muslims lived largely in the central regions and in the hinterland of the far northern towns; Christians largely in the south.

Archaeological research has also disproved the long-held view that the Frankish settlers lived exclusively in the towns of the coastal region. On the contrary, extensive farming and settlement of the interior of the kingdom has been established (Ellenblum 1996: 41–210; Boas 1999: 60–90). Here again, however, a model of spatial distribution can be applied. Frankish rural settlement seems mainly to have been within the bands of settlement of indigenous Christian villages; for example, 'Abud, St George near Tiberias, and Darum (Ellenblum 1996: 119–44). Typically, as in the West, the Frankish landlords built manor houses on hilltops, which were sometimes fortified, and from which they could dominate the villages over which they exercised lordship. Although the Franks in the countryside did not usually live among the indigenous Christians, but maintained separation from them on the same ethnic basis as we find, for example, in the crusader laws, there is evidence of some pragmatic interaction between Franks and natives. Agricultural resources, such as water and irrigation, might be shared. This made sense, because the Frankish rural settlers had no interest in denying the indigenous peasants the means by which they could provide the agricultural surpluses that formed the income from land.

Another kind of sharing was the double use of rural churches. Because the number of Franks in the countryside was relatively small, it made no sense to build new parish churches for them, when every Christian village had its own. The archaeological evidence from parishes such as 'Abud and St George indicates that the Orthodox churches built in the Byzantine period were enlarged or altered in the twelfth century, to provide for the observance of Roman as well as Orthodox liturgy. The literary record confirms this: for example, a letter from a Byzantine theologian, Theorianus, to the Orthodox priests of Beth Zechariah ('Ain Karim), a few miles west of Jerusalem, deals with the practical consequences of having to share the church with Latin clergy (Jotischky 1997: 218). The village, which as the birthplace of John the Baptist was the centre of a minor cult, was given to the Hospital of St John, but the Orthodox priests were evidently expected to continue using the church alongside the Hospital's clergy.

The subject peoples: resistance or docility?

It is striking that the literary evidence for hostility in Frankish attitudes towards the eastern Christians begins to become more noticeable after the shrinking of the kingdom of Jerusalem as a result of the conquest of 1187–91. The population of Acre, for example, doubled from the influx of refugees from the countryside, many of who were eastern Christians. At the same time, most of the hinterland was lost, which meant that indigenous communities were split between Frankish and Muslim rule. James of Vitry's suspicion of the Syrian Christians on the grounds that 'for a little money they sell the secrets of the Christians to the Saracens' makes sense in a context in which indigenous Christians were moving across political frontiers (Stewart 1895: 67). As James observed, the Syrian and Greek Orthodox Christians were Arabic speakers, and shared a common culture with the indigenous Muslims.

Indigenous Christian attitudes to the Franks are difficult to ascertain, partly because of the paucity of sources dealing with the question, and partly because, in the absence of any political or cultural leadership among the indigenous communities, such evidence as can be found shows no consistent pattern. Local Christians gave help and moral support to the crusaders in 1099, and the Armenian chronicle of Matthew of Edessa welcomes the crusade in terms of a liberation. In 1144, however, the indigenous population of Edessa saw no reason to support a Frankish régime that had been at best negligent of local interests, and at worst oppressive; a Syrian Christian chronicle of the period describes the joyful welcome given to Zengi when he entered the city in triumph. On the whole, however, the indigenous Christians of the kingdom of Jerusalem seem to have felt more investment in a Frankish rather than a Muslim government. Saladin's conquest must have reinforced such calculations, for in 1183 his raid in the Galilee damaged the Orthodox monastery on Mt Tabor, and in 1187 the monks of St Euthymius, in the Judaean desert, were enslaved by the Turks (Jotischky 2001: 85). Although one contemporary Arabic Christian source attributes Saladin's capture of Jerusalem in 1187 to the espionage of an Orthodox Christian merchant who had known the sultan in Egypt, western sources indicate that Orthodox senior clergy supported Richard I's crusade. The abbot of Mar Sabas, for example, showed Richard where a piece of the true cross could be found (Jotischky 1999: 189–91).

The destruction, exile, or forced resettlement in the countryside of a large section of the urban middle class between 1099 and 1110, when the Franks were engaged in capturing the coastal towns, meant that those Muslims who survived lacked effective leadership. This doubtless explains their apparent docility throughout the crusader period. Although there were sporadic revolts against Frankish overlords, these occurred mainly when the army of a neighbouring Muslim power was in the region (Kedar 1990: 155). On the other hand, collaboration with the Franks was also rare. It is particularly striking that there was no attempt by the Frankish crown to establish a military levy from among their Muslim subjects, as the twelfth and thirteenth-century kings of Sicily did, and as the Franks did from their indigenous Christian subjects. As Kedar has pointed out, the Muslims living under Frankish rule, like the indigenous Christians, were subject to both the Frankish authorities and to the neighbouring Muslim powers (1990: 158). When it came to war, they were simply ignored by both sides.

The Church in the Crusader States

The Latin Church

The first offices created by the crusaders in the East were ecclesiastical. Latin clergy filled empty sees as the army progressed from Antioch to Jerusalem. By the time Jerusalem itself was taken, the patriarchate of the city was itself vacant, but since Symeon II, the Orthodox patriarch who had been in correspondence with Adhémar of Le Puy, died at about the same time as the holy city was captured, the crusaders could not have known this. The only senior cleric left alive in the army, a south Italian bishop, was thoroughly unsuitable, so the office was filled by Arnulf of

Choques, a Norman canon. It is likely that Urban had intended to return Symeon, who had been exiled by the Fatimids, to his office, but in any case the pope himself died before the news of the seizure of Jerusalem reached Rome. The election of Arnulf represents the abandonment of Urban's policy regarding the eastern Churches by the crusaders (Hamilton 1980: 11–12).

The legate dispatched to the East to replace Adhémar, Daimbert, archbishop of Pisa, arrived in September 1099. Among his responsibilities was to assist the new settlers in establishing an ecclesiastical framework. A good deal of organisation was necessary, for the Orthodox episcopal and parochial framework, which had been set up in the sixth century, no longer applied in practice. Many of the sees in the old *Notitia* were no more than names, for the cities of the Roman provinces on which they were based had either vanished or lay geographically too far beyond the influence of the Orthodox world. Some sees did have resident bishops, who were either appointed direct from Constantinople or had their elections confirmed by the patriarch of that city. Although the evidence is patchy, most seem to have been replaced by Latins, sometimes in fractious circumstances, once their sees came under crusader control. Thus, a bishop was appointed to Lydda in June 1099, but Sidon had an Orthodox bishop as late as 1110, because it was still in Muslim hands. The supreme ecclesiastical authority in the kingdom of Jerusalem was the patriarch of Jerusalem. Under him were the four archbishops – of Caeserea, Tyre, Nazareth and Petra – and the ten bishops (Stewart 1895: 32–5). In the twelfth century the patriarch of Antioch fought to retain jurisdiction over the archbishopric of Tyre, which had in the Byzantine period been its suffragan. Jerusalem won, however, because it would have been inconvenient for ecclesiastical jurisdiction to cut across political frontiers (Edbury and Rowe 1988: 116–23). This kind of pragmatism was also evident in the exchange effected in the positions of Ascalon and Bethlehem. Presumably because it lay so close to Jerusalem itself, Bethlehem had never been a bishopric in the Orthodox system, whereas Ascalon had; but since Ascalon was in Fatimid hands until 1153, and because Bethlehem was so important a shrine, the Franks reversed their positions (Mayer 1977: 56).

Daimbert's first actions, however, were to overturn the crusaders' own religious policies, by refusing to recognise the election of Arnulf, and by making himself patriarch in his place. As patriarch, he then tried to wrest control of Jerusalem from the secular power. It may have been Urban's intent that Jerusalem should become an ecclesiastical lordship, but Paschal II evidently feared that Daimbert's political activities would destabilise the new state, and deposed him in 1101. Subsequent patriarchs tended to be pious servants of the crown, notable for conventional rather than spectacular piety and for competence in administration rather than theological learning. With the exception of Stephen of Chartres (1128–30), who was elected while on pilgrimage to Jerusalem, Fulcher (1145–57) and Aimery II (1197–1202), none of the patriarchs before 1220 were monks, and few of the bishops. None of them disgraced their office, and some were deeply pious, even to the point of naivety. There were of course exceptions, such as Baldwin, the first abbot of Notre-Dame de Josaphat, who apparently extorted money from the faithful by branding a cross on his forehead and claiming that the mark had been made by an angel (Riley-Smith 1986: 82). But the qualities required of prelates in the Crusader States were those of the court rather than the cloister, and only a handful, such as Patriarch Heraclius, William, archbishop of Tyre and Gerard, bishop of Laodicea, were scholars. More typical of twelfth-century bishops was Ralph, bishop of Bethlehem (1155–74), an Anglo-Norman who served as chancellor of the kingdom and not only accompanied the kingdom's armies but was wounded while carrying the True Cross. Perhaps the outstanding prelate in the Crusader States in the twelfth century was Aimery of Limoges, patriarch of Antioch (1142–90), a skilled politician, diplomat and administrator who for long periods in his term of office effectively ruled the principality. Although he established a friendship with the Syrian Orthodox patriarch and brought the Maronites to obedience to Rome, his interest in the eastern Churches was pragmatic rather than theological (Hamilton 1999: 1–12).

During the period of reconstruction after the Third Crusade, the papacy's involvement in the affairs of the Church in the Crusader States became more intensive. Innocent III directly appointed bishops from his own circle, among them Albert of Vercelli to the patriarchate of Jerusalem in 1205, Peter of Ivrea, a Cistercian, to Antioch in 1209, and James of Vitry to Acre in 1216. These three alone stand out for their learning and for the vigour of their pastoral ministry. But what really distinguishes them, and other thirteenth-century prelates such as James Pantaleon and Thomas of Lentino, patriarchs of Jerusalem successively 1255–61 and 1272–7, or Albert of Rizzato, patriarch of Antioch (1227–46), from their predecessors, is the closeness of their programme to that of the papacy (Bolton 2000: 154–80). In the absence of a strong monarchy, or in a period in which secular authority was contested by the representatives of the Holy Roman Emperor and local barons, the patriarchs provided leadership and a means by which papal policy could be implemented. Thus, for example, during Frederick II's crusade in 1229, the patriarch of Jerusalem, Gerold of Lausanne, upheld papal instructions rather than the authority and prestige of his own church when he placed the holy city under an interdict.

One important element in the religious life of Outremer in the thirteenth century was the arrival of the friars. The Dominicans established their first house in the Holy Land in 1229, the Franciscans at some point in the 1220s. Although their mission was largely the same as in the West – supplementing the work of the parish clergy by preaching, hearing confessions and offering the sacraments – the East had the added frisson of the potential for new converts to Christianity. Over the course of the thirteenth century the friars attempted what the Latin Church had been reluctant to do before, namely to tackle the problem of the strength of Islam by conversion. William of Tripoli, the Dominican Master-General, was surely too optimistic when he declared in 1273 that almost all the Muslims were ready to convert to Christianity, but some headway was made and, given the much smaller numbers of Muslims under Latin rule in the thirteenth century, it was perhaps not so outlandish a notion (Kedar 1984: 180–2). But friars also continued to support the crusades by preaching the cross, and there is little trace of tolerance, let alone pacifism, in their approach to Islam (Maier 1996). The friars were particularly useful as agents of the papacy, and many of them were appointed legates to the East. Innocent IV, notably, used the Franciscan Lorenzo da Orte to try to bring about a closer relationship between the Latin and Orthodox Churches, based on the direct submission of Orthodox clergy to Rome, without the intermediary authority of the Latin hierarchy in Outremer (Hamilton 1980: 322–4). The friars were also instrumental in bringing about closer relations with the separated eastern Churches. In 1237, for example, the Dominican prior-general appeared to have secured the personal conversion to Rome of the Jacobite patriarch of Antioch.

It would be unfair to criticise the twelfth-century Latin Church in Outremer too heavily for its 'pre-Gregorian' character (Hamilton 1980: 134–5). Its most important task, as far as Christendom was concerned, was to safeguard and to service the shrines of the Holy Land. In the words of one chronicler describing the Council of Clermont, the Holy Land was itself a relic. Another chronicler described the crusade itself as a holy theft, in which the Holy Sepulchre had been snatched from the Muslims. The Latin clergy were the guardians of this relic. The Church of the Holy Sepulchre was substantially redeveloped by the Franks during the first half of the twelfth century. After the damage inflicted during the anti-Christian persecution of al-Hakim in 1009, the rebuilding, largely financed by the Byzantine emperor Constantine IX, resulted in a more compact church. This was enlarged by the Franks, with the addition of a cloister and conventual buildings for the canons. The choir of the church, which lay at the centre, was surrounded by a complex of cave-chapels, each associated with an episode in the story of the Passion, and each of which would have been by itself the centre of an important cult in any western church. These chapels created what has been called an 'enclosed stage set' for pilgrims, for whom the visit to the Holy Sepulchre marked the climax of an elaborately choreographed ritual. The uniqueness of each moment of the drama of the Passion was preserved by the architectural role of the different components of the whole complex. Although the tenor of the redecoration was to supplement rather than to replace what was already there, the costs must have been enormous. The chapel of Calvary was entirely redecorated with mosaic, the edicule of the Sepulchre itself was gilded, and the choir was substantially improved. Much of the energies of the chapter must have been consumed up to 1149, when the rebuilding was completed, by the business of soliciting and administering the grants of money necessary for such work (Hamilton 1977: 105–16; Folda 1995: 175–245).

The Holy Land was studded with shrines commemorating the lives of Jesus, the Apostles and the prophets of the Old Testament. Moreover, the potential for identifying and exploiting new shrines, and thus of extending the pilgrim itinerary, was enormous. In 1114, for example, the canons of Hebron discovered a cave underneath their cathedral containing bones that were immediately identified as belonging to the patriarchs Abraham, Isaac and Jacob, who had long been associated with the site but whose tombs had never been located. The clergy, monks and nuns of the Holy Land were, as Peter the Venerable reminded the Cluniacs of Mt Tabor, specially favoured by their guardianship of the holy sites on behalf of Christendom. Lay Christendom showed this favour in the traditional manner, through the bequests of land and other gifts. Because donors gave what was within their gift, the shrine churches, particularly the Holy Sepulchre, the Church of the Nativity and the abbey of Notre-Dame de Josaphat, became extensive landowners in the West.

James of Vitry, writing in the 1220s, describes the Holy Land as a 'garden of delights' that abounded in monasteries, churches and the dwellings of hermits:

From all parts of the world, all peoples and tongues, and from every
nation under heaven, pilgrims devoted to God, and religious of all orders,
drawn by the sweet scent of the holy places, rushed to the Holy Land.
Ancient churches were repaired and new ones were built; monasteries
were constructed in those places from the donations of great men and the
alms of the faithful. (Stewart 1895: 26–7)

The most significant of these foundations, mostly those at shrine churches, pre-dated the First Crusade. Notre-Dame de Josaphat, in the valley of Jehosaphat outside the eastern wall of Jerusalem, was founded over the tomb of the Blessed Virgin, probably by western pilgrims in the last quarter of the eleventh century. Royal and noble patronage from early in the twelfth century assured its prosperity, as is attested by its cartulary. Like the other houses whose cartularies survive, the Holy Sepulchre and Mt Sion, Notre Dame's properties, estates and villages enabled it to become not only one of the largest landowners in the kingdom of Jerusalem, but throughout the Mediterranean – with properties in Cyprus and Sicily – and even in the West (Mayer 1977: 258–372). Another foundation by western pilgrims, Saint Mary Latin, was as close to a conventual community at the Holy Sepulchre as the Roman Church could reach before the First Crusade. Although its monks, as far as we know, did not serve the shrine itself, it lay only a stone's throw from the Holy Sepulchre. Mt Sion was refounded by Godfrey de Bouillon as a house of regular canons, as also was the Church of the Ascension on the Mount of Olives. The Qubbat as-Saqra, or 'Dome of the Rock', built by the Muslims in 638, was reconsecrated as the 'Templum Domini', because it stood on the site of the Jewish Temple. It also became a house of regular canons. As was the practice in the West, all cathedral churches, whether at shrines or not, were served by regular canons, though in the case of the Holy Sepulchre itself, the canons beneficed after the First Crusade did not adopt the Augustinian rule until 1114.

Some Latin houses, as James of Vitry indicates, were re-foundations of disused churches. St Anne's, a Benedictine convent in Jerusalem, was built adjacent to the ruined Byzantine church that commemorated the house of the mother of the Blessed Virgin. In Sebastia, the burial place of John the Baptist, a Latin monastery replaced a disused Orthodox one. The number of completely new monastic foundations, however, was surprisingly small. Two houses of Premonstratensian canons, St Samuel and St Habbakuk, commemorated Old Testament sites, but the reluctance on the part of Frankish landowners to convert their subject peoples probably discouraged further such foundations.

The character of the Latin Church, centred as it was on the shrines and on pilgrim itineraries, ran contrary to the currents in contemporary monastic spirituality in the West. The most important new foundations in the twelfth-century West were marked by their eremitical character – not necessarily in the sense of promoting individual solitude, but rather of settlement detached from existing ties of landownership or of populated centres. There was plenty of wilderness available in Outremer, and ample opportunity for independence from feudal relationships, but such places presented security problems, and in any case the new orders did not as a rule warm to the prospect of new foundations in a land so marked by fixed associations with biblical events.

The Cistercians are a case in point. It is no coincidence that the first Cistercian foundations in Outremer occurred only in 1157, after the death of Bernard of Clairvaux. Bernard's own lukewarm attitude to monastic settlement in the Holy Land is exemplified in the letters he wrote in 1124 concerning the proposed pilgrimage to Jerusalem of a fellow Cistercian abbot, Arnold of Morimond, and some of his monks. Bernard argued that cloistered monks who made a pilgrimage to the earthly Jerusalem were missing the whole point of being a monk, which was that the cloister was the earthly Paradise, the foretaste of heaven. The physical shrines of the Holy Land needed to be defended by righteous warriors, but they were an irrelevance for monks (Jotischky 1995: 2–4). Until the arrival of the Cistercian Peter of Ivrea as patriarch of Antioch in 1209, there were only two Cistercian monasteries in Outremer, and one of those, Salvatio, was probably lost in 1187 (Hamilton 1976: 406). Peter applied well-known Cistercian strategies in his new office, persuading unaffiliated reforming foundations in the Black Mountain near Antioch to become Cistercian houses. A similar thing happened elsewhere in the patriarchate of Antioch, when in 1231 Cistercians took over the abandoned Orthodox monastery of St Sergius at Gibelet.

One of Peter of Ivrea's 'new' Cistercian monasteries, St Mary Jubin, had a chequered past in the twelfth century that demonstrates some of the difficulties of monastic reformers in the crusader states. Gerard of Nazareth, who before becoming bishop of Laodicea in c.1140 had himself been a hermit, describes the Black Mountain as a centre of eremitical monasticism of a type familiar from the same period in the West (Kedar 1983: 55–77; Jotischky 1995: 17–46). Some of the monks of Jubin, a monastery founded in the 1120s, tried to adhere to a strict interpretation of the Rule of St Benedict, including an abandonment of property. The monastery experienced a schism over the prior's refusal to countenance this, as a result of which one of the reformers, Bernard of Blois, 'stirred up like a gadfly, ran through the woods, prepared rather to die of hunger rather than to allow such an unspeakable thing'. Bernard and many of the monks left for another house, Machanath, as a result of which Jubin fell on hard times. Another example of Gerard of Nazareth's reforming monks, Elias of Narbonne, demonstrates the attraction that the Holy Land exercised for some westerners. Elias came to Jerusalem on pilgrimage, but remained behind to live as a monk, first in a reforming community and later in a cave. He eventually entered the cloister at Notre-Dame de Josaphat, but he later became abbot of another reforming community, Palmaria in Galilee. Here he angered the monks by trying to import Cistercian customs; he fled to Jerusalem, and Palmaria eventually became a Cluniac house (Jotischky 1995: 29–35).

The Holy Land inevitably attracted extremes of ascetic behaviour from some individuals. We know of a hermit who preceded St Francis' more celebrated devotion to the nursing of lepers; of another who incarcerated himself in a cell in the city wall of Jerusalem; of another who was so detached from the world that he did not even know whether Antioch was still held by the Franks or Muslims (Jotischky 1995: 25–29). One of the characteristics of Latin monasticism in Outremer was the apparent ease with which individuals moved from the cloister to the wilderness and vice versa. Although we know of no direct evidence that connects western monasticism in the East to Orthodox practices, it is probable that the presence and traditions of Orthodox monks and anchorites had an influence on the character of Latin monasticism.

The Orthodox and the eastern Christians

If the election of Arnulf of Chocques as patriarch in 1099 was the consequence of an anti-Orthodox policy stemming from the crusaders' disillusionment with the Byzantine role on the crusade, the arrival of Daimbert presaged a return to cooperation. One of Arnulf's first actions had been to expel the Greek Orthodox monks from the Church of the Holy Sepulchre; a year later, however, they returned when it became apparent that the Easter Fire miracle (at which fire descended spontaneously from heaven to light the lamps that had been hung around the Sepulchre) would not work without their expertise. The description of this ceremony at Easter 1107 by the Russian pilgrim Daniel – even allowing for the author's slant – leaves no doubt that the Orthodox monks were a vital component of the liturgy (Wilkinson 1988: 166–71). The twelfth-century rebuilding of the church, moreover, left the Greek Orthodox and other eastern Christians in possession of altars and rights of worship.

The hierarchy of the Orthodox Church, however, did not fare so well. The patriarch, Symeon II, had been exiled by the Fatimids in 1098, and many of the sees seem to have been vacant in 1099 or long since to have fallen into disuse. But the coastal towns, with their communities of indigenous Christians, probably did have bishops, and these were replaced by Latins as their towns fell into the crusaders' hands in the early years of the twelfth century. This policy conformed to ecclesiological theory, as was later explained in a canon of the Fourth Lateran Council. The reasoning was that the Orthodox were part of the same Catholic Church as the Latins, and the Church, which was a single body, could not have two heads. Consequently a diocese could only have a single bishop, and it was natural for the ruling group to appoint their own rather than a Greek. However, both in the mainland Crusader States and in Cyprus in the thirteenth century, the Latins found it necessary to use Greeks as suffragan or 'coadjutor' bishops, with particular responsibility for the Orthodox flock of a diocese. Orthodox bishops are known to have been appointed to Lydda, Gaza/Eleutheropolis and Sidon, and probably to other dioceses that had substantial populations of Orthodox laity. In this respect, the separated eastern Churches fared better than the Orthodox. Because the Monophysite Syrian Orthodox (Jacobite), Armenian and Nestorian Churches were not in communion with Rome, the question of multiple bishops did not arise, and these Churches continued to appoint their own to sees under crusader rule in accordance with their own hierarchical arrangements; thus, there was a resident Jacobite patriarch of Antioch and bishop of Jerusalem, as well as Coptic and Nestorian bishops.

The Orthodox clergy was left to minister to their flocks as before the Latin conquest. Thus Orthodox liturgical life continued as before: priests said Mass using leavened bread, married, observed the fast on Saturday, and so on. In practice, the replacement of Greeks with Latins probably made little difference to many parochial clergy in the patriarchate of Jerusalem, because the Orthodox bishops had tended to be Greeks appointed from Constantinople, whereas the Orthodox clergy and laity were Arabic-speaking. The indignation of James of Vitry at the latitude given to Orthodox clergy in Acre in 1216 not only signals a change of direction under an increasingly interventionist papacy; it also reveals how tolerant the Latin authorities had become in the twelfth century of the continuation of Orthodox practices. In theory, the submission to Latin bishops meant that the Orthodox clergy were recognising the authority of the papacy, but, as Oliver of Paderborn remarked in the 1220s, who could tell what they were really thinking, even if they did observe the correct forms when required by their bishops? Other evidence confirms James' impressions. Gerard of Nazareth complained that the Orthodox clergy of his diocese (Laodicea) were consecrating cemeteries and practising confirmations – ceremonies that in the Roman Church could only be carried out by a bishop (Jotischky 1997: 223–4). In Cyprus, there were complaints about Orthodox encroachments on Latin episcopal rights (Coureas 1997: 251–317). The Frankish laity in the East sometimes adopted the cults of eastern saints, such as the Syrian Bar Sauma. Such tendencies can only have been encouraged in the kingdom of Jerusalem by the practice of sharing churches, particularly in rural areas, between Orthodox and Latin communities. Nor were complaints all on the Latin side: a letter from the Orthodox clergy of 'Ain Karim, near Jerusalem, for example, raises the question of whether altars that had been used for Mass in the Latin rite needed to be purified before they could be used again for Orthodox services (Jotischky 1997: 218).

Contact with the other eastern Christians seems to have been determined partly by political conditions. Thus, Aimery of Antioch's friendship with Michael, the Jacobite patriarch of Antioch, intensified as he found himself the victim of Byzantine pressure to replace him with the Orthodox patriarch Athanasius (1165–70). In 1237, the Jacobite patriarch Ignatius II submitted a statement of faith to the Dominicans in Jerusalem while on a pilgrimage to the holy city – an act of friendship that was hailed in the West as the patriarch's conversion to Rome (Hamilton 1980: 349–53). In the 1240s Innocent IV may even have tried to create a uniate Jacobite Church, in which the clergy and hierarchy could retain their own liturgical traditions and become dependent directly on Rome, rather than submitting to the local Latin Church. Some pilgrims to the Holy Land remarked on the diversity of religious traditions with tolerance or even approval. There is an implication, for example, in the account of Burchard of Mt Sion, that such diversity of forms of worship from Christians whose origins are so widespread – not only Greeks, Armenians and Syrians but Egyptians, Ethiopians, Georgians and even Indians – is part of the special quality of the Holy Land itself (Stewart 1896: 102–11).

The monastic life of the Orthodox Church experienced a significant revival in the crusader period. This was not so much because of the policy of the Latin Church, but because political circumstances made it possible for patronage to be extended to the monasteries by the Byzantine emperor. No better account of this revival can be given than a comparison between the pilgrimage accounts of Daniel (1106–7) and John Phokas, a Cretan monk who visited the Holy Land in 1185 (Wilkinson 1988: 120–71, 315–36). Daniel visited St Sabas, in the Kidron Valley southeast of Bethlehem, which was still functioning, but found many others, including St Theodosius, also near Bethlehem, St Euthymius, between Jerusalem and Jericho, and St Mary Kalamon by the Jordan, in ruinous states. By 1185, these and others had been rebuilt and filled with monks again. Others, such as St Aaron on Mt Horeb, and St John Prodromos by the Jordan, were functioning soon after 1110. Phokas remarked on the prosperity of a group of Orthodox monasteries by the Jordan near Jericho: 'the land, divided and shared out among these holy monasteries, is full of woodland and vineyards, as the monks have planted trees in the fields and reap rich harvests from them' (Wilkinson 1988: 328). Some monasteries, among them St John Prodromos by the Jordan and St Elijah near Bethlehem, are known to have been restored as a result of gifts from Emperor Manuel Komnenos, which locates the physical revival of Orthodox monasticism between 1143 and 1180. There is no evidence for the patronage of Orthodox monasteries by the Latin rulers, as was the case, for example, in Norman Sicily, or even for certain monasteries in Lusignan Cyprus. But that the rulers of the kingdom of Jerusalem were content for Manuel to exercise such patronage is even more remarkable, for it suggests that they were colluding in a Byzantine imperial policy to promote the image of the emperor as the guardian of Orthodoxy. Even when the ownership of a monastery lay in Latin hands – as for example was the case with St George's at Beth Gibelin, which was a Hospitaller property – prayers were said for the emperor (Jotischky 1995: 83–8).

St Sabas and St Catharine's on Mt Sinai are the only two Orthodox monasteries that enjoyed an uninterrupted history from their early Christian foundations. But equally as striking as the physical revival of ruined monasteries is the revival of traditions associated with them. For example, when Gabriel, a monk of St Sabas, had to be disciplined for his attempted murder of another monk while under demonic possession, he was sent to the monastery of St Euthymius to carry out a régime of manual labour. This not only parallels a case from the early eleventh century, but shows that the original relationship between the *laura* of St Sabas and the *cenobium* of St Euthymius in the sixth century had been re-instituted (Jotischky 2001: 89–90). Similarly, the Orthodox hermit whom Phokas met in the ruins of St Gerasimus by the Jordan in 1185 is described as consciously reproducing the practices of the founder of the monastery in the sixth century (Jotischky 2000: 110–22). Such traditions must have been

1 The 'Dome of the Rock'/Qubbat as-Saxra and al-Aqsa mosque, Jerusalem: view from Mount of Olives

2 *Church of the Holy Sepulchre, Jerusalem: south façade*

3 *Chapel of the Ascension, Mount of Olives*

4 *Carved capital, Chapel of the Ascension, Mount of Olives*

5 *Monastery of Mar Sabas, Kidron Valley*

6 *Tariq al-Wad, Old City of Jerusalem (one of the main thoroughfares of the crusader city)*

7 *Church of St Anne, Jerusalem*

8 *Citadel of David, Jerusalem: crusader gateway*

9 *Aqua Bella: Hospitaller fortified estate*

10 *Caeasarea: fortified gateway*

11 Church of the Nativity, Bethlehem: Syro-Palestinian mosaics of 1160s on north nave wall

12 *Hospitaller castle at Kolossi, Cyprus*

13 *The edicule of the Holy Sepulchre, Church of the Holy Sepulchre, Jerusalem.* © CORBIS

14 *Hospitalier castle of Krak des Chevaliers, Syria.* © *Michael Nicholson/CORBIS*

15 *St John leading crusaders into battle, from a 14th century French manuscript of the Apocalypse. Ms Royal 19.B.XV, fol 37. Akg-images/British Library*

16 *Joust between Crusader and Muslim warrior, from the Luttrell Psalter (circa 1340). Ms 42130, fol 82. Akg-images/British Library*

17 *The Dome of the Rock (Qubbat as-Saqra), Jerusalem, which during the 12th century became the Templum Domini, a house of Augustinian canons. © A. Griffiths Belt/CORBIS*

preserved in literary form, in the works of Cyril of Scythopolis, John Moschus and others. Among the functions of the Orthodox monasteries in the Holy Land was the preservation, by copying, of canonical, hagiographical and theological texts. The surviving manuscripts produced at the scriptoria of the Orthodox monasteries of the Holy Sepulchre, St Sabas, St George Choziba, St Euthymius and others, demonstrate, through the range and extent of these works, the vitality of monastic life (Pahlitzsch 2001: 213–24). Besides the copying of the Greek Fathers and the liturgical works needed for services, there were original compositions, including theological treatises on the errors of the Latin Church. In the absence of an Orthodox hierarchy, the monasteries provided a voice and a focus for the maintenance of Orthodox traditions. Far from being detached centres of contemplation, they were at the centre of the life of the Orthodox community: when Emperor Manuel Komnenos provided money for the rebuilding of the monastery of St Elias, between Jerusalem and Bethlehem, it was at the request of the local Orthodox villagers.

Art and architecture

Nowhere is Fulcher of Chartres' picture of an assimilated society more realistic than in the area of visual culture. In the religious architecture and the decoration of shrine churches, in painting and manuscript production, the influence of indigenous artists increased throughout the twelfth and thirteenth centuries. Not only were local artists employed in workshops and scriptoria, but western artists borrowed Byzantine and indigenous methods and iconographical programmes, with the result that a distinctively hybrid artistic culture evolved in the Crusader States.

In the early years of the Crusader States this was not yet apparent. The earliest extant crusader church, in Tarsus, and the new church of St Anne in Jerusalem, built before 1110, echo the architectural forms of southern French Romanesque (Boase 1977: 73–4), and surviving fragments and reconstructed ground plans suggest that this remained the standard model for crusader ecclesiastical architecture. The Church of the Holy Sepulchre was a more complex site, however. Constantine's churches of the Anastasis and Sepulchre, linked by a courtyard, were destroyed in 614 by the Persian invasion, and in 1009 al-Hakim, the Fatimid khalif, ordered the seventh-century rebuilding to be torn down. A Byzantine-sponsored rebuilding from 1048 onward produced a domed rotunda over the Sepulchre, with a polygonal apse extending to the east and housing the Anastasis church. The crusader rebuilding of the twelfth century was designed to house a greater complex of shrines in a single building. The site of the crucifixion occupied a chapel in the south transept, a new arch was cut from the rotunda to the Anastasis church, and chapels commemorating the events of the Passion radiated from the apse. Separate chapels commemorating Calvary and the Empress Helena's miraculous discovery of the whole site opened off an ambulatory that encircled the apse. But the Franks also extended the conventual buildings, in order to house the Augustinian canons who served the church from 1114 onward. The style of this complicated church has been a matter of debate, but despite the retention of some elements of the eleventh-century Byzantine work, the basic models seem to have been northern French, Aquitainian or Provençal (Boase 1977: 78; Lyman 1992: 63–80), and the lay-out both of the architecture and the iconographic programmes corresponded to churches along the great pilgrimage routes of southern France and Spain (Folda 1995: 177–245).

The architectural sculpture, such as that in the column capitals of the south façade, bears comparison with classical Syrian models, though the surviving lintel relief, with its unique iconography of Passion Week, is purely western in influence – possibly Provençal, and in particular deriving from La Daurade in Toulouse (Barasch 1971: 107; but see Borg 1969: 25–40; Boase 1977: 82). Other sculpture from the Crusader States has similarly western origins. The extraordinary figural sculpture from the twelfth-century cathedral at Nazareth, for example, bears comparison with contemporary Burgundian work such as that at Vézelay, Autun and La Charité-sur-Loire, but there are also traces of southern influence (Barasch 1971: 109; Folda 1986: *passim*). Column capitals in the Temple Mound area and on the Church of the Ascension on the Mount of Olives similarly derive from French models (B. Kühnel 1977: 41–50; Folda 1995: 253–73). There is little trace in any surviving sculpture from Outremer of indigenous influence or workmanship.

In contrast to architecture and sculpture, the fields of monumental and panel painting, mosaic and manuscript illumination show the exceptionally strong influence of indigenous artistic production. The Church of the Nativity in Bethlehem is a good example of cultural synthesis. Unusually among shrine churches, the Constantinian church remained almost intact in 1099, and therefore retained its basilica shape. The nave columns were painted in the mid-twelfth century with a series of saints' portraits. The mixture of eastern and western styles and the eclectic choice of saints demonstrates the presence of artists of varying traditions, and may also signify a sophisticated iconographical programme. The paintings may, however, represent votive offerings from pilgrims, and therefore also reflect the diversity of religious traditions that gathered at this important shrine (G. Kühnel 1988: 83–8; Folda 1995: 364–71).

In the 1160s, moreover, a decorative programme sponsored by Emperor Manuel Komnenos, King Amalric I and Ralph, bishop of Bethlehem, added a series of mosaics to the walls of the nave arcade and the apse. Mosaic work is in itself an indication of local artistic production, since knowledge of the craft was relatively unknown in the West. The iconography and style of the mosaics is wholly Byzantine, and separate inscriptions record the names of two artists: the Orthodox monk Ephraim and Basil, who was probably also Orthodox (Hunt 1991: 69–85; Folda 1995: 347–64; but see Boase 1977: 119–21). The nave mosaics are of particular interest because they also reveal a distinctly Orthodox theology. Although it was once thought that some of these mosaics dated from the eighth century (Stern 1936: 101–52, 1938: 415–59; Boase 1977: 121), it is now clear after recent restoration that the whole programme is twelfth century. The subject, the ecumenical and selected provincial councils of the early Church, is rendered largely through inscriptions, mostly in Greek, and deriving from Palestinian Orthodox written sources. The use of Latin and Syriac in occasional inscriptions in the scheme points to the involvement of indigenous and western artists, and to the use of the church by varied religious traditions. But the selection of texts, for example in the inscription recording the profession of faith of the Council of Constantinople (381), which omits the *filioque*, seems like a deliberate ploy to promote Orthodox theology at the expense of Latin, and in the context of a Latin cathedral and the co-sponsorship of the Latin bishop, is a remarkable statement not only of the role of local Orthodox monks as artists, but of the role of Orthodoxy itself in the religious culture of the crusader kingdom (Hunt 1991: 69–85; Jotischky 1994: 207–24).

The Church of the Nativity is only one example of such artistic co-operation. In the Church of the Holy Sepulchre itself, the mosaics and wall paintings – no longer extant – show a similar situation. Byzantine influence is also noticeable in the wall paintings at Abu Ghosh, a Hospitaller church to the west of Jerusalem (Carr 1982: 215–44). It is clear that not only did western artists learn new techniques from Byzantine or indigenous Orthodox artists, but that a distinctly new artistic style evolved in the Crusader States in the twelfth and thirteenth centuries. How did this come about? Manuscript illumination from the mid-twelfth century already reveals the existence of workshops of Italian, French, English and local artists, and the cross-over of techniques and styles was a natural consequence. The

'Melisende Psalter', a manuscript made at the scriptorium of the Holy Sepulchre during the 1130s, has been identified as the work of several artists, among them the Byzantine-influenced Basil (who may even have been the same Basil who signed his name to the mosaic angels in the nave at Bethlehem), a south Italian and another western artist using an English liturgical calendar (Folda 1977: 252–3, 1995: 137–63).

Such a synthesis must also have reflected – or, in the first place, prompted – the tastes of patrons. It is not surprising to find a workshop with such accomplished artists as those of the Melisende Psalter attached to the Holy Sepulchre, the most prestigious of crusader churches. But the Byzantine influence – whether from Byzantines or indigenous Orthodox artists – also suggests that court patronage favoured such works as a matter of taste, or even in an attempt to rival aristocratic culture at Constantinople: as Folda remarked, 'Byzantine' was synonymous with 'aristocratic' (1995: 159). This should not be surprising at a court whose queen was herself half-Armenian, or in a royal house that consistently throughout the twelfth century favoured marriages with Armenian or Byzantine princesses. A political agenda may also underlie the Melisende Psalter, if, as Folda (1995: 155) suggests, the manuscript was commissioned by Fulk as a reconciliation gift for his queen after the rebellion of 1134. In this case, the creation of a book so clearly indebted to eastern artistic inspiration says a good deal about the desire of the Franks to balance the diverse ethnic and religious components under their rule.

The Byzantine elements in crusader manuscript illumination intensified from the 1160s onward. Byzantine artists may have supervised the work of Latin illuminators in the scriptorium of the Holy Sepulchre (Buchthal 1957: 26–8), but there is also stylistic evidence that Armenian scribes were working there (Folda 1995: 341). The Abu Ghosh frescoes, which date from the 1170s, suggest Byzantine artists being employed by the Hospitaller patrons of the church (Folda 1995: 387–9). It is surely no coincidence that the period between c.1158 and 1180 was a period of political alliance between the kingdom of Jerusalem and the Byzantine Empire, and that both Baldwin III and Amalric I married Byzantine princesses. The establishment of queens' households may have provided opportunities to introduce artists directly from Constantinople (Carr 1982: 221–4), though the Bethlehem mosaics indicate that indigenous Orthodox workmanship of very high quality was available. With the passage of time, such relationships between Byzantine-trained and western artists became unnecessary. A 'crusader style' of painting developed in the thirteenth century that incorporated both eastern and western elements, exemplified in the remarkable group of manuscripts produced in the scriptorium of Acre from 1275 onward (Folda 1976).

Perhaps the most remarkable evidence for Byzantine tastes in crusader art is the production of a large number of icons for western patrons. The icon, a cult image integral to Orthodox worship, was more or less unknown in the West during the crusader period. Yet a genre of icons exhibiting signs of western artistic influence, and sometimes including western as well as Orthodox saints, survive from the monastery of St Catherine on Mt Sinai, where there may even have been an atelier of western artists supervising their production (Weitzmann 1963: 179–203, 1966: 51–83), and from the thirteenth-century crusader kingdom of Cyprus (Mouriki 1985/6: 9–112). It is in Byzantine icon painting, mediated through the culturally diverse Crusader States, that Italian panel painting of the thirteenth and fourteenth century has its origins.

The existence of a distinctive genre of such art raises questions – as yet unanswered – about the use of icons by western settlers in the Crusader States. Did the settlers merely admire them as portable works of art, or collect them as sophisticated pilgrim emblems associated particularly with St Catherine's? Or do the existence of these icons suggest that devotional practices as well as artistic tastes were borrowed from the indigenous Orthodox? These are questions that may be answered in the future, as the fruits of art historical research are more fully absorbed by religious historians of the Latin East. Yet the synthetic nature of the visual culture of the Crusader States, even if reduced to a question of aristocratic taste, is sufficient by itself to make us sceptical of the idea that crusader society was entirely segregationist in its treatment of the native peoples.

Economy and society

Historians who regarded crusader society as essentially based on segregation between Franks and native peoples tended to do so partly because they also thought that the Franks were exclusively urban dwellers and the native peoples rural peasants. Patterns of settlement, according to this model, reinforced the tendency towards separation of peoples. Research over the past generation, however, has thrown considerable light on Frankish settlement in the interior of the Crusader States, particularly in rural areas of the kingdom of Jerusalem. It can no longer be doubted that conscious attempts were made in the twelfth century to create 'model settlements' for Franks in the countryside (Ellenblum 1998: *passim*; Boas 1999: 60–90).

A recent estimate numbers 235 Frankish settlements among the 1,200 or so villages in the kingdom of Jerusalem before 1187 (Boas 1999: 63–5). Taxes were levied on Frankish and native settlers alike by feudal overlords on land and livestock, but also on natural products such as olives and honey, and on rights to use mills and olive presses. This was in fact simply the extension of the western custom of monopolising control over the means of production of food by the landowners. The structure of rural life under crusader rule appears to have been little altered: such villages as have been excavated resemble those of pre- and post-crusader periods. But two types of Frankish village are discernible: those, such as at 'Abud, that indicate settlement alongside native Christian communities (Ellenblum 1998: 128–35), and planned villages such as al-Bira, north of Jerusalem. In one planned village, al-Kurum, peasant freehold seems to have been the norm, and many of the settlers were probably westerners who were deliberately encouraged to emigrate on the promise of favourable terms (Boas 1999: 63). At al-Bira we know of southern French, Italians and Spaniards, numbering 142 families in total by 1187 (Benvenisti 1970: 223–4). These were probably people of peasant origin who, as William of Tyre suggests, found urban life too expensive for their means (WT 1986: 937). Excavations at al-Qubaiba, like al-Bira a new village owned by the canons of the Holy Sepulchre, have yielded a bakery, olive-press and pressing-floors. Both villages lay within a radius of a dozen miles of Jerusalem, and were located on hilly land with soil conducive for olive cultivation and certain kinds of animal husbandry, but not for cereal agriculture.

Most of the evidence suggests that planned villages were largely established on lands belonging to the crown, the major religious institutions or the Military Orders. They demonstrate, therefore, serious attempts by the major landowners to exploit the natural resources of the region: wine, olives and sugar cane, and livestock such as goats, sheep and pigs. In the case of viticulture and pig-farming (Prawer 1980: 133–4, 185–6), such activity was intended to supplement their diet, but sugar-cane production, which was revolutionised under the Franks – perhaps because it demanded a high degree of servile labour – became an important industry for export to the West (Prawer 1980: 196–7; Boas 1999: 74–87). Despite attempts to support rural agriculture, however, contemporary sources suggest that grain, poultry and wine were sometimes imported. This must have been especially so when a large crusading army from the West was present, and during the course of the thirteenth century, after the loss of much of the rural hinterland, increasing amounts of food were imported.

The countryside and town in Outremer did not represent completely different spheres of economic activity. The natural resources of the hinterland included sand for glassmaking and ashes for soap production – both of which were primarily urban industries – and sugar cane production was moved from the Jordan Valley to the coastal region near Acre in the thirteenth century (Prawer 1980: 160). There is also evidence of sophisticated craft-based industries in crusader towns, such as metalwork and pottery production (Boas 1999: 143–92). All towns also had their own markets for the sale of produce, as in the West. But what distinguished most crusader towns from those in the west – with the obvious exception of Jerusalem itself – was their location by the sea, and thus their place in an existing pattern of trade between East and West. Geography provided the Crusader States with the capacity to become the agents of exchange in the long-distance trade between East and West. From the West, timber, iron, copper, furs, amber and textiles were sold in exchange for gold, spices and silk. In this way, the crusader ports, though never as important as Egypt in the luxury trade, became a medium to serve the increased demand for luxury goods in the West. In this respect, the crusader occupation changed nothing of the pre-existing economic structures in Palestine, in which the coastal towns had, since classical antiquity, been the main centres of population and commerce. Some of the coastal towns – Tyre, Acre, Beirut, Tripoli, Antioch, perhaps Ascalon – had populations that far outnumbered most cities in the West before the mid-thirteenth century (Benvenisti 1970: 27; Prawer 1980: 181–2).

Under the Franks, however, two new elements greatly benefited from the economic potential of the coastal towns. One was the crown, as income from the passage of goods through Acre, Jaffa, Tyre, Sidon, Tripoli and Beirut, and the markets established in these towns, gave the kings a wider range of resources than those customarily available in the West. Another was the western mercantile community, primarily Italian and Provençal in the twelfth century, with the addition of Catalans in the thirteenth century, which established economic colonies in the ports. These were autonomous quarters, exempt from local laws and governed instead by the laws of their parent city. According to the agreement made between the Venetians and the patriarch of Jerusalem – who was deputising for the king – in 1123, within the Venetian quarter of Tyre, not only the mercantile community itself, but anybody who lived within the defined quarter, was subject to Venetian jurisdiction (Prawer 1980: 221–5). Moreover, the Venetian quarter included some of the surrounding rural area, which included villages and their inhabitants. This land could also be used to enfeoff Venetian noble families who had initially been attracted to the Levant by trade. The grant of land as well as property and jurisdictional rights inside the town indicates a transition from a purely mercantile to a colonial community. Venice acquired more privileges than its commercial rivals Genoa and Pisa, despite a rather sporadic participation in crusading from 1095 to 1124. But cities such as Pisa, Amalfi, Genoa, Marseilles and Barcelona, which had bargained military assistance in conquering and defending the towns in the first place in return for favourable trading privileges, were also greatly enriched by assured access to eastern markets.

The extent of the early twelfth-century grants reflects the urgency of the crown in securing naval and military assistance. In the agreement made in 1110 at the siege of Sidon, Baldwin I gave Venice a street in Acre, known as the *vicus Venetorum*, but only in 1123, in the more comprehensive agreement known as the *Pactum Warmundi*, did crown give up its seigneurial rights to the bakery, mill and control of weights and measures, even in this district. Initially, the Italian mercantile communities were interested primarily in commercial rights such as warehousing and market spaces, rather than jurisdiction, and the nature of their presence in the port towns was determined by the seasonal nature of long-distance trade. Not all ports were equally attractive: Tyre (from 1123), Acre, Sidon and Tripoli were preferred, probably because they already enjoyed dominance in preexisting trading patterns. From about 1111 onward, coinciding with the capture of Sidon, the Italian presence took on a more permanent look, with fixed rather than seasonal markets, and eventually the emergence of permanent merchant communities in ports.

The *Pactum Warmundi* (1123) reflects the superior bargaining position of the Venetians over the crown at the time. Tyre had been invested but there was little prospect of capturing it without Venetian naval support, and that could be secured only at the price demanded in the *Pactum*. Moreover, Baldwin II was in Muslim captivity at the time, and negotiations were therefore conducted by Warmund, patriarch of Jerusalem. In order to safeguard the agreement, the Venetians even insisted that future kings of Jerusalem should be required to agree to uphold the *Pactum Warmundi* before their coronation (Prawer 1980: 224). The agreement gave for the first time genuine autonomy within the area defined as 'Venetian', as well as commercial privileges. In return for a rather negligible knight service owed to the crown, Venice secured sovereignty over not only her own nationals but over anyone living in the Venetian quarter in Tyre. This meant that Venice had the right to administer justice within its enclave, and to retain the financial benefits of justice in the form of fines – another drain of income away from the crown. The range and type of seigneurial assets granted to the Venetians in the *Pactum Warmundi* indicate that a new phase of Italian colonisation had begun. Henceforth, the Italians became urban settlers in Outremer.

The ways in which Venice and Genoa exploited the financial potential of its quarters in Acre can be reconstructed from surviving thirteenth-century documentation. A Venetian inventory divided the quarter into four districts, at the centre of which lay the *fondaco*, a kind of warehouse and retail complex which also included lodgings. In addition to smaller houses, there was also a palace belonging to the *bailli*, another belonging to the *fondaco* and a fortified tower. The palace of the *bailli* was a two- or three-storey building with space for six shops on the ground floor, and apartments that could be rented out on the other floors. Next to the palace stood the quarter's parish church, dedicated to St Mark. Around the palace on ground level were pitches for stalls, which were also rented out by the commune. The great palace of the *fondaco* was the largest building owned by the Venetian commune. It comprised sixteen shops on the ground floor space, and apartments on its upper floors, some of which were set aside for commune officials such as the parish priest and the court clerk. We even know the level of customary rental prices for rooms in the palace, which ranged from 2 to 15 besants per month depending on location and season (Prawer 1980: 233–4). A similar picture is painted by a Genoese inventory of 1249 for Acre: the Genoese maintained four houses and six palaces, which were divided into apartments to be rented out seasonally for short periods, and in addition six houses rented out for a year. Income came partly from rentals by the commune itself, but partly from other buildings in the quarter whose occupants paid a property tax to the commune for the land they occupied.

Even allowing for the fact that the crown had little bargaining leverage when it made the initial agreements with the Italians, it does seem as though a great deal was given away in the long term for short-term military gains. The full effect of the mercantile quarters was to stifle the development of a Frankish commercial class by allowing foreign merchants to monopolise retail as well as wholesale trade. Prawer has suggested that at the time when the agreements were made, neither side fully appreciated the implications of the concessions made by the crown (1980: 219). However, a more positive assessment of the agreements from the crown's perspective has been made by Riley-Smith, who argues that so much income flowed through the ports in the form of taxable goods and the taxes levied on ships that the losses to the Genoese, Venetians and other mercantile communities were minimal (Riley-Smith 1973b: 109–32). The mutually beneficial relationship between the crown and the western merchants was one of the defining characteristics of the kingdom of Jerusalem, as well as the main source of its economic stability in the twelfth century. In the thirteenth century, however, and particularly as the Hohenstaufen government ran into difficulties in the 1230s–40s, the economic strength of the mercantile colonies strained the kingdom's resources, while their autonomy only added to the process of political fragmentation. As early as the reign of Amalric I (1163–74) the crown attempted to claim back rights of jurisdiction over Italians who had settled permanently and owned property in the kingdom of Jerusalem, but although further attempts were made in the thirteenth century, the crown simply had insufficient authority to make headway. The War of St Sabas, so called because of the parish in Acre that formed the centre of a bitter quarrel over jurisdictional rights, polarised the Italians and the Military Orders in a civil war in the 1250s that severely damaged the kingdom's ability to function as a single unity. In the absence of central authority during a long interregnum, there was no balancing force to hold in check the political and economic ambitions of the Italian colonists and their home governments.

CHAPTER 7

Recovery in the East, new challenges in Europe: crusading, 1187–1216

A fter the fall of the kingdom of Jerusalem in 1187, immediate efforts were made in the West for its recovery. The years covered in this chapter saw the launch of new crusades, which achieved varying degrees of success. But at the same time, the papacy began to apply crusading concepts to problems nearer home such as the rise of heretical movements. This chapter examines the reasons for the limited success of the Third Crusade and the failure of the Fourth, then proceeds to focus on Pope Innocent III's development of crusading ideas and practice.

The Third Crusade

The Third Crusade ultimately failed in its objective, for Jerusalem remained as firmly in Muslim hands when Richard I and Saladin drew up the treaty to end hostilities in 1192 as it had been when the crusade was launched. A fuller assessment of its significance, however, may be reached by asking whether the recovery of Jerusalem was in fact necessary for the continued life of the Crusader States. When Gregory VIII issued his crusading bull *Audita tremendi* in October 1187, Tyre was the only one of the kingdom of Jerusalem's towns still in Frankish hands. Tripoli was unable to protect itself, and Antioch was threatened by Saladin in 1188. Yet only ten years after Hattin, the coastline had been reconquered almost in its entirety from Gaza to the Gulf of Attalia; Antioch, though now faced with a powerful new neighbouring kingdom of Armenia, was safe, and Tripoli's security had been assured by absorption into Antioch. Much of the work itself was completed after 1192, but the groundwork was laid by the Third Crusade, and in particular by the vigour and military skill of Richard I.

Audita tremendi was issued after Hattin but before news of the fall of Jerusalem had reached the West. Predictably, the response was immediate and massive. The themes of the papal appeal – the need for penitence and internal peace in Christendom – struck a chord that Eugenius III's bull in 1145 had been unable to reach. Even without news of the capture of Jerusalem itself, the loss of the true cross at Hattin provided symbolism sufficient to bring conflicts between kings to a hasty end so that resources could be deployed instead towards the crusade. For moralists, the lesson was clear enough: the West's refusal to listen to repeated appeals for aid from Outremer had left no alternative but for God to allow Saladin victory but, as always, true penitence could bring about both the restitution of the Holy Sepulchre and forgiveness for individuals. In fact, the situation was more complex than this, for western dynasties had been intricately involved in the political life of Outremer since the Second Crusade (Phillips 1996: *passim*). Moreover, it could be argued that Hattin presented openings for western kings that were not unwelcome. Henry II of England and Philip II of France allowed Joscius, archbishop of Tyre, to negotiate a truce in a war that neither appeared to be capable of ending. Philip had little choice but to take the cross once Henry – and indeed his heir Richard – had publicly done so (Tyerman 1988: 59); but for the German emperor Frederick Barbarossa, who had crusaded with his uncle Conrad III in 1147–9, the expedition offered opportunities for military glory denied him since his defeat in Italy in 1176, for reconciliation with rebel barons, and for a demonstration of imperial prestige. The Second Crusade had set the precedent for royal participation, and Henry II had taken the cross as early as 1177 – as in 1188, as part of a settlement with the Capetians. Both he and Philip II had dynastic connections in Outremer; and Baldwin IV's will in 1185 had appealed to the three principal rulers of the West to safeguard the kingdom in the event of his heir's death.

The Third Crusade, however, depended on others besides the main leaders. The first action in response to *Audita tremendi* was committed by a Sicilian fleet that handicapped Saladin's Egyptian navy, and by the time Richard and Philip arrived in the Holy Land an army of over 15,000 Frisians, Danes and Flemings, supported by some French contingents led by Henry, count of Champagne, had gone far towards the recapture of Acre. Genoese, Pisans, Venetians and Bretons had also joined what was indeed a truly international expedition.

The successes enjoyed by the crusade were partly due to extensive preparations. Henry II and Philip both levied the 'Saladin tithe', a tax of a tenth on all movables on laity and clergy alike, which assured that at least the Angevin contingent remained well resourced throughout the crusade. Philip II encountered such opposition from his barons that he had to abandon the tax, but the strength of Angevin government lay in its ability to organise the efficient collection of money. The tax in England was levied parish by parish, by committees that included representatives of local landowners, the bishop, the crown and the Military Orders (Tyerman 1988: 75–85). Moreover, the message of sinfulness and the need for penitential discipline was useful to secular government in organising the campaign. The Geddington ordinances promulgated by Henry II in February 1188 included restrictions on ostentatious clothing, swearing and gambling (Powicke 1981: 1025–9). The moral discipline required of Christendom by the papacy also served the needs of military strategy, and Henry's decrees indicate a close partnership between crown and Church in the Angevin realms.

The death of Henry II in 1189 left the Angevin crusade in the hands of his heir Richard. Although Henry had initially considered travelling overland, Richard decided instead to go by sea. Moreover, unlike Philip II, he provided his own ships – some built specially, others chartered; a lesson that a decade later the Fourth Crusade failed to note. Naval mastery was to prove decisive in Richard's military strategy; not only did it enable him to ensure a supply line independent of the land campaign, but it also provided him with the conquest of Cyprus from Isaac Komnenos in May–June 1191. Richard's delay in the Mediterranean (he also stopped over in Sicily to enforce payment of compensation by Tancred of Lecce for the dowry for his sister Joanna's marriage to William II) has led to suspicions that he was using the crusade as an opportunity to serve wider dynastic ambitions. But in fact the Cypriot adventure, useful though it was subsequently to prove, was probably the result of simple chance, and in Richard's defence, a stable succession in Sicily was crucial to the long-term defence of Outremer.

The recapture of Acre in July 1191, soon after Philip and Richard had arrived in the East, masked underlying problems in the Frankish army. In fact the Third Crusade suffered from the start from the continuing political divisions that had characterised the last years before Hattin. The siege of Acre had begun in 1189 because Guy of Lusignan, after abjuring his oath to Saladin not to take up arms against Islam again, had been refused entry to Tyre by Conrad of Montferrat and had nowhere else to go. With the help of the Pisans, who began a naval blockade, and steady reinforcements from the surviving barons in Tyre, the siege grew in significance until Conrad could no longer afford to ignore it. Guy's position, however, was ruined by the death of his queen Sybil and their daughters in autumn 1190, for he had no claim to the throne in his own right. Indeed, his right to wear the crown in 1186 had been suspect because he had not been crowned by the patriarch but by Sybil herself. The heir to the throne was now Isabel, Sybil's half-sister, the daughter of Amalric I and his second wife, the Byzantine princess Maria Komnena. Guy had done much to restore a reputation damaged by the events of 1186–7, but Conrad still had considerable support from the native barons, now led by Balian of Ibelin, who had married Maria Komnena after Amalric I's death. Maria's influence was critical, for she sanctioned the annulment of her daughter Isabel's marriage to Humphrey of Toron, so that Conrad could marry her and thus wear the crown himself on the same grounds that Guy had worn it in 1186. They probably thought that Humphrey, who had declined to allow himself and Isabel to become figureheads of a rival coronation to that of Guy and Sybil in 1186, would prove equally pliant on this occasion. Baldwin, archbishop of Canterbury, refused to cooperate on the grounds that Isabel's half-sister Sibyl had previously been married to Conrad's brother William Longsword, and Isabel and Conrad were thus, in canon law, related within the prohibited degrees. Besides this, as Baldwin knew, the marriage infringed the interests of the Angevins, whose vassals the Lusignans were. The papal legate, however, seems to have had no such scruples, and when Baldwin died that winter the marriage duly took place, apparently against the express wishes of Humphrey and Isabel themselves. This effectively disinherited Guy of whatever claim he might still have enjoyed to the crown; furthermore, it meant that when Richard and Philip arrived in 1191, they found themselves embroiled in a succession dispute.

The settlement of the dispute by Richard I and Philip II, according to which Guy was confirmed as king for his lifetime, but forced to recognise Conrad as his heir, seems to have satisfied neither party. Richard's support for Guy, his Lusignan vassal, cost him the cooperation of Conrad, and most of the native Frankish barons, for the duration of the crusade. The Hospitallers supported Guy, the Templars – in a change of allegiance – Conrad. Gerard de Ridefort, the Templar Grand Master who had been such an important confederate of Guy in 1186–7, had died during the siege of Acre. Conrad was subsequently to open separate negotiations with Saladin in the hope that the Muslims might recognise his new lordship of Tyre, Sidon and Beirut.

Philip II's premature departure after the fall of Acre, to ensure that his interests in France were safeguarded after the death of the count of Flanders, was probably less serious a matter, for he left 10,000 troops behind with the duke of Burgundy, and besides, his withdrawal left Richard a freer hand. But the internal divisions were not fully resolved by the agreement in July 1191, and in February 1192 Richard had to divert resources to defend Acre against an attempt by Conrad to seize it for himself. Finally, in the spring of 1192, Richard was forced to abandon Guy and recognise Conrad as king. Then, suddenly, Conrad was murdered in the streets of Tyre. Because it happened so soon after the succession had been settled in Conrad's favour, and under such mysterious circumstances, Richard has been suspected by some of engineering the murder. Fortuitous though it might have been for Richard, the murder was in fact the work of the Assassins, who had a grievance against Conrad because he had impounded a ship carrying goods belonging to them (Gillingham 1978: 206–7). But it left the succession question wide open again.

The reluctance of the barons of the rump of the kingdom to cooperate cannot wholly explain why the Third Crusade enjoyed only limited military success. Another reason was the disaster that befell Barbarossa and his German army. Although Leopold of Austria and some others took the sea route, the emperor himself, accompanied by his son, Frederick of Swabia, nine prelates and 32 great magnates, chose to march overland, following the traditional route taken by German crusaders. This was a provocative move, for since 1185 the Byzantine empire had been an ally of Saladin, and when he heard of the German expedition Saladin demanded that Isaac II Angelus deny access to Barbarossa's army. But the Byzantines were militarily too weak to keep their side of the treaty, and the Germans' threats were sufficient to secure access to Asia Minor in April 1190. Barbarossa expected an easier time in Asia Minor than the dreadful mauling he had experienced in 1147 when accompanying his uncle on the Second Crusade, for he had taken the precaution of making an alliance with Kilij Arslan III, who was for his part only too pleased to be able to undermine Saladin's power in Syria. But by 1190 real authority in Asia Minor had passed to Kilij's son Qutb ad-Din, who had married one of Saladin's daughters, and knew his obligations. In May 1190, therefore, Barbarossa had to face a hostile Seljuq army. This time, he was able to avenge the defeat of half a century earlier, and Qutb was brought to terms. An expedition that had promised so much was reduced to a shambles, however, by an unfortunate accident. Barbarossa, impatient to cross the river Saleph in Seleucia, drowned when he was dragged away by the strong current, possibly after suffering a heart attack. Frederick of Swabia was unable to keep the army together, and the contingent that eventually reached Acre in October 1190 was scarcely significant enough to make a difference to the crusade.

The siege of Acre was a triumph of persistence on the part of Guy and his followers and of the early arrivals, and a testament to the siegecraft of the Capetian and Angevin armies. Though not the last military success of the crusade, it was to prove the most significant. For although Richard showed, between summer 1191 and autumn 1192, that Saladin was far from invincible and that the apparent unity of the Islamic world for the purpose of the *jihad* was a façade, his campaign also exposed the limitations of traditional crusading.

If Richard inherited a winnable situation when he arrived at Acre at the beginning of June, his victory at Arsuf was of his own making. In late August he began the march south along the coast past Mt Carmel towards Jaffa, supplying the army from the accompanying fleet. As they reached Caesarea, Saladin began to harass the crusaders' rearguard, but he did not commit himself to battle until Richard reached Arsuf, a few miles north of Jaffa. Here Saladin hoped to use the forest as a screen for the advance of his main forces, and to squeeze the crusaders on the plain between forest and sea. The fighting march became a pitched battle on 7 September, after a cursory attempt to come to terms had failed. Richard's success at Arsuf was due to his ability to control the disparate army, made up of native barons of the kingdom, Flemish, French, Angevins, Poitevins and the Military Orders. The Hospitallers, bringing up the rear of the army on the march, had already been goaded almost beyond endurance by the harassment of the Turkish horse archers. Now they suffered the brunt of the initial Turkish attack as Saladin tried to separate them from the rest of the army. Richard refused to let them charge, for fear the whole battle order would be lost, but when at last the Hospitallers gave way to temptation and unleashed the heavy cavalry charge, he timed to perfection the moment to commit the rest of his forces to the charge, and the Turks were driven from the plain.

The victory at Arsuf (September 1191), although once again demonstrating the effectiveness of heavy cavalry against lightly armed Muslim forces on ground favourable to them and under skilled leadership, recovered no territory. It cleared the coastline, enabling Richard to secure the defences as far south as Darum, but also dissuaded Saladin from risking his army in the field against Richard. Yet only decisive victory in battle could bring the crusade to a successful conclusion. Richard's caution in securing and fortifying Jaffa as a base before marching on to Jerusalem meant that he was faced with the prospect of a winter siege that he was unwilling to risk. Once again, in early June 1192, he hesitated a few miles from Jerusalem, this time in the face of French impatience, for Hugh of Burgundy was keen to begin. But Saladin had refortified the walls, and Richard, a canny general, could see little point in a long siege that might be unsuccessful. Moreover, he realised that even if he succeeded in recovering the city there was little likelihood of defending it against an enemy that, even without the distant allies of 1187–91, could still call upon reinforcements from Syria and Egypt. He preferred an aggressive operation against Egypt, but this, too, failed to materialise.

Instead, at the end of July 1192, Richard found himself having to recover Jaffa from a surprise attack by Saladin, in an episode that did much for his reputation for personal valour but little to take the crusade farther than it had advanced by the autumn of 1191. Ambroise, the Anglo-Norman chronicler who recorded the crusade in verse, described how Richard, who had sailed down the coast from Acre as soon as he heard of Saladin's advance, leapt from his ship and waded ashore just before the last remnants of the garrison surrendered. The Muslims, surprised by the ferocity of the king's charge from the beach, fled in confusion. Only days later, Richard's courage and example swung the tide again in an engagement at Ascalon: 'On that day his sword shone like lightning and many of the Turks felt its edge . . . He mowed down men as reapers mow down corn with their sickles . . . He was an Achilles, an Alexander, a Roland' (cited in Gillingham 1978: 215).

Had Saladin succeeded, as he nearly did, in recapturing Jaffa, the Frankish coastline recovered by the crusade would have been cut in half. It may have been this close shave, rather than the failure to recover Jerusalem, that persuaded Richard to abandon aggression in favour of consolidation. For it must have been clear to him that the objectives of the crusade as articulated in 1188 – namely, to recover the Holy Sepulchre and the True Cross – were at odds with the military imperatives. Christendom could expect little more than the defence of what had been won back since 1189. Richard himself could not stay in the East indefinitely, and by spring 1192 was already keen to turn his attention to the problems that had been developing in his western territories during his absence.

Moreover, the dynastic prospects for the kingdom of Jerusalem were looking more settled than at any time since 1174. Isabel had successfully prevented Hugh of Burgundy, who was sympathetic to Conrad, from taking over Tyre. Richard had already compensated Guy of Lusignan for the loss of his crown with the island of Cyprus, and wisely did not try to bring him back. Instead, overlooking the fact that in his eyes Isabel was still canonically married to Humphrey of Toron, Richard proposed that the uncrowned queen now marry Henry, count of Champagne, who had proved himself a valuable ally during the crusade. The wedding took place with the same haste as Isabel's enforced marriage to Conrad in November 1190. Between 1192 and 1197, when he died, Henry showed himself a vigorous defender of his kingdom, though he never accepted the title of king.

What had the Third Crusade regained from the disaster of 1187? Richard's treaty with Saladin granted access to the Holy Sepulchre for Christian pilgrims, the right of clergy to officiate at the shrine, and a corridor of land from Jaffa to Jerusalem to maintain a pilgrim road. In terms of territory, the rich coastal plains around Acre and Tyre, the two cities that formed the heart of the royal demesne. Much rural land had been lost, of course, but the crown in the thirteenth century was still able to grant fiefs en bezant from increased revenues in the towns. The towns themselves, particularly Acre, increased in size and population after 1187 as both Franks and Arabic-speaking Christians who had previously lived in the countryside withdrew into the new frontiers. The continued existence of the kingdom of Jerusalem would thus depend, in the thirteenth century even more than had been the case in the twelfth, on a successful urban economy. The loss of land in 1187, of course, meant the loss of many agricultural resources, but in fact the coastal plain retained by the Franks was highly conducive to the intensive farming of sugar cane and olives. The demographic balance in the 'new kingdom' was also different, for the large population of Muslims who lived in the rural areas were now once again under Turkish rule. For the Franks this represented a loss in forced manpower for building work, but also removed the danger – admittedly never a serious one in the twelfth century – of an uprising against their authority. Against this loss must be balanced the acquisition of Cyprus, to which Guy and his brother Aimery attracted many of the Frankish nobility of the East to replace the Byzantine ruling class.

The treaty of 1192 was the result not only of fifteen months of warfare between Saladin and Richard, but also of sporadic negotiations since September 1191. At one point it looked as though Saladin might accept a deal to restore the former extent of the kingdom, which would then be ruled by his brother al-'Adil, who would in turn marry Richard's sister Joanna. Implausible though this might seem to us – and unacceptable as it proved to Joanna – this would in fact have represented simply the transfer of standard western techniques of conflict resolution to the East, and the fact that the plan was proposed demonstrates that both sides accepted the inevitability of a Frankish presence in the Near East. The fate of Jerusalem itself is also instructive. Saladin, understandably, was not prepared to abandon his greatest conquest. For their part the Franks realised that the kingdom could function without it, but that it made no sense to occupy the city itself unless Kerak, to the east, and Ascalon, on the Mediterranean coast, were also in Frankish hands to secure its flanks. This consideration was to make a Frankish military recovery of Jerusalem an impossibility throughout the medieval period.

Problems of leadership and logistics: the Fourth Crusade

The idea that Constantinople might have to be taken into Frankish hands as a preliminary to the successful defeat of the Turks in the East had been mooted by crusaders in 1096, 1147 and 1190, but on none of these occasions had it been seriously attempted. In 1204 it actually happened, but as the result of a chain of circumstances rather than as the original intent of a new crusade. Widespread western suspicion of Byzantine culture and society, the Byzantine alliance with Saladin, the role of Venice in eastern Mediterranean commerce, and the presence on the crusade of 1202–4 of individuals who had personal interests in Byzantium have led some historians to suppose that the Fourth Crusade was from an early stage disposed to seize Constantinople, whose wealth could then be deployed for the recovery of the Holy Land (Brand 1968: v; Nicol 1988: 126, 130; Harris 2003: 147). How far one gives credence to this probably depends in part on the extent to which one subscribes to a 'conspiracy theory' model of historical causation.

The chronicler Geoffrey de Villehardouin's explanation for the fate of the crusade is one of logical cause and effect. The crusaders could not meet the terms of their treaty with the Venetians, so they were compelled to accept Venetian transport on revised terms that included the restoration of Alexios IV to the Byzantine throne (Shaw 1963: 50–6). Once installed, Alexios would guarantee their further passage to the Holy Land or Egypt. When Alexios was unable to deliver, the crusaders had no alternative but to sack the city in order to secure sufficient resources to fulfil their vows; having conquered the city, however, they found that they were not strong enough both to hold their new territories and to mount an assault on the Holy Land, and so stayed put. However, Villehardouin's attempt to justify the actions of the crusaders, of whom he was one, begs important questions both about the planning and organisation of the crusade before it ever left Italy, and about decisions made once it was under way.

The crusade announced by Innocent III on his election in 1198 can be seen as an attempt to continue the momentum built up by the crusade of the emperor Henry VI in 1197. Henry died in Sicily in September, leaving the advance contingents of his army, together with the native Frankish barons of the East, to face an Egyptian army that had been summoned in anticipation by al-'Adil. Preachers were commissioned in 1198 and attempts were made to reconcile the warring Richard I and Philip II. The main impetus for recruitment, however, seems to have developed in north-eastern France, Flanders and Germany, perhaps after a group of knights meeting for a tournament in the Ardennes in late 1199 had been shamed by the papal legate Peter Capuano into taking the cross instead. The French barons formed a committee which, early in 1201, travelled to Venice to negotiate transport across the Mediterranean. In May 1201 Innocent III approved an agreement according to which the Venetians undertook to provide transport for 4,500 knights and their horses, 9,000 squires and 20,000 foot soldiers, plus food for nine months, at a cost of 85,000 marks. The details of the agreement are recorded by Villehardouin, a member of the committee who later wrote one of the fullest and most vivid accounts of any crusade (Shaw 1963: 32–3).

Problems developed even before the crusade had set out. The count of Champagne, the nominal leader, died in 1201, and after others had refused the leadership, Boniface, marquis of Montferrat was chosen. This was a critical choice. The Montferrats were related both to the Hohenstaufen and the Capetians, and had been involved, since 1176, in the dynastic affairs of the kingdom of Jerusalem. The marquisate, an imperial fief situated in the wealthy Piedmont region of northern Italy, could provide large numbers of knights. But the family also had interests in Constantinople, for Boniface's younger brother, Renier, had married the daughter of Manuel Komnenos, before falling victim to the anti-western sentiment manipulated by Emperor Andronikos I (1182–5), while Conrad of Montferrat had followed Renier to Constantinople before sailing for Jerusalem in 1187. This Byzantine connection has been seized upon by historians who argue that the crusade fatefully changed course from the Levant to Constantinople even before leaving Venice, with the election of Boniface as leader (Nicol 1988: 129–30). In autumn 1201, Boniface met the pretender to the Byzantine throne, Alexios (IV), who had been exiled when his father Isaac II was deposed by his brother Alexios III in 1195. There was good reason for Alexios to seek help in the western Empire, for his sister Irene was married to Philip of Swabia, Boniface's own suzerain. Queller and Madden suggested that the possibility of taking the crusade by the overland route via Constantinople, where Alexios could be installed as emperor in place of Alexios III, must have been considered, but warned against the danger of assuming that leadership of the army necessarily meant that Boniface could use the army as he pleased (1997: 37–9). For one thing, Boniface was not in a position to change the treaty drawn up with Venice in spring 1201, according to which the crusaders were bound to travel by sea; for another, he could scarcely impose the ambitions of his own dynasty on other magnates, such as the count of Flanders, who did not share them.

The Venetians have often, particularly by Byzantinists, been cast as the villains of the piece. Venice developed from a Byzantine protectorate in the early Middle Ages into an independent commercial power, but by the last quarter of the twelfth century the former colony enjoyed a dominant role in the eastern Mediterranean. The expulsion of Venetian merchants and their families from Constantinople in 1171 – according to one contemporary source, in retaliation for a practical joke in which Venetian sailors dressed up a monkey in imperial robes – was a serious blow in the commercial rivalry between Venice and Genoa for Black Sea and Aegean trade. While Genoa suffered the anti-western backlash of the 1180s in Constantinople, Venice profited from the dynastic instability in the Byzantine Empire, and after the *coup* of 1185, Isaac II retained the throne with Venetian naval help, in return for a virtual trading monopoly in the waters around Constantinople (Brand 1968: 195–206).

Some historians have portrayed Venice as a chronically exploitative state whose involvement in the crusade was necessarily a cynical piece of commercial enterprise: the crusade was 'bad for business' (Nicol 1988: 124). But this is unfair. Venetians, along with the other Italian maritime powers, had been enthusiastic crusaders from the early twelfth century, and their commercial prosperity enabled the crown in the crusader kingdom to build up a secure financial base. In 1202 Enrico Dandolo and many Venetians took the cross themselves. It is true that Alexios III's *coup* of 1195 had threatened Venetian dominance in the Bosphorus once again, but this need not mean that the Venetians saw the crusade preached in 1198 as a means to deposing him. The treaty negotiated by the doge, Enrico Dandolo, and the baronial committee in spring 1201 specified Egypt as the target of the crusade. Until the emperor's nephew Alexios IV arrived in the West, in September 1201, having escaped from his uncle's custody, there was no pretext for deviating from this plan. Even after

Alexios turned up to complicate matters, Venice stood to gain far more economically from the conquest of Egypt than from restoring the monopoly in the Bosphorus. Damietta, the principal Mediterranean port of Egypt, was a terminus of the trade route across the Red Sea to India, and the prospect of establishing a commercial quarter there was more enticing than anything in Byzantine waters. It was only from the mid-thirteenth century onward that the Black Sea could rival the Levant as an artery of trade. The rumour that Venice had been bribed by the sultan of Egypt to divert the crusade to Constantinople has no contemporary basis.

The crusade began to falter in summer 1202, when it became apparent that insufficient numbers were turning up to meet the terms of the Venetian treaty. Villehardouin, seeking to explain the direction taken by the crusade in 1202–3, blamed those who had taken the cross but either never set out from home or, like many of Baldwin of Flanders' forces, chose alternative routes. These, he implies, betrayed the crusade; however, he offers no convincing explanation for why they should have done so (Shaw 1963: 41). Another eyewitness to events, Robert of Clari, whose account represents the point of view of the 'average' crusader rather more than does Villehardouin's, leaves no room to suppose that knights made different arrangements because they already suspected that the crusade was to be diverted by Boniface of Montferrat and the Venetians to Constantinople. Even the original destination of Egypt was not revealed to Robert until he was already at sea. It seems, then, that the baronial leadership was simply unable to impose the conditions to which they had agreed on disparate and inchoate groups of crusaders. The failure of the crusade was one of unclear leadership and poor organization.

By late August 1202 only some 11,000–12,000 of the 33,500 crusaders estimated in the treaty had assembled in Venice, and it must have been apparent that they would be unable to meet the terms agreed in 1201. Despite Villehardouin's accusations, it is far from certain that, even if all those who did eventually set sail had turned up in Venice, the crusaders would have been able to raise the money required. The estimate of about 2,100 extra men who went directly to the Holy Land would only have raised the total number in Venice slightly (Queller and Madden 1977: 48). Even after Baldwin count of Flanders, Boniface of Montferrat and other leaders had borrowed to their limits to make up the amount, they fell short by 34,000 marks. The implication must be that the baronial committee of 1201 was either gulled by the Venetians into agreeing exploitative terms, or that they gravely overestimated the numbers who would take the cross and fulfil their vows. The figure agreed in 1201 was 85,000 marks, or four marks for each mounted knight, and two for a squire or sergeant, but the sum was to include provisions for a year. Comparison with the contract made by Philip Augustus with Genoa for transport to Acre in 1190 demonstrates that the Venetians were asking hardly more than the going rate (Queller and Madden 1997: 11–12). The problem, therefore, must lie with the numbers estimated. No wonder, Queller and Madden conclude, Villehardouin does not lay blame on the treaty itself, 'for he himself stands out among those responsible for the ruinous blunder' (1997: 48).

Could the Venetians have done more to preserve the crusade at this stage? Time was running short, for if they were going anywhere the crusaders needed to set sail before the autumn set in. Meanwhile food was scarce, and neither the crusaders themselves nor the Venetians could feed the army indefinitely. The doge must have feared – as Byzantine emperors had found in 1096 and 1147 – that crusaders kept hungry and waiting would eventually turn against their hosts. However, he could not let them disperse and return home without writing off huge losses against the Republic. Robert of Clari reports that Venice had suspended commercial shipping for eighteen months prior to 1202 so as to be able to build and equip the fleet (McNeal 1996: 38). Ordinary commercial vessels could not be used to carry horses in the numbers required; specialised transport ships had to be built. The Venetians, whose existence depended on maritime trade, were taking a risk in agreeing the contract in the first place, and now had to find a way of covering that risk.

One of the most controversial aspects of the crusade, for contemporaries as well as for later historians, is the diversion to Zara. Villehardouin and Robert of Clari concur that most of the crusaders were ignorant of Dandolo's proposal to save the crusade, but Gunther of Pairis, whose account relies on the eyewitness report of his abbot Martin, hints that some knew and made their displeasure known. Martin himself asked the papal legate to be released from his vow rather than attack a Christian city, but Peter Capuano refused him. The papal desire to salvage the crusade outweighed the moral consideration of shedding Christian blood. At this stage, there was no reason to suppose that the crusaders would not eventually have the opportunity to spill Muslim blood also, so the argument that evil must be accepted for good to prevail must have carried some weight. The problem was that although the same crusaders – even, in some cases, clerical ones – would under most circumstances have had little compunction about attacking a city held by a Christian enemy, especially when, as in the case of Zara, the claim to the city was disputed, knights did see a distinction between what one did when signed with the cross and what one might do when not under a crusading vow. Those who tried to prevent the attack, such as the Cistercian Abbot Guy of Vaux-de-Cernay, Enguerrand de Boves and Simon de Montfort, were very clear in their understanding of what crusading should and should not involve. Innocent III's role at this point is ambivalent. He certainly condemned the attack, but only when the army was already outside Zara's walls, and the presence of both Boniface and Peter Capuano in Rome during the attack itself suggests that he accepted the inevitability of a policy that he could not condone, and from which he sought to disassociate his legate and the official leader of the crusade (Andrea 2000: 41–5, 46–8, 55–7, 58–9, for Innocent III's letters to the crusaders and their replies).

The entanglement of the pope in Alexios IV's plans is also a murky issue. By the time the crusaders had left, Innocent had already received an embassy from Alexios III appealing to him not to allow the crusade to be diverted to Constantinople, to which he replied favourably (Andrea 2000: 35–9). But Innocent's failure to prevent the sack of Zara, even at the pain of excommunication, must cast doubt on his ability to influence the course of the crusade any further. The doge had persuaded the crusaders to attack a city in the possession of a king, Emeric of Hungary, who had himself taken the cross, and from whose walls crosses had been hung by the inhabitants; what chance did the pope have of turning them away from a city that was widely regarded as the enemy of crusading? Moreover, the papacy had every reason to support the deposition of Alexios III in favour of his nephew. Although his father, Isaac II, had been an ally of Saladin, the young Alexios not only undertook to supply the crusaders with sufficient ships and provisions and troops to complete the crusade, in return for putting him on the throne, but further promised to ensure the submission of the Orthodox Church to Rome. Papal/imperial negotiations over the next century and a half were to show the futility of such a promise, but in 1202–3 it must have looked to Innocent like a short cut to the long-desired union of the Churches. A further complication was that Philip of Swabia, who was Innocent's preference for the western imperial throne, openly supported Alexios IV (Brand 1968: 190–1, 225) and when the plan was revealed to the army at Zara in January 1203, the German crusaders naturally argued in its favour. Yet Innocent, having confirmed the excommunication of the Venetians for the Zaran affair, could scarcely approve the crusade leaders' decision to accept Alexios IV's proposal. In fact, the crusaders had little choice but to go along with what Dandolo decided, for although some individuals, notably de Montfort and Guy of Vaux-de-Cernay, took their small contingents directly to the Holy Land, the transport of over 10,000 men and horses still depended on the Venetians. Villehardouin's justification of the course of the crusade is correct in so far as the crusaders as a body had, literally, no control over their own movements.

The crusaders made short work of restoring the blinded Isaac II to occupy the throne jointly with his son; Alexios III fled as soon as an assault was launched in July 1203. But Alexios IV, dazzled by the prospect of the throne, had promised far more than he could deliver. The crusaders had restored him to the city of Constantinople, but not to the whole empire. He had also reckoned without the hostility of his own people for an emperor – albeit a legitimate one – so clearly dependent on Latin arms. Caught between his allies and his subjects, Alexios wavered so much that historians have been unable to agree whether he persuaded the crusaders to stay beyond summer 1203 (Richard 1999: 249–50) or turned suddenly against them (Queller and Madder 1997: 149–50). In any case, he and his father were murdered in January 1204 when the Byzantines revolted. Villehardouin's account of being sent into the hostile city to confront Alexios with his faithlessness (Shaw 1963: 82–3) suggests that the crusaders had already decided to attack the city if the new emperor did not fulfil his promises. Once Alexios Mourtzouphlos had installed himself in his place on a tide of anti-Latin feeling, they had in any case to defend themselves against attempts to burn their fleet. A second assault on the city was inevitable. On 13 April the city fell to the crusaders, and for three days suffered a vicious and vindictive sack. One reason why Greeks still regard the Fourth Crusade with such horror is the very nature of the crusaders' victory. Gunther of Pairis' proud list of the relics stolen by his abbot Martin, which included 'a considerable part of St John, an arm of St James, a foot of St Cosmas, a tooth of St Lawrence' as well as a piece of the True Cross, could doubtless be amplified many times over by unrecorded acts of looting (Andrea 1997: 125–7).

The division of the empire between the crusade leaders and Venice had already been agreed in a concord drawn up in March. The loser turned out to be Boniface of Montferrat, whose stock had fallen since the death of his protégé Alexios IV. However, the new emperor, Baldwin of Flanders, soon found that he had exchanged a strongly centralised county in the West for a paper empire. Without the capacity to distribute fiefs – a privilege reserved for a joint Venetian/French committee – his authority was limited from the start, and by the 1220s the Latin Empire was already doomed as a political entity. Venice, on the other hand, held on to its new island possessions in the Aegean and Ionian seas until the sixteenth century. As far as the crusade itself was concerned, there was now no possibility of reaching the Holy Land or Egypt, for even if the will had been there, the crusaders had relied since the beginning of 1203 on promises of Byzantine resources that they now found to be worthless.

The Fourth Crusade is often thought to have been a failure, but in military terms its achievement was almost as impressive as that of 1097–9. Moreover, if no impact had been made on the Holy Land, the crusaders could, as Villehardouin attempted, rationalise their actions by arguing that they had furthered the cause of future expeditions. As it turned out, Constantinople was never again to be pivotal to a crusade for the recovery of the Holy Land, but with the exception of the period between 1158 and 1180, the Byzantines had always represented an obstruction to crusaders, so the argument was by no means frivolous. The manner of the victory, however, proved fatal to any attempt to treat the imposition of a Venetian patriarch and canons in Constantinople as the genuine union of the Churches so desired by the papacy. In the history of crusading as a phenomenon, the Fourth Crusade's chief significance is twofold: first, it demonstrated how poor organisation before departure could wreck an expedition, and second, it gave weight to the notion that crusades could legitimately be launched against other enemies of Christendom than the Muslims.

Innocent III and the idea and practice of crusading

On his election as pope in 1198, Innocent III took as the text on which to base his first encyclical the verse 'O Jerusalem, if ever I forget you' (Psalm 137: 5). The preaching of a crusade to the East was undertaken immediately. The patriarch of Jerusalem, Aimery the Monk, was asked to submit a report to the curia describing the state of the Holy Land, as an *aide-mémoire* for planning a crusade. Innocent, moreover, involved himself in the affairs of the Latin Church in the East more than any previous pope. After the death of Aimery in 1202, the papal legate Soffredo, who had been sent to supervise a new election, was himself elected, and when he died after only a year in office another Italian in Innocent's circle, Albert, bishop of Vercelli, was sent to the East. Innocent persuaded yet another of his associates, the Cistercian Peter of Ivrea, to take up the patriarchate of Antioch. The Holy Land was central to the piety and spirituality of the pope and his intimate circle at the beginning of the thirteenth century (Bolton 2000: 154–80).

It is ironic, given this interest in the Holy Land, that Innocent has the reputation of being the pope who, more than any other, widened the scope of crusading. Other regions than the eastern Mediterranean – the Iberian peninsula and the Baltic in particular – had long been targets of crusades. But Innocent's pontificate saw a conceptual shift as well as a geographical one, with the extension of crusading privileges to those who took the cross against Christian enemies of the papacy in Sicily and against Cathar heretics in southern France. As is often the case in papal history, this was simply the logical extension into practice of ideas implicit in previous actions or statements. Gregory VII had encouraged the use of force to remove 'unreformed' clergy, and canny rulers, like Henry II in his proposed invasion of Ireland in 1170, had seen how the idea of the crusade could be exploited to further his territorial aims. Whether Innocent intended, however, to place crusades in other directions than the Holy Land and against enemies other than the Muslims in the same category, as some historians have argued or implied (Housley 1992: 2–6; Riley-Smith 2002: *passim* – but *contra* see Tyerman 1998: 37) depends partly on one's interpretation of papal practice.

Because so much of lasting significance was articulated in Innocent's pontificate, it is easy to see him as an idealist who created new ways of formulating papal primacy. But Innocent's crusading policy can just as convincingly be interpreted as a series of pragmatic reactions to problems as they arose. Some of these problems were chronic, others the result of fortuitous circumstances; but all were inherited, rather than being of Innocent's own making.

Heresy and Christian unity

Perhaps the most problematic inheritance was the increasingly strong hold taken by heretical beliefs in Christendom. The regions particularly susceptible to the dissemination of heresy, urban northern Italy and southwestern France, though vastly different in character, had one crucial feature in common: the lack of a powerful central government in whose interests it was to stamp out heresy as a potentially dangerous subversion of society. In southern France by c.1200, the parochial Church seems to have lost the loyalty of large sections of society. Lack of education among the clergy meant that parishioners were often vague about what constituted orthodox beliefs, and the provision of pastoral care depended largely on the competence of individual bishops. Catharism, a dualist religion that presupposed the existence of two equal powers, good and evil, and that in the extreme forms that had taken hold in southern France, also denied the divinity of Christ and the validity of the sacraments, had by Innocent's pontificate become organised into bishoprics and offered a coherent set of alternative beliefs to orthodox Christianity (Lambert 1998: 61–81; Barber 2000: 68–106). The organisation of the religion made it particularly difficult to detect, because with the exception of the Cathar priesthood (the 'perfect'), Cathar believers were not required to observe ways of life that marked them out as observably different from their neighbours.

The application of the crusade to the Cathar problem was a last resort on Innocent's part. Repeated demands to the dominant magnate of the region, Raymond VI of Toulouse, to stamp out Catharism, failed because they were impossible for him to meet. Raymond was the largest land-owner, with rich and varied lordships in the Toulousain, Gascony and to the east of the Rhone, but he lacked the machinery of central government available to northern magnates. He was suspected of Cathar sympathies himself, but whether or not this were true – and the charge rests largely on the indictment of the Cistercian Peter of Vaux-de-Cernay – he was simply in no position to drive the Cathars out of his territories, even if he had wished to do so. By the time he inherited the county in 1194, at the age of thirty-eight, Catharism was an accepted condition of life in the Languedoc (Sumption 1978: 64–5; Costen 1997: 25–51). What the count himself was unable to achieve, the lesser magnates could certainly do no better, and the promise of indulgences equal to those for the great pilgrimages was insufficient to induce them to rid their lands of heretics. Raymond already had a history of antagonism towards the clergy – he was excommunicated for the first time in 1196 – but Innocent's preferred method of resolving the heretic problem, which was to send legates with absolute powers to enforce obedience to Rome, turned the local bishops against them too. The legates, the Cistercians Peter of Castelnau and Arnald-Amaury of Citeaux, although themselves southerners, were resented by the bishops whose authority they overrode. In consequence, the bishops of Béziers, Toulouse and Viviers were all replaced within months of the arrival of the legates. Arnald-Amaury, who was also to lead a crusading army in Spain, had a propensity for bloodshed: it was he who allegedly advised at the siege of Béziers: 'Kill them all; God will know his own'. The legates found – though perhaps never fully understood – that the presence of Catharism was a symptom, rather than the cause, of the disaffectedness of so many Christians in the Languedoc from the Church. Between 1203 and 1208 teams of Cistercians, supplemented by a Spanish bishop, Diego of Osma, toured the Languedoc preaching and disputing publicly against the Cathars and trying to reform the local clergy.

By 1207 Peter of Castelnau's patience had been exhausted; having negotiated a truce among Raymond's vassals without his knowledge, he excommunicated the count himself for failing to keep his promises. The problem with excommunication as a weapon, however, was that it relied on a local Church willing to respect it. Innocent thus tried to back it up by appealing to Philip II of France to enforce it. His appeal of November 1207 is in some ways a logical application of the joint encyclical of Pope Lucius III and Frederick Barbarossa of 1184, in which the principle of using the secular sword to enforce the submission of heretics is articulated. But Innocent went further by offering the same indulgence as for a crusade against the Muslims. Raymond was finally scared into negotiating, but in January 1208 Peter of Castelnau was murdered by one of Raymond's officials, and the crusade was duly preached throughout France. Raymond's attempt to save himself by taking the cross against the heretics may be seen as cynical opportunism, but in any case it did not work, and by the summer of 1209 he had been forced to hand over control of the administration of his lands to the papal legates. Although after initial losses in 1211–12 he successfully invaded the Languedoc and retook Toulouse, the crusade found new vigour in the leadership of Louis VIII of France in 1224, and by 1229 Raymond VI had been defeated and humili-ated. The county of Toulouse, halved in extent during Raymond's own lifetime, would effectively be annexed to the French crown.

The course of the crusade, which was largely a territorial war fought by northern French nobles led initially by Simon de Montfort and sub-sequently by Louis VIII, against the families and properties of the southern nobility, whether or not they were themselves Cathars, can obscure the ideals that had initiated it. The crusade was not an impetuous act of vengeance for the murder of the legate, for Innocent had already offered indulgences both to those who expelled heretics from their own lands and, in November 1207, to Philip Augustus and anyone who took the cross under him. Nor was Innocent, by using force against the Cathars, attempt-ing anything new. The theological justification for acts of violence in defence of the Church had been developed in the curia of Gregory VII for application against those who threatened the integrity of Christian society, and the Third Lateran Council in 1179 had approved the use of force to deal with heresy (Tanner 1990: 216). Heresy was regarded throughout the twelfth century as a social crime as well as a theological error, and lay rulers were quicker than the Church to punish heretics by force where necessary (Moore 1987: 8–9). It was not, therefore, the resort to war that troubled contemporary critics of the Albigensian Crusade, so much as the harnessing of crusading ideals to that war.

The problem, as critics saw it, lay not in Innocent's use of crusading language *per se*, but in the fact that by the early thirteenth century the language of crusading was beginning to have specific and practical implications. Since the Geddington Ordinances of 1188, a precedent had been set for the assumption of the cross as equivalent to a religious vow, which in turn enacted a change in status for the vow-taker and, in consequence, set in train practical measures by which he could raise money, defer repayment of loans, evade jurisdiction, and so on. Preaching any crusade, even one limited in its geographical application to one part of Europe, unlocked the machinery by which money could be collected from the faithful throughout Christendom. Although there was as yet no coherent canonical theory of crusading, the term 'crusade' had begun to appear after the Third Crusade, both in the Latin form 'crucesignatus' (lit.: someone signed with the cross, hence, a crusader) and, more telling in the context of the Albigensian Crusade, in the vernaculars of northern and southern France (*croisier*, *crozeia*) (Tyerman 1998: 49–50). This implies that perceptions were hardening about the meaning of an activity that had since 1095 had no specific terminology.

Most of the contemporary criticism of the Albigensian Crusade comes from the Languedoc itself; in other words, from the victims. The landowning classes of the Languedoc, whether they were sympathetic to the Cathars or not, could see that a crusade against heretics was in practical terms impossible to carry through. In current terminology, it was an 'asymmetric' war. War could be waged against individual lords and their property, or against towns, castles and armies put into the field to defend them, but less easily against religious beliefs or organisations. To Innocent and the crusaders, however, the property of Cathars, and places where Cathars lived or were sheltered, were no less legitimate targets than Aleppo or Damascus might be for crusaders to the East. Although it is counterintuitive for us to equate the Muslim threat in the East, which was largely territorial in nature, to that posed by heretical beliefs within Christendom, it was not so for most medieval Christians. For one thing, Islam was itself largely thought of as a Christian heresy rather than as a separate systematic religious tradition. Then again, Cathars were not simply the disenfranchised urban poor or peasants; they could be found in all social classes, from Raymond VI's second wife downward, and in positions of authority in towns in both the Languedoc and northern Italy. Cathars in Orvieto even succeeded in expelling the bishop.

More profoundly, however, the existence of Catharism was understood by the Church as a threat to its own validity. In denying the efficacy of the sacraments, Catharism made the priesthood irrelevant, and in so doing, it cut through at a stroke the complex social and spiritual ties that held Christian society together. In an age in which religious choice has become axiomatic as an indicator of individual freedom, it is difficult for us to acknowledge how deeply this struck at the heart of the social order. The structuralist interpretation of the Church's defence of its authority, favoured by sociologists, is that the clergy needed to enforce orthodox Catholic beliefs and practices in order to preserve their mystical power as an élite over the illiterate masses (Kaelber 1998: 106–17). Innocent and his circle, however, did not think in these terms. They thought, simply, that maintaining universal Catholic practices was necessary for the salvation of society. 'The Church' was not a separate entity from 'society'; in all senses the Church *was* Christian society, meaning all people baptised into the Church at birth. The Church, which had been instituted by Christ with St Peter at its head, was a single body, and as a single body must be kept safe and delivered up to Christ at the end of time. The pope, the successor of St Peter, was thus responsible for the salvation of society by preserving and maintaining the sacraments through which individuals could gain assurance of Paradise.

The key to Innocent's thinking is the idea of unity. Within the Church there would always be those who in their manner of life rejected salvation – as Augustine had envisioned it, the elect and the damned lived side by side. But to establish an alternative Church was to upset the unity implied by the single body created through the agency of the Holy Spirit at the first Pentecost. There could be only one Church; coexistence with another that offered different doctrines and sacraments was impossible because the presence of another invalidated the first.

The crusade was the logical weapon for Innocent to use in the defence of Christendom for practical reasons too. Since 1187, four crusades had been preached; it had become a more normative and recognisable means of raising armies than ever before. Despite the appeal to Philip Augustus, Innocent was aware that he was hardly more likely in 1207 than he had been in 1199 to persuade him to act as the enforcer of papal policy. But by commissioning preachers to accept crusading vows from the knights of Christendom, Innocent was making a virtue out of necessity. The capacity to attract knights through the offer of a crusading indulgence was a visible demonstration of the pope's overriding authority as the leader of Christendom.

After the experience of the Fourth Crusade, in which papal authority over the crusaders had been so hollow, it is hardly surprising that Innocent wished to play a more direct role in the crusade against the heretics. But he was to be equally frustrated here by the extremism of his legates, particularly Arnald-Amaury, and by the ambitions of Simon de Montfort. Innocent seems to have wanted Raymond VI to keep not only his title as count but most of his possessions in the Languedoc, but he had to give way in the council he himself summoned in 1214 and recognise the reality of the battlefield (Sumption 1978: 171–81). The Albigensian Crusade, in the end, accomplished only the violent transfer of the lands of the southern nobility to northern French crusaders; the real work of combating heresy was achieved by a combination of the Inquisition and better preaching and more uplifting example on the part of a new generation of preachers (Lambert 1998: 215–29). Yet this brutal episode, like the Investiture Contest for Gregory VII and his successors, provided an arena of conflict in which new and more coherent ideas about crusading and its role in Christian society could be forged.

Towards a centralised crusade

It has been said that hierarchical ideas formed the basis of Innocent III's thinking about crusading (Sayers 1994: 166, differing in interpretation from Mayer 1988: 217). When we look at crusading activity in his pontificate, Innocent certainly appears as the organising principle. It is striking that none of his crusades was led by, or even featured to any great extent, royal figures. Although individual nobles, such as Philip of Flanders in 1177, had mounted crusades in the past, since 1145–6 the papacy had aimed bulls proclaiming crusades at kings. The problem was that in the period 1198–1216 none of the dynasties on whom popes had previously relied was any longer in a position to answer an appeal. The death of Emperor Henry VI in 1197 was followed in 1199 by that of the Angevin Richard I, whom Innocent had in fact approached a year earlier; in both cases, the consequence was a succession dispute. The German throne was claimed by three candidates: Henry's infant son Frederick II, his uncle Philip of Swabia and Otto of Brunswick, neither of whom wished to jeopardise their candidature by leaving Germany. John, who succeeded Richard to the throne of England, had to fight off a claim from his nephew Arthur, who was supported by Philip II. Philip himself, who had not relished his crusade ten years earlier, would not take the cross until the Angevin succession had been resolved; he was even reluctant to lead a crusade into the Languedoc.

Circumstances may have suited Innocent's ideals well. Although in 1199 and in 1212 Innocent tried to reconcile Philip and John, the absence of royal figures also provided him with an opportunity to direct the crusade more personally than any pope since Urban II. For, as Richard has pointed out, a dangerous precedent had been set by Henry VI's crusade (1999: 236–7). Henry had taken the cross as a matter of personal conscience in 1195, rather than in response to a papal appeal. But by 1197 a fortuitous set of circumstances had altered his stature in the East. When Richard I fell into the hands of Leopold of Austria, Henry, as Leopold's overlord, took Richard into his custody and demanded as part of the exorbitant ransom suzerainty over Cyprus. As overlord Henry then sent Amalric II a crown to wear; Cyprus thus became an imperial vassal. Moreover, when in September 1197 Henry of Champagne died when he fell from a window in Acre, Aimery was chosen to marry the thrice-widowed Isabel and thus become king of Jerusalem as well as Cyprus. Although the kingdoms were in no way constitutionally linked, it was obvious that Henry VI, as Aimery's suzerain in Cyprus, would have unparalleled influence for a western monarch in the crusader states. Innocent III realised that any successful claimant to the imperial throne would enjoy the same influence.

Yet by 1212 it was apparent that committees of barons, as had operated in 1202–4, lacked sufficient direction, and that papal legates could dictate crusading policy on the ground more easily than the pope himself. Two decrees, *Novit ille* (1212) and *Quia maior* (1213) were instrumental in formulating a papal theory of crusading, and in paving the way for the decree *Ad liberandam* of the Fourth Lateran Council, which was the most detailed statement of how crusading as an institution was to be managed. The first of these was only peripherally connected with crusading. *Novit ille* was a decretal letter sent to Philip Augustus in which the pope claimed the right of arbitration in the war between the Capetian and Angevin dynasties (Tierney 1964: 134–5). No pope, not even Gregory VII, had so explicitly articulated the rhetoric of papal primacy so as to claim to judge the actions of kings in the purely secular sphere. But Innocent did so not, he says, as a feudal judge, but rather as a judge over matters of sinfulness. In this he was simply following the Petrine commission. 'No sane man could deny our right to judge and punish any Christian in matters of sin', he argued, and he further claimed that the matters in dispute, concerning as they did breaches of oaths and of treaties, were spiritual, not secular affairs. *Novit ille* is a far-reaching and high-minded ideology of papal authority. At the same time, it is a document born out of a pragmatic desire to end a war that jeopardised the possibility of launching another crusade. Its significance for crusading lies in the assumption that the crusade is the highest aspiration not only of papal policy but also of Christian kingship, and that kings' secular policies ought thus to be subordinated to serving the needs of the crusade.

Quia maior, which outlined Innocent's plans for a new crusade for the recovery of Jerusalem, contained little new; instead, it summarised and articulated more coherently than before a century of previous papal policy. Most importantly, the plenary indulgence, never yet precisely defined, was now established as a fixed inducement to crusaders. All who took the cross, whether they actually fought or simply paid for others to take their place, were forgiven the sins they had confessed to a priest. One problem evident to contemporaries was that of equivalence: did the penalty meet the gravity of the sin? It was not until the 1230s that theologians devised a system, known as the 'Treasury of Merits', according to which the substitute for an earthly penance that might be deemed to have fallen short of the sin committed was notionally provided by the extra store of grace 'in the bank' from the inexhaustible goodness of Christ and the saints.

Furthermore, the grant of indulgences to those *crucesignati* who commuted their vows to money payments, and that could then be used to resource substitute crusaders, was controversial at the time (Throop 1940: 91). In fact, it was not Innocent's invention; he was merely trying to make sense of a precedent implied in a practice that had developed during the preaching of the Third Crusade. Later critics of the abuse of indulgences, thinking of caricatures such as Chaucer's pardoner, might see *Quia maior*'s policy of vow commutation as cynical. But, like so much of Innocent's policy, it was pragmatic in origin. Crusading as an *effective* activity was largely limited to the social élite: those who could afford to equip themselves and who could contribute in a militarily useful way to campaigns. But crusade preaching and recruitment was not limited to royal and noble households: especially from the early thirteenth century onward, preachers sought to recruit from among the growing urban populations. Research on crusade recruitment in England shows blacksmiths, tanners, cobblers and merchants taking the cross (Tyerman 1988: 168–75). In 1196 the pope instructed that *crucesignati* who had failed to fulfil their vow should be forced to do so. But Innocent III realised that it was impossible to force, for example, a poor blacksmith who had taken the cross as an emotional response to an eloquent crusade sermon, to fulfil a vow that should not have been accepted in the first place. Some of the English *crucesignati* who turned up in the records of the archbishop of Canterbury's survey into vow-breakers had set out but turned back when they ran out of money; others had become invalid or had families to support. Besides, although urban crusaders proved invaluable in the sieges of Lisbon in 1147, Acre in 1189–91 and Damietta in 1217–19, on the whole such crusaders could be a liability. Thus in 1200/1 Innocent ordered that the *crucesignati* identified as vow-breakers should instead redeem their vows for cash. In *Quia maior* this measure becomes official policy, by providing a procedure for commuting a vow to a money payment which was to be collected in a crusade chest placed in the parish church. Although officially vow redemption was still distinct from commutation, in practice the two were probably often confused.

In hindsight it is easy to see how this policy can be interpreted as the manipulation of people's sensibilities to encourage them to take vows that they were not in a position to fulfil; or, more cynically still, as the indirect taxation of the laity to fund crusades. But Innocent's policy was at the same time practical and profoundly idealistic. He was trying on the one hand to ensure that crusades could be made more self-sufficient, so that they would never again be blown off course as had happened in 1202–4. On the other hand, he wanted to make the crusade an activity controlled by the head of Christian society but in which the whole of that society could participate. If the loss of Jerusalem was the consequence of human sinfulness, then all Christians should play a part in its recovery. By taking the crusade vow, even non-combatants could affirm public support for the ideal, and by commuting it, they could support a campaign financially.

Innocent III's legislation had the effect of linking the ideal of crusading to financial realities. The attempt to sort out anomalies in previous practice gave crusading a book-keeping aspect; thus, for example, indulgences were graded in proportion to whether the vow was kept or commuted. The same attention to practical detail characterises *Ad liberandam*, the final – and lengthiest – decree of the Fourth Lateran Council (Tanner 1990: 267–71). In overall vision and grasp of detail, the conciliar decrees of Fourth Lateran were the most mature regulatory document yet produced by the medieval Church. They provided a blueprint for how Christian society should function, from baptism to last unction, from parish to archdiocese. Yet by drawing to a conclusion with the specific arrangements for the new crusade first referred to in *Quia maior*, Fourth Lateran makes it appear as though the entire programme for the reform of society, from union with the Greek Church to the visitation of monasteries, was to serve the end of the crusade. The decree begins by asserting that the recovery of the Holy Land is the 'ardent desire' of the Church. Although another decree of the Council offered the same indulgence to those combating heresy as to those fighting the Muslims in the Holy Land, it is – as traditionalists among crusade historians have pointed out – by reference to the Holy Land crusade that the scale and nature of indulgences is conceived (Tyerman 1998: 37–8; cf. Mayer 1972: 312–13). *Quia maior*, indeed, offered the indulgence for the Spanish Reconquista only to Spaniards and Provençals. This suggests that Innocent wished to subordinate, at least for the duration of this expedition, all other military activity against the enemies of Christendom to the crusade for Jerusalem.

Although the measures in *Ad liberandam* were specific to the new crusade, they called on earlier precedents. A general peace was declared throughout Christendom for four years, and tournaments were banned for three; bishops were instructed to check that *crucesignati* left by the declared departure date of 1 June 1216; crusaders' debts were to be deferred, and interest payments rescinded; materials of war such as timber and iron were prohibited from trade in Muslim-held ports, indeed all shipping to Muslim ports was to be suspended for four years. These measures were to be repeated in the crusade decrees of the First and Second Councils of Lyons in 1245 and 1274 (Tanner 1990: 297–301, 309–14), and featured in crusade planning well into the fourteenth century. More contentious in future crusade planning was to be the measure regarding taxation of the clergy. A rate of one-twentieth was levied for three years, in addition to which the papacy would itself contribute £30,000. The principle of clerical taxation having been thus enshrined in canon law, rulers planning crusades came to expect to be granted the faculty by future popes. The rate varied: in 1225 a tenth was levied in France but a fifteenth in England; in 1245 the rate was again a twentieth but in 1274 a tenth. Once again, Innocent was merely codifying recent precedent: in 1199 his legate Peter Capuano had persuaded the French bishops to pay a thirtieth in lieu of sending knights on crusade. But secular rulers had always looked to the Church for financial help for crusades, sometimes heavy-handedly. Taxation meant that systems had to be put in place to collect the money and gifts in kind. Innocent III envisaged this being done by bishops, but it was not long before the task was allocated to papal legates. In England at least, this caused bitter resentment, as for example in 1236 when the legate was assaulted by students in Oxford (Luard vol III 1876: 481–4).

Innocent III's crusades were, on the face of it, a failure. This is not to say that they achieved little, but that their achievements were rather different from what he had expected. In both the eastern Mediterranean and the Languedoc, the principle of Christian unity was enforced at the point of the sword, rather than by voluntary submission to papal authority. He found few takers for the indulgence offered to resist the Hohenstaufen in Sicily (see Chapter 8, below). The crusade heralded by *Quia maior* and *Ad liberandam*, that he did not live to see, despite painstaking planning and initial success in the end accomplished nothing. Yet Innocent's contribution to crusading is immense. In letters to bishops and friends he articulated a pastoral and devotional theology based on the Holy Land (Bolton 2000: 154–80). His appointments re-energised the Latin church in the Crusader States. Above all, he clarified existing practices in preaching, recruitment and financing crusades. Whether through idealism or in reaction to practical problems, Innocent succeeded in making the crusade a central feature of the daily life of ordinary Christians. He may be criticised for opening the way for abuses; nevertheless, in the century after his pontificate far more people took the cross and far more money was raised for crusading than in the previous hundred years.

CHAPTER 8

Varieties of crusading from the eleventh to the thirteenth centuries

Crusading took many forms in the Middle Ages. The success of the First Crusade had an important effect on the way in which Europeans perceived the campaigns of reconquest in Spain, which began to assume an overtly spiritual character. Similarly, territorial wars of expansion against pagan peoples in the Baltic came to be seen as crusades and, in the thirteenth century, so did the wars of attrition fought by the popes against political enemies in Italy. After the Fourth Crusade, the defence of the Latin Empire of Constantinople also attracted crusading privileges. This chapter examines each of these examples as a case study and offers interpretations of the similarities and differences between them.

Crusades and holy wars

Even while preaching the first crusade in 1096, Urban II realised that his use of holy war against the Turks in the East could have similar application elsewhere. As we have seen, the late Roman idea of the just war was revived by the papacy in the later eleventh century in a political climate – conflict with the Empire – that required the threat, if not the actual use, of force. Alexander II sanctioned the invasion of England in 1066 on such grounds; Gregory VII allowed that the bearing of arms could in certain circumstances act as a penance; and in 1089 Urban II offered the campaign against Muslim-held Tarragona, in Catalonia, as a substitute for penance (Mayer 1972: 29–32; Bull 1993: 97). In 1096, Urban discouraged Catalan counts from leaving their own territories to join the armed pilgrimage to Jerusalem: fighting the Muslims on their own doorstep was just as efficacious a means of doing penance (Riley-Smith and Riley-Smith 1981: 40).

The logic of the holy war implied a wide diffusion of actions against the enemies of Christ, wherever they might be found. Opposition in principle to such diffusion – for example, against heretics in Italy and southern France, or pagans in northeastern Europe – was limited, perhaps because contemporary opinion had during the course of the eleventh century already become accustomed to the idea that a holy war, with some form of spiritual privilege, could reasonably be applied to a wide range of opponents. In the cases of the *Reconquista* in Spain and the crusades against the Baltic peoples, holy war served the needs of an existing drive to conquest and the colonisation or exploitation of lands occupied by non-Christians. In the case of the so-called 'political' crusades of the thirteenth century, the theory of the crusade was put to the test in service of papal claims to political sovereignty in Christendom.

Were these campaigns 'crusades' in the same sense as was understood of the expeditions to the East? Did those who undertook them assume that they were engaging in the same meritorious process, and did the leadership of the Church regard them as equal in merit? I leave questions of definition aside for two reasons: first, because contemporary attitudes to crusading were never frozen into a universally understood set of terms – the terminology of crusading was inexact and fluid, at least until the late thirteenth century – and second, because the arenas of conflict in which crusading terminology came to be applied were geographically and historically different. What marked out the crusades discussed in this chapter from those in the rest of the book was that they were either existing conflicts, or potential conflicts that would in all probability have continued, even if the events of 1095 had never taken place. The forms of such conflict, the rhetoric and expectations of participants may have differed, but they were all essentially territorial wars that came to be defined by the application of a theological rationale. The purpose of this chapter is twofold: first, to demonstrate how the language and ideas of crusading could be manipulated both by the papacy and by those doing the fighting; and second, to compare the results of crusading with respect to settlement and institutions of government in different regions.

The *Reconquista* in Spain, *c.*1057–1212

Before the First Crusade, the Iberian peninsula was the only part of Latin Christendom to have endured lasting conquest by the Islamic world. But although towns as far north as Barcelona and Santiago were sacked at the end of the tenth century, the period of conquest was largely over, and the political map fixed, by *c*.900. The dominant power in Spain was the khalifate of Cordoba, whose influence stretched into the foothills of the Pyrenees. Most Christians in Spain lived under Islamic rule; only in the valleys of the Pyrenees and the Basque regions of northwestern Spain and Galicia did Christians exercise authority. Although these tiny regions – and especially the kingdom based in Oviedo under Alfonso III in the 880s – began to engage in aggressive warfare against the khalifate, it was the internal collapse of the khalifate of Cordoba in 1031, and the subsequent political vacuum in Spain, that made the *Reconquista* possible.

In the wake of the khalifate, independent Muslim states, known as 'taifas', emerged based on the major cities, such as Zaragoza, Valencia, Seville, Toledo and Badejoz. Because they had no institutional legitimacy in Islamic law, but relied on the strength of dynastic rulers, they were vulnerable as much to internal discontent as to external threat. Thus a system developed in the eleventh century according to which taifa rulers paid subsidies, known as 'parias', to stronger or more aggressive neighbours for protection – both from the neighbour himself and from other rivals, whether Christian or Muslim.

The Christian powers of northern Spain in the mid-eleventh century – Leon, Navarre, Aragon and the counties of Catalonia – though tiny in size, had nevertheless developed a military aristocracy and a military technique that made them dangerous to their larger Muslim neighbours. They were essentially geopolitical structures, with no basis in tribal or ethnic distinctions between Christian peoples under their rule. They could be divided and reunited, as happened to Leon–Castile in the 1060s, or partitioned, as happened to Navarre. As elsewhere in early medieval Europe, kingship in Spain was fixed on the personality and possessions of a ruling dynasty.

By 1054, the early hegemony among the Christians of Navarre under Sancho Garcia III (1000–35) had given way to that of the kingdom of Leon–Castile, united by his son Fernando I. This was the first Christian power with the military capacity to intervene in Islamic affairs. In 1063 his raid into al-Andalus induced the taifas of Seville and Badajoz to begin paying him parias; such payments, naturally, enhanced his own military capacity. Fernando's heir, Alfonso VI, continued to raid into Islamic territory in order to acquire the wealth that would enable him to dominate his Christian rivals. Thus, for example, Granada paid him an annual tribute of 10,000 dinars. As the taifa ruler of Granada, 'Abd Allah, observed: 'I knew that the payment . . . protected me from his misdeeds, and was

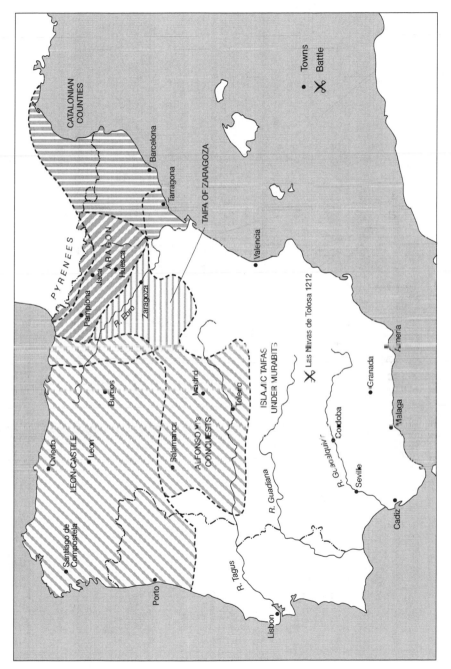

Iberian peninsula during the Reconquista, c. 1109

preferable to incurring losses and the ravaging of my country' (cited in McKay 1977: 18). The paria system enabled a Christian hegemony to be paid for out of Muslim wealth. In 1074, Alfonso VI enjoyed an annual income of 74,000 dinars from parias – incomparably more than could have been acquired from taxation of his subjects or the income from crown estates.

In 1076, two events assured Alfonso VI's supremacy among the Christians: the murder of Sancho Garcia IV of Navarre, who was pushed off a cliff in a bid for the throne by his brother and sister; and the parias paid by Zaragoza in return for colluding with that taifa's conquest of Valencia. The resources in coin and land, including the fertile Rioja valley taken from Navarre, enabled Alfonso VI to campaign in the south again in 1079–80, and to extend a protectorate over the taifa of Toledo. In 1083 he raided Seville, symbolically riding his horse into the Mediterranean to demonstrate his intended mastery of the peninsula. The surrender of Toledo in 1085, by arrangement with its ruler, al-Qadir, demonstrated not only his military strength but also his ability to redraw the political map, for al-Qadir was imposed on Valencia as a compensation, thus interfering with Zaragoza's conquest.

The annexation of Toledo not only increased Leon–Castile's size, but also added the largest and most cultivated city in Spain. But it had the effect of frightening the other taifas into inviting the Murabit leader Yusuf from North Africa for their protection. Yusuf, who enjoyed an unofficial authority as a spiritual as well as a political leader, and his Murabit successors, came over the next half-century to regard Spain as a natural extension of their African territory. The introduction of the Murabits instilled a sense of Islamic identity into the taifas' resistance to Christian expansion. In 1086 Yusuf defeated Alfonso, in 1093–4 his cousin conquered the taifa of Badajoz, and in 1102 the Murabits repulsed the Catalans at Lérida and captured Valencia.

Alfonso VI's death without a male heir, and the civil war that erupted as a result of the marriage between his daughter Urraca and Alfonso I of Aragon, meant that Leon–Castile was, for the first quarter of the twelfth century, eclipsed as an active proponent of *Reconquista*. The struggle for power demonstrates how far the Christian kingdoms were from any sense of a united front against Islam in Spain, even along purely territorial lines. But it also shows how resilient Leon–Castile had become by 1100 that it was able to withstand such internal pressure without succumbing.

Meanwhile the initiative passed to Aragon. From its origin as a tiny state based on the Pyrenean town of Jaca, Aragon inched south through the valleys. If the state's independent survival was assured by the acquisition of Pamplona as its share of the partition of Navarre in 1076, the conquest of Huesca by Peter II in 1096 sealed a generation of accumulation and consolidation. The conquest doubled Aragonese territory and provided it with a city of some 2,000–3,000 souls – about three times the population of Jaca, Pamplona, or for that matter Leon. Alfonso I, 'the Battler', assured Aragonese greatness with his conquest of Zaragoza in 1119. Huesca doubled Aragon's size, but Zaragoza, together with its satellite cities such as Tudela, engorged it. The resources of Zaragoza made possible an extended *razzia* (raid) deep into al-Andalus in 1125–6.

Alfonso's death in 1134 led to an implausible but ultimately successful merger with the county of Barcelona. Alfonso had been married briefly and disastrously to Urraca of Leon–Castile, but the marriage had no issue, and he neither remarried nor produced any illegitimate heirs. In an astonishing gesture, he left his kingdom jointly to the Holy Sepulchre in Jerusalem, the Hospitallers and the Templars. This impossible will was never enforced; instead, Alfonso VII of Leon–Castile grabbed Zaragoza, while Garcia Ramirez, a descendant of Sancho I, revived a tiny kingdom of Navarre based in Pamplona. But the Battler's younger brother, Ramiro Sanchez, who had become a monk at Huesca, emerged from his monastery, contracted a marriage with a daughter of William IX of Aquitaine, begat an heiress, Petronilla, and betrothed her to the heir of Barcelona, Ramon Berenguer IV. The dynastic future of his kingdom assured, he retired once more to his monastery, and was heard of no more.

The union of Aragon and Barcelona meant that in 1137, for the first time, there was now a Christian kingdom able to compete on equal terms with Leon–Castile. The county of Catalonia itself was more or less the creation of Ramon Berenguer, who moulded together a number of old Carolingian counties through marriage, diplomacy and fortunate bequests. At times it looked as though Catalonia's orientation would lead north and east, to Toulouse and southern France, rather than south, but Alfonso I's conquest of Zaragoza freed the cities of the Ebro plain for conquest or the payment of parias to Barcelona. Meanwhile an independent kingdom emerged in the west. The county of Portugal, consisting of the southern part of Galicia and Oporto, and briefly of Lisbon, was detached from Leon–Castile by Alfonso VI and bestowed on his daughters Urraca and Teresa and their spouses, Raymond and Henry of Burgundy. After Urraca's disputed accession to the throne of Leon–Castile, Teresa managed to cling to enough of Portugal to pass on to her son Afonso Henriquez, who eventually assumed a crown (Reilly 1992: 201–4).

On the eve of the Second Crusade, therefore, three kingdoms had become dominant enough to represent Christian expansion in Iberia. Up to this point, the war of conquest had only fleetingly been coloured by anti-Islamic ideals. This is not to say that the *Reconquista* was not justified to contemporaries by religious rhetoric. Alfonso III of Leon had in the ninth century seen his war against the Muslims as part of the duty of a Christian monarch. But the ideal to which he aspired was the return of a dimly remembered Visigothic past, of which the expulsion of Islam was a by-product. In the eleventh century, when the *Reconquista* began to become a reality, Christian kings fought and schemed against each other quite as much as against the taifa kings. A consensus has now emerged among historians against the orthodoxy of an earlier generation of Spanish scholars, such as Menendez Pidal, who believed that the defeat of Islam in Spain was the enactment of a Spanish religious and national destiny (Fletcher 1987: 31–3; Linehan 1992: 1–21).

A particularly useful example in this argument has been the career of Rodrigo Diaz Vivar, known as the Cid. Menendez Pidal, a literary scholar, interpreted the *Poem of the Cid*, a late twelfth-century source, as though it bore witness to the conditions of the eleventh century. The Cid, in his hands, became an authentic national hero of the *Reconquista*. In reality, however, the Cid was an ambitious operator of a type particularly recognisable in the eleventh-century Mediterranean, comparable to Bohemond or Roger I of Sicily (Fletcher 1987: 36). Exiled from Leon–Castile by Alfonso VI for exceeding his authority, the Cid took service with the taifa king of Zaragoza before conquering and establishing himself as autonomous ruler of Valencia. His achievement was based on the ruthless manipulation of religious and political rivalries, coupled with a tactical brilliance in warfare. There is very little in the *Poem*, in fact, that suggests any religious inspiration behind the Cid's self-advancement. He was an outstandingly successful warlord in a golden age for freebooters in Spain, and the wealth he amassed from war is admiringly described in a passage in the *Poem* where he filled a boot with gold and silver and sent it as a gift to Santa Maria, Burgos, for masses to be said for his soul (Merwin 1959: 94–5).

Nevertheless, the eleventh-century period of the *Reconquista* opened the way for a series of wars that did come to be defined by crusading language, ideals and methods. Geography helped in some respects. Jaca, in Aragon's heartland, lay at the foot of the only pass in the western Pyrenees negotiable for pilgrims en route to Santiago de Compostela. Aragon and Catalonia in particular were open to cross-Pyrenean influence. Whether Cluny, through its network of daughter-houses along the pilgrim route, was really as responsible for promoting holy war ideals as historians once thought, is far from certain (Cowdrey 1973: 285–311, Constable 1997: 179–93); however, there is no doubt that Christian Spanish rulers were susceptible to monastic reforming initiatives. As early as 1149–53, Fernando I granted 1,000 dinars annually to Cluny, while Saint Victor, Marseilles was a beneficiary in Catalonia, and Saint Pons de Thomières in Aragon. Personal contacts were crucial: for example, Bernard, archbishop of Toledo after 1085, was a Cluniac monk, and Alfonso VI's second wife was the niece of Abbot Hugh of Cluny (Reilly 1992: 66–7; but Linehan 1992: 174–5, 260–1 is more sceptical). But Roman influence was probably more critical. In 1068, Sancho Ramirez, while on pilgrimage to Rome, offered his crown to the pope in a gesture of vassalage, one result of which was the replacement of the Mozarab rite by the Roman in Aragon, to be followed eventually by the same process in Leon–Castile. Alongside reform came the influence of papal interest in holy war. Alexander II encouraged Norman participation at the siege of Barbastro in 1064; Gregory VII proposed an expedition to Spain to be led by Ebles, count of Roucy, whose sister Ramiro Sanchez had married, in 1073; although this came to nothing, French knights fought at Tudela in 1086, and the capture of Huesca, in the same year as the first crusaders left for the East, was aided by the presence of the archbishop of Bordeaux and other southern French bishops.

Although French knights had been involved in the *Reconquista* for a generation before the First Crusade, no crusade as such was preached in Spain until 1118, and even then the campaign in question, the siege of Zaragoza, would have taken place in any case. It could be argued that the involvement of French knights is no more significant of proto-crusading ideals than the activities of Normans in southern Italy: there was land and booty to be won, and kinship loyalties to be observed. But the participation at Zaragoza of knights like Gaston of Béarn, who had been on the First Crusade, suggests that contemporaries increasingly saw the *Reconquista* as an equivalent exercise, while that of Rotrou of Perche, a Norman, shows that its appeal was more than regional. Calixtus II's bull of 1123, emphasising that the *Reconquista* was to be seen as identical to wars against the Muslims in the East, may have been intended to clarify confusion among contemporaries as to the status of the Spanish wars (Fletcher 1984: 297–8). The presence of Spanish clergy at papal councils is as significant as that of French knights in Spain; particularly so in the case of Oleguer, bishop of Tarragona, who was at Toulouse and Rheims in 1119, First Lateran in 1123 and Clermont in 1130, and who instituted the Templars in Catalonia in 1134 (Fletcher 1987: 43).

It was the Second Crusade, however, that located the *Reconquista* firmly within the wider crusading context. The archbishop of Toledo was in Rome in 1145 when news of the fall of Edessa arrived, and may have been at Vezelay in 1146 when Alfonso Jordan, count of Toulouse, who was Alfonso VII's half-brother, took the cross. Eugenius III named Iberia as a target for crusade preaching in 1147, and both Leon–Castile and Aragon–Barcelona sought Genoese help for the conquests of Almeria in 1147 and Tortosa in 1148. But the most spectacular result of crusading in Iberia was the capture of Lisbon by Afonso Henriquez and a mixed group of Flemish, German and English crusaders. Most historians have attributed it to the coincidence of crusaders en route for the East by sea arriving in Portugal and being persuaded by the bishop of Oporto to take service under the king, but Phillips (1997: 485–97) and Livermore (1990: 1–16) have shown that Afonso Henriquez's campaign had been endorsed by Bernard of Clairvaux, and that Bernard had probably recruited crusaders from Flanders and Germany for this venture during his preaching tour in 1146. The eye-witness account of the siege, *De expugnatione Lyxbonensi*, uses language strikingly similar to that of Eugenius III's bull *Quantum praedecessores* in the regulations made for the conduct of crusaders, and Cistercian influence may be detected in the provisions for weekly confessions and chapters for each shipload (David and Phillips 2001: 56–7). Not all historians are convinced: Tyerman, for instance, argues that the *De expugnatione* lacks ideological coherence and resorts to old Augustinian justifications for warfare (1998: 17). Such distinctions are important, for if we accept that the siege of Lisbon was an integral part of the crusade preached in 1146, then there is no reason not to see Bernard's preaching of the Danish campaign against the Wends in the same light, and we are thus presented with a three-pronged attack against the enemies of Christendom, of which the campaign to the east was only a part. This, in any case, was the effect that a contemporary, Herman of Bosau, observed (Constable 1953: 213–79), though there is room for scepticism about the extent to which such a strategy was planned by the papacy, rather than being the result of Christian rulers jumping on a crusading bandwagon.

The death of Alfonso VII (1157) and the increasing involvement of Aragon–Barcelona in the Languedoc in the later twelfth century meant that the *Reconquista* ran out of steam. Enough had been done in the previous century, however, to ensure that the Christian kingdoms would remain dominant: the forces of Islam could win battles without being able to occupy territory, whereas Christianity could suffer losses without having to yield territory. The Christian victory at Las Navas de Tolosa (1212), which assured Christian hegemony over the peninsula, was the result of a crusade preached by Innocent III and promoted by Alfonso VIII of Castile, but in which Sancho VII of Navarre, Peter II of Aragon, the lay rulers of Provence and the archbishops of Bordeaux and Narbonne also participated. The crusade was preached in northern France, Italy and Germany as well as Spain and Provence, and almost 70,000 non-Spanish crusaders appeared (O'Callaghan 2003: 66–76). The internationalism of the campaign, however, while on the surface it appeared to place the *Reconquista* within the papal concept of crusading, also revealed the particularism of the Spanish wars against the Muslims and the different conception of the crusade by Spanish and northern crusaders. The French, applying the same rules to this campaign as to any crusade, were disgusted by the toleration shown by the Spanish to their defeated Muslim enemies and their possessions, and deserted en masse before the final victory (O'Callaghan 2003: 71).

The dissension at Las Navas points to one distinctive feature of the *Reconquista*: it was a war of domination over, rather than extermination of, Islam. Christian rulers were always prepared to allow a degree of self-determination to the defeated Muslims, because their aim was never to kill or expel them, but to rule over them. On the whole, this was also true of the Franks who settled in the Levant, though not necessarily of western crusaders to the east. Territorial conquest brought with it new lands and new wealth from captured towns, but also new populations that had to be absorbed into unfamiliar structures – in twelfth-century, Aragon, for example, Muslims comprised 30–35 per cent of the population (McKay 1977: 37). The *Reconquista* essentially was the victory of a small rural society based on a military aristocracy over a largely urban one. Despite its relatively large number of towns, Spain was lightly settled even in the thirteenth century. Unlike England and France, it did not experience a dramatic increase in population and therefore there was little incentive to bring marginal lands into cultivation or to develop more efficient farming methods so as to exploit more land. Because manpower was always short, rulers offered privileged conditions of settlement in newly conquered areas: in the 1170s, for example, settlers in the Ebro valley were offered exemption from taxation by the Aragonese crown. This meant that the *Reconquista* extended the conditions of a frontier society over increasingly large parts of the peninsula. Most of the colonists who settled in newly conquered regions, for example, were smallholders, and Spain thus retained the character of fairly small estates held 'in chief' of a lord without becoming fragmented by the demands of land shortage. The emergence of a Christian nobility in the north was also to some extent shaped by the demands of the frontier. Because war was endemic, military service was one route to nobility. One of the themes expressed in the *Poem of the Cid* is social mobility: after the fall of Valencia to the Cid, 'Those who had been on foot now became *caballeros*. And the gold and silver – who could count it?' (Merwin 1959: 128–9). The Cid himself, though of noble birth, was not of the highest nobility, but proved the greater value of knights who earned nobility through military service than worthless heirs of great families.

Just as the crusader states in the East had their indigenous Christian populations, reconquered Spain had its Mozarabs – Christians who lived under Islamic rule. On the eve of the First Crusade, these greatly outnumbered those living in the Christian kingdoms. The Mozarabs, largely isolated from religious developments across the Pyrenees, had developed a monastic tradition, liturgy and art of their own. One aspect of the *Reconquista* was the relentless imposition of the Roman rite on the Spanish Church, a process that accompanied military aid. By the 1070s Leon–Castile, Aragon and Catalonia had all accepted the Roman rite, and despite the resistance of the Mozarabs of Toledo, the appointment of a Cluniac, Bernard, to the archbishopric after its conquest by Alfonso VI made the further extension of Roman conformity inevitable. As the territorial expansion of the north into the rest of the peninsula acquired a more religious character, colonisation tended to be accented by the rhetoric of reform. Thus one of the beneficiaries of the cheap new lands available in the twelfth century was the Cistercian Order, whose first foundation at Fitero in 1140 was swiftly followed by two others in Catalonia and another in Portugal before 1153. The Cistercians, and likewise the Military Orders, who thrived as frontier settlers, were a conduit for the flow of papal influence, and thus for the absorption of the Mozarabs into mainstream Roman Christianity (Linehan 1992: 261–4).

The *Reconquista* continued to attract crusaders and crusading privileges until the end of the Middle Ages. Despite the virtual dominance enjoyed by James I of Aragon and Frederick III of Castile over the peninsula in the thirteenth century, the *Reconquista* remained incomplete. Quasi-feudal status was imposed on the remaining Muslim rulers of al-Andalus, a region that could not be subdued by military conquest until the fifteenth century. Indeed, al-Andalus remained well into the fourteenth century a foothold from which the Marinid rulers of North Africa could still cause concern in Spain, as for example in 1275, when Alfonso X of Castile was defeated by Abu Yusuf Ya'qub, and the siege of the Muslim city of Algeciras in 1342–4 attracted crusaders from all over Europe, at a time when recruitment for crusades to the East was proving more difficult.

The 'political' crusades of the thirteenth century

The label 'political crusade', though it enjoys widespread currency among crusader historians, begs certain questions about how we, and about how contemporaries, understand the term. As has been remarked, all crusades were political, in so far as they all aimed at the conquest of territory (Strayer 1962: 343–4). Moreover, a sharp distinction between one 'type' of crusade and another is impossible for the period before crusading had an institutional apparatus. However, some holy wars preached or sponsored by the papacy from the eleventh to the fourteenth centuries were ostensibly 'political' in the sense that they had aims that could only by some rhetorical justification be said to fulfil the spiritual mission of the Church. Popes from Gregory VII onward sometimes made such justifications, and offered spiritual incentives for those who engaged in such wars on their behalf – what else, after all, could they offer? As the conceptual and legal apparatus of crusading developed, it was natural that those spiritual incentives should conform to the privileges offered to crusaders. It was not until the very end of the twelfth century, however, that a pope for the first time made an explicit parallel between a war in defence of papal rights and war against the Muslims.

Popes, like all rulers, maintained their political position through a series of alliances. Throughout the twelfth century, they could normally count on the opposition of the emperor to the Norman kings of Sicily, whose legitimacy they denied. But Frederick Barbarossa had married his heir, Henry VI, to the king of Sicily's sister Constance. The match was probably intended as little more than a diplomatic gesture, but William II of Sicily died in 1189 and Henry pressed his claim to Sicily in his wife's name, a claim given greater weight by the birth of an heir, Frederick, in 1194. It must have looked in 1197 as though Henry's premature death had left all the cards in the hands of the papacy (Sayers 1994: 49); but the emperor had shown great foresight. In a masterstroke of *realpolitik*, he left his heir as a ward of the papacy. By recognising that Sicily was a papal fief, he publicly forced Innocent to defend Frederick's interests, thus reversing the support that Innocent's predecessor had given to the rival claimant Tancred of Lecce. As Sayers has observed (1994: 83), Innocent had little choice but to behave as a feudal overlord, which meant mobilising armies in defence of his ward against Henry VI's regent, Markward von Anweiler.

Markward invaded Sicily in 1199 and took the young heir under his protection – in fact, as a hostage against papal attempts to impose overlordship. It was in some ways analogous to the struggle between the papacy and Barbarossa in northern Italy in the 1160s–70s, in so far as the papal aim was to prevent the extension of imperial influence in Italy. In a letter of 1199, Innocent threw the weight of crusading rhetoric into the struggle, denouncing Markward as 'another Saladin' – on the grounds that he had won over the Muslims of western Sicily – and claiming that he was hindering efforts to mount a crusade to the East (Kennan 1971: 231–51). Some troops, under the leadership of the French noble Walter of Brienne, were recruited under a form of plenary indulgence. Although the expedition came to nothing in the end, an important precedent had been set in Innocent's willingness to apply the language of crusading to conflict in an arena that – despite his sophistry – really had nothing to do with the expeditions he had launched against the Muslims in 1198.

It is instructive to contrast the speed with which Innocent adopted crusading parallels in the campaign against Markward – even if in fact he used the 'barest minimum' of resources (Strayer 1962: 346) – with the length of time that elapsed before he preached a crusade against the Cathars and their supporters: in 1209, after all, the Cathar problem had been recognised for thirty years. Innocent regarded control over territory in Italy, and suzerainty over Sicily, as essential for the survival of the papacy as a political institution. The case of Markward, however, was more significant as a precedent than for anything it achieved; and that precedent was not to be followed for forty years.

Although Gregory IX in the 1230s raised armies to clear the papal territories of central Italy of ghibellines (imperial supporters), he stopped short of using the same terminology as that for the crusade to the East, or even of offering the same indulgence. Frederick II was guilty of threatening the Church's liberty, and thus, in the papacy's view, lost his claim to the fidelity of its subjects, but the pope did not deny the emperor's right to rule over the lands for which he had been anointed. It is significant that the first mention of a crusade came only in 1240, when Frederick's army threatened Rome itself (Strayer 1962: 351). Crusade preachers were then commissioned in Milan and Germany, and in 1241, Gregory specified that the same indulgence be offered for those who took the cross against the Hohenstaufen as for those who went to the Holy Land. Hungarians who had taken the cross were even told not to go to the East, but to join the papal armies instead (Strayer 1962: 352). The defence of Rome against an aggressive emperor could be identified as a crusade, because crusading rhetoric from Clermont onward, and canon law in its wake, had always seen the wars against the Muslims as essentially defensive wars to protect land that in theory belonged to Christians. If the Holy Land was the patrimony of Christ, Rome was the patrimony of St Peter. Both Gregory IX (1227–41) and Innocent IV (1243–54), moreover, argued that any lay ruler who hindered, or even refused to help, the 'business of the Holy Land', was guilty of a crime against Christian society itself (Weiler 2003: 9–10).

How did contemporaries receive such arguments? In England at least, there was opposition to papal demands for subsidies from clerical incomes to pay for the crusade, and the fact that arrears were still arriving several years later is as eloquent as Matthew Paris' reasoned rejection of the papal case against the emperor (Siberry 1985: 132–8). It is unlikely that the grant of a twentieth promised by the French clergy all reached its intended destination (Strayer 1962: 353), and in Germany, where crusade preaching was targeted most extensively, the emperor forbad the clergy to pay anything. At first sight it seems curious that crusade preaching was concentrated so heavily in Germany, but it was here that lay the papacy's best chance of loosening Frederick's grip on the imperial throne. After the First Council of Lyons (1245), the papal legate preached successfully in the northwest of the Empire, where the anti-king William of Holland was active. Italian participation was determined by local and feudal relations, while Louis IX's crusade to Egypt, for which he took the cross in 1244, preoccupied potential crusaders from France.

At the root of the popes' problems throughout the 1240s, however, lay not so much opposition in principle – although this was certainly expressed by some – but rather the unwieldiness of the crusade as a means of winning the war against the Hohenstaufen. This was a war of attrition, in which Frederick had no intention of committing himself in the field against a papal army, but preferred to reduce the guelf (papalist) towns in central and northern Italy one by one (Waley 1961: 142–53). Individual campaigns, such as William of Holland's capture of Aachen in 1248, featured *crucesignati*. But crusade preaching had always presupposed a campaign with clearly defined limits and a firm objective, and the war against the Hohenstaufen had neither. Moreover, Innocent IV wavered about the relative place of his war against Frederick in the wider context of crusading. Although canon lawyers regarded the crusade against internal enemies of the papacy as equally valid to those against Muslims (Russell 1975: 205; Tyerman 1998: 38), Innocent was reluctant to let crusading vows taken anywhere outside Germany – even by William of Holland's subjects in Frisia – be commuted to service against the Hohenstaufen.

The crusade launched by Innocent III against the Cathar heretics set a different precedent, but one that was to become relevant to 'political' crusades. The Third Lateran Council of 1179, which sought to tackle the increasing problem of heresy, offered indulgences to those who defended Christendom against heretics (Tanner 1990: 216). The experience of the Albigensian Crusade taught the Church that it was easier to attack those who supported or provided a home for heretics than it was to identify and make war against heresy itself. Thus, for example, in 1212 Innocent III threatened the use of force against the Milanese on the grounds that they had let Cathars have free rein in the city, just as in the Languedoc the lands of lords who failed to prevent the spread of Catharism were subject to confiscation. In Italy, as Housley has observed, there was a strong link between heterodoxy and anti-papalism, with the result that heresy often had political overtones (1982a: 193–4). Heretics, persecuted by the Church, naturally found common cause with ghibelline opponents of the papacy, while in turn the imperial cause became identified by guelfs as 'heretical', or supportive of heresy, even though the Hohenstaufen themselves were entirely orthodox. The ghibelline Ezzelino of Romano, for example, pursued blatantly anti-clerical policies, seizing Church property and taxing the clergy. He also defied the Inquisition and harboured Cathars, though there is no evidence that he himself was a Cathar. But the pope could treat him as one on the grounds that whoever was not with him was against him (Housley 1982a: 199–200). In the 1230s indulgences were offered to groups of laity who had formed into anti-heretical confraternities, such as the Society of the Blessed Virgin in Milan, or the Militia of Jesus Christ in Parma. In Florence in 1245, the Dominican inquisitor turned his pursuit of Cathars to the political advantage of the papacy by citing the ghibelline *podestà* of the city as a defender of heretics; in defending him, a ghibelline mob thus aligned itself with the Cathars.

The crusade against the Hohenstaufen did not end with Frederick II's death in 1250, but it changed significantly in nature. After 1254, when Frederick's heir Conrad IV died, the focus of attention turned to Sicily, where the emperor's illegitimate son Manfred was proving an effective ruler. Now it was possible to preach a crusade with a clear objective – the conquest of Sicily. Full crusading indulgences were extended to support the offer of the throne of Sicily to the junior members, first of the English, then the French royal houses. In 1255 Henry III was allowed to commute his crusading vow for the Holy Land to the campaign against Manfred. In the same year Pope Alexander IV preached a crusade against the ghibelline Ezzelino of Romano in Lombardy.

In contrast to the sparing use of the crusade indulgence by Gregory IX and Innocent IV, subsequent popes turned to it instinctively. Urban IV, who in the words of one historian 'perfected the technique of the political crusade' (Strayer 1962: 363), was particularly lavish. In 1263 he offered a full crusading indulgence to Charles of Anjou – and with it the fruits of a tenth on French clerical incomes – for his conquest of Sicily; in the same year he also preached crusades against ghibellines in Sardinia, and against the Byzantine Empire. After Manfred's death in battle against Charles in 1266, the crusade was preached once again in 1268 when Frederick II's grandson Conradin tried to claim his Italian inheritance. Yet it is important to note that in the papal discourse promoting these campaigns, the Holy Land still has a central place. Even if it were treated as empty rhetoric – and it is far from certain that it was – Popes Alexander IV in 1255, 1259 and 1263, Gregory X in the 1270s and Boniface VIII in 1302 justified such use of the crusade indulgence in support of political protégés on the grounds that it benefited the cause of the recovery of the Holy Land (Weiler 2003: 10–11).

The extension of crusade privileges into the arena of secular politics made it possible for rulers to carry out their political ambitions under the guise of a crusade. Because a ruler who took the cross was granted a subsidy from clerical incomes in his kingdom, there was great incentive to do so. Charles of Anjou proved especially skilled at using crusading vows as an instrument of policy. From his conquest of Sicily in 1266 onward, he tried continually to use his crusade vow to mount an invasion of the restored Byzantine Empire. Between 1271 and 1280 his ambition was stifled by the insistence of Gregory X and Nicholas III that the Holy Land should take priority, but with the election of a French pope, Martin IV, in 1281, he was granted a crusading privilege for his conquest. Officially, this was for a war in defence of the Holy Land, but at least one contemporary observer, the Sicilian Bartholomew of Neocastro, was not convinced, sneering that the cross Charles bore was not Christ's, but that of the unrepentant thief (Strayer 1962: 369–70).

Perhaps the most far-fetched use of the crusade privilege in support of a political objective came about as a result of the revolt against his rule, known as the Sicilian Vespers, that ended Charles' dream of a new Mediterranean empire. The revolt, carried out by the largely Greek-speaking Sicilians, was fomented by a secret alliance between Michael VIII Paleologus, the Byzantine emperor, and Peter III of Aragon, who had a claim to the Sicilian throne through his marriage to Manfred's daughter (Runciman 1958: 201–13). In retaliation for the revolt, Pope Martin IV preached a crusade against Aragon, which was undertaken in 1285 – disastrously, as it turned out – by Philip III of France. The purpose of the crusade, from the papal point of view, was to prevent Peter from seizing the Sicilian throne, and thus to protect the papal protégé, Charles. The rationale was that only if his throne were secure could Charles' proposed crusade to the East be salvaged, and thus the crusade against Aragon was, indirectly, a contribution towards the longed-for crusade to save the Holy Land. But although Philip III seems to have taken the crusade quite seriously as a religious obligation, there was by 1285 no longer any prospect of an Angevin crusade to the Holy Land – if, indeed, there had ever been one – because all Charles' resources were needed in the attempt to save his throne.

The result of the papacy's use of the crusade to further its aims in Italy during the course of the thirteenth century must be analysed on different levels. The Sicilian Vespers produced a kingdom of Sicily partitioned between Angevins and Aragonese, with the result that there was no single power in Italy strong enough to threaten the papal states. Especially when combined with the turmoil that followed the end of Hohenstaufen rule in Germany, this was an outcome that would have satisfied papal policy since Innocent III. Financially, however, the 'political crusades' were a disaster, for once the precedent had been set for kings who had taken the cross to collect subsidies from clerical incomes, there was nothing to prevent a king from taking the cross, collecting the tax and using it for whatever immediate purpose he needed, then regretting that he had run out of money and was unable to fulfil his vow. The original idea of funding crusades through clerical taxation was sensible, but it depended for its credibility on transparency – not only kings, but popes also were prone to raising money for one purpose and spending it on another. More intangibly, papal authority was weakened as a result of popes' involvement in warfare in Italy. The 'political crusades' recruited well on certain occasions, but they were generally unpopular if only because the cause of the Holy Land remained popular, and critics could see that the one detracted from the other. By c.1300 the German Empire could no longer claim leadership of Christendom, but its heir was the crown of France, not the see of St Peter. When attempts were made in the first half of the fourteenth century to launch crusades against the Turks, it was the French king who pulled the strings.

The northern crusades

The first indication that the papacy would come to regard the wars waged by Scandinavians and Germans against the pagan peoples of the Baltic coastal regions as crusades came in 1147, when St Bernard persuaded Pope Eugenius III that the German campaign against the Wends was a holy war for the conversion of non-Christians and the extension of Christendom, and thus analogous to the situation in Spain. This was in effect the recognition of a *fait accompli*. Bernard had found, while touring Germany to preach the crusade to the East, that the northern Germans were much more interested in the Wendish campaign than in Edessa, and the bull *Divina dispensatione* thus looks rather like a rationalisation of the refusal of some Christian knights to toe the line, by finding a religious justification for a war of territorial conquest that would have taken place in any case.

Such justifications were far from new; nor were the wars. Carolingian and Ottonian attempts at the conquest of pagans to their northeast had been accompanied by mass conversion of the defeated enemy. By the beginning of the twelfth century missions to convert the Wends had been established, but raids into their territory continued, as part of the process by which the north German aristocracy sought new lands for exploitation. In 1140–3 a group of north German nobles drove the Wendish chieftain from his base, and a bishopric was established at Oldenburg. This was the first time that the traditional Wendish structures of authority had been displaced, and therefore marks the beginning of an era of conquest.

The Danes and Germans on whose behalf St Bernard obtained the bull may have been gratified to secure an indulgence for their expedition of 1148, but it is doubtful whether Bernard's exhortation to continue the crusade until all the pagans were converted or killed suited their intentions. They were content, in the end, with the surrender of the Wendish stronghold, Dobin, and the token baptism of Nyklot, the Wendish prince, and his warriors. They wanted to colonise and exploit the conquered lands and people, rather than devastate them.

Perhaps because the Wendish crusade of 1148 failed in the objective of conversion, the papacy appeared to lose interest in the northern wars against pagans. There was neither papal preaching nor indulgence for a generation, until Alexander III's bull *Non parum animus* (1172) placed the war against the Estonian and Finnish pagans on the same level as that in the Holy Land. Yet the subjugation of the Wends by the Saxons and Danes continued, and continued to be understood by contemporaries in religious terms. Saxo Grammaticus, the Danish chronicler, praised Absalom, bishop of Roskilde (1158–92) and Lund (1178–1202) for his military prowess, 'for it is no less religious to repulse the enemies of the faith than to uphold its ceremonies' (Christiansen 1980: 61). Self-defence, that necessary component of the just war, was another justification. Saxo's chronicle, the

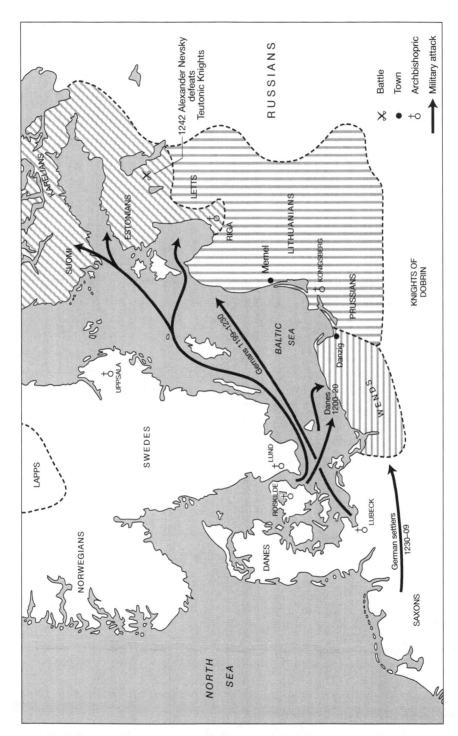

The Baltic crusades

Gesta Danorum (1185–1215), which portrays the development of the Danish state as a product of an epic struggle with paganism, justifies king Valdemar's conquest of Rügen as retaliation for Wendish raids on Danish shipping. What may appear today as the cynical manipulation of pious sentiment was in fact more complex. It would be wrong to imagine, for example, that contemporaries thought their wars against non-Christians freed them from any concern about the fate of the Holy Land. Henry the Lion, duke of Saxony, made an armed pilgrimage to Jerusalem in 1171 – only a year before Alexander III's bull recognised the Danish war in Estonia as equal in merit to a crusade to the east (Phillips 1996: 214).

One reason why the northern crusades have received relatively little attention from crusader historians is that Scandinavian and Baltic historians themselves have been reluctant to see them as part of the same impulse that governed the wars against Muslims in the Near East. The recent historiographical tradition has instead preferred socio-economic explanations for the conquest of the non-Christian Baltic, perhaps in reaction to earlier generations of German nationalist historians who saw in the northern crusades the manifest destiny of a superior against inferior peoples (Johnson 1975: 547). This is understandable, given the ambivalence that the twelfth-century northern crusaders themselves showed about conversion, as opposed to conquest and colonisation. It is true that crusaders and settlers in the East were also largely unconcerned about the conversion of Muslims who came under their rule, but the issues of conquest in the East were clearer because of the universally held conviction that the Holy Land was rightfully a Christian possession. Nobody could argue that the pagans of the Baltic had taken land that was rightfully Christian, and the theoretical justification for the conquest of pagan lands was thus uncertain. In practical terms, mass conversion was implausible, and wholesale destruction as an alternative – as St Bernard had advised in 1147 – counter-productive. The colonising families wanted to derive an income from newly conquered lands, and this could not be achieved by extirpating the subject population that worked the land.

Henry the Lion at first followed a policy of dispossessing the Wendish aristocracy in order to force the colonised people directly under Saxon rule, but a revolt in 1164 led to the establishment of a puppet ruler, Pribislav, who had converted to Christianity in 1160 (Christiansen 1980: 68–70). In contrast, King Valdemar of Denmark, after his conquest of Pomerania in 1185, was content to leave the native prince of Riga in place as a tributary. Both missionary bishops and Christian rulers realised that the conversion of a pagan population was easier to accomplish if the local ruler of that population set the example. Once this happened, there was little incentive to undertake wholesale conversion. In this respect, the northern crusades differed little from those in the East, or for that matter from the *Reconquista* in Spain.

Just as in the East and Iberia, the Military Orders established a strong presence in the Baltic. Their role here, however, was to be far more controversial. The Teutonic Knights, founded with the patronage of Henry VI from those German knights left in Palestine after the siege of Acre in 1191, were modelled on the Templars. Like the other large military orders, their primary commitment was to the Holy Land; unlike them, the strong links to the German imperial throne also drew them to Prussia and Livonia, where the German colonial presence was strong. It was not until the second half of the thirteenth century, however, that the Teutonic Knights began to focus on the north as an area of military activity, rather than simply as a resource from which men and income could be drawn and redirected towards the Holy Land. The shift in emphasis came about partly as the Crusader States in the East became more precarious, but also because of the composition of the order. The recruits themselves were Germans, and as the frontiers of German-held territory spread to the east of the Elbe, so also did the catchment area from which the recruits were drawn. The Teutonic Knights were pre-empted, however, by two orders specifically founded as the private armies of colonising bishops – the Sword Brothers in Riga, founded by Albert of Buxtehode, bishop of Riga, in 1202, and the Knights of Dobrin, founded by the Cistercian Christian of Prussia in 1206. Both German and Danish military activity in the Baltic, particularly in the thirteenth century, was driven by commercial pressures. The defence of Riga, a German trading centre, lay at the heart of these orders' activities. Above all, the orders provided a permanent garrison of knights who could, as Christiansen observed, sit out the winter year after year (1980: 81). Both the Sword Brothers and the Knights of Dobrin were eventually absorbed by the Teutonic Knights. The Sword Brothers had acquired a particularly unsavoury reputation. In 1205–6, during the first phase of conquest, even a peaceful attempt to convert the Livonians by performing a miracle play failed when the natives were terrified by the realism of a battle scene. In 1222 they were rebuked by Honorius III, and in 1230 the papal legate recommended that Gregory IX suppress them altogether; in retaliation for this, the Brothers captured and imprisoned him (Christiansen 1980: 94, 102). Their methods in dealing with the local population were so harsh that the colonists they were supposed to be protecting were frequently threatened by revolt.

Between *c*.1200 and 1292, the territory added to Latin control in the Baltic had extended up to 300 miles east of its previous frontier at Danzig, and comprised as many as a million new inhabitants. The Livonians, who had been tributaries of the Lithuaninas and the Russians of Novgorod, were conquered with the help of crusading indulgences offered in 1195 and 1198, initially by Hartwig, archbishop of Bremen, then by his nephew Albert of Buxtehode and his Sword Brothers. The mission to the Prussians, begun in *c*.1200, turned into conquest by the Teutonic Knights, but this was completed only by *c*.1272, and only after Urban IV had diverted a German crusade against the Mongols to help them. Further to the north, the conquest of Estonia was a Danish preserve. The crusading tradition of Valdemar I was followed by Cnut VI (1187–1202) and Valdemar II (1202–41), and in 1218 Honorius III granted the Danes whatever they could conquer from the Estonians. The key to Danish success, not only against the indigenous pagans but also in their rivalry with Sweden for the northwest coast, lay in their command of the Baltic Sea. Unlike German expansion to the east, Danish expeditions were controlled by the crown, and conquests treated as economic investments. There was little Danish settlement, and the purpose of Danish involvement was to secure trading privileges so as to control the supply of furs, amber and timber from the northeast (Jensen 2001: 164–79)

Likewise, Swedish raiding in the eastern Baltic was largely a means of supplementing the crown's income. Swedes settled in western Finland throughout the twelfth and thirteenth centuries, but although Innocent III authorised the conversion of the Finns in 1209 and placed them under the guidance of the archbishop of Lund, it was impossible to find a suitable candidate. The Swedes also came up against the entrenched interests of Novgorod; in 1240 Alexander Nevsky defeated a Swedish crusade intended to assimilate the pagan Tavastians of eastern Finland by force. This was a blatantly economic conflict, as Novgorod sought to preserve its monopoly over the Karelian fur trade. Similarly, the Swedish kings Eric XI in 1249 and Birger I in 1292 secured indulgences for crusades whose intention was to protect Swedish trade routes. Swedish kings had little intention of occupying land, and unsurprisingly, the knighthood was disinclined to settle in lands so barren and frozen for much of the year. However, a religious principle was at stake, at least as far as the papacy was concerned. If the Russians succeeded in dominating the Finns, they would become absorbed into Orthodox Christianity, and linked by the great rivers via Novgorod and Kiev to Constantinople. This explains why popes were prepared to treat Swedish economic wars as crusades, and why the mixed communities of Latins and Orthodox that had coexisted in Baltic trading centres such as Novgorod, Riga and Gotland became unsustainable. The logic that governed the dominance of the Greek Orthodox under Latin rule in the Mediterranean was also applied to the north; thus, in 1222 Honorius III forbad observance of the Greek rite in Latin-controlled regions. After the Byzantine reconquest of Constantinople in 1261, the lines were drawn even more clearly.

The character of the northern crusades, particularly in Wendish Pomerania, Prussia and Livonia, was determined by the conviction that the continued existence of paganism was a genuine threat to Christianity. Pagans, just as much as Muslims, were the forces of darkness. This belief, which at the beginning of the crusading period appears largely eschatological, acquired increasingly theoretical justification in the hands of thirteenth-century canonists. If the lands and possessions of heretics were subject to confiscation by Christians, why not also those of pagans? By the mid-thirteenth century the canonist Hostiensis could argue that pagans, by their refusal to accept manifest truth, had lost their rights to liberty, and that it was the obligation of Christendom to subjugate them (Muldoon 1979: 15–18).

There were also practical and political reasons why, at certain times, the papacy found it convenient to offer indulgences or to sanction conquest in the north. Many Germans had taken the cross for a crusade to the Holy Land in the 1190s but were left stranded by the death of Henry VI; given the succession crisis in the Empire, it was unlikely that their vows would be fulfilled unless they could be commuted to parallel service. Especially after the conflict with the Hohenstaufen intensified from the 1230s onward, the Teutonic Knights became the beneficiaries of the papacy's desire to outbid the emperor for their support. Although Prussia was claimed by the papacy in a bull of 1234, Gregory IX and his successors feared that the attempt to enforce it would alienate a powerful order that could be enlisted by the emperor in Italy.

Innocent III's concept of papal primacy made intervention in the northern world irresistible. The thirteenth century saw not only crusades against the pagans of the Baltic, but also the heightened activity of papal legates – among them James Pantaleon, legate to Prussia in 1247–9, who was later to become patriarch of Jerusalem and ultimately pope as Urban IV. Another of the most prominent legates was Baldwin of Aulne, a Cistercian sent to Livonia during the period 1231–4. As in frontier regions from Spain to Scotland to the eastern Mediterranean, the foundation of Cistercian monasteries was a crucial feature of Latin expansion. The Cistercians, like the legates and the friars, were instruments of theocratic government, but conquest and mission did not always converge. Legates and friars all disputed the military orders' claims to sovereignty in conquered lands. The Teutonic Knights' suzerainty in Livonia was checked by the presence of the Dominicans and Franciscans; indeed, the Dominican Master-General Humbert of Romans urged the papacy to abandon all support for their conquests in 1274.

Although such disputes were aligned along institutional lines, this may be misleading. German settlement and colonisation in particular had a family character. The work of the initiator of the Livonian conquest, Hartwig of Bremen, was continued by his nephew, Albert, bishop of Riga, who from 1204 onward returned to Germany annually to recruit knights from his family's *mouvance*. One of Albert's brothers became abbot of a new Cistercian foundation in Livonia, a brother-in-law was enfeoffed with large estates in the region, and a cousin became bishop of Dorpat (Tartu). The Sword Brothers, whom Albert founded, were also recruited partly from among his kinship. In this respect, the character of the German settlement in the east Baltic shares much with the northern French and Lotharingian settlement of the kingdom of Jerusalem in the early twelfth century.

Contemporary attitudes to the northern crusades are difficult to assess. There is a striking contrast between the idealism of Henry of Livonia's *Chronicle* of 1225/9, in which he argues that military conquest was merit-orious provided that Christian rule gave the natives of the Baltic peace and prosperity in greater measure than they had known before, and the German *Livländische Reimchronik* of the end of the thirteenth century, which reflects bitterly that the only way to guarantee genuine conversion was at the point of the sword. Such attitudes reveal the despair of almost a century of failure in enforcing the subjugation of a people who took every opportunity to cling to what was left of their social, religious and political traditions. But contemporaries' horror at the brutality of some of the methods used to enforce conquest does not mean that they regarded the northern crusades as less meritorious than those to the Holy Land. There were complaints about the diversion of subsidies to the northern crusades, as there were also of crusades against the Hohenstaufen. But participants doubtless took papal bulls that equated participation in the northern crusades with the pilgrimage to Jerusalem at face value. It is true that Honorius III reminded the bishop of Riga of his continuing responsibility to donate to the proposed recovery of Jerusalem, but this suggests not so much that the northern crusades were a 'second-best, cut-rate enterprise'

(Christiansen 1980: 127) as that even outlying parts of Christendom were seen as linked to, and able to contribute to, the central cause. Interest in one variety of crusading did not necessarily mean a lack of concern for another. As a parallel example, the English crusaders who took Tortosa, in Catalonia, in 1148, were buried in a cemetery subsequently given to the canons of the Holy Sepulchre in Jerusalem, and even in this Christian frontline against Islam, a strong link was maintained through regular donations to the Holy Sepulchre (Jaspert 2001: 90–110).

The northern crusades were wars of expansion, conquest and colonisation. One historian has recently used the process of conquest and the organisation of a hierarchy of power in the Baltic to supply a definition of a more general movement, 'the expansion of Latin Christendom' (Bartlett 1993: 18). This process was made possible not only by superior military technology but also by the prevailing ideology of the crusade that placed at the disposal of the conquerors the full machinery of the Church – the ability to recruit through preaching, the offer of spiritual rewards for fighters and the administrative means to establish structures of government in conquered territories.

Frankish Greece

Following the conquest of Constantinople by the Fourth Crusade, a substantial new Frankish presence was added to those in Syria, the Holy Land and Cyprus. Rule was established by those crusaders who remained in Constantinople over the city and adjoining territories, and those areas that had still been under the control of the Byzantine government – Thrace, Greece and the extreme northwest of Asia Minor, together with the Aegean and Ionian islands.

The new emperor was elected by a council comprising six Venetians and six Frankish crusaders. The choice of Baldwin, count of Flanders, rather than Boniface of Montferrat, was a genuine surprise, at least to Boniface, but may be explained by Baldwin's more conciliatory approach to the Byzantines (Lock 1995: 43–5). Another committee decided upon the division of spoils. The Venetians retained one-eighth of the city of Constantinople, the Adriatic coast, the Ionian and Aegean islands (the latter of which were to become the Duchy of the Archipelago), Euboea (Negroponte) and Crete. They established a genuinely colonial maritime empire, ruled through a *podestà* by the Republic of Venice itself. Most of these possessions were lost to the Ottomans in the fifteenth century, though Crete remained in Venetian hands until 1669. The emperor

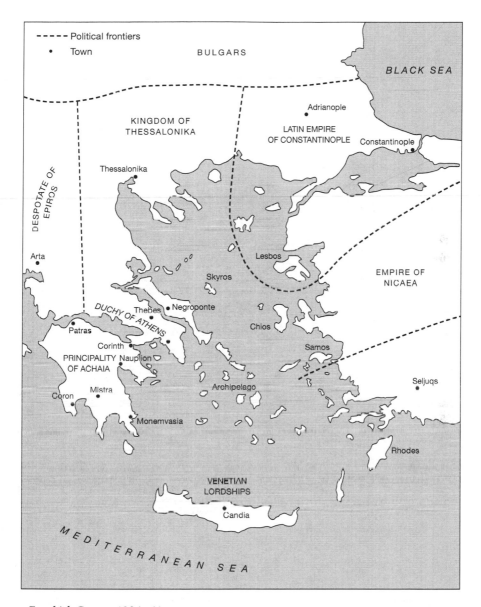

Frankish Greece, 1204–61

received the rest of the city of Constantinople, together with Thrace, the northwest region of Asia Minor and the islands of Chios, Lesbos and Samos. He soon found, however, that he had exchanged his strongly cent- ralised county in the West for little more than a glamorous title. Although the partition of fiefs was probably based on recent Byzantine tax registers

(Oikonomides 1976: 3–28), Baldwin had little say in the distribution, which meant that he had little opportunity to establish a dynastic base in support of his title. In recognition of his family's claim to titles held by his brother Renier before 1185, Boniface was granted the barony of Thessalonika. Central Greece and the Peloponnese were carved up between the Frankish crusaders, and eventually settled into the two main power bases of the Principality of Achaia and the Duchy of Athens.

A narrative account of the fortunes of the Latin Empire makes unedifying reading. The Empire faced enemies on three fronts: in east and west, the two rival Byzantine successor states of Nicaea, ruled by the Lascarids (who also swallowed up the Komnenos state of Trebizond) and the Despotate of Epiros, ruled by the Angeloi; and the Bulgar kingdom to the north. Baldwin and Boniface were both killed fighting the Bulgars, in 1205 and 1207 respectively. Had the Franks been prepared to recognise the Bulgar kingdom, they might have been able to count on their help against their Byzantine rivals, but Emperor seems to have been deluded by his title into a false sense of his power. In 1208, his successor, Henry II (1206–16) married the daughter of the Bulgar tsar Kalojan, which gave him breathing space to concentrate on defending his empire against the Lascarids. In 1211 he won a victory in Asia Minor that secured his Asian possessions, and when he died in 1216 the Latin Empire was probably at its height.

The 1220s, however, proved disastrous. The 'kingdom' of Thessalonika, a Montferrat dynastic possession, fell to the Despotate of Epiros in 1224, despite the launch of a crusade from the West by William IV of Montferrat to save it. In 1225 the Nicene Emperor John III Vatatzes (1222–54) drove the Franks out of Asia Minor and was prevented from reconquering Constantinople itself only by the opposition of Epiros. A further stay of execution came in 1230 when Epiros itself was destroyed by the Bulgar tsar John II Asen (1218–41). The Empire had a minor, Baldwin II (1228–61) at its helm, and even the appointment of the experienced former king of Jerusalem, John of Brienne to the position of co-emperor in 1229 could not turn the tide. When John died in 1237, after an adventurous career that had included the captaincy of papal armies in Italy, marriage to the heiresses of Jerusalem and Armenia and the leadership of a crusade, the Latin Empire consisted of little more than the city of Constantinople itself. The end came only in 1261, when the Nicene Emperor Michael VIII Paleologus (1259–82) realised just how weak the Franks were. By that stage, the unfortunate Baldwin II had been reduced to selling or pawning anything of value, including the remaining relics, the lead from the palace roof, and even his own son.

One corner of Frankish Greece, however, provided a model of strong and successful feudal colonisation. The Villehardouin dynasty established itself in Morea (Peloponnese) and between 1209 and 1259 enjoyed secure and peaceful rule over the Principality of Achaia. William II of Villehardouin (1246–78) conquered the whole of Morea, and made himself overlord of Negroponte, the duchies of Athens and the Archipelago. The constant warfare that weakened the emperors in the north also acted as a screen to protect the more remote Morea, with the result that the Villehardouin princes were able to consolidate centralised feudal authority with little external threat. Alongside their Frankish fief-holders they also ruled over the remaining Byzantine *archontes* (landowners), using French feudal customs enshrined in a written code of law. After 1266, following the defeat of William II by the Lascarids in 1262, the Principality survived only as a vassal state of Charles of Anjou's Sicilian kingdom. In the fourteenth century it disintegrated under the growing strength of the new Palaeologan regime in Constantinople, and as a consequence of the seizure of the duchy of Athens by the Catalan Company (1311), a group of mercenaries from northern Spain. During its zenith, however, under Geoffrey II (1229–46) and William II, the Frankish court of Achaia enjoyed a reputation as a centre of traditional chivalric culture. William II, who participated in Louis IX's crusade in Egypt (1249–50), built a palace at Mistra whose grandeur can still be appreciated today.

The weakness of the Latin Empire must be seen in the context of wider collapse in the region. That the crusaders had triumphed in 1204 with a small army (about 20,000) over the defences of the largest city in Christendom says more about problems within the Byzantine imperial system than about Frankish military strength. The Byzantine Empire had suffered five changes in regime between 1182 and 1204; loyalty to the emperor was so loose by 1204 that the Greeks of Thrace were quite prepared to recognise a Latin as just another in a succession of emperors. The Franks never had sufficient military strength, nor could they mobilise western colonisation as the settlers in the Holy Land after 1099 had been able to. But if the Franks were hopelessly weak, their rivals were scarcely stronger. The Latin Empire was therefore 'an additional element in the regional mosaic of princelings' (Lock 1995: 55); they were distinguished only by their possession of the 'queen of cities' herself. The Latin Empire was a symptom of the breakdown in power in the northeast Mediterranean.

The Latin Empire may have been weak, but that does not mean it was unimportant in the political life of Europe. Popes until the mid-thirteenth century saw it as a vital component in the Crusader States. There is no reason to suppose that Innocent III's reference to the Frankish settlers after 1204 as 'pilgrims' was simply conventional phrasing: like Gregory IX (1227–41), he was probably sincere in the belief that the best hope for the Holy Land lay in securing a bridgehead to control the passage from West to East. Thus Gregory saw a crusade against Nicaea (1237–9), which never in fact materialised, as part of the wider effort to protect the Crusader States. Here the papacy seems to have been out of step with western opinion (Barber 1989: 111–28). Although individual rulers such as Charles of Anjou (1266–85) were prepared to invest in Frankish Greece, the cause of the Latin Empire had little resonance among potential crusaders. Richard, earl of Cornwall, for example, resisted papal pressure to commute his vow for the Holy Land to the Aegean in 1239. The failure of the Latin Empire to appeal to western crusading instincts can also be seen in the large numbers of Franks who took service in the Byzantine armies of Nicaea or the Despotate of Epiros after 1204. Frankish mercenaries were prepared to commit horrifying acts of violence against the ruling Franks on behalf of their employers, as for example in 1210 when Adamée de Pofoy and a group of his knights were crucified in Thessaly by western knights fighting for the Despotate. Papal policy changed, perhaps to conform to public opinion, under Innocent IV (1243–54). Seeing that there would never be sufficient western interest in propping up the failing régime in Constantinople, he recognised the legitimacy of the Orthodox patriarch of Constantinople in exile, and initiated a more conciliatory policy towards Orthodox Christendom in general. This change in papal policy surely contributed to the revival of Byzantine authority and the eventual downfall of the Latin Empire.

Relations between Franks and the indigenous Greek population were, as may be supposed, often strained. Yet it may have been the confrontation with unfamiliar and – according to the Byzantine view – erroneous religious customs that proved most offensive to the Greeks. As in the Crusader States in the Levant, Orthodox monasteries and the parochial system continued to function. The difference between the twelfth-century Crusader States and the situation in Frankish Greece and Cyprus in the thirteenth century, as well as in Syria and the Holy Land, was the more intensive level of papal oversight. In part, this may have been caused by the suspicion that Greek monasteries were collaborating with the rival régimes of Nicaea and Epiros (Lock 1995: 227), but on the whole this changed situation reflected new directions in papal policy under Innocent III and his successors. Where the Latin Church in the twelfth-century Crusader States

seems to have made little trouble over the observance of Orthodox customs that ran contrary to Latin norms, in Frankish Greece there was
less tolerance of difference in customs relating to fasting, the Eucharist,
consecration and holy orders. Latin oversight was aided by the arrival of
new religious orders, notably the Franciscans and Dominicans, who were
valuable agents on behalf of papal policy. A dozen Cistercian monasteries
and nunneries were also founded in Frankish Greece, reflecting the
involvement of the order in the Fourth Crusade (Brown 1958: 63–120;
Panagopoulos 1979). Nevertheless, the number of Franks was always
small compared with Greeks, and consequently there was little attempt
outside Constantinople and Thessalonika to impose a Latin parochial
system on conquered territory. This meant that there must often have been
little alternative for isolated Frankish communities to sharing churches
or even attending Orthodox services, and even in the early years of the
settlement Pope Innocent III worried about Franks adopting Orthodox
religious customs. By the fourteenth century, Venetians in Crete who 'went
native' were a serious concern for the papacy. It is probably going too
far to describe the Frankish settlement as a process of acculturation. As
one historian has observed, the occupation was too brief and too limited
geographically to be 'anything other than a curdling, rather than a true
intermingling' (Lock 1995: 266). Because Frankish and Greek communities lived largely separate different existences, however, it does not follow
that they were necessarily hostile to each other. Although historians,
naturally enough, dwell on violent incidents such as the anti-Latin riots of
1182, and examine the polemical discourse against Latins in the Orthodox
tradition, there is a danger in overplaying this type of evidence. There was
little anti-Latin resistance in Frankish Greece in the thirteenth century, and
although polemical writing continued to be produced in monasteries and
schools, it was not representative of how the societies interacted. Once 'a
caricature became a person with a face and a name, cooperation, friendship and conjugal fidelity were thought possible' (Lock 1995: 274).

We may even question the extent to which 1204 really marks a cata-
clysm in Byzantine history. For one thing, creeping 'westernisation' during
the twelfth century had accustomed cosmopolitan Byzantines to the sight,
sound and customs of Franks. It has been estimated that before the last
quarter of the twelfth century there were between ten and twenty thousand
western merchants and their families in Constantinople; marriage between
Italian merchants and Byzantine women was, moreover, encouraged by
imperial policy (Magdalino 1993: 27–108). At the middle and lower levels
of society in particular, the Latin conquest offered possibilities of social

and professional advancement, and it may be in this largely undocumented hinterland that something like a hybrid society developed. Even for those Greeks who were unwilling to cooperate actively, or in the provinces where Franks were thinly spread, the resentment against them may have been no greater than that felt for any ruling élite. In the provinces, land was redistributed, but the impact of this was greater for the pre-1204 élites than for those who actually worked on the land. The Frankish effect on institutional life was minimal. Although some French feudal terminology crept into the everyday speech, few placenames seem to have been changed. Debates over whether the Franks introduced feudal landholding arrangements have been inconclusive. The age-old Byzantine system of *pronoia* – a grant of land or privilege in return for services to the grantor – looks on the surface rather like western feudalism anyway, with the exception that the grant was typically for the lifetime of the recipient and not hereditary. After 1261, Emperor Michael VIII seems to have made some grants of *pronoia* hereditary, which may indicate residual Frankish influence. Even this, however, may simply have been a stage in the process of westernisation that can be seen in the twelfth century, in Emperor Manuel Komnenos' adoption of chivalric concepts such as the tournament and dubbing to knighthood. The Frankish occupation of Greece, therefore, may be seen as a stage in a process, rather than as the sole cause of the eventual decline of the Byzantine Empire.

CHAPTER 9

Crusading and the Crusader States in the thirteenth century, 1217–74

Crusading saw some new departures in the thirteenth century, among which was a shift in the focus and target of large-scale expeditions. The Fifth and Seventh Crusades were launched against Egypt rather than the Holy Land, although the intention was still the recovery of Jerusalem. This shift represents both new military strategies and a more subtle change in the western perception of crusading. In this chapter, three crusading expeditions are examined in turn and the reasons for their varying fortunes analysed. The main themes that emerge are:

- conflicting interests of western crusaders and Frankish settlers in the East;
- the narrow margin between military success and failure;
- the continuing problem of dynastic succession in the Crusader States.

The challenge of Egypt: the Fifth Crusade, 1217–21

The crusade that Innocent III had announced in his bull *Quia maior* in 1213, and over the planning of which Fourth Lateran had taken such pains, did not begin until 1217. There were signs in the intervening years of the pope's frustration at the delay – almost to the point of ending the Albigensian Crusade in 1212 so that attention could be turned to the East.

But Innocent died in 1216, and problems over leadership were to dog the progress of the crusade throughout.

Part of the difficulty, indeed one reason for the delay, was that the kingdom of Jerusalem had assured its immediate survival by negotiating a series of truces that expired only in 1217. Moreover, the very ability of the restored kingdom to occupy a place once again in the political society of the eastern Mediterranean, may have made the cause that Innocent found so emotive – the recovery of Jerusalem – appear less immediate to Europeans. In 1217 there was no immediate challenge to the security of the kingdom, other than the fortification of Mt Tabor, in the Galilee, by the Muslims. As the patriarch of Jerusalem informed the new pope, Honorius III, the Ayyubids were in no position to repeat the successes of 1187–91. All but one of Saladin's heirs had been disinherited by their uncle, al-'Adil. Although he was an experienced statesman who maintained a secure grip on Egypt and Syria, al-'Adil was incapable of preventing the revival of the kingdom of Jerusalem between 1194 and 1217. Either as a result of his experiences in confronting the Third Crusade in 1191–2, or out of genuine disinclination for war, he left the Franks in the East alone. Above all, al-'Adil lacked the moral authority that his brother had been accorded in the Islamic world, and was thus unable to form the alliances that had delivered such success to Saladin in 1187–91. Innocent III seems to have believed that he might even be willing to concede Jerusalem (Richard 1999: 297), and when the first phase of the crusade began, al-'Adil's policy was one of withdrawal rather than confrontation.

The first contingents of crusaders arrived in the East as a trickle of national groupings rather than a series of coordinated armies. In part, this reflected the process of preaching and recruitment. The preaching campaign was more widespread and 'professional' than ever before. By the early thirteenth century, preachers such as Robert of Courçon, Fulk of Neuilly, Oliver 'Scholasticus' of Paderborn and James of Vitry, had developed specialist crusade sermons. Following the Bernardine model of the mid-twelfth century, preachers focused their message on the responsibilities of individual Christians and the opportunities presented by the crusade for salvation (Powell 1986: 51–63; van Moolenbroek 1987: 251–72; Maier 2000: 82–99, 100–27, esp. 94–9). Alongside the 'feudal' message of recovery of the Holy Land there had always since 1095 also been another element in crusade preaching: the ideal of the imitation of Christ. This came to the fore in the mouths of the thirteenth-century preachers. Moreover, Innocent III's bull *Quia maior*, and the reforms in the organisation of recruitment during his pontificate, had the effect of

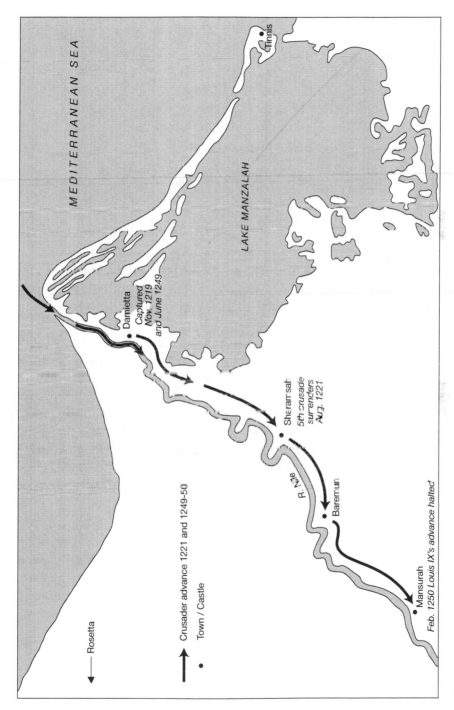

MEDITERRANEAN SEA

Rosetta

Crusader advance 1221 and 1249–50

Town / Castle

LAKE MANZALAH

Tinnis

Damietta
Captured Nov. 1219 and June 1249

Sharamsah
5th crusade surrenders Aug. 1221

R. Nile

Baremun

Mansurah
Feb. 1250 Louis IX's advance halted

Crusading in Egypt, 1217–21 and 1248–50

widening participation in crusading. That this was effective is attested by a survey of named crusaders between 1217 and 1221 that includes counts and carpenters, bishops and porters, from France, Italy, England, Flanders, Frisia, Germany and Austria (Powell 1986: 209–46). Taking the cross was not only an increasingly formalised and ritual event, it was also more available to a wider range of people than ever before. In consequence, the Fifth Crusade was characterised by waves of recruitment and by the periodic arrival of fresh crusaders as earlier ones left from 1217 to 1221. This undoubtedly had an effect on the nature of the crusade itself, and perhaps also on its failure.

The Fifth Crusade marked a new departure in that the preaching extended to the Latin East itself. The king of Jerusalem, John of Brienne, King Hugh I of Cyprus and Bohemond IV of Antioch all took the cross, with many of their leading barons. Although this reflected Innocent III's conception of the crusade as a penitential enterprise to be undertaken by the whole of Christendom, it was to raise thorny problems of leadership. John, a French baron who had become king through marriage to Maria, the daughter of Isabel and Amalric II, was eventually elected leader by the combined army in 1218, after the initial phase of campaigning in the Holy Land itself was over. Throughout the crusade, the aspirations of John and the native barons were to heighten tensions with the western crusaders. One reason may have been that, although he wore a crown, John of Brienne's pedigree was less impressive than that of many of the western crusaders. Besides King Andrew II of Hungary, who had taken the cross as early as 1195, the crusaders numbered among them the dukes of Austria and Burgundy, the counts of Nevers and La Marche, the earl of Chester and the prince of Bavaria.

The first phase of the crusade, in which the Hungarians and Leopold of Austria's Germans dominated, was inconclusive. The fortifications on Mt Tabor were assaulted, which removed the threat to Acre, but a raid in the direction of Damascus was less successful, and Andrew of Hungary returned home early in 1218, while Hugh of Cyprus died. Most assessments of the Fifth Crusade have been critical of the conduct of King Andrew and concluded that nothing was achieved in 1217–18, but more might have been done had the expected Rhenish–Flemish fleet not been detained by action in the Iberian peninsula (Powell 1986: 125–6).

Now the crusade turned its attention to Egypt, in the hope that an assault on the Ayyubid power base of Cairo would weaken al-'Adil's hold on Jerusalem. In summer 1218 the siege of Damietta began. The city, which lay at the mouth of the Nile Delta, was itself guarded by an island tower that cut the mouth of the river into two channels, one of which was unnavigable, and the other barred by chains hung across it. In effect, the Tower of Chains, as it was called, protected all access upriver towards Cairo. The tower was successfully taken with the help of a siege engine designed – as he tells us himself in his *Historia Damiatana* – by Oliver of Paderborn (Peters 1971: 64–5). Al-'Adil died soon after, leaving the Ayyubid territories in the hands of his three sons: al-Kamil in Egypt, al-Muazzam in Damascus and al-Ashraf in Mesopotamia. Damietta itself, however, held out for another fourteen months, until early November 1219. During this period two new waves of crusaders had arrived, including in autumn 1218 the papal legate Pelagius, cardinal-bishop of Albano. It was Pelagius' refusal to accept the negotiated settlement offered by al-Kamil in 1219 that prolonged the crusade. At the beginning of 1220, his stance seemed prescient, for the way to Cairo lay open, al-Muazzam's assaults on Caesarea and Athlit were unsuccessful, and Frederick II's imperial forces were finally being prepared for departure to reinforce the crusading army.

In the end, however, 1220 turned out to be a wasted year. Al-Kamil built a new fortification farther up the Nile at Mansurah to protect the route to Cairo. John of Brienne, who had enjoyed uncontested leadership before the arrival of Pelagius, returned to Acre to pursue his claim to the throne of Armenia, which looked precarious after the death of his second wife Stephanie. His departure looks like pique or, at the least, disagreement over who was really in charge, exacerbated by the dispute over al-Kamil's offer. More crusaders arrived during the course of the year, including a Venetian fleet, but this only offset the departure of others. Meanwhile, prophecies relating to a mysterious Christian king of the East who, it was anticipated, would attack the Islamic world in unison with the crusaders, and circulated around Damietta by Pelagius, kept hopes of a decisive victory alive. The imperial forces, which only amounted to about 500 knights under the command of the duke of Bavaria, did not arrive until late spring 1221. Once again, disagreements broke out among the crusaders: should they wait for Frederick II to arrive with the bulk of his army, or begin offensive operations at once on Mansurah before the annual Nile floods rendered the Delta impassable for armies? Louis of Bavaria was anxious to fight, despite instruction from Frederick II to wait for him, but John of Brienne argued that there was insufficient time. Pelagius, whose reputation has suffered as a result (Mayer 1988: 218), refused to wait, and the army set off in mid-July. John's misgivings were proved right: the Nile swelled to the point where the Egyptians could use a secondary canal to encircle the crusaders and cut off all supplies from Damietta; then, opening the sluice gates, they flooded the countryside and trapped them on a narrow island of land opposite Mansurah. By the end of August, the crusaders' fighting retreat had been defeated, and Pelagius accepted a dishonourable peace. The crusaders were forced to leave Egypt altogether, surrendering Damietta, and to accept an eight-year truce. Everything achieved so painfully since 1218 had been lost.

It is easy to ascribe the defeat of the Fifth Crusade to a failure of military leadership, and in particular to blame Pelagius for the fateful decision to advance. Western ignorance of local geography and climate, and a refusal to listen to the advice of those who had local knowledge, certainly contributed. But, as at Hattin in 1187, the margin between defeat and victory was narrow – perhaps to be measured in terms of only a few weeks. The crusade had already wasted a year waiting for Frederick II, and Pelagius may have sensed the army's mood correctly when he insisted on decisive action rather than waiting out another summer in Damietta. It is curious that he should have left such a large portion of the troops available to him in Damietta, and this may indicate that the campaign of July 1221 was intended simply to drive al-Kamil from Mansurah, before the arrival of a fresh army made a larger-scale advance possible (Powell 1986: 184–5). In any case, although it is true that the Egyptians had been reinforced by the arrival of al-Muazzam from Syria, the defeat at Mansurah was not the result of superior forces. The leadership question is similarly thorny. As Powell has pointed out, the decisions taken in July–August were not simply Pelagius' own, but represented the popular will among the crusaders (1986: 188–9). The role of John of Brienne is also complex. He certainly clashed with Pelagius over the negotiations in 1219, but his departure in 1220 can be explained on other grounds. For one thing, al-Muazzam was trying to divert attention from Egypt by threatening Caesarea. Besides, the insecurity of John's own position as king of Jerusalem – underlined by the fact that his wife Maria, through whom his claim was made, had died in 1212 – was exploited by Pelagius, who refused to recognise the king's claim of Damietta as part of his lordship. One source says that John returned to Acre to press a claim to Armenia through his second wife, Stephanie, the daughter of Leo II, who had died in 1219. Her death and that of their heir in 1220 cannot have helped his cause. John, who ended his life as Latin emperor of Constantinople, is an example both of the possibilities open in the East to resourceful but landless western knights, and of the limitations on their authority if the lineage on which their claims were based failed. That ultimate leadership of the crusade was never determined may reflect the wishes of Innocent III that crusading be the collective enterprise of Christendom; on the other hand, Honorius III seems to have placed his hopes on the arrival of Frederick II, and it can be argued that the emperor's failure to appear betrayed the crusade. Perhaps more action might have been taken in 1220 had the crusaders known that he would never in fact arrive.

In hindsight, the refusal to accept the terms offered by al-Kamil in 1219 – the exchange of Damietta for Jerusalem – was particularly frustrating. But this, too, is understandable if the interests of different parties are taken into account. John of Brienne and the barons of the crusader kingdom were in favour, naturally, because al-Kamil's offer would have restored the kingdom's territories to their extent before 1187, with the exception only of the Transjordan castles of Kerak and Montreal, for which compensation would be paid. Al-Kamil was even, by late August 1219, prepared to rebuild the walls of Jerusalem and to return the True Cross. But the Military Orders baulked at the exception of the Transjordan castles, recognising that as long as Egypt and Syria were in alliance, the defence of Jerusalem hinged on them. Others had their own reasons: for the Venetians, for example, Damietta, with its control of the Red Sea trade, was the real prize, not Jerusalem. In any case, much of al-Kamil's offer may have been bluff. The location of the True Cross was unknown, perhaps even to al-Kamil, and in any case, his offer was empty without the compliance of al-Muazzam: what if he refused to play along with the return of the Holy Land?

The contemporary narratives of the crusade indicate a constant preoccupation with reinforcements. The fourteen-month siege of Damietta exhausted both sides, and by 1220 both money and troops were insufficient for further large-scale operations. The crusaders may have numbered as few as three thousand knights. However, the numbers of fresh arrivals never compensated for those who departed, and after the autumn of 1219 the peak of new crusaders had passed (Powell 1986: 165–72). The total number of crusaders between 1217 and 1221 was high, but most had already arrived before the end of 1218, and the effect of staggered contributions to the crusade was to stretch resources during crucial moments. The crusaders were simply unable to press home their advantage in 1220, when it might have counted. A comparison with the First Crusade may help make the point. The Fifth Crusade outlasted all save the First in length of time; both crusades involved long sieges followed by periods of uncertainty and recuperation, but in 1099 the final assault on Jerusalem was made possible by the timely arrival of the Italian fleet, whereas in 1221 the expected reinforcements did not arrive.

The policy of attacking Egypt rather than pursuing the campaign in Syria in 1218 was ultimately a failure. As a strategy, however, it is understandable. As early as 1192, Richard I had questioned whether it was possible for a force that was to be in the East for a limited duration to recapture and secure Jerusalem. The objective of the Fifth Crusade was to strike at the heartland of Ayyubid power, and to recover the former extent of the kingdom of Jerusalem as a result of a decisive military victory. There was every reason to be hopeful of such an outcome: had Richard and Henry of Champagne not shown what could be done in the 1190s? Crusade planners knew that if they wanted to engage the Ayyubids in such an encounter, they would have to find a target in the defence of which al-'Adil and his sons would be prepared to risk everything. Jerusalem was not such a target: it might fall to the crusaders, but not at the cost of the Ayyubids' real military muscle. Moreover, by 1217 the Egyptian strategy was already traditional. Egypt had apparently been the intended target in 1204, and according to Oliver of Paderborn, the Fourth Lateran Council had determined an attack on Cairo as the policy of the new crusade (Peters 1971: 61). The first phase of the crusade in 1217 should thus be seen as a diversion with the limited intention of freeing Acre of the threat posed by the fortification of Mt Tabor. Significantly, even the disappointment of defeat did not undermine this strategy in the future. Although some crusades in the future (in 1239–41 and 1270, for example) did target the Holy Land directly, the Fifth Crusade set a precedent for a wider military strategy that was to be adopted by most major expeditions from the West in future.

Frederick II and the kingdom of Jerusalem

Frederick II's failure to appear in Egypt in 1221 may have cost the crusade dearly, but it enabled both emperor and pope to avoid the potentially embarrassing question of authority. When Frederick did eventually fulfil his vow, in 1229, it was on his own terms, in pursuit of his own objectives and, ultimately, without the cooperation of the papacy. This was not how his crusade began, for it was initially Honorius III who took the initiative, by insisting that Frederick fulfil the vow taken in 1215 and persuading him to agree to a departure date in 1225. By the time he arrived in the East, however, there could be no doubting that the emperor alone was in charge of his crusade.

Frederick's status had materially altered since 1221. In 1225 he married Yolanda (also known as Isabella II), the daughter of John of Brienne and Maria, and thus gained precisely the same claim to the throne of Jerusalem through his wife as John had enjoyed. Although in the marriage negotiations it had been agreed that John of Brienne should remain king for life, once the marriage had been celebrated – by proxy – Frederick had himself crowned king of Jerusalem and insisted on the homage of the barons of the kingdom who had brought his bride to Sicily. Fulfilling his crusading vow would thus enable him at the same time to exercise lordship over distant possessions; but it would also provide a triumphal accolade for the most powerful ruler in the West. To add Jerusalem, the symbolic centre of Christianity, to his imperial title, was to wield a new kind of authority that combined the spiritual with the secular. Yet the marriage must have been at the very least approved by the pope, for Yolanda, through her Montferrat ancestry, was a cousin of Frederick, and a dispensation was needed. Either Honorius was outmanoeuvred by an unexpectedly proprietorial attitude to the crown on Frederick's part, or he simply regarded John's fate as a necessary sacrifice for the future of the kingdom of Jerusalem. The luckless King John may have been the victim of his own ancestry, for his father, Walter, had fought to put Tancred of Lecce on the throne of Sicily instead of Frederick in 1198. He was now deprived of all but an empty title.

The situation in the Near East also changed in the decade following the Fifth Crusade. Al-Kamil and al-Muazzam fell out with each other, and when, in 1226, the sultan of Egypt courted Frederick as a lever against his brother, the emperor seized the opportunity to form an alliance. Al-Kamil doubtless knew that Frederick was planning a crusade, and wanted to ensure that it was directed against al-Muazzam in Syria rather than against Egypt. Frederick may, as some historians have argued, have approached the problem in the role of a king of Jerusalem rather than a crusader (Abulafia 1988: 171–2), but it did not take a strategic genius to realise that the best hope for crusading success lay in keeping the Ayyubids apart rather than united. The mutual jealousy of the sons of al-'Adil presented the most encouraging opening for a crusade since the death of Saladin.

Frederick finally embarked in September 1227, but fell ill and turned back to recuperate on land. Gregory IX was furious at the further delay, caused by what he thought was Frederick's incompetence in embarking his troops before disease broke out in the camp, and promptly excommunicated him for breaking the treaty of 1225. It is unlikely that Frederick was dissimulating, for some of his barons, including the landgrave of

Thuringia, died of illness at the same time. But the excommunication reveals the pope's frustration at the wider situation. Gregory's accusations of tyranny against the Sicilian Church, and his insistence that Sicily be governed as a papal fief, tell the real story of his fears of imperial domination. He now prohibited Frederick from sailing for the East until he had been reconciled to the Church, and the conditions for reconciliation were set as the submission of Sicily to its feudal overlord, the papacy. These demands probably explain Frederick's further delay in setting sail, until June 1228: he wanted to allow sufficient time to come to reasonable terms. When this proved impossible, he surprised Gregory by setting out anyway. Different interpretations of the affair are possible. One can see Gregory as an excessively legalistic old man with a predetermined view of the dangers Frederick posed to the papacy (Abulafia 1988: 164–74). On the other hand, Frederick had delayed so long since 1215 over fulfilling his vow that the pope may genuinely have believed that this was yet another pretext. Once Frederick had been excommunicated, moreover, the pope had no choice but to insist on reconciliation to the Church before embarking on his expedition again, for to do otherwise would make the same mockery of the crusade as had occurred in 1202–3.

The delay, in fact, was not at all to Frederick's advantage, for in April 1228 – by which time he had expected to be firmly entrenched in his kingdom of Jerusalem – his constitutional claim to the throne disappeared with the death of his wife Yolanda. He was now, like John of Brienne before him, only regent for their infant son, Conrad. This was to prove damaging to Frederick's expectations of wielding real authority in the East. Moreover, Frederick's crusading strategy was jeopardised by the death of al-Muazzam. Al-Kamil had welcomed Frederick as an ally against his brother, but al-Muazzam's death, and the absence of an adult male heir, removed the need for the emperor's presence in the East. Indeed, it made his crusade an embarrassment to both Frederick and al-Kamil, for he now had nobody to fight for the recovery of Jerusalem but his ally.

Perhaps this is one reason why Frederick headed straight for Cyprus rather than the Holy Land. There was work to be done here, for since 1195 the island had technically been an imperial fief, and the king, Henry, was only a child. Frederick thus had every right to assure himself that the regency was in responsible hands. Moreover, according to western custom he, as suzerain, was due the revenues accrued during the regency, and he may have intended to use them to pay his soldiers (Riley-Smith 1973a: 161). In fact the real ruler of the island was not Alice, the king's mother, but John of Ibelin, lord of Beirut, whom she had appointed *bailli*. Like most of the higher nobility in the kingdom of Jerusalem, he also held lands in Cyprus. This was a potentially messy situation, because the two kingdoms were constitutionally separate. Because neither kingdom had an adult king in the 1220s, it was inevitable that the boundaries between the two should become rather fluid. In fact, Frederick seems to have become mired in a rivalry between competing noble families in Cyprus, represented on the one hand by the Ibelins, and on the other by Aimery Barlais and his adherents, who were largely the 'new' nobility created by the Lusignans on the island.

Frederick's solution to the problem posed by the overlap of power in the hands of barons such as John of Ibelin was to play the same game himself. He bullied John into submission at Limassol, and threatened him with the seizure of his fief on the mainland. But here he blundered, for, as John pointed out, even if he had the authority to act as regent on behalf of his son Conrad, the matter had to go before the kingdom's High Court, and could not be decided by the emperor simply in his capacity as suzerain of Cyprus. The baronial view was that John's conduct in Cyprus, even if judged culpable, could not be punished by confiscation of his mainland inheritance, because the offence related to a different kingdom, and thus came under a different authority. Frederick seems to have been taken aback by the strength of resistance from the native barons of Jerusalem to his claims. Assuming that his imperial title was sufficient to make him appear as a universal secular ruler, he underestimated the gravity with which constitutional problems were treated in the kingdom of Jerusalem. The resistance of the Ibelins and their supporters to his attempt to throw his weight around masks the real question that he had raised in Cyprus – namely, the abuses of power and privilege by John of Ibelin as regent (Edbury 1997: 40).

On the mainland, Frederick was caught between the expectations of the barons and Military Orders on the one hand and the fear of what might be happening in Sicily on the other. Gregory had threatened seizure of Sicily if he defied the prohibition on leaving Italy without seeking reconciliation. But his army was not large enough to take on al-Kamil – most of the army that had sailed without him in 1227 had by now returned to Italy – and in any case Frederick did not like direct military confrontation if it could be avoided. He could hope only to extract from al-Kamil what had been promised in return for fighting al-Muazzam, but since there was no longer any fighting to be done, al-Kamil had no reason to give it to him. The fact that Frederick succeeded in persuading the sultan to restore anything at all represents something of a diplomatic triumph, in a situation in which he had no leverage over al-Kamil. In fact, as his letter to Henry III of England makes clear, the territories regained were not negligible (Peters 1971: 162–5). Besides Jerusalem and Bethlehem, and the narrow corridor linking them to the coast, Frederick also secured Nazareth and part of the Galilee, Toron and the immediate hinterland of Sidon. This meant that some of the most productive and populous areas of the old kingdom were back in Frankish hands. As Frederick pointed out, the return of Sidon, which had been split before 1228, was crucial for both commercial and strategic reasons. Frederick had succeeded through persistent pleading and a token show of strength – a march down the coast with his small army, followed by the Military Orders – where the meticulous planning and superior forces of the crusades of 1189–92 and 1217–21 had failed.

Why, then, was everyone – the native barons, the Military Orders, the Church in the East and the papacy – so dissatisfied with the liberation of Jerusalem in 1229? Part of the reason is undoubtedly the same attitude that had been shown by Pelagius in 1219. Taking the cross was a penance that conferred immense spiritual privileges. Where was the penitential exercise in Frederick's diplomacy; where the sacrifice and heroism? In his desperation to have something to show for his journey, Frederick had resorted in his diplomacy to compromising his own dignity and that of the holy city itself. All he wanted, he wheedled to al-Kamil, was an undefended and empty hill town with no strategic value, that happened to be the birthplace of Christianity; all he needed was some assurance that his reputation would not be lost. No matter how he dressed it up in his triumphant crown-wearing ceremony in Jerusalem in 1229, the whole affair looked shabby and rather pathetic. Besides, al-Kamil had lost very little. The Temple Mount, including the 'Templum Domini', was retained by the Ayyubids, for of course they would not give up the al-Aqsa or the Dome of the Rock. As al-Kamil told the Muslims, all he had given up were 'some churches and ruined houses', while Islamic worship continued as before in Jerusalem. There was disagreement over whether the terms of the treaty allowed Frederick to rebuild the city walls, but even with them, Jerusalem would be very difficult to defend without more territory, and especially without the Transjordan castles. Most devastating was al-Kamil's complacent assurance that once the ten-year truce was over, he would easily be able to retake the holy city at his leisure (Gabrieli 1984: 271). In the wider context of his ambitions to dominate Syria as his uncle had done, Frederick must have appeared to al-Kamil as a mere irritant.

The native barons and Military Orders were naturally disappointed that Frederick was unwilling to challenge al-Kamil for more than the return of a city they had long realised was only of symbolic importance. If the barons had been suspicious of his motives in Cyprus, they must now have realised that the emperor had no intention of extending the kingdom further; indeed, that he was interested in the crown primarily to advance his own authority in the West. Even the Galilean lands around Nazareth won back by the treaty were not restored to their former lords but given to the Teutonic Knights. Frederick might have expected the patriarch of Jerusalem, at least, to be pleased at the restoration of his church, but Gerold, who had studiously avoided him, followed the papal line (Peters 1971: 165–70, for translation of his account). He even placed an interdict on Jerusalem on the grounds that it had been polluted by the presence of the excommunicate emperor. If this sounds like cutting off one's nose to spite one's face, it demonstrated the contempt felt in the Church for Frederick's methods. Given the reports by the Arab chronicler Ibn Wasil about Frederick's favourable behaviour towards the Muslims in Jerusalem, it is hardly surprising that he did not win the confidence of the Church in Outremer (Gabrieli 1984: 271–3). Neither the patriarch nor the canons of the Holy Sepulchre, nor the monks and canons of St Mary Latin, Mt Sion or Notre-Dame de Josaphat thought it worthwhile to return to Jerusalem during the period covered by the truce.

Frederick's crown-wearing – which was not a coronation at all but a ceremonial act such as occurred on great feasts – was designed to publicise his achievement. He wrote a boastful letter announcing his achievement to Henry III of England, but the main source from the Latin East, Philip of Novara, scarcely mentions the emperor's presence in Jerusalem. An Ibelin supporter, Philip did not regard him as eligible to wear the crown. Frederick's own letter to Henry suggests that the crown he wore was in fact the imperial one, which may indicate either that he was respectful of the constitutional niceties of his position, or that he saw no real difference between his role as universal emperor and as regent for his infant son in Jerusalem. The rest of the crusade was no less ignominious. Frederick was forced to give way to the High Court when the Ibelins threatened to withdraw their vassalage because the emperor had dispossessed another baron, Alice of Armenia. When he left Acre in May 1229, a riot broke out against him and he was pelted with offal in the market. The appointment of two native barons as his *baillis*, Garnier the German and Balian of Sidon, shows how far he had been forced to climb down since 1228.

The legacy of Frederick's crusade was a decade of civil war between those who upheld and those who rejected imperial claims to rulership of the kingdom of Jerusalem, in person or by proxy. Frederick cannot be blamed entirely for this. It is true that his own behaviour in 1227–9 was politically both naïve and heavy-handed. But the situation created by the failure of male succession to the throne was impossible. Fundamentally, this problem had been the same (with a hiatus in the mid-twelfth century) since the 1130s – the need to import from the West husbands as consorts for female heirs to the throne. Where such an import was, like John of Brienne, without prospects or ambitions in the West, he could identify his interests with those of the kingdom of Jerusalem. But Frederick never intended to rule personally in the East, and seems deliberately to have blurred the distinctions between his authority as regent in the kingdom of Jerusalem and imperial overlord in Cyprus. Moreover, both Frederick and Honorius III underestimated the implications of constitutional develop-ments in the kingdom of Jerusalem since 1187. To understand why the barons of the kingdom took his feudal *faux pas* so seriously, we must turn aside to examine the internal workings of the kingdom since the 1180s.

Crown and barons in the kingdom of Jerusalem from Hattin to St Louis

The barons of the kingdom of Jerusalem in the thirteenth century have not, on the whole, had much sympathy from historians. They were, Riley-Smith concluded, a dislikeable lot: prone to 'donnish pedantry' and the 'mantling of political actions in self-justificatory and sometimes spurious legal terminology' (1973a: 229). It is tempting to view them as argument-ative and lacking the breadth of vision to suspend their constitutional jealousies for the greater good of the kingdom. A western observer, James of Vitry, bishop of Acre, was disgusted by the superficial rhetoric in which they indulged (Stewart 1895: 65). But it was this very quality of legal expertise and the ability to plead a case in court that the barons themselves prized. Ralph of Tiberias, for example, became a heroic figure among the thirteenth-century baronage for the constitutional grasp he showed in his resistance to Aimery I in 1198. Accused of plotting the murder of the king, and faced with the seizure of his fief, Ralph demanded that he be judged not by the king but by the kingdom's fief-holders in the High Court, as he was entitled by provision of the *Assise sur la ligece*, a law dating from the 1160s. When Aimery refused to grant him this right, the barons, persuaded by Ralph's oratory, withdrew their feudal service from the king. The fact that in the end Ralph's resistance was futile, and that he went into exile until Aimery's death, was less significant in baronial memories than the principle on which he had stood.

Significantly, the same principle was applied by John of Beirut in 1229 and 1231 when he persuaded many of the barons to withdraw their service from Frederick II, in the first case because of the arbitrary seizure of Alice of Armenia's fief, and in the second in response to the attempt by the imperial lieutenant Richard Filangieri to seize his fief of Beirut. The latter instance, however, showed how ineffective the recourse to twelfth-century law was in a case when the aggressor was not dependent in any case on the Palestinian barons but rather on the Italian army he had brought with him (Riley-Smith 1973a: 179–80).

Ralph of Tiberias was the stepson of Raymond of Tripoli, and after Hattin an adherent of Conrad of Montferrat; it is therefore not surprising to find him opposing the Lusignan king Aimery. In the early years of the thirteenth century he was an ally of the Ibelins. On the surface, his career appears to show how the alignments formed among the barons in the years before 1187 could continue in the next generation. This is misleading. Ralph's brother Hugh, who had fought with Raymond of Tripoli at Hattin, became an adherent of the Lusignans in 1189–90 and later of Henry of Champagne (Edbury 1997: 25). Moreover his marriage to Margaret, the daughter of Balian of Ibelin and Maria Komnena, the widow of Amalric I, suggests that baronial 'factionalism' was less significant than dynastic interests. In any case, however they pursued their interests, the barons' habits of resistance to the crown, formed in the late twelfth century when regalian rights were on the wane (Riley-Smith 1973a: 147), had already become traditional by the time of Frederick II's arrival in the East.

It was not only obstructionist opponents of the crown who took jurisprudence in the application of feudal law so seriously. Aimery I himself, and Bohemond IV of Antioch, were highly regarded for their detailed knowledge of the law; and another of the great feudal jurists of the 1220s and 1230s, Balian of Sidon, turned out to be an upholder of Frederick II's rights. Theoretical knowledge of feudal law, and the ability to turn it to practical use in the courts, seems to have been a fundamental requirement of the kingdom's barons in the thirteenth century. Moreover, the examples of Philip of Novara and Renier of Gibelet, a burgess, show how legal expertise could be a path to higher status in the kingdom. The second generation of 'lawyer-barons', such as Philip himself and John of Jaffa, the nephew of the lord of Beirut who opposed Frederick II, produced sophisticated theoretical treatises on the laws of the kingdom. By the 1260s, interest in the law had become the chief characteristic of a literate and cosmopolitan baronage in Jerusalem and Cyprus (Edbury 2003).

Why was study and application of feudal law so important, even to the point of apparently throwing away the possible advantages of western military support in 1228 and indulging in a damaging civil war from 1229 to 1243? First, it is important to note that the Palestinian barons were very unlike western lords in their manner of life. The primarily urban money economy and the fact that there were very few rural fiefs after 1187 meant that the barons tended to live much more like mercantile princes or senior clergy than like their counterparts in the West. The towns, larger and wealthier in any case than most western ones outside Italy, developed a literate aristocratic culture in which the barons played an important role. But the political conditions of the crusader kingdom also contributed to the legalistic character of its barons. The disaster of 1187 did more than remove political control of the Holy Land from the Frankish nobility. The loss or destruction of the body of written law that had developed since 1099 meant that in the thirteenth century the kingdom of Jerusalem followed a path diametrically opposite to that in western Europe. In England, for example, the period between c.1160 and 1260 is characterised by the steady growth of written assizes and a resulting body of case law to replace customary law. But the Palestinian barons had already constructed a body of written law before 1187, and its loss resulted in the reversion to customary law mediated by orally transmitted memory. Philip of Novara expressed this problem when he acknowledged that 'We know [the laws] rather poorly, for they are known only by hearsay and usage . . . and we think that an assize is something that we have seen used as an assize . . . in the kingdom of Jerusalem [the barons] made much better use of the laws and acted on them more surely before the land was lost' (Riley-Smith 1973a: 133–5). As Philip realised, it was uncertainty about the law that led to overriding concern about how it was to be applied.

The reliance on orally transmitted legal custom naturally led to the creation of a mythical past in which the dimly remembered laws could take on a constitutional significance quite different from that which had originally been intended. For example, the *Assise sur la ligece*, which Amalric I had introduced in the 1160s in order to strengthen the crown by binding rear-vassals more closely to it, came to be interpreted as a measure to limit the crown's authority. The first step was the insistence that because all vassals had done homage directly to the king, they were thus all members of the High Court; the second was the argument that sovereign authority rested not in the king alone but in the king's judgement in consultation with the High Court – and thus with all fief-holders. In the case of the *Assise*, it was not only the dislocation of 1187 that made this development possible, but, perhaps even more important, the consistent failure of adult male succession in the kingdom from 1185 onward. From 1186 to 1286 kings of Jerusalem were either minors or enjoyed limited authority because they owed their title to marriage. In such circumstances, and particularly from 1225 onward, when there was no resident king, it is hardly surprising that the barons should have interpreted feudal law as they did. In fact, the idea that sovereignty rested with the king in the High Court was not so much a limitation of royal prerogative as an idealisation of it, and an attempt to prevent imperial *baillis* or regents from assuming the authority of the crown.

The crusade of Frederick II marked only the beginning of the attempt to impose direct rule on the kingdom of Jerusalem from the West. The Ibelins' refusal to accept the authority of Frederick's *baillis* led to the seizure of their estates in Cyprus; in retaliation John of Beirut led an invasion of Cyprus that succeeded in 1229–30 in restoring Ibelin control over the island. But in 1231 Frederick sent a fleet under Richard Filangieri, an imperial official, to the kingdom of Jerusalem. Richard's Italian army besieged Beirut and took control of Tyre, but the Ibelinist barons were able to retain control of Acre, and in 1232 formed a commune to protect their interests. Up to this point it was possible for most barons to remain, like the imperial *bailli* Balian of Sidon, moderate supporters of Frederick's régime. But, just as in Cyprus in 1228, Frederick's aggression in attacking John of Beirut forced such moderates into the Ibelins' camp. There was now a principle at stake: what right had the emperor to destroy an individual baron, and if he could do this to one baron, why not to others? (Edbury 1997: 45). The Commune of Acre was thus more widely representative of baronial interests than John's personal resistance to Frederick had been.

The Commune acted as a surrogate parliament, in circumstances in which no High Court could be summoned – because for that a representative of the crown was needed – and in which withdrawing service from the crown, according to the baronial understanding of the *Assise sur la ligece*, was pointless in any case because Filangieri was using his Italian army rather than feudal service from the kingdom. Although it probably never exercised any governmental functions, the Commune did determine baronial policy until Conrad, Frederick's heir, came of age in 1243 (Prawer 1980: 54–76). During the 1230s, the kingdom remained divided between Filangieri's imperial régime in Tyre and Jerusalem, and the communards, who were loyal to Balian of Sidon and after 1239 his replacement Odo of Montbéliard. The situation would have been more dangerous had the Ayyubids not been bound by the truce of 1229.

Frederick II's truce with al-Kamil guaranteed the security of Jerusalem for ten years, ten months and ten days, which according to Islamic custom was the maximum length for a treaty with infidels. As early as 1234 Gregory IX began to alert western rulers and barons to the need to prepare to defend the holy city, and although some crusaders were diverted to the Latin Empire of Constantinople, Theobald, count of Champagne, together with the duke of Burgundy and a number of other Burgundian nobles, left for the Holy Land in 1239 (Weiler 2000, 192–206). They were followed in 1240 by an English expedition led by Richard, earl of Cornwall, the king's brother. The timing for a crusade was auspicious, because al-Kamil's death a year earlier had left the Ayyubid empire divided between his sons as-Salih and al-'Adil II. Despite a defeat near Gaza, Theobald's crusade succeeded in reinforcing the military presence in the kingdom, and between 1240 and 1242 Richard of Cornwall refortified Ascalon and negotiated a treaty with as-Salih to evict the Muslims from Jerusalem again. The treaty of 1241 reconstituted the kingdom of Jerusalem to its greatest extent since 1187: it now included the whole of the Galilee and the plain of Jaffa and lacked only the Samarian region between Jericho and Bethsan, the Judaean desert and Transjordan and the Gaza region. Even more might have been accomplished had Richard, in particular, been able to count on the support of a united baronage and the Military Orders (Jackson 1987: 32–60).

Part of the trouble was that Richard of Cornwall was Frederick II's brother-in-law and seems to have been regarded by both as the emperor's representative in the East. Certainly the Palestinian barons felt hard done by when the most important of the newly acquired territories, Ascalon, was handed over to Filangieri, strengthening his hand in the kingdom. If they had expected anything to change when Conrad entered his majority in 1243, they were disappointed, for Frederick continued to grant rights and privileges as though still king. They now developed a new constitutional argument, namely that since the rightful king – Conrad – had not come to claim his throne, he should still be regarded as a minor, and as represented through the regency of Alice, the daughter of Henry of Champagne and Isabel, who was the next in line to the throne after him. This was in fact the same claim that Raymond of Tripoli had made in claiming the regency in 1174.

Briefly the barons seemed to have gained the upper hand, when in 1243 they wrested Tyre away from the Italians and entered Jerusalem. For the first time since 1187, the Templum Domini was restored to the kingdom, through an alliance with Damascus. At the same time Galilee and the lordship of Sidon were resettled, and the future must have looked promising.

But the barons' policy of alliance with Damascus, though understandable, proved tactically disastrous. In a sense they were caught on the wrong side. Their resistance to Frederick meant that as-Salih had no need to observe any treaty, because, as he told the pope, he was bound only by treaty made with the imperial representative. As soon as the barons exercised their autonomy from the emperor, therefore, they lost his protection against the Egyptians. As-Salih, moreover, was strengthened by his own alliance with a central Asian Turkish people recently displaced by the Mongols, the Kharismians. A joint Egyptian–Kharismian invasion in 1244 destroyed the Damascene–Frankish army at La Forbie, near Gaza. Jerusalem had already been reconquered earlier in the year, with considerable savagery towards the indigenous Christian population, by the Kharismians; now many of the territories gained by the treaty of 1241 were reoccupied by as-Salih. The battle of La Forbie, though less famous today than Hattin, was no less important. The slaughter was far more comprehensive, and this marked the last occasion on which the Palestinian nobility was able to operate as an effective military force. Of the Military Orders, only thirty-six out of 348 Templars, twenty-six out of 351 Hospitallers and three out of 400 Teutonic knights escaped alive. The patriarch of Jerusalem put the total losses at 16,000 men (Richard 1999: 329–30). Frederick II blamed the recalcitrant barons and the Templars, but in fact even a baronage united behind the new imperial *bailli*, Thomas of Accerra, might have been similarly defeated, and a continued policy of non-engagement with the Ayyubids was probably no longer realistic.

St Louis, the 'king over the water'

Louis IX became king of France as a boy of seven in 1226, but the regency of his mother Blanche, and the early years of his personal rule, were shaped by internal upheavals arising from the acquisition of vast new territories by the crown since 1204. Louis may have taken the cross after hearing the news of the devastation of Jerusalem by the Kharismians, but the defeat at La Forbie was not yet known in the West, and it seems equally likely that his vow, taken in December 1244, was the result of his recovery from serious illness. In hindsight the crusade looks like the obvious means by which Louis could enact his ideology of Christian kingship. Already known for a singular piety, Louis had in 1239 acquired the crown of thorns, a relic symbolising Christ's kingship.

Despite the awful situation in the kingdom of Jerusalem, it was by no means easy to recruit for a new crusade after 1244. The pope, while

approving Louis' vow, tried to persuade him to fulfil it by joining the
crusade he preached against Frederick II at the Council of Lyons in 1245,
rather than in recovering Jerusalem. Rather petulantly, Henry III of
England at first refused to allow the crusade to be preached in England,
before giving way in the face of baronial enthusiasm – Simon de Montfort,
earl of Leicester, himself the grandson of a crusader, and Henry's two
half-brothers (both of them, confusingly, with identical names to twelfth-
century crusaders, William Longsword and Guy of Lusignan!) took the
cross (Tyerman 1988: 108–10). Elsewhere in Europe there was less enthusiasm: Haakon of Norway, despite having already taken the cross,
declined to join Louis, and neither Frederick II himself nor the pope was
keen for German barons to enlist (Richard 1999: 340).

The result was that Louis' crusade was largely a French affair, in which
the organisation and financing was undertaken by the French royal government (Jordan 1979: *passim*). Louis was given authority to tax ecclesiastical revenues to pay for the crusade, and in addition he imposed a tax
of his own on the French towns. A new port, Aigues-Mortes, was built
on the south coast specifically for embarkation, to obviate the need to rely
on Italian interests. The material preparations for the crusade, based on a
successful governmental machine, were thorough enough to ensure that
Louis could afford to take into his household nobles and their retinues
who, like Jean de Joinville, seneschal of Champagne, had run out of their
own money (Shaw 1963: 198).

Louis sailed in August 1248 and arrived in Cyprus in September. One
of the puzzles in his strategy is why, instead of making straight for the
Holy Land, he delayed so long that he had to over-winter on the island.
Not all of his forces had been able to leave at the same time; his brother
Alphonse of Poitiers, who had encountered difficulties in recruiting, did
not leave France until August 1249. It is also possible that he was as yet
undecided about the best strategy to adopt; certainly his preferred option
of a frontal assault on Egypt was only one of a number offered by the
Cypriot and Palestinian barons (Richard 1992: 117–19). But Ayyubid
power was still, as in 1218, based in the heartland of Egypt. Louis realised,
moreover, that there were many ways in which his forces could be frittered
away in alliances with this or other rulers in Syria, but that the best hope
of lasting success lay in inflicting a single decisive military blow to the
rulers in whose control the Holy Land lay. If this coincided with Louis'
own traditional view of what a crusade ought to be, it was still first of all a
strategically sound policy.

The seizure of Damietta, which thirty years earlier had taken fourteen months, was accomplished in 1249 in a single beachhead landing.

Joinville, whose autobiographical memoir of his king provides the fullest and most vivid account of the campaign, captures the excitement and intensity of the occasion. The arrival of John of Ibelin, count of Jaffa, evidently impressed him. As for Louis, at his first sight of the enemy on disembarking, he was fully prepared to charge at them (Shaw 1963: 202–5). In the event, there was scarcely any need for a fight, for the Egyptians simply abandoned Damietta. News had spread of the sultan's illness, and it was assumed that he was already dead. The crusade could scarcely have had a more auspicious beginning.

Because it was already June, there was no time to continue towards Cairo before the Nile flooded. Louis consolidated his position in Damietta, first beating off an attempt to recapture it, then securing its future as a French conquest. Louis seems to have regarded it as a personal – or perhaps a national – conquest; there was certainly no thought either of annexing it to the kingdom of Jerusalem or of using it as a bargaining counter with the sultan for the return of Jerusalem. A Frenchman was appointed archbishop and the cathedral, founded in 1220, was restored. Joinville was troubled by the king's insistence on retaining control over all captured property, including food, rather than dividing it up according to crusading custom among the army (Shaw 1963: 206–7). Louis was adopting the same fundamental view as Pelagius had in 1219, that territories conquered by crusaders were within the gift of crusaders. But because he was the undisputed leader of his crusade, there was no repetition of the infighting that had followed the capture in 1219. This makes it easy to miss the significance of what he was doing. Throughout crusading, a fault-line ran through crusaders' perceptions of the status of conquered territories. On the Second Crusade, Louis VII's intention to enfeoff his own vassal with Damascus lost him the support of the Palestinian barons; the proposed conquest of Egypt in 1177 may have foundered for the same reasons, and the disagreements between Richard I and his followers and the native barons over the fate of restored territory jeopardised the success of the Third Crusade.

It was still not decided whether Louis would follow the exact route of the Fifth Crusade. The more cautious strategy of first seizing Alexandria, the other Ayyubid access to the Mediterranean, was favoured by most of the barons. But the king's brother Robert of Artois attacked this plan on the grounds that it was better to attack the head rather than the tail of the serpent (Shaw 1963: 210), and Louis favoured this advice. Joinville leaves little doubt that fraternal affection played a role in the decision, but in fact Robert's advice was sound: it made sense to attack Cairo while the sultan was still ailing, and besides, the route south from Alexandria led through desert, so the army would have had to return to Damietta in any case before marching south. Louis' decision was consistent with his preference for as immediate a confrontation as possible with an Ayyubid army.

The problem with such a strategy, however, is that it required for success not only a measure of luck but good generalship. Louis' previous campaigning experience in Poitou had been largely successful, but there he was up against the inept Henry III and on his own ground; he may have been a competent soldier but he was certainly no Lionheart in his grasp of tactics. The route south to Cairo led first to the fortress of Mansurah, built by al-Kamil in 1219. Before they could attack it, in February 1250, the crusaders had first to cross a branch of the Nile. Louis, who must have known this, had his engineers build a rampart to cross the river, but this had to be abandoned in the face of fierce resistance. The fording place eventually chosen for the crossing was not ideal, because it entailed splitting the army. Instead of waiting for the king with the bulk of his troops, Robert, leading the vanguard, stormed the Egyptian camp outside Mansurah and, emboldened by easy success, blundered into the town itself. Neither the Grand Master of the Templars – who was incensed that Robert, rather than he, had been given the honour of leading the army across the ford – nor the king could stop him. In the narrow streets of Mansurah, Robert, with William Longsword and many of the English, were slaughtered.

At this point the crusade might yet have achieved something, for as-Salih had died in November 1249, and the new sultan, not yet secure in his inheritance, was clearly unnerved by the presence of a Frankish army in the Delta. While negotiations with the sultan stalled, however, the Egyptians outmanoeuvred the crusaders by navigating the lower reaches of the Nile and depriving them of fresh supplies from Damietta. Disease, according to Joinville caused by eating the river fish that had themselves fed off corpses, spread rapidly through the camp, until perhaps only a few hundred knights remained in fighting condition. The retreat, in early April, was hampered both by the illness of the king himself and by simple mistakes – the failure to cut bridges behind them, for example, made the Egyptians' pursuit easier. A day short of Damietta, the crusaders capitulated. Louis himself was captured. The humiliation of the crusade was complete. Damietta was given up as ransom for Louis himself, and the figure of 800,000 bezants set for the army.

Louis' failure was all the more disappointing because the preparation had been so meticulous and the conditions so generally favourable. Despite superficial similarities with the Fifth Crusade, the military campaign failed for different reasons. There was no precipitate dash for Mansurah ahead of the floods, no division in the leadership. Louis' strategic decisions were eminently sensible. It was at the tactical level, rather, that mistakes were made. Joinville, out of loyalty for his lord, declined to blame the one figure whom the ample evidence of his chronicle accuses for the débâcle at Mansurah, Robert of Artois. As a soldier, Joinville knew that a commander was only as good as his subordinates, and the ultimate blame for failure to control his army must rest with Louis himself. But the assault on Mansurah had been compromised even before Robert had forded the Nile, by the failure to cross in a position where the whole army could operate together. Despite his foresight in bringing engineers to build a dyke, Louis was defeated by the geography and topography of the Nile: the Egyptians simply dug away the bank on their side to widen the river. The final defeat was the result of being trapped without supplies between an army in Mansurah and a fleet between themselves and the safety of Damietta.

The great irony of Louis' failure was that the final débâcle took place against the background of the collapse of the Ayyubid régime in Egypt. As-Salih's successor, Turanshah, was murdered by his own Mamluks just days before the king was released from captivity, probably because they resented the fact that he had brought advisers with him from Baghdad. The political union between Egypt and Syria forged by Saladin in the last quarter of the twelfth century was broken; the Ayyubids were limited to Syria, while the Mamluks kept Egypt. This situation naturally benefited the kingdom of Jerusalem, and Louis remained in the East for a further four years in order to guide its affairs. To an extent this was forced on him by the need to extricate something from the crusade; he had left most of his army in captivity when he sailed to Acre in May 1250, and only managed to arrange their release by a new treaty with the Mamluks in 1252. His presence in Acre made him *de facto* ruler of the kingdom of Jerusalem; that he was recognised as such by neighbouring states is evidenced by the Ayyubid invitation that he should join them in the reconquest of Egypt in July 1250. At one point it looked as though Louis might have been able to restore Jerusalem as the price of his alliance with the new Mamluk régime against Ayyubid Damascus. But he was understandably cautious, and the chance was lost when the Khalif patched up the Mamluk/Ayyubid division in 1253. Louis spent his time in refortifying the kingdom's towns, notably Jaffa, Caesarea, Sidon and Acre. A further ten-year-and-ten-month truce was agreed for most of the kingdom, excluding Jaffa, before Louis returned to France in 1254. One of the most significant results of the crusade was the permanent presence of a small force of knights paid by the French crown, under the command of Geoffrey de Sergines, who became seneschal of the kingdom. This was to be a commitment maintained until Louis' death in 1270.

Crusading resources between the Fourth Lateran Council and the Second Council of Lyons (1274) were overwhelmingly directed against Egypt. There were good strategic reasons for this, both in the long and the short term. Egypt, though subject to violent changes of régime in 1171 and 1250, was fundamentally a politically stable country. Its geographical isolation from the other centres of power in the Islamic world meant that its conquest would not necessarily involve the defence of difficult frontiers: if it was difficult to conquer, it would also be comparatively easy to defend. Moreover it was agriculturally self-sufficient, because of the natural fertility of the Nile Delta. The spice trade from the Indian Ocean fetched up in Alexandria and Damietta before being filtered through the Syrian markets. Economically, Egypt was a rich prize. In both 1218 and 1248 Egyptian campaigns also made political sense. Al-'Adil's refusal to give battle in Palestine in 1217–18 forced the crusaders into a diversionary attack. In 1248, as-Salih's war with Aleppo offered the prospect of stretched Ayyubid resources, and the sultan's decline and death in 1249 only confirmed this judgement.

It should not be forgotten, however, that during the period up to 1274 crusades were still launched to other destinations than Egypt. Frederick II's crusade and that of 1239–41, which was a natural consequence of the treaty of 1229, showed that a role was evolving in crusading policy for a balance of defensive warfare and diplomacy in Palestine. In the 1260s Sultan Baibars' assault on the Crusader States returned the Holy Land to the forefront of crusade policy. James I of Aragon set sail for Acre with high hopes in 1269, but turned back when he fell sick, and the few hundred of his knights who reached the Holy Land were able to do little more than help in the defence of Acre against Baibars. Edward, the heir to Henry III of England, also landed in Acre in 1271, but his force was so small that he dared not risk battle and contented himself with a raid or two and the repair of Acre's walls. Both James' and Edward's campaigns were intended to form part of a much larger crusade, led again by Louis IX. A combination of Louis' own misjudgement and the self-interest of his brother Charles of Anjou diverted the major part of the crusade to Tunis, where in August 1270 Louis died. The crusade splintered into ineffective smaller operations, the anticipated alliance with the Mongols failed, and the opportunity to recover what Baibars had conquered was squandered. It was to be the last chance before the final sunset fell over the mainland Crusader States.

CHAPTER 10

The later crusades, 1274–1336

W hy did crusading fail so often in its military aims, and what did Europeans make of such failure? The period from 1274 is distinguished by an impressive deeper level of reflection on crusading and how to make it work; yet in the end the fall of Acre in 1291 brought no response comparable to the Third Crusade after Hattin. In this chapter the final disintegration of the Crusader States is set against a background of continuing failures in papal attempts to organise crusades. The new genre of 'recovery treatises' is used to examine some of the structural flaws in crusade planning in the late-thirteenth and fourteenth centuries. Finally, some new approaches to the Holy Land and the problem of Islam are considered.

Mongols, Mamluks and the Crusader States

Even while wintering in Cyprus in 1248, St Louis had looked farther afield, to the world beyond Islam. By the time he arrived in the East, the military threat posed to the civilised world by the Mongols was well known. Matthew Paris, the chronicler of St Albans, captured the mood of apprehension in northern Europe when he reported that in 1239 even the annual Baltic herring fleets did not put in at Yarmouth for fear that the Mongols were on the move, hundreds of miles to the east (Luard, vol. III 1876: 488) Like many contemporaries, he believed that the Mongols were the descendants of the hordes of Gog and Magog, who according to tradition had been confined to the frozen wastes of Asia by Alexander the Great. Whoever they were, they were a sufficient threat for the heir to the German throne, Conrad IV, to mobilise a German army to defend northern Europe against them in the spring of 1241, and German bishops began to preach a crusade against the Mongols (Jackson 1991: 1–18). In that year, after sweeping in an arc of destruction through southern Russia, Poland and Hungary, a Mongol army stood poised to seize Vienna, before abandoning the European campaign because of the lack of sufficient pasture and the need to elect a new khan. In 1243 a separate Mongol force defeated the Seljuqs of Asia Minor, and tribute was demanded of both Aleppo and Antioch. The pope, alerted to the threat to the Christian lands of the East, sent an embassy to the new khan, Guyuk, in Karakorum, in 1245. Innocent IV asked by what authority the Mongols attacked the lands of Christians who had offered them no threat. The reply was chilling. 'We do not understand your words', Guyuk told the Franciscan envoy, John of Plan Carpino, and he continued, 'Through God's power, all lands, from east to west, have been granted to us' (Dawson 1955: 85–6). The khan demanded nothing less than the total submission of all rulers, temporal and spiritual, Christian and Muslim, to himself as universal world ruler.

From what had been observed so far of Mongol military capability, this was a very real threat. But it was also logical to those Europeans who had been tempted to identify the Mongols with the mythical Christian ruler of the East, 'Prester John'. The tradition of a powerful eastern ruler who would march west to smash the forces of Islam from behind had been alive since the Second Crusade. After a period of dormancy, expectations had been revived in 1220 during the Fifth Crusade, when contact had actually been made with the Mongols, albeit inadvertently, for the first time. The experience of the 1240s seemed to support the legend's claims of the might of this new power in the East, but not necessarily of its goodwill towards western Christendom. In fact the expectation that the Mongols might be sympathetic to Christendom was more than wishful thinking. The Church of the East, which began to spread throughout Persia in the sixth century when the followers of the heretical patriarch of Constantinople, Nestorius, were exiled from the Byzantine Empire, had successfully proselytised in central Asia, and some of the tribes who made up the Mongol federation were members of this Church. At the time of Louis' capture of Damietta, rumours spread as far as England of the conversion of the khan by an Indian monk (Luard 1880: 87). The Syrian chronicler Bar Hebraeus believed that Guyuk was himself a Christian. One of the wives of Mongka, who became khan in 1255, was Christian, and hopes were high that he would himself convert. Both the pope's envoys in the 1240s, John of Plan Carpino and Andrew of Longjumeau, and the Dominican sent by Louis from Acre in 1253, William of Rubruck, pursued the conversion of the Mongols alongside their request for alliance. William even took with him a specially designed portable chapel with little statues to represent the main characters of the Gospels (Shaw 1963: 197–8). The Mongols regarded this as, at best, puzzling, at worst, insulting: they had their own perfectly good religious traditions, and besides – remembering eastern Europe – why should they adopt the faith of people whom they had conquered so easily? But the Mongol khans were shrewd enough to recognise that their own ambitions of conquering the Near East would be enhanced by acting in concert with Islam's western enemies, and in 1262 Hulagu, the Mongol leader in the Near East, offered an alliance to Louis IX. An uneasy series of temporary alliances with the Mongols followed in the second half of the thirteenth century, but it was always an unequal relationship, and nothing substantial came of any of them.

The Mongol project of world domination began formally in 1253, with the targeting of three arenas of conquest – Europe, China and the Near East. Hulagu, the il-khan responsible for the conquest of the Near East, caused tremors throughout the East by his sack of Baghdad in 1258. The Abbasid khalifate, which had been in existence since the eighth century, was abolished. Early in 1260 Aleppo and Damascus were conquered: Ayyubid Syria seemed to have been destroyed in a single short campaign. While Pope Alexander IV and the Franks in the East watched nervously, Bohemond VII of Antioch and Hethoum I of Armenia, his father-in-law, both submitted to the Mongols: Antiochenes even took part in the sack of Damascus, thereby finally achieving what had been intermittently attempted since the beginning of the Crusader States. The pragmatic attitude of Antioch, though deplored by the patriarch of Jerusalem, was rewarded by the restoration of former territories taken by the Ayyubids. But any hopes of accomplishing the restoration of the Holy Land on the backs of the Mongol cavalry were dashed in September 1260 at one of the most decisive battles to take place in the Near East since the first Arab invasion in the seventh century. At 'Ain Jalud, in Galilee, the Mamluks of Egypt routed Kitbuqa, Hulagu's general, and the Mongols were forced to withdraw across the Euphrates. Those Christians who had sided with the Mongols would pay dearly for choosing the wrong side.

The architect of Mamluk success during the 1260s was Baibars, who not only reconfigured the territorial empire fashioned by Saladin a hundred years earlier, but rediscovered the religious zeal of the later twelfth century (Thorau 1992). His victories were built on the foundation of governmental reform in Egypt that delivered ultimate power to a military élite, the *mamluks*, or slaves, usually Christians from Asia Minor, who were educated in Islam and given military training and their freedom. In 1263 he failed to take Acre, but succeeded in pillaging Nazareth and Mt Tabor, signalling his deep-rooted hostility to Christianity. The balance of power that had been created between al-Kamil and Frederick II and renewed by the treaty of 1241 and by the diplomacy of Louis IX was upset. The Franks of Outremer were caught off guard by this new political configuration, but in any case the barons of the kingdoms of Jerusalem and Cyprus and the Antiochenes and the military Orders lacked the resources to deal with an enemy who had no other distractions and who was determined to put an end to western settlement in Outremer. In 1265 Caesarea and Arsuf fell; in 1266 Safed, the great Templar castle in Galilee. Equally worrying was the destruction of an Armenian army, which left Antioch defenceless, and two years later the city itself fell amid scenes of anti-Christian savagery. Baibars himself boasted of the massacre of monks and priests and the firing of the cathedral of St Peter (Gabrieli 1984: 310–12). The destruction was so complete that the city was never again to amount to anything more than a provincial village. To the south, Jaffa surrendered in the same year, and the process of reducing the Military Orders began with the seizure of their castles: Krak des Chevaliers and Montfort both fell in 1271. The systematic nature of Baibars' conquest is demonstrated by the destruction of everything that he did not intend to reuse himself; he left no room for accommodation because he was not interested in coming to terms with the Franks. It is also significant that he treated the indigenous Christian population in the same way as he did the Franks, massacring the peasantry near Acre and the inhabitants of Qara, near the Orontes. Perhaps the only thing that saved the Franks in Outremer for another twenty years was that Baibars was as virulently anti-Mongol as anti-Christian, and during the last few years before his death in 1277 he concentrated on trying to reduce their power in the Near East.

Just as in 1190–2, the Franks were unable to put aside constitutional wrangles even at times of extreme danger. The 'war of St Sabas' between Venetians and Genoese in Acre had dragged in the Military Orders and local baronage as well from 1256 to 1258, but at least the external threat had been less grave. In 1277, however, the Franks 'managed to make things infinitely worse for themselves by indulging in the luxury of two kings' (Mayer 1988: 271). This was because the crown was disputed between two descendants of the daughters of Isabel I – Hugh III of Cyprus, the grandson of Alice, daughter of Isabel by her marriage to Henry of Champagne, and Maria of Antioch, Isabel's granddaughter from her final marriage to Amalric II of Jerusalem. Most barons preferred Hugh's claim, doubtless in part because he was a man, but on the principle that he was descended from the more senior line of Isabel. In 1277, however, Maria, realising she would get nowhere with her claim, sold it to Charles of Anjou, king of Sicily. Although the Hohenstaufen line had finally been extinguished in 1268, Charles seemed to regard the claim to the crown of Jerusalem as belonging to the kingdom of Sicily, which he had been granted by the pope, rather than to a particular dynasty. Charles never attempted to claim what was left of the kingdom in person, but he interfered sufficiently, by sending Roger of San Severino as *bailli* to represent him, to divide the loyalties of the remaining barons. The Templars, aligning themselves behind Charles, prevented Hugh from entering Acre. Hugh returned to Cyprus, and the kingdom was left once more kingless.

By the time that Charles' administration collapsed in 1282, in the wake of the revolt against his rule in Sicily, individual barons had already begun to make separate treatises with the Mamluks for their own security. Palestine and Syria were returning to the situation of *c*.1099–1110, albeit in which Christian and Muslim positions were reversed, with individual lordships preserving a precarious autonomy for as long as possible in the face of a heavy military superiority. Division of territories was common: for example, at Chateau Pèlerin, on the coast south of Haifa, the town and its fortifications, with their cultivable land, gardens and vineyards and sixteen surrounding villages, remained Frankish, while eight villages belonged to the sultan Qalawun and the remainder of the lordship was divided between Franks and Muslims (Richard 1999: 111). The notion of divided ownership required a joint Frankish and Mamluk administration to collect revenues and give justice. Although the odds were weighed heavily in favour of the Mamluks, who insisted on Frankish neutrality in such areas in the event of a crusade, it was the best that the remaining Franks could hope for.

The death of Abaga, the strong pro-western Mongol commander in the Near East, in 1285, left the way open for a fresh assault on the last Frankish lordships. This began in the north, with the seizure of the Hospitaller castle of Margat in 1285 and Latakia in 1287. Qalawun's campaign was aided by chronic warfare between Pisa and Genoa, in which Venice joined the Pisan side. Italian possessions in the Levant were dragged into the war, with the result that in 1289 Tripoli, from which the Genoese had expelled the comital administration, fell to Qalawun. The city was completely destroyed and the population either killed or enslaved; naturally, the smaller towns on the coast fell in their turn or capitulated. Amid concern over the fate of Acre, western help began to arrive, and Pope Nicholas IV launched a crusade early in 1290. Promises of immediate contingents for the city's defence were secured from James II of Aragon and Edward I of England. The truce covering Acre was broken at Easter when a group of Italian crusaders killed some defenceless Muslim peasants trying to go to market in the city. Even Qalawun's death in the summer of 1290 could hardly slow the preparations now clearly under way for a massive Mamluk assault on Acre.

Acre had been the heart of the Frankish presence in the East for more than a century. The population, which had probably doubled in the thirteenth century, stood at around 20,000 – easily the largest in the Crusader States. The walls had recently been reinforced, and the city could be supplied with extra manpower and food from the sea. Moreover, at last the kingdom had an adult male king whose authority was unquestioned – Henry II, son of Hugh III of Cyprus, whose coronation in 1286 had been fêted with two weeks of festivities in Acre. Henry arrived in the city early in May with a detachment from Cyprus and refused the sultan's terms of surrender. But his confidence was misplaced. By the middle of May the Mamluk sappers had undermined a tower in the outer wall, and the Templar Grand Master William of Beaujeu was killed in the fighting. The evacuation of non-combatants began as the fighting spilled into the streets, and resistance coalesced around the Military Orders' fortified bases. Towards the end of May the Templars' keep fell; the Franciscans, Dominicans and Claresses were massacred and many women and children who had not been able to leave with King Henry were enslaved; others, including the patriarch of Jerusalem, drowned when the boats used to evacuate them were overloaded and sank in the harbour. A moving description by Ricoldo of Monte Croce, the Dominican missionary in Baghdad, of lines of prisoners being led off into slavery brings us full circle from the crusaders' massacre of the inhabitants of Jerusalem in 1099 (Röhricht 1884: 258–96).

Tyre, Beirut and Sidon, observing from afar the carnage at Acre, saw no point in resistance; not only could they expect no immediate help from the West, but the Mamluks' siegecraft was simply too effective to withstand. The Frankish and Syrian Christian population of Cyprus expanded hugely from refugees, but very many Franks simply disappeared into the Mamluk labour force. Christian pilgrims in Egypt, such as the Franciscan Angelo of Spoleto in 1303, visited prisoners, and the diplomatic exchanges between James II of Aragon and the sultan indicate that there were Templars in captivity as late as the 1320s (Jotischky 2004). Angelo also reported the presence of 'renegade Christians' – some of them friars who had converted to Islam.

The Mamluk campaign against the Frankish settlements, begun by Baibars in the 1260s, was not simply a continuation of the intermittent warfare between western Christians and Muslims that had characterised the region since 1098. Baibars and his successors Qalawun and al-Ashraf succeeded in accomplishing what, uniquely before them, only Saladin had attempted – the utter eradication of a Frankish power on the mainland. From 1099 until 1187, and then from 1193 to the accession of Baibars, Muslim military leaders – whether Seljuq, Ayyubid or Mamluk – had been prepared to acknowledge the reality of Frankish settlement in the Levant. But the destruction of captured towns like Antioch, Acre, Sidon and Tripoli was a deliberate policy to ensure that even the physical memory of the Franks would disappear. As a result, the continuity of an urban coastal civilisation in West Asia that stretched back into antiquity was disrupted. The Franciscan pilgrim Leopold von Suchem, travelling in the Holy Land in the 1330s, describes wandering along a desolate coastline where once cities and palaces had stood (Stewart 1895: 49–50, 60, 63).

The papacy, crusading and the Holy Land, c.1274–91

It is tempting to assume that because no major expedition landed in the East between 1249 and the fall of Acre, the West lost interest in crusading. This is entirely mistaken. Particularly from 1274 onward, papal attention on the fate of the Crusader States in the second half of the thirteenth century was intensive, and major expeditions were either planned or launched in 1269–70, 1276 and 1290. That none of them had the intended result was due to a combination of bad luck and divided strategic interests, but above all to the much greater complexity of European politics in the second half of the thirteenth century. Paradoxically, the period from 1274 onward was also one of increased clarity and articulacy in the planning and theorising about crusades, both in the development of a canonical underpinning for crusading and in the wide-ranging practical measures designed to enable crusades to become more effective.

One outcome of the failed crusade of 1270 was the election of a pope who would once more put the Holy Land, rather than either national interest or the problems of Italy, at the heart of his policy. The installation of the papalist Charles of Anjou, brother of Louis IX, in Sicily, and his execution of the last Hohenstaufen heir, Conradin, in 1268, had offered the hope of a Mediterranean unity that might make a joint crusade possible. But Charles' self-interested diversion of Louis' army to Tunis, his own failure to rescue his brother, and the inability to coordinate Aragonese and English forces with Louis' army, meant that the opportunity was lost. Tedaldo Visconti was in the Holy Land with the Lord Edward's forces when he was elected pope as Gregory X in 1271. The Second Council of Lyons, summoned by Gregory in 1274, was self-consciously modelled on Fourth Lateran. Like Innocent III, Gregory saw the crusade as the pinnacle of papal policy, to be supported by a substructure of institutional reform within Christendom. But the very breadth of his vision, which included the reunion of Latin and Orthodox Churches, may have distracted the energies of Christendom from the single-minded pursuit of the crusade to the East. Gregory saw no merit in hurrying into a crusade; instead he sent out requests for discussion papers to be tabled at the council so that the wise heads of Christendom could tackle the underlying organisational problems that had so beset the 1270 expedition. This approach was to signal the arrival of a new literary genre – the 'recovery treatises', or proposals for the best way to recover the Holy Land.

Some of these treatises were couched as complaints, like that of an anonymous Franciscan who wrote a *Compilation of the Scandals of the Church* (Throop 1940: 69–104). Others were more balanced. The Dominican Master-General, Humbert of Romans, introduced his proposal (*The Three-Part Work*) by reviewing current criticism of crusading (Throop 1940: 147–83; Schein 1991: 28–36). Such documents give us insights into why contemporaries thought crusading was failing – for example, poor leadership, abuses of the taxation of clerical income, and God's disfavour towards the use of violence – but they do not tell us how widespread or representative such views were. Scepticism about crusading is certainly more widespread in the sources in the later thirteenth century: there is evidence of criticism of some kind against most thirteenth and fourteenth century crusades (Tyerman 1985: 105–6). But, as Humbert's memorandum shows, fundamental doubts about the merits of crusading were mixed with displeasure at the policies that had led to defeat and to general despair that God had abandoned the crusading ideal. A good deal of the thirteenth-century criticism of crusading really consisted of complaints about how they were financed, which is not the same thing as scepticism about the project itself (Siberry 1985: 111–49).

One of the problems that had dogged crusades since the early thirteenth century, and that the Second Council of Lyons was designed to address, was the diffusion of targets. According to Matthew Paris, Richard of Cornwall's crusaders took a public oath to set out for the Holy Land, '[L]est they be prevented from fulfilling their vows by the delaying tactics of the Roman Church, or turned aside from their vows to shed the blood of Christians in Greece or Italy' (Luard 1876: 620). Gregory IX had in 1239 authorised a crusade against the Byzantine 'empire of Nicaea', in order to protect the failing Latin Empire of Constantinople. In 1245, Innocent IV tried to persuade Louis IX to fulfil his crusading vow in the papal crusade against Frederick II rather than going to Egypt. To popes, there was no conflict of interest here. For one thing, the war against the Hohenstaufen could be rationalised as a necessary action to bring about the unified conditions in the west in which a crusade to the East could succeed. Besides, it was a matter of the logical application of the principle of a justifiable war for the defence of the Church. But to the English crusaders in 1239 – or perhaps to Matthew Paris – it looked as though their sense of priorities – in which the Holy Land represented the height of crusading – was not shared by the papacy.

Some of the doubters, such as William of Tripoli, the Dominican provincial in the Holy Land, whose *State of the Holy Land* advocated peaceful conversion instead of war, represented a minority view among friars. Roger Bacon, the English Franciscan natural philosopher, had as early as 1250 argued that war would never succeed in recovering the Holy Land because each crusade simply instilled greater resentment among a new generation of Muslims, and thus contributed to an endless spiral of violence (Siberry 1985: 207–8). This was itself a reflection of the fundamental critique of crusading in Ralph Niger's *Military Affairs* (1187), but by the mid-thirteenth century Ralph's gloominess had been succeeded by a more positive alternative approach to crusading through conversion. If Roger's views seem hauntingly prescient today, they did not command sufficient respect in 1274. Humbert's response was that, desirable though the peaceful conversion of Muslims would be, it was impossible without securing territory first. The thirteenth-century kingdom of Jerusalem, lacking the rural hinterland as far east as the Jordan, where the Muslims had largely lived before 1187, had a much higher proportion of Christians to Muslims. Conversion, therefore, could only take place through missionary work in hostile territory, and the experience of Franciscans and Dominicans was that this led to the martyrdom of the missionaries (Siberry 1983: 103–10).

Against this general background of conflicting priorities and mixed signals about the purpose and efficacy of military expeditions to the East, Gregory X attempted to place Jerusalem once more at the forefront of crusading policy (Schein 1991: 20–50). The crusade was preached and revenues collected throughout Christendom, from Frankish Greece to

Greenland. But familiar problems arose: Philip III of France was still fighting the king of Castile in 1275, while the new German emperor-elect, Rudolph, did not succeed in defeating the rival claim of the king of Bohemia until 1278. The novel feature of Gregory's crusade was to have been an alliance with the Mongols in the Near East, cemented in an embassy sent to the Council of Lyons. One of the Mongols' ambassadors was a Dominican missionary, David of Ashby, who had first-hand knowledge of their military tactics. From the surviving portions of a treatise he wrote for the pope, it is probable that the alliance was premised – from the Mongols' perspective – on western forces fighting under Mongol command and following Mongol strategy (Richard 1949: 291–7).

Gregory X's crusade had every chance of success in inflicting a military defeat on Baibars, but this alone would not have saved the Crusader States. A more permanent commitment to the Frankish remnant in the East was needed. Some strategists, including the heads of the Military Orders, recommended abandoning the traditional 'massed crusade' in favour of smaller, more focused and short-term strategic expeditions that would accomplish specific aims. This kind of crusade, which in the fourteenth century would come to be called the *passagium particulare* as opposed to the *passagium generale*, has of course always been part of the history of the Crusader States. In 1110, the capture of Sidon had been made possible by the armed pilgrimage of the king of Denmark; in 1122, the Venetians had rendered similar service in the siege of Tyre, and the expeditions of Henry the Lion in 1170 and Philip of Flanders in 1177 also fall into this category. St Louis' policy of funding a small permanent force of knights in the Holy Land from 1254 onward (Jordan 1979: 79) provided a model for William, patriarch of Jerusalem, to demand in 1267 the commitment of European rulers to a standing army in the east (Schein 1991: 18). The experience of trying to coordinate different national forces, all leaving at different times, which had proved unmanageable in 1270, doubtless added weight to such opinion. There is evidence that the policy was tried, but although small contingents were sent east in 1272, 1273 and 1275, Gregory regarded these as holding forces to prepare the way for his 'passagium generale'.

Gregory's death in 1276, and the succession of a series of short-lived popes over the next ten years, put paid to his *passagium generale*. The collection of tithes for Gregory's crusade continued, but papal policy, whether by design or accident, once again subordinated the needs of the Franks in the East to the Church in the West. Nicholas III (1277–80) was accused of spending the Lyons tithes on rebuilding the Vatican (Schein 1991: 57), while Martin IV (1281–5), himself a Frenchman, openly colluded with the French king Philip III's war against first Castile and then Aragon, which both king and pope were disposed to view as itself a crusade. But the major obstacle to the realisation of Gregory's crusade was the posturing of Charles of Anjou, king of Sicily. The revolt against his rule in Sicily in 1282, known as the 'Sicilian Vespers', resulted in a flurry of crusading activity, but all directed against Christians – the Byzantine emperor Michael VIII Paleologus, who had fomented trouble among the Greek-speaking population of Sicily, and Peter III of Aragon, who took advantage of the revolt to invade the island.

The Vespers themselves were the consequence of Charles' own ambitions. Determined to create a Mediterranean empire on the axis of Sicily and Constantinople, he persuaded Martin IV to authorise a crusade in 1281 against the 'schismatics and usurpers' of the Byzantine Empire. Martin's nationalist chauvinism thus overturned the policy espoused at Second Lyons of pursuing union between Latin and Orthodox Churches. If we look at the complicated situation from Martin's perspective, it is possible to credit him with believing that Charles really would use his imperial ambitions for the good of the Crusader States – why else, after all, had he bought the title of king of Jerusalem in 1277? Perhaps Martin was in fact far-sighted enough to realise that the only hope for the kingdom of Jerusalem was the protection that could be offered by a single Christian ruler holding sway throughout the Mediterranean. This, in effect, was what Honorius III had hoped for back in 1225 when he sanctioned the marriage of Frederick II to Yolanda. But in the 1280s it all went spectacularly wrong. Michael VIII responded to the threat of the crusade against him by precipitating the Sicilian revolt, and in January 1283 the crusade was preached against the Sicilian rebels. In 1285 Philip III died leading a disastrous French crusade against Aragon.

Even after the fall of Tripoli in 1289, the new pope Nicholas IV (1288–92) was still issuing indulgences for the Sicilian crusade while making plans for the defence of Acre. Anti-papal sentiment at the diffusion of crusading policy showed its disgust in the speech put into the mouth of a Templar ambassador to the West by the Sicilian chronicler Bartholomew of Neocastro: 'You could have saved the holy land [by using] the might of kings . . . but instead you chose to attack a Christian king and the Christians of Sicily, arming kings against a king to recover the island' (Housley 1982b: 77). It would be unjust, however, to blame Capetian interests or papal policy alone for delaying a crusade to the East. Edward I of England, asked by Martin IV in 1284 to lead a crusade, frustrated papal attempts to levy a tithe until 1288, and then became embroiled in wars in

Wales and Scotland. As Housley has observed, crusading could not take place, either in the minds of planners or in reality, in a political vacuum (1995: 262). Kings always had other commitments, and part of the responsibility of popes had always been, since the 1170s if not earlier, to help to sort out dynastic and territorial conflicts so that the business of crusading could continue.

In fact Nicholas IV was to prove more effective than any pope since Gregory X in organising a *passagium generale* for the Holy Land. In summer of 1290 he sent a small force of Italians to assist in defending Acre; the following spring, the crusade itself was proclaimed for 1293. As in 1270 and 1274–6, this was to be an international enterprise, involving Edward I, Alfonso III of Aragon and his brother James of Sicily, and Charles II of Anjou. Plans for the crusade continued to proceed after the fall of Acre in May 1291. As far as western Christendom was concerned, all that had changed was the need for a new military strategy, since no bridgehead remained from which an attack on Qalawun could be launched. Promised participation widened even further, with the expectation of an end to the Venetian–Genoese war. Of the major western powers, only Philip IV of France refused to become involved. But his stance had a knock-on effect on the crusade, for the threat he posed to English Gascony meant that Edward could not afford to leave the West. Coincidental deaths also spoiled the pope's plans. Rudolph of Habsburg, Alfonso III and the Mongol il-khan Arghun all died in 1291: there would be no single successor in Germany until 1298, while the accession of James of Sicily to the throne of Aragon meant the eruption of the Sicilian–Angevin war once again, and Arghun's death meant that the opportunity to use Mongol military power in the East, so tantalising a possibility since 1264, had disappeared. In 1292, Nicholas IV himself died. Although he achieved nothing, he was widely recognised at the time, and has been by modern historians, as a crucial figure in the history of crusading (Atiya 1938a: 45–6; Schein 1991: 90–1). The policy he pursued during his pontificate – notably the economic blockade of Egypt and the attempt to reform the Military Orders – not only stemmed directly from the recovery treatises, but were to form the basis for crusade planning in the fourteenth century.

The difficulties of crusading in the fourteenth century

Briefly, in the early months of 1300, it looked as though Jerusalem might be recovered for Christendom. The Mongol il-khan, Ghazan, won a dramatic victory over the Mamluks at Homs, in northern Syria. When the news reached the West, diplomatic efforts were launched to revive the Frankish–Mongol alliance, this time with Armenian help. This was the best opportunity for recovering the holy city since the 1240s, and it seemed particularly apt that in 1300, the year proclaimed by Pope Boniface VIII as a 'Jubilee Year' for Christendom, Christian forces – albeit Armenians rather than Franks – should have entered Damascus and Jerusalem in triumph (Schein 1979: 805–19). According to western chroniclers, Ghazan wrote asking for westerners to come and resettle the Holy Land. Henry II of Cyprus, eager to press the advantage, sent sixteen galleys to raid Alexandria. But if the episode showed how easily the Mamluks might be defeated in battle, its aftermath demonstrated how ephemeral the results of such victories were in reality. After a brief and largely symbolic occupation of Jerusalem, Ghazan withdrew to Persia, and a year later was unable to coordinate his forces with the Cypriots. In 1303 he once again invaded Syria, but this time he was defeated by the Mamluks, and a year later he died. In fact, what looked in western chronicles like an apocalyptic liberation through the instrument of God was nothing more than an episode in the long-running struggle between Mongols and Mamluks for mastery of the Levant (Schein 1991: 167–75). The irony is that the instrument of God, Ghazan, had in 1295 converted to Islam. His 'conquest' was motivated by territorial rather than religious aspirations.

Crusade planning continued regularly throughout the first half of the fourteenth century, but it became increasingly difficult to put well-conceived projects for the recovery of the Holy Land into practice. In part, this was the inevitable consequence of the passage of time, for the longer the Holy Land was in Muslim hands, the easier it was for Christendom to become accustomed to the idea of not having possession of its spiritual treasures. By 1336, when the last serious attempt by the French monarchy to launch a crusade collapsed, only people in middle age could remember a time when any part of the Holy Land had been in western possession, and one would have had to be elderly to recall a flourishing state in the Levant. And of course, the Mongols' failure to dominate western Asia as they did eastern, and the collapse of Armenian Cilicia as an effective independent power, meant increasingly that all the work of conquest would have to be done from the West. This had been the case also in 1095, but the First Crusade had taken advantage of a moment of unique dislocation in the political climate of the Near East – and it had also been singularly lucky in succeeding even in such conditions. Turkish political unity more or less since the collapse of Fatimid Egypt in 1169–71 meant that those conditions were never again replaced.

There were also reasons quite external to the politics of the Near East why crusades never reached far beyond the planning stage in the fourteenth century. Simply, the political society of the West was more complex than it had been in the late eleventh century; and although governmental bureaucracy, both ecclesiastical and secular, was far more effective, this did not mean that it was easier either for Clement V (1305–14) or John XXII (1316–34) to launch a crusade than it had been for Urban II, or for kings or great lords to respond to an appeal. It is true that financing and recruitment of crusading, which had been haphazard in 1095–1101, was in contrast efficient and well organised by the 1270s. But the process that had made such efficiency possible was the growth of political bureaucracies that had their own ideologies and their own national interests. The increased power of European sovereign states made it less likely that Christendom could react spontaneously or with any degree of unity. Moreover, the papacy itself had changed. The long struggle against the Hohenstaufen in Italy and the subsequent entanglement of the popes from the 1260s onward with the Sicilian succession had the effect of making the institution appear unable to rise above the interests of national politics in the way in which an Urban II, Eugenius III or Innocent III had done. The domination of the office by French prelates, and from 1305 the residence of the popes themselves in Avignon, only exacerbated such perceptions. This is not to say that the Avignon papacy was uninterested in crusading (Housley 1986), but rather that its voice was less universally heard than had been the case up to the late thirteenth century.

It may appear paradoxical that a system of financing crusades that had become better organised since 1274 should have been correspondingly unable to deliver results in respect of armies in the field. Thirteenth-century popes realised that the best way of ensuring a steady income stream to pay for crusading was to tax clerical incomes. Despite attempts in England and France, the imposition of a regular tax on the laity on the model of the 'Saladin tithe' of 1188 proved impossible to enforce. Taxing the Church, however, was a way of taxing the whole of Christian society, because clerical incomes were made up, at the lowest level, of the tithes paid to their parish by every parishioner. Moreover, taxing clerical incomes was also a means of retaining control of crusading by the Church: according to Innocent III's doctrine, what was the business of the Church ought to be organised by the Church. Gregory X's bull *Cum pro negotio* (1274) laid down the way tithes were to be assessed and collected throughout Christendom. Although setting up an institutional strategy does not guarantee that the money will flow in, the fact that money from tithes was spent by popes from the 1270s to *c*.1300 on a variety of crusading projects is testimony to the effectiveness of the system. Sources reflecting both those taxed and those implementing taxation indicate the level of the practical problems to be overcome. First, there was the difficulty of assessing levels of taxation. Since each bishopric had a different income, which was made up from a variety of sources and would fluctuate yearly, this was a complex matter. Then there were the practical problems of collection. In 1277, the papal collectors in England found that travelling around the country to do their job was dangerous because of the brigandage that increased whenever the king was fighting outside the kingdom. There was evasion on a massive scale: from delays, appeals and claims of exemption by monastic and other institutions to open defiance. Some clergy were even claiming that after Gregory X's death in 1276 they were no longer under any obligation to pay. The most telling part of the letter from which these complaints are taken is the statement 'We see no way in which this can be remedied other than by calling on the assistance of the secular arm' (Lunt 1917: 66-9; Housley 1996: 21-5).

Even when the correct amount was finally collected, it had to be accounted for. At one level, this meant record-keeping on a huge scale; at another, the far more difficult process of ensuring that the money collected was actually spent on crusading. As we have seen, opinions about the most legitimate use of tithes could be fiercely critical of papal policy, but a more serious problem in the fourteenth century was the difficulty of holding crusading kings to their vows so that tithes collected under the terms of their vow were indeed spent on the crusade and not on some other purpose. For example, when Charles IV succeeded his brother Philip V as king of France in 1322, the pope took up with him the crusade project that Philip had been pursuing. Philip had been granted a tax of one-tenth in 1321 in addition to some of the proceeds from the one-tenth levied for six years at the Council of Vienne (1314), but Charles found that this had already been spent. Pope John XXII's reply more or less accuses the French crown of only feigning interest in crusading as a pretext for enjoying the fruits of clerical taxation (Housley 1980: 166-9). A king could take the cross, receive a grant of tithes – and perhaps an additional papal subsidy – then find reasons for delaying an expedition, while all the while using the money to fund other projects – as, for example James II of Aragon had done from 1305 onward when he used the tithes granted him by Clement V to fund his conquest of Sardinia and Corsica. This is why, in 1321, John's negotiations with Philip V involved the clause that the king should agree to go on crusade in person before a set date, rather than sending a proxy commander in his place.

Ten years later, in 1331, the boot was on the other foot, as Philip VI accused John XXII of spending money set aside for the Holy Land on other purposes. Now the pope had to defend not only his specific conduct – the use of the money for the defence of the Church against schism (in this case the imperialist anti-pope Nicholas V in 1328) – but also the more general principle that tithes and subsidies for crusading could in emergencies be spent on other campaigns. He gave as instances the cases of Philip IV and Philip V in France, and implied that since 1295 the practice had been widespread (Housley 1996: 64–7). He might also have mentioned that even forces raised for crusading were diverted for service elsewhere: in 1319, for example, the ten war galleys equipped at papal expense for a *passagium particulare* by Philip V were instead lent to Robert, king of Naples, when bad weather prevented them from sailing to the East. What such examples indicate is that popes and kings alike, as figures on an international stage, had finite resources that had to be deployed to meet immediate demands. This does not necessarily mean that they were insincere in their concern for the fate of the Holy Land, or more generally of the Christians in the East. But the very processes that made the collection of taxes possible in England, France or Castile, for example – in other words, more effective governmental methods – also made it less likely that rulers would allow those taxes to be used by other kings who might be rivals. Philip IV clashed with Boniface VIII when he refused to allow French tithes to be used by the king of England, with whom he was at war – for how was Philip to know that they would not be used to finance a campaign in Gascony, rather than a genuine crusade?

National interests may have prevailed in different ways over the desire to reconquer the Holy Land. Yet nationalist ambitions and crusading were no less incompatible in the fourteenth century than they had been for St Louis. The last attempt to mount a 'traditional' crusade for the recovery of the Holy Land was the projected expedition of Philip VI in 1336 (Tyerman 1985: 25–52). The preaching began in France at the end of 1331, but it took a further sixteen months (February 1332–July 1333) before the details of who was to go, how much money was to be collected and how, and what indulgence was offered, were agreed. Philip wanted control of tithes collected outside France, which was impossible for the pope to deliver. By the summer of 1336, the pope was complaining that there was no evidence that even a single galley had been built. In fact a fleet was amassed, because when the crusade was eventually cancelled, the sailors waiting in Marseilles staged a mock sea-battle in which they hurled oranges at each other (Tyerman 1985: 47). Philip would not leave until a safe peace had been concluded with Edward III of England over the future of Gascony, England's last Angevin possession. It is easy to accuse Philip of insincerity in his planning, but the diversion of crusading taxes to the start of war against England (1336–7) in fact represented the failure of French policy. Philip knew that his political aspirations would have been better served by going on crusade than by fighting the English.

Other national interests than those of defence could also interfere with crusading plans. James II of Aragon (1291–1327) declared his intent to go on crusade in c.1289–91, and in 1293 launched a diplomatic mission to construct alliances in the East to prepare the way for a crusade. During the 1290s he was seen by the papacy as the natural leader of a crusade because of Aragon's maritime power (Schein 1991: 151–2, 188–92). But although he periodically remembered his vow – in 1296, in 1300 and in 1305, for example – he also exchanged friendly embassies with an-Nasir, the sultan of Egypt, on five occasions between 1303 and his death in 1327 (Atiya 1938a, 1938b: 510–17). Flouting Nicholas IV's ban on trade with Egypt – for which he was excommunicated by Boniface VIII – James' embassies sought to establish commercial relations with the most important power in the eastern Mediterranean. In fact, although James legislated in his own kingdom against trade with Egypt in 1302, this law was never enforced, and in 1326 he was remitting fines against illegal traders in return for pay-offs to the crown (Housley 1986: 203).

On the surface, it looks as though James simply paid lip-service to commitments expected of him, while pursuing a policy diametrically opposed to the interests of crusading. But it would be simplistic to assume that there was an equation of 'either the crusade or commerce'. In the first place, James was certainly not opposed to crusading in itself; in 1305 he linked the recovery of the Holy Land with a crusade against the Muslims of Granada, and Ramon Lull, the influential Dominican crusade planner, was impressed by the king's zeal (Schein 1991: 190). Besides, James would have been more than willing to bring about the recovery of the Holy Land if it could be achieved without jeopardising the profitable relations with Egypt that were so important to all Mediterranean maritime powers. As we shall see, his interest in the Holy Land and the fate of eastern Christians was profound. Aragonese policy in this regard was similar to that of Genoa and Venice; and James' successor Alfonso IV continued to play the same hand, to the extent that Aragon was later accused by the French of sabotaging the mission of Peter de Palud, the patriarch of Jerusalem, to Egypt in 1329–30 (Dunbabin 1991: 164–73).

Problems of finance and national policy provide two fundamental and interlinked reasons why plans for the recovery of the Holy Land never materialised into action in the fourteenth century. A third is the nature of warfare in general, and of the specific form of warfare required in the eastern Mediterranean in particular. Urban II and his twelfth-century successors saw participation in crusades as a voluntary activity, albeit one in which a strong sense of moral obligation ought to figure. We cannot assume that even in 1095, this meant that every *crucesignatus* made entirely individual decisions: recruitment always followed patterns set by family interests, networks of patronage and political alliances. Even in the mid-thirteenth century, recruitment to crusades seems to have been largely traditional. The process described by Joinville, for example, does not seem to have differed far from that in the chronicles of the First Crusade: lords formed individual military households by taking vassals with them, and if they ran short of the means to support them, they and their knights had to hope they could be taken on by a greater lord (Shaw 1963: 191–2). But by 1291 most European kings no longer recruited armies according to older feudal patterns. Charles IV and Philip VI, for example, negotiated the terms of their proposed crusades with the papacy on the assumption that they would provide given numbers of ships and men. These quotas could not be met by the haphazard method of sending out preachers and waiting for crusade vows to yield manpower and money. Instead, taxation of clerical income brought with it, by the end of the thirteenth century, a more systematic method of contractual recruitment. If a king knew roughly how much he could expect to raise for his crusade from the grant of a tithe or a papal subsidy, he could estimate the numbers of knights, foot soldiers and archers that could be supported by that amount, and for how long.

This more streamlined and efficient process of raising armies ought perhaps to have made it easier for kings to implement their vows. At least it ensured that the fiasco of 1202, when an over-estimate of manpower scuppered the crusade, would not be repeated. But at the same time it created a distance between vow and participation. Crusading vows were still made, but except in the case of those at the top of society, they had little effect in determining the shape or nature of an expedition. Popes could no longer, like Innocent III, expect to create little communities of Christendom to recover the Holy Land; even Gregory X realised that the knighthood was no longer in a position to direct crusading, and that he had to deal with kings. The real value of the crusading vow, by the end of the thirteenth century, was to raise extra money through redemptions sold to people who would never in any case have been participants. But should this necessarily have meant that crusades to the East were less likely to take place? The benefits of central organisation of recruitment and financing by royal governments are undoubted; conversely, however, greater systemisation resulted in a narrowing of the executive authority that could make crusading actually take place. Royal governments were more likely to recruit and finance successfully, but they were also more prone to subordinate the crusade itself to national policy. The centralisation of political life in the West may have made a crusade for the recovery of the Holy Land more effective if it ever took place; the problem was that it was by the same measure less likely ever to take place.

The Holy Land in the later medieval mind

Just as methods of finance and recruitment became more efficient from 1274 onward, so also military and geopolitical strategy took a leap forward through the 'recovery treatises'. The problem of the Holy Land was now seen in a fuller economic and political context. The treatises dating from after the fall of Acre, in particular, are characterised by, one the one hand, a high level of theoretical abstraction, and on the other, a profound knowledge of the political and military problems to be overcome (Leopold 2000: *passim*). In the first category belong those written by public servants such as Pierre Dubois (*c*.1306–7) and William of Nogaret (1310–11); in the second, those by the Armenian prince Hethoum (1307), the Hospitaller Grand Master Fulk of Villaret (1305), the Venetian merchant Marino Sanudo Torsello (1309), Henry II of Cyprus (1310–11), and the Dominican missionary William Adam (1316–17 and 1332). Fidenzio of Padua, the Franciscan provincial in the Holy Land at the fall of Acre, belongs in both categories, because although he makes practical recommendations, his work is rooted in a moral paradigm of sinfulness and judgement. Ramon Lull's work is so voluminous, and even his practical writing so coloured by philosophical interests, that it is difficult to fit him into any category.

One effect of the recovery treatises, taken as a whole, was to diffuse the western view of crusading through a wider lens. Even after the fall of Acre, the goal of a crusade was by no means clear. Fidenzio, writing before the fall of Acre, wanted to link an amphibious assault on the Holy Land to an economic blockade of Egypt. This strategy, which was to prove highly influential (it was also adopted by Ramon Lull, Marino Sanudo and Hethoum, amongst others) was a continuation of the 'Egyptian policy' of thirteenth century crusades, in so far as it was based on the idea that Egypt was the key to Mamluk power. But it also demanded the subjection to western Christendom of Asia Minor, Syria and North Africa. Marino Sanudo, Dubois and Henry II concurred: if the Holy Land was to be not only recovered but held for Christendom, it was not just the Mamluks, but Islam in general, that had to be defeated.

This thread had been unravelling almost since crusading began. The Egyptian policy, whether it involved invasion or blockade, was an implicit rejection of the narrow focus of Urban II's crusade for the recovery of a specific holy place, the shrine of the Lord. The widening of crusading during the twelfth century acknowledged a conceptual link between the holy places and the more intractable problem of Islam. Even when the recovery of the holy land was the target, by the early thirteenth century other strategic imperatives either took priority – as for Richard I in 1191–2, for the Military Orders in 1219 and St Louis in 1249 – or, as in some of the recovery treatises, subsumed Jerusalem in a wider geopolitical dimension. From 1274 onward, alliances with Mongols and Armenians, the problem of Sicily, reunion with the Orthodox Church, the remnants of Frankish Greece and the future of Cyprus all featured in crusade planning.

One way in which the thread unravelled was through the evolution of a papal theory on non-Christian peoples. From the pontificate of Gregory IX (1227–41) onward, popes developed a canonical position in which Muslims were seen as occupying the same relationship to Christendom as schismatics and Jews. It is no coincidence that the popes who initiated the mendicant missions in the Near East and to the Mongols – Gregory himself and Innocent IV (1243–54) – were also those who thought most deeply about the theoretical implications of crusading. Innocent IV developed a justification of the crusade starting from Christendom's right to the ownership of certain property – the holy places – and concluding that Christians could also legally seize lands other than the Holy Land from Muslims. At the heart of this thesis was the question of whether infidels could have *dominium* – legal lordship – over any territory. Innocent IV thought that they could, because they enjoyed the rights of natural law. On the other hand, he argued that they were in theory subject to Christ's *dominium*, and thus to the pope's authority. This was a relatively conservative position, compared to that of the canonist Hostiensis (*c*.1270–1), who argued that those who rejected Christ lost the right to *dominium* altogether (Muldoon 1979: 5–7, 15–17, 47–8). Even the more conservative position might be taken to justify the defeat of Islam as a whole, rather than simply the recovery of the Holy Land.

Such ambitious theories of Christendom's entitlement over Islam were a long way from reality. Islam might lack authority as a religion, but its armies were mighty. Fidenzio of Padua, on the eve of the fall of Acre, conceded that an invasion of Egypt was implausible because the Mamluks were simply too strong. But a growing perception of crusading as a war against Islam and Muslims rather than simply for the recovery of a specific territory was necessitated precisely by the imbalance of forces in the eastern Mediterranean. In 1318, John XXII recognised that the recovery of the Holy Land had to be secondary to the defence of Cyprus and Armenia (Housley 1980: 168) and a fleet was collected for that purpose in 1319. The Armenian–Mamluk truce of 1323 may have resolved the problem temporarily, but it also revealed European impotence. The background to Philip VI's crusade preparations of the 1330s was Turkish raiding on Venetian island possessions in the Aegean, and Philip's reluctance to be diverted from his 'passagium generale' by such side-issues only demonstrates the widening gap between western crusading aspirations and the needs of the Christians in the eastern Mediterranean (Tyerman 1985: 36–7, 52). The advance of the Ottoman Turks from the 1330s onward changed crusading utterly. The second half of the fourteenth century saw specific expeditions to defend territory under threat – most famously the disastrous Nicopolis crusade of 1396 – but no more general crusades for the recovery of the Holy Land were planned after 1336.

This does not mean, however, that the holy places themselves were forgotten. A quite different approach to the Muslim occupation of the holy places was 'peaceful liberation'. In 1322 James II of Aragon, during the course of one of the regular embassies he sent to Egypt, asked an-Nasir to entrust the custody of the Holy Sepulchre to a group of Aragonese Dominican friars who would take up residence in Jerusalem. Although the sultan agreed readily, the Dominicans stayed only a year before returning to the West. Undeterred, James asked for the same concession for the Franciscan Order in 1327. This time, an-Nasir prevaricated, and James' own death stalled the plan. The project was revived, however, by Robert the Wise, king of Naples, and his queen Sancia. In 1333 a group of Franciscans from Provence, led by Roger Garin, occupied the site of the Cenacle on Mt Zion, where they built a priory, and from which they provided a permanent Latin staff for the Holy Sepulchre. Queen Sancia funded all of this activity. The Franciscan 'custody of the Holy Land', as it became known, soon extended to the Church of the Nativity in Bethlehem and the tomb of the Blessed Virgin in the garden of Gethsemane.

James' approach to an-Nasir did not come out of the blue. As early as 1300 he had pressed the sultan to grant access to Christian pilgrims to the holy places; in 1303 he asked for churches in Egypt to be reopened; while in 1305 and 1314 he presented himself as the protector of all Christians under Mamluk rule. Alongside these demands he secured the release of prisoners of war, presumably from the fall of Acre or Ruad (1302) (Atiya 1938b: 18–33). James' diplomacy operated in parallel to his and other rulers' plans for crusading. In 1327 his embassy to Cairo coincided with that of Charles IV of France, who had papal authority to negotiate with the sultan. But whereas French dealings with an-Nasir between 1327 and 1331 were full of thinly veiled bombast about a crusade, James never coupled his requests with the threat of force. The return of the holy places was to him, and to the Angevins of Naples in the 1330s, a separate matter from the more general problem of Islam.

Jerusalem and the Holy Land did not cease to appeal to western Christians even when hopes of their recovery must have dimmed. The Franciscan custody made western pilgrimage easier, but accounts of pilgrimages, mostly by friars, survive from 1302, 1320, 1322–3, 1327 and 1330–1, as well as from 1335, 1336, 1345, 1346–7 and 1384. The liveliest of these, by James of Verona (1335) and Niccolo da Poggibonsi (1346–7) fully demonstrate the emotions that could still be raised in western Christians by being in the presence of the Holy Sepulchre. Pilgrimage to Jerusalem continued to be popular in the fifteenth century and even beyond, though by this time an element of antiquarian curiosity can be seen merging with an older devotional tradition (Sumption 1975: 257–302).

Yet, precisely because it was so iconic in western piety, the Holy Land had long been present in Europe. Relics of the Passion were located in the West; for example through the discovery of the holy blood of Mantua in 1048, or the Oviedo relics, including bread from the Last Supper, in 1075 (Morris 2000: 95–6). The Lateran palace itself housed an impressive collection of relics from the Holy Land on the eve of the First Crusade (Cowdrey 1997: 740–2). More remarkably, symbolic representations of the Holy Sepulchre were constructed in western churches. In the Carolingian Church these served a purely liturgical purpose for Easter celebrations, and were not intended to be architectural copies. But by the eleventh century, a new wave of 'precise architectural reminiscences' of the Holy Sepulchre swept across Europe (Morris 2000: 101). The most ambitious example is San Stefano in Bologna (Morris 1997: 31–59), but many others were built as a result of the crusades, such as Holy Trinity (the 'Round Church') in Cambridge (Krautheimer 1942: 1–33; Bresc-Bautier 1974: 319–24). These constructions were an attempt to confer some of the spiritual charge of the original through a reproduction of certain elements, such as shape, proportion or decoration. Pilgrims returning from the Holy

Land, similarly, might build models of the shrines at which they had worshipped, in an attempt to evoke emotional experiences of the holy. Entire urban landscapes could be configured to represent pilgrimages in the Holy Land, as for example in Suzdal in Russia (Price 2000: 255–7). London was decked out to represent Jerusalem in Henry V's victory parade after the battle of Agincourt (Housley 2000: 239).

Such cultural translations originally had the effect of rendering the Holy Land a place 'where no Christian could be a foreigner' (Smail 1977: 38). But they could have a different effect when the holy places themselves were no longer in Christian hands. Already in the thirteenth century – the period in which, according to one historian, 'the recovery of the Holy Land . . . pervaded the minds of men . . . [it was] inseparable from the air they breathed' (Powicke 1962: 80) – the Holy Land was being allegorised by followers of the apocalyptic tradition of Joachim of Fiore. According to some Joachites, St Francis, who was 'another Christ', had made Italy a new 'holy land' (Wesley 2000: 181–91). In a different vein, Clement V's bull *Rex gloriae* (1311), proclaimed that France enjoyed a status in Christendom similar to that of the Holy Land. This bull, issued under extreme pressure from Philip IV, shows how the self-assurance of the French crown might be theorised into an abstract statement of divine favour on the nation. Emerging concepts of national identity brought with them notions of the special quality, even holiness, of the national homeland. Where did this leave the Holy Land itself? One of the most thoughtful of the recovery theorists, Pierre Dubois (c.1306–7), suggested that, once reconquered, it ought to be divided up among the western nations who had assisted in the conquest (Brandt 1956: 84–5). As Housley has observed, this was a far cry from Fulcher of Chartres' perception of the Holy Land as a new *natio* forged in the fires of migration, conquest and settlement (2000: 241; Murray 1995: 59–73).

Crusading as an activity had also become a cultural concept by the end of the fourteenth century. Chaucer's knight, the model of late fourteenth-century chivalric ideals, had served on a crusade; indeed, crusading was part of his chivalric credentials. But his crusade was not for the recovery of the Holy Land, or even the defence of the Christian Mediterranean against the Turks. Like Henry IV of England, Duke Albert III of Austria, Gaston-Phoebus, count of Foix, Henry duke of Lancaster and many others, Chaucer's knight had been a guest of the Teutonic Knights on one of their perpetual expeditions against the pagans in Prussia. Here, in the frozen terrain of the north, the European knighthood acted out its ideals and fantasies in a drama of hunts and elaborate feasts that enlivened the often inconclusive fighting (Keen 1984: 172–4). One result of such activities was the strengthening sense of a common chivalric culture among the knightly classes that cut across national identities. English knights met their counterparts from France, Germany and Bohemia, and engaged with them on a common cause that linked them, however tenuously, to the great crusading expeditions to the East of the twelfth and thirteenth centuries. In this vein, Froissart tells the story of two French knights trapped in a siege during the Hundred Years War who were rescued from a dishonourable death at the hands of English crossbowmen by a party of English knights whom they recognised from an expedition they had shared with the Teutonic Knights only a few years previously (Jolliffe 1967: 134–5).

A readiness to embark on a crusade against the Turks was a mark of chivalric virtues among fifteenth-century knights. Crusading vows came to have a symbolic quality, as for example when Henry V took the cross in grave illness in 1422. He may have been echoing St Louis' vow of 1244; more likely he was simply making a gesture that had long been customary. Even when vows were part of a genuine plan for crusading, as in Duke Philip of Burgundy's intended rescue of Constantinople from the Turks in 1453, they were set in a lavish ceremonial that evoked and glorified individual martial prowess above all. The vow of the *crucesignatus* had evolved into a complex and formulaic culture that represented the more general knightly ideals of the day. In a sense, the revolution brought about by Urban II was complete. In 1095, warfare against the Turks had been offered to knights by the Church as one among many acceptable forms of the penance required of all knights. By the close of the Middle Ages, declaring a readiness to fight the Turks had been adopted by the knighthood as a necessary part of chivalric culture. Whether Urban II would have recognised his intention in the festivities at Philip of Burgundy's Feast of the Pheasant in Lille, we can only speculate.

Brief biographies

Baldwin I

The youngest of the three brothers of the Boulogne–Rethel dynasty, all of whom took the cross in 1095–6. Unknown before the crusade, Baldwin made his reputation through the acquisition of Edessa in 1097–8. In 1100 he succeeded his brother Godfrey in Jerusalem, and had himself crowned first king of Jerusalem. He expanded the frontiers of the kingdom and defended it against Egyptian and Syrian assaults, but died without direct heirs in 1118.

Bernard of Clairvaux (1090–1153)

A Cistercian monk from a knightly family in Burgundy, Bernard founded the monastery of Clairvaux in 1115, and became a prominent champion of monastic reform. He probably met Hugh of Payens, founder of the Templars, in 1129, and at his request wrote the treatise *In Praise of the New Knighthood*. In 1146 he was commissioned by Pope Eugenius III, his former pupil, to preach the Second Crusade. His preaching reveals a characteristically spiritual understanding of the crusading vow, and extended crusading practice to the Iberian peninsula and the Baltic.

Bohemond

The eldest son of Robert Guiscard, head of the Hauteville family from Normandy that settled in Apulia in the mid-eleventh century. With Guiscard he attempted an invasion of the Byzantine Balkans in the 1080s. He may have taken the cross in order to secure a Byzantine military governorship at a time when his dynastic future in Italy looked uncertain. He played the decisive role in the capture of Antioch and defeat of Ridwan of Aleppo in 1098, but his refusal to give up the city to the Byzantines ruptured relations for half a century. In 1106 he launched a crusade against Byzantium but was defeated and forced to assent to the Treaty of Devol (1108); he died in 1111.

Emperor Frederick II (1194–1250)

Heir to both the Norman kingdom of Sicily, through his mother Constance, and the Holy Roman Empire, Frederick took the cross in 1215 on his imperial coronation, but failed to participate in the Fifth Crusade (1217–21). In 1225 his marriage to Yolanda, daughter of Maria and John of Brienne, gave him a claim to the throne of Jerusalem, which he pursued in 1228–9 with a crusade that saw no fighting but resulted in the return of Jerusalem by treaty. His attempts to rule the kingdom through his representative Richard Filangieri in the 1230s led to a damaging struggle for power in Jerusalem.

Emperor Manuel Komnenos (1142–80)

The grandson of Alexios Komnenos, the initiator of the First Crusade. Manuel was noted for his pro-western sympathies, and he pursued a policy of close alliance with the Crusader States after 1158. During the 1160s he supported a joint Franco-Byzantine invasion of Egypt. His reign also saw attempts to regain prestige by recovering imperial authority in Italy. He promoted Orthodox monasticism in the Crusader States, and was a generous patron to the holy sites. After his death, however, the Komnenos dynasty fell apart and his successors allied with Saladin rather than the Franks.

Guy of Lusignan

A member of a prominent Poitevin noble family, Guy followed his brothers Geoffrey and Aimery to Jerusalem, and in 1180 was surprisingly chosen to marry the heiress to the kingdom, Sybil. He was given charge of the army to face Saladin in 1183 by Baldwin IV, but dismissed for incompetence, and rebelled against the king. In 1186 he and Sybil staged a *coup* to seize the throne, but in 1187 he led the kingdom to disaster at Hattin. Captured by Saladin, he was soon released and initiated the siege of Acre in 1189. When Sybil died in 1191, however, he lost his claim to the throne, and was compensated by Richard I with Cyprus.

Isabel

The daughter of King Amalric I (1162–74) by his second marriage to Maria Komnena, Isabel became an important political figure at her half-sister

Sybil's death in 1190. As heiress to the kingdom she was persuaded to marry Conrad of Montferrat, but at his assassination in 1192 she married Henry of Champagne. After Henry's death in 1197 she married Amalric II, Guy's brother. Her children from these marriages formed separate ruling dynasties in Jerusalem and Cyprus. Isabel's life provides an example of how prominent women were expected to marry successive husbands in order to further the dynastic line.

John of Brienne, king of Jerusalem (1210–25)

The son of a French baron who had been a prominent papal supporter in Italy, John became king of Jerusalem through marriage to the heiress Maria, daughter of Isabel and Conrad of Montferrat. After her early death, his position became insecure, and he married Stephanie, heiress to Armenia. In 1218 he was elected leader of the forces on the Fifth Crusade, but despite initial success at Damietta he quarrelled with the papal legate and did not take part in the final advance in 1221. In 1225 he lost his claim to the throne when his daughter married Emperor Frederick II, but in 1231 he became co-ruler of the Latin Empire of Constantinople. He died in 1237.

Louis IX, king of France (1226–70)

Louis took the cross in 1244 and spent four years assembling a predominantly French army to attack Egypt. After initial success in capturing Damietta (1249), Louis marched down the Nile towards Cairo, but his advance was halted by a defeat at Mansurah (1250) and during the retreat to Damietta he fell ill and was captured. Although the crusade was a disaster, Louis' personal prestige was enhanced by his conduct and by his stay in the kingdom of Jerusalem 1250–4, where he provided much-needed leadership in the absence of royal authority. He died before the walls of Tunis leading a Second Crusade, which had been diverted from its original target of Egypt.

Melisende

The eldest daughter of Baldwin II and his Armenian wife Morphia, Melisende married Fulk V of Anjou in 1129, and succeeded to the throne jointly with her husband and infant son Baldwin III in 1131. In 1134 she may have encouraged the revolt of Hugh of Jaffa, her father's protégé, in protest at Fulk's attempt to marginalise her from government. After Fulk's death in 1143 she acted as regent for her son, but provoked civil war from 1150 to 1152, when she refused to renounce power on his majority. She chose to be buried by the steps leading to the tomb of the Virgin Mary in the Valley of Jehosaphat.

Nur ad-Din

The son of Zengi, Nur ad-Din came to power as ruler of Aleppo, but in 1154 assumed control in Damascus during a power vacuum in the city. He was thus the first Seljuq leader since the 1090s to unite north and south Syria. He mobilised *jihad* ideals in his struggle against the Franks, particularly the importance of the recovery of Jerusalem. He won important victories against the Franks in 1149 and 1164, but was also active in fighting Muslim rivals, especially the Shi'ites of Damascus, and in 1171 he suppressed the Fatimid khalifate in Egypt.

Pope Innocent III (1198–1216)

Innocent made crusading the cornerstone of his pontificate, launching more than any previous pope. He extended crusading privileges to conflicts against non-Muslims, notably the Cathar heretics of southern France in the Albigensian Crusade (1209–29). He is also credited with preaching the first 'political' crusade against papal enemies in Italy. However, he lost control of the Fourth Crusade (1202–4) and was unable to prevent it from being diverted to Constantinople. Innocent developed new systems for preaching and financing crusades, which are epitomised in the Fourth Lateran Council (1215).

Pope Urban II (1088–99)

A former Cluniac monk, Urban was part of the reforming circle of Pope Gregory VII. At the Council of Piacenza (May 1095) he received an appeal for military aid from the Byzantine emperor, which at the Council of Clermont (November 1095) he turned into a more general call for an armed pilgrimage to seize the Holy Sepulchre from the Turks. Papal primacy lay at the heart of his policy, but he was conciliatory towards the Orthodox Church and established relations with Symeon II, the Orthodox patriarch of Jerusalem. He died before news of capture of Jerusalem reached Rome.

Raymond III of Tripoli

A grandson of Baldwin II through his mother Hodierna, Raymond succeeded his father as count of Tripoli, but was captured in 1164 and imprisoned for ten years. After his release he married Eschiva, heiress to Tiberias, and became regent of the kingdom of Jerusalem for Baldwin IV, but was dismissed in 1180 amid suspicions about his loyalty. He became a focal point for dissent against Guy of Lusignan and Sybil in 1185–7, but was reconciled to Guy in time to fight at Hattin. He survived the battle, but died soon afterward.

Reynald of Chatillon

A French noble who came to the East on the Second Crusade and remained behind; in 1152 he married Constance, heiress to Antioch, but as prince of Antioch he overreached his authority, raiding Byzantine territory and provoking invasion from Manuel Komnenos in 1157–8. He was captured and imprisoned by the Muslims (1160–76), but on his release became lord of Oultrejourdain in the kingdom of Jerusalem. His bold raiding in the Transjordan and the Red Sea incited Saladin to invade the kingdom of Jerusalem in 1183 and 1187. After Hattin he was executed by Saladin.

Richard I 'Lionheart' (1189–99)

Richard took the cross to replace the vow of his father Henry II in 1189. His expedition to the East was logistically well managed and achieved considerable success in restoring parts of the coastline to the kingdom of Jerusalem after the disaster of Hattin. He finished the siege of Acre in July 1191, and defeated Saladin in battle at Arsuf in September 1191 and at Jaffa in 1192. However, although he was within a few miles of Jerusalem, he never succeeded in recapturing it. He was the first western crusader to propose an assault on Egypt instead of Jerusalem, a policy that was taken up in the thirteenth century.

Saladin

A Kurd who first came to prominence in 1169 as part of Nur ad-Din's protectorate over Egypt. Saladin used brutal methods to suppress the Fatimid khalifate in 1171 on Nur ad-Din's orders, but on his master's death in 1174 he usurped the authority of his heir as-Salih in Damascus. Between

1174 and 1186 he pursued a war for the mastery of Syria, capturing Aleppo in 1183 and Mosul in 1186, while periodically also attacking the Franks. He continued Nur ad-Din's policy of using *jihad* ideals to legitimise his rule. In 1187 he defeated the Franks at Hattin and captured Jerusalem, but failed to drive them out completely, and in 1191–2 he saw many of his gains reversed by the Third Crusade. He died in 1193.

Zengi (1127–45)

The atabeg of Mosul, Zengi came from a prominent Seljuq noble family. He was noted among contemporaries for the savagery of his conduct. Although initially active in Iraq, he became ruler of Aleppo in 1128, but never succeeded in his aim of conquering Damascus. He introduced the *madrasa* (religious school) into western Syria and succeeded in creating the alliance between the Seljuq military ruling class and Arab scholars. For this reason he is often seen as the first Islamic leader to have fought a *jihad* against the Franks, though his conquest of Edessa in 1144, which ended the independent existence of the first of the Crusader States, was a by-product of a campaign against Islamic rivals.

Bibliography

Primary sources

Andrea, Alfred J. (1997) trans. *The Capture of Constantinople: The Hystoria Constantinopolitana of Gunther of Pairis*. Philadelphia: University of Pennsylvania Press.

Andrea, Alfred J (2000) ed. and trans. *Contemporary Sources for the Fourth Crusade*. Leiden: Brill.

Benton, John F. (1984) trans. *Self and Society in Medieval France: The Memoirs of Abbot Guibert of Nogent*. Toronto: University of Toronto Press.

Berry, Virginia (1948) ed. and trans. Odo of Deuil, *De profectione Ludovici VII in orientem*. New York: Columbia University Press.

Brandt, Walther (1956) trans. Pierre Dubois, *The Recovery of the Holy Land*. New York: Columbia University Press.

Broadhurst, Ronald J.C. (1952) trans. *The Travels of Ibn Jubayr*. London: Cape.

Constable, Giles (1967) ed. *The Letters of Peter the Venerable*, 2 vols. Cambridge, MA: Harvard University Press.

Cusimano, R. and J. Moorhead (1992), trans. *The Deeds of Louis the Fat by Suger*. Washington DC: Catholic University of America Press.

David, Charles W. (2000) trans. *The Conquest of Lisbon*. New introduction by Jonathan Phillips. New York: Columbia University Press.

Dawson, Christopher (1955) ed. and trans. *The Mongol Mission: Narratives and Letters of the Franciscan Missionaries in Mongolia and China in the Thirteenth and Fourteenth Centuries*. New York: Sheed and Ward.

Edbury, Peter (2003) ed. *Le Livre des Assises*. Boston: Brill.

Fink, H. and Ryan, F. (1969) *A History of the Expedition to Jerusalem, 1095–1127 by Fulcher of Chartres*, trans. F. Ryan, edited with an introduction by H. Fink. Knoxville: University of Tennessee Press.

France, John (1989) ed. and trans. *Historiarum libri quinque: The Five Books of the Histories by Rodolfus Glaber*. Oxford: Clarendon Press.

Gabrieli, Francesco (1984) trans. *Arab Historians of the Crusades*. Berkeley: University of California Press.

Greenia, Conrad (1977) trans. Bernard of Clairvaux, *In Praise of the New Knighthood*. Kalamazoo, MI: Cistercian Publications.

Hill, Rosalind (1962) ed. and trans. *Anonymi Gesta Francorum et aliorum Hierosolimitanorum*. Oxford: Clarendon Press.

Hitti, Philip (1929) trans. *An Arab–Syrian Gentleman and Warrior in the Period of the Crusades*. Beirut: Khayats.

Hubert, Merton J. and La Monte, John (1941) trans. *The Crusade of Richard the Lion-Heart by Ambroise*. New York: Columbia University Press.

Jolliffe, John (1967) ed. and trans. *Froissart's Chronicles*. London: Harvill Press.

Levine, Robert (1997) trans. *The Deeds of God through the Franks. A Translation of Guibert de Nogent's Gesta Dei per Francos*. Woodbridge: Boydell and Brewer.

Luard, H. (1876–1880) ed. *Matthaei Parisi Chronica Majora*. 6 vols. London.

McNeal, Edgar (1996) trans. *The Conquest of Constantinople by Robert of Clari*. Toronto: Medieval Academy of America.

Merwin, W.S. (1959) trans. *The Poem of the Cid*. New York: New American Library.

Peters, Edward (1971) ed. *Christian Society and the Crusades 1198–1229: Sources in Translation including The Capture of Damietta by Oliver of Paderborn*. Philadelphia: University of Pennsylvania Press.

Richards, Donald (2001) trans. *The Rare and Excellent History of Saladin by Baha ad-Din Ibn Shaddad*. Aldershot: Ashgate.

Röhricht, Reinhold (1884) ed. 'Lettres sur la prise d'Acre, 1291' *Archives d'Orient latin*, 2: 258–96.

Sewter, E.R.A. (1969) trans. *The Alexiad of Anna Comnena*. Harmondsworth: Penguin.

Shaw, Margaret (1963) trans. *Chronicles of the Crusades: Villehardouin and Joinville*. Harmondsworth: Penguin.

Stewart, Aubrey (1895) trans. *The History of Jerusalem by Jacques de Vitry*. London: Palestine Pilgrims Text Society (repr. 1971).

Stewart, Aubrey (1896) trans. *Burchard of Mt Zion*. London: Palestine Pilgrims Text Society (repr. 1971).

Tanner, Norman (1990) ed. and trans. *Decrees of the Ecumenical Councils I: Nicaea to Lateran V*. Washington DC: Georgetown University Press.

Tierney, Brian (1964) trans. *The Crisis of Church and State 1050–1300*. Englewood Cliffs, NJ: Prentice-Hall.

Upton-Ward, Julia (1992) ed. and trans. *The Rule of the Templars*. Woodbridge: Boydell and Brewer.

Wilkinson, J. (1988) ed. *Jerusalem Pilgrimage 1099–1185*. London: Hakluyt Society.

WT (1986) William of Tyre, *Chronica* ed. R.B.C. Huygens. 2 vols. Turnhout: Brepols.

Secondary sources

Abulafia, David (1988) *Frederick II, a Medieval Emperor*. London: Allen Lane.

Alphandéry, Paul and Alphonse Dupront (1954–9) *La chrétienté et l'idée de croisade*. 2 vols. Paris.

Asbridge, Thomas (2000) *The Creation of the Principality of Antioch*. Woodbridge: Boydell and Brewer.

Asbridge, Thomas (2003) 'Alice of Antioch: a case study of female power in the twelfth century', in Peter Edbury and Jonathan Phillips (eds) *The Experience of Crusading, 2: Defining the Crusader Kingdom*. Cambridge: CUP, pp. 29–47.

Atiya, Aziz (1938a) *The Crusade in the Later Middle Ages*. London: Methuen.

Atiya, Aziz (1938b) *Egypt and Aragon: Embassies and Diplomatic Correspondence Between 1300 and 1330 AD*. Leipzig: F.A. Brockhaus.

Bachrach, David (2003) *Religion and the Conduct of War c.300–1215*. Woodbridge: Boydell.

Baldwin, Marshall W. (1936) *Raymond III of Tripolis and the Fall of Jeruslaem (1140–1187)*. Princeton, NJ: Princeton University Press.

Baldwin, Marshall W. (1958) 'The Decline and Fall of Jerusalem, 1174–89' in K. Setton (gen. ed.) *A History of the Crusades. I: The First Hundred Years*. ed. Marshall W. Baldwin. Philadelphia: University of Pennsylvania Press, pp. 590–621.

Barasch, Moshe (1971) *Crusader Figural Sculpture in the Holy Land*. New Brunswick, NJ: Rutgers University Press.

Barber, Malcolm (1989) 'Western Attitudes to Frankish Greece', in B. Arbel, B. Hamilton and D. Jacoby (eds) *Latins and Greeks in the Eastern Mediterranean after 1204*. London: Frank Cass, pp. 111–28.

Barber, Malcolm (1994a) *The New Knighthood: A History of the Order of the Temple*. Cambridge: Cambridge University Press.

Barber, Malcolm (1994b) 'The Order of St Lazarus and the Crusades', *Catholic Historical Review*, 80: 439–56.

Barber, Malcolm (2000) *The Cathars: Dualist Heretics in Languedoc in the High Middle Ages*. Harlow: Pearson Education.

Bartlett, Robert (1993) *The Making of Europe: Conquest, Colonization and Cultural Change 950–1350*. Harmondsworth: Penguin.

Becker, Alfons (1997) 'Le voyage d'Urbain II en France', in *Le concile de Clermont de 1095 et l'appel à la croisade*. Rome: Ecole française de Rome, pp. 127–40.

Becker, Alfons (1988) *Papst Urban II (1088–99)*. Vol 2. Stuttgart: Monumenta Germaniae Historica.

Benvenisti, Meron (1970) *The Crusaders in the Holy Land*. Jerusalem: Israel Universities Press.

Berry, Virginia (1958) 'The Second Crusade', in K. Setton (gen. ed.) *A History of the Crusades. I: The First Hundred Years*. ed. Marshall W. Baldwin. Philadelphia: University of Pennsylvania Press, pp. 463–512.

Blake, Ernest (1970) 'The Formation of the "Crusade Idea" ', *Journal of Ecclesiastical History*, 21: 11–31.

Blake, Ernest and Morris, Colin (1985) 'A Hermit Goes to War: Peter and the Origins of the First Crusade', *Studies in Church History*, 22: 79–109.

Boas, Adrian (1999) *Crusader Archaeology*. London: Routledge.

Boase, T.S.R. (1977) 'Ecclesiastical Art in the Crusader States in Palestine and Syria', in K. Setton (ed.) *A History of the Crusades. Vol 4: The Art and Architecture of the Crusader States*, Madison: University of Wisconsin Press, pp. 69–116.

Bolton, Brenda (2000) ' "Serpent in the Dust: Sparrow on the Housetop": Attitudes to Jerusalem and the Holy Land in the Circle of Innocent III', *Studies in Church History*, 36: 154–80.

Borg, Alan (1969) 'Observations on the historiated lintel of the Holy Sepulchre, Jerusalem', *Journal of the Warburg and Courtauld Institutes*, 32: 25–40.

Bouchard, C.B. (1987) *Sword, Miter and Cloister: Nobility and the Church in Burgundy 980–1198*. Ithaca, NY: Cornell University Press.

Brand, Charles M. (1968) *Byzantium Confronts the West, 1180–1204*. Cambridge, MA. Harvard University Press.

Bresc-Bautier, G. (1974) 'Les imitations du St-Sépulcre de Jérusalem: archéologie d'un devotion', *Revue d'histoire de la spiritualité*, 50: 319–42.

Brown, Elizabeth A.R. (1958) 'The Cistercians in the Latin Empire of Constantinople and Greece, 1201–76', *Traditio*, 14: 63–120.

Brundage, James (1960) 'An Errant Crusader: Stephen of Blois', *Traditio*, 16: 380–95.

Buchthal, Hugo (1957) *Miniature Painting in the Latin Kingdom of Jerusalem*. Oxford: Clarendon Press.

Bull, Marcus (1993) *Knightly Piety and the Lay Response to the First Crusade*. Oxford: Clarendon Press.

Bull, Marcus (1997) 'Overlapping and Competing Identities in the Frankish First Crusade', in *Le concile de Clermont de 1095 et l'appel à la croisade*. Rome: Ecole française de Rome, pp. 195–211.

Cahen, Claude (1940) *La Syrie du nord à l'époque des croisades et la principauté franque d'Antioche*. Paris: P. Guenther.

Carr, A. Weyl (1982) 'The Mural Paintings of Abu Ghosh and the Patronage of Manuel Comnenus in the Holy Land', in J. Folda (ed.) *Crusader Art in the Holy Land*. Oxford: British Archeological Reports, pp. 215–44.

Carr, A. Weyl (1995) 'Art in the Court of the Lusignan Kings', in N. Coureas and J.S.C. Riley-Smith (eds) *Cyprus and the Crusades*. Nicosia: Cyprus Research Centre, pp. 239–74.

Charanis, Peter (1958) 'The Byzantine Empire in the Eleventh Century', in K. Setton (gen. ed.) *A History of the Crusades. I: The First Hundred Years*. ed. Marshall W. Baldwin. Philadelphia: University of Pennsylvania Press, pp. 177–219.

Chazan, Robert (1996) *In the Year 1096: The First Crusade and the Jews*. Philadelphia, PA: Jewish Publication Society.

Christiansen, Eric (1980) *The Northern Crusades: The Baltic and the Catholic Frontier 1100–1525*. Basingstoke: Macmillan.

Cole, Penny J. (1991) *The Preaching of the Crusades to the Holy Land, c.1095–1270*. Cambridge, MA: Medieval Academy of America.

Constable, Giles (1953) 'The Second Crusade as Seen by Contemporaries', *Traditio*, 9: 213–79.

Constable, Giles (1982) 'Financing the Crusades in the Twelfth Century', in B.Z. Kedar, H.E. Mayer and R.C. Smail (eds) *Outremer. Studies in the History of the Crusading Kingdom of Jerusalem Presented to Joshua Prawer*. Jerusalem: Yad Izhak Ben Zvi, 1982, pp. 64–88.

Constable, Giles (1997) 'Cluny and the First Crusade', in *Le concile de Clermont de 1095 et l'appel à la croisade*. Rome: Ecole française de Rome, pp. 179–93.

Constable, Giles (2001) 'The Historiography of the Crusades', in A.E. Laiou and R.P. Mottahedeh (eds) *The Crusades from the Perspective of Byzantium and the Muslim World*. Washington DC: Dumbarton Oaks, pp. 2–25.

Costen, Michael (1997) *The Cathars and the Albigensian Crusade*. Manchester: MUP.

Coureas, Nicholas (1997) *The Latin Church in Cyprus, 1195–1312*. Aldershot: Ashgate.

Cowdrey, H.E.J. (1968) 'The Papacy, the Patarenes and the Church of Milan', *Transactions of the Royal Historical Society*, 18: 25–48.

Cowdrey, H.E.J. (1970) 'Pope Urban II's Preaching of the First Crusade', *History*, 55: 177–88.

Cowdrey, H.E.J. (1973) 'Cluny and the First Crusade', *Revue Bénédictine*, 83: 285–311.

Cowdrey, H.E.J. (1982) 'Pope Gregory VII's "Crusading" Plans of 1074', in B.Z. Kedar, H.E. Mayer and R.C. Smail (eds) *Outremer: Studies in the History of the Crusading Kingdom of Jerusalem Presented to Joshua Prawer*. Jerusalem: Israel Academy of Sciences, pp. 27–40.

Cowdrey, H.E.J. (1993) 'Pope Gregory VII and the bishoprics of Central Italy', *Studi medievali*, 34: 51–64.

Cowdrey, H.E.J. (1997) 'The Reform Papacy and the Origin of the Crusades', in *Le concile de Clermont de 1095 et l'appel à la croisade*. Rome: Ecole française de Rome, pp. 65–83.

Cowdrey, H.E.J. (1998) *Pope Gregory VII, 1073–1085*. Oxford: Clarendon Press.

Cushing, Kathleen (2004) *Reform and the Papacy in the Eleventh Century: Spirituality and Social Change*. Manchester: Manchester University Press.

Delaruelle, Etienne (1944) 'Essai sur la formation de l'idée de Croisade', *Bulletin de literature écclesiastique*, 45: 13–46, 73–90.

Dodu, G. (1914) *Le royaume latin de Jérusalem*. Paris.

Duby, Georges (1971) *La societé aux XIe et XIIe siècles dans la region maconnaise*. Paris: SEVPEN.

Duby, Georges (1977) *The Chivalrous Society*. London: Edward Arnold.

Dunbabin, Jean (1991) *A Hound of God. Pierre de la Palud and the Fourteenth Century Church*. Oxford: Clarendon Press.

Duri, A.A. (1982) 'Bait al-Maqdis in Islam', in A. Hadidi (ed.) *Studies in the History and Archaeology of Jordan. I*. Amman: Department of Antiquities, pp. 351–5.

Edbury, Peter W. (1993) 'Propaganda and Faction in the Kingdom of Jerusalem: The Background to Hattin', in M. Shatzmiller (ed.) *Crusaders and Muslims in Twelfth-Century Syria*. Leiden: Brill, pp. 173–89.

Edbury, Peter W. (1997) *John of Ibelin and the Kingdom of Jerusalem*. Woodbridge: Boydell and Brewer.

Edbury, Peter and J.G. Rowe (1988) *William of Tyre, Historian of the Latin East*. Cambridge: CUP.

Edgington, Susan (2001) 'Albert of Aachen, St Bernard and the Second Crusade' in J. Phillips and M. Hoch (eds) *The Second Crusade: Scope and Consequences*. Manchester: Manchester University Press, pp. 54–70.

Ehrenkreutz, Andrew (1972) *Saladin*. Albany: State University of New York Press.

Elisséeff, Nikita (1967) *Nur ad-Din: un grand prince musulman de Syrie au temps des croisades (511–569 H./1118–1174)*. 3 vols. Damascus: Institut français de Damas.

Ellenblum, Ronnie (1996) 'Three Generations of Frankish Castle-Building in the Latin Kingdom of Jerusalem' in M. Balard (ed.) *Autour de la première croisade*. Paris: Sorbonne, pp. 517–52.

Ellenblum, Ronnie (1998) *Frankish Rural Settlement in the Latin Kingdom of Jerusalem*. Cambridge: Cambridge University Press.

Erdmann, Carl (1977) trans. Marshall W. Baldwin and Walter Goffart. *The Origin of the Idea of the Crusade*. Princeton, NJ: Princeton University Press. (First published as *Die Entstehung der Kreuzzugsgedankers* (1935) Stuttgart.)

Favreau-Lilie, Marie (1995) 'The German Empire and Palestine: German Pilgrims to Jerusalem between the 12th and 16th Centuries', *Journal of Medieval History*, 21: 321–41.

Ferzoco, George (1992) 'The Origin of the Second Crusade', in M. Gervers (ed.) *The Second Crusade and the Cistercians*. New York: St Martin's Press, pp. 91–100.

Fletcher, Richard (1984) *St James' Catapult: The Life and Times of Diego Ramirez of Santiago de Compostela*. Oxford: Clarendon Press.

Fletcher, Richard (1987) 'Reconquest and Crusade in Spain, c.1050–1150', *Transactions of the Royal Historical Society*, 37: 31–47.

Flori, Jean (1997) *La première croisade 1095–99: l'Occident chrétien contre l'Islam*. Brussels: Editions complexes.

Flori, Jean (1999) *Pierre l'hermite et la première croisade*. Paris: Fayard.

Flori, Jean (2001) *La guerre sainte. La formation de l'idée de croisade dans l'Occident chrétien*. Paris: Aubier.

Folda, Jaroslav (1976) *Crusader Manuscript Illumination at Saint Jean d'Acre, 1275–91*. Princeton, NJ: Princeton University Press.

Folda, Jaroslav (1977) 'Painting and Sculpture in the Latin Kingdom of Jerusalem, 1099–1291' in K. Setton (gen. ed.) *A History of the Crusades. IV The Art and Architecture of the Crusader States*, ed. Harry W. Hazard. Madison, WI: University of Wisconsin Press, 1977.

Folda, Jaroslav (1986) *The Nazareth Capitals and the Crusader Shrine of the Annunciation*. University Park PA: Penn State Press.

Folda, Jaroslav (1995) *The Art of the Crusaders in the Holy Land, 1098–1187*. Cambridge: Cambridge University Press.

Forey, Alan (1971) 'The Order of Mountjoy', *Speculum*, 46: 250–66.

Forey, Alan (1994) 'The Second Crusade: Scope and Objectives', *Durham University Journal*, 55: 165–75.

Forey, Alan (1995) 'The Military Orders, 1120–1312', in Jonathan Riley-Smith, ed. *The Oxford Illustrated History of the Crusades*. Oxford: OUP, pp. 184–216.

France, John (1971) 'The Departure of Tatikios from the Crusader Army' *Bulletin of the Institute of Historical Research*, 44: 137–47.

France, John (1994) *Victory in the East: A Military History of the First Crusade*. Cambridge: Cambridge University Press.

France, John (1996) 'Patronage and the Appeal of the First Crusade', in Jonathan Phillips (ed.) *The First Crusade. Origins and Impact*. Manchester: MUP, pp. 5–20.

France, John (1999) *Western Warfare in the Age of the Crusades, 1000–1300*. Ithaca, NY: Cornell University Press.

Gibb, Hamilton (1958) 'Zengi and the Fall of Edessa', in K. Setton (gen. ed.) *A History of the Crusades. I: The First Hundred Years*. ed. Marshall W. Baldwin. Philadelphia: University of Pennsylvania Press, pp. 449–62.

Gibb, Hamilton (1958) 'The Career of Nur ad-Din', in K. Setton (gen. ed.) *A History of the Crusades. I: The First Hundred Years*. ed. Marshall W. Baldwin. Philadelphia: University of Pennsylvania Press, pp. 513–27.

Gieysztor, A. (1948) 'The Genesis of the Crusades: The Encyclical of Sergius IV', *Medievalia et Humanistica*, 5: 3–23.

Gil, M. (1992) *A History of Palestine 634–1099*. Cambridge: CUP.

Gilchrist, John (1985) 'The Erdmann Thesis and Canon Law, 1083–1141', in P.W. Edbury (ed.) *Crusade and Settlement*. Cardiff: University College Cardiff Press, pp. 37–45.

Gillingham, John (1978) *Richard the Lionheart*. London: Book Club Associates.

Grabois, Aryeh (1985) 'The Crusade of King Louis VII: A Reconsideration' in P.W. Edbury (ed.) *Crusade and Settlement*. Cardiff: University College Cardiff Press, pp. 94–104.

Grousset, René (1934–6) *Histoire des croisades et du royaume franc de Jérusalem*, 3 vols: vol 1, 1934; vol 2, 1935; vol 3, 1936. Paris.

Hamilton, Bernard (1976) 'The Cistercians in the Crusader States', in Basil Pennington ed. *One Yet Two. Monastic Tradition East and West*. Kalamazoo, MI: Cistercian Studies Institute, pp. 405–22.

Hamilton, Bernard (1977) 'Rebuilding Zion: The Holy Places of Jerusalem in the Twelfth Century', *Studies in Church History*, 14: 105–16.

Hamilton, Bernard (1978) 'The Elephant of Christ: Reynald of Châtillon', *Studies in Church History*, 15: 97–108.

Hamilton, Bernard (1980) *The Latin Church in the Crusader States: The Secular Church*. London: Variorum.

Hamilton, Bernard (1999) 'Aimery of Limoges, Latin Patriarch of Antioch (*c.*1142–*c.*1196) and the Unity of the Churches', in K. Ciggaar and H. Teule (eds) *East and West in the Crusader States II Context, Contacts, Confrontations. Acta of the Congress held at Hernen Castle May 1997*. Leuven: Peeters, pp. 1–12.

Hamilton, Bernard (2000) *The Leper King and his Heirs: The Reign of Baldwin IV of Jerusalem*. Cambridge: Cambridge University Press.

Hamilton, Bernard (2001) 'Three patriarchs at Antioch, 1165–70', in M. Balard, B.Z. Kedar and J.S. Riley-Smith (eds) *Dei Gesta per Francos: Etudes sur les croisades dédiées a Jean Richard*. Aldershot: Ashgate, pp. 199–207.

Harris, Jonathan (2003) *Byzantium and the Crusades*. London: Hambledon.

Head, Thomas (1987) 'Andrew of Fleury and the Peace League of Bourges', *Historical Reflections*, 14: 513–29.

Hendy, Michael (1970) 'Byzantium, 1081–1204: An Economic Reappraisal', *Transactions of the Royal Historical Society*, 20: 31–52.

Hiestand, Rudolf (2001) 'The Papacy and the Second Crusade', in J. Phillips and M. Hoch (eds) *The Second Crusade: Scope and Consequences*. Manchester: Manchester University Press, pp. 32–53.

Hillenbrand, Carole (1994) 'Jihad Propaganda in Syria from the Time of the First Crusade until the Death of Zengi: The Evidence of Monumental Inscriptions', in K. Athamina and R. Heacock (eds) *The Frankish Wars and their Influence on Palestine*. Jerusalem: Birzeit University, pp. 60–9.

Hillenbrand, Carole (1999) *The Crusades: Islamic Perspectives*. Edinburgh: University of Edinburgh Press.

Hillenbrand, Carole (2001) ' "Abominable Acts": the Career of Zengi', in J. Phillips and M. Hoch (eds) *The Second Crusade: Scope and Consequences*. Manchester: Manchester University Press, pp. 111–32.

Hitti, Philip (1929) trans. *An Arab-Syrian Gentleman and Warrior in the Period of the Crusades*. New edition. London: Tauris.

Hoch, Martin (1992) 'The Crusaders' Strategy against Fatimid Ascalon and the "Ascalon Project" of the Second Crusade', in M. Gervers (ed.) *The Second Crusade and the Cistercians*. New York: St Martin's Press, pp. 119–28.

Hoch, Martin (1996) 'The Choice of Damascus as the Objective of the Second Crusade: A Re-evaluation', in M. Balard (ed.) *Autour de la première croisade*. Paris: Sorbonne, pp. 359–70.

Hoch, Martin (2001) 'The Price of Failure: the Second Crusade as a turning-point in the History of the Latin East?' in Jonathan Phillips and Martin Hoch (eds) *The Second Crusade. Scope and Consequences*. Manchester: MUP, pp. 180–200.

Holt, Peter (1986) *The Age of the Crusades. The Near East from the Eleventh Century to 1517*. Harlow: Longman.

Housley, Norman (1980) 'The Franco-Papal Crusade Negotiations of 1322–3', *Papers of the British School at Rome*, 48: 166–85.

Housley, Norman (1982a) 'Politics and Heresy in Italy: Anti-Heretical Crusaders, Orders and Confraternities, 1200–1500', *Journal of Ecclesiastical History*, 33: 193–208.

Housley, Norman (1982b) *The Italian Crusades: The Papal–Angevin Alliance and the Crusades against Christian Lay Powers, 1254–1343*. Oxford: Oxford University Press.

Housley, Norman (1986) *The Avignon Papacy and the Crusades*. Oxford: Oxford University Press.

Housley, Norman (1992) *The Later Crusades From Lyons to Alcazar 1274–1580*. Oxford: Oxford University Press.

Housley, Norman (1995) 'The Crusading Movement, 1274–1700', in Jonathan Riley-Smith, ed. *The Oxford Illustrated History of the Crusades*. Oxford: OUP, pp. 260–93.

Housley, Norman (1996) ed. and trans. *Documents on the Later Crusades*. Basingstoke: Macmillan.

Housley, Norman (2000) 'Holy Land or Holy Lands? Palestine and the Catholic West in the Late Middle Ages and Renaissance', in R.N. Swanson (ed.) *The Holy Land, Holy Lands and Christian History*. Woodbridge: Boydell and Brewer, pp. 228–49.

Hunt, Lucy-Anne (1991) 'Art and Colonialism: The Mosaics of the Church of the Nativity in Bethlehem (1169) and the Problem of "Crusader" Art', *Dumbarton Oaks Papers*, 45: 69–85.

Irwin, Robert (1995) 'Islam and the Crusades, 1096–1699', in J. Riley-Smith (ed.) *The Oxford Illustrated History of the Crusades*. Oxford: Oxford University Press, pp. 217–59.

Jackson, Peter (1987) 'The Crusades of 1239–41 and their Aftermath', *Bulletin of the School of Oriental and African Studies*, 50: 32–60.

Jackson, Peter (1991) 'The Crusade Against the Mongols', *Journal of Ecclesiastical History*, 43: 1–18.

Jacoby, David (1973) 'The Encounter of Two Societies: Western Conquerors and Byzantines in the Peloponnesus after the Fourth Crusade', *American Historical Review*, 78: 873–906.

Jaspert, Nikolas (2001) '*Capta est Dertosa, clavis Christianorum*: Tortosa and the Crusades,' in Jonathan Phillips and Martin Hoch (eds) *The Second Crusade. Scope and Consequences*. Manchester: MUP, pp. 90–110.

Jensen, K.V. (2001) 'Denmark and the Second Crusade: the Formation of a Crusader State?' in Jonathan Phillips and Martin Hoch (eds) *The Second Crusade. Scope and Consequences*. Manchester: MUP, pp. 164–79.

Johnson, Edgar (1975) 'The German Crusade on the Baltic', in K. Setton (gen. ed.) *A History of the Crusades. III: The Fourteenth and Fifteenth Centuries*. ed. H. Hazard. Madison, WI: University of Wisconsin Press, pp. 545–85.

Jordan, William C. (1979) *Louis IX and the Challenge of the Crusade: A Study in Rulership*. Princeton, NJ: Princeton University Press.

Jotischky, Andrew (1994) 'Manuel Comnenus and the Reunion of the Churches: The Evidence of the Conciliar Mosaics in the Church of the Nativity in Bethlehem', *Levant*, 26: 207–24.

Jotischky, Andrew (1995) *The Perfection of Solitude: Hermits and Monks in the Crusader States*. University Park, PA: Penn State Press.

Jotischky, Andrew (1997) 'Gerard of Nazareth, Mary Magdalene and Latin Relations with the Greek Orthodox in the Crusader East in the Twelfth Century', *Levant*, 29: 217–26.

Jotischky, Andrew (1999) 'The Fate of the Orthodox Church in Jerusalem at the End of the Twelfth Century', in T. Hummel, K. Hintlian and U. Carmesund (eds) *Patterns of the Past, Prospects for the Future: The Christian Heritage in the Holy Land II*. London. Melisende, pp. 179–94.

Jotischky, Andrew (2000) 'History and Memory as Factors in Greek Orthodox Pilgrimage to the Holy Land under Latin Rule', *Studies in Church History*, 36: 110–22.

Jotischky, Andrew (2001) 'Greek Orthodox and Latin Monasticism around Mar Saba under Crusader Rule', in J. Patrich (ed.) *The Sabaite Heritage in the Orthodox Church from the Fifth Century to the Present*. Leuven: Peeters, pp. 85–96.

Jotischky, Andrew (2004) 'The Mendicants as Missionaries and Travellers in the Near East in the Thirteenth and Fourteenth Centuries', in R. Allen (ed.) *Eastward Bound: Travel and Travellers in the Medieval Mediterranean*. Manchester: Manchester University Press, pp. 88–106.

Kaelber, Lutz (1998) *Schools of Asceticism: Ideology and Organization in Medieval Religious Communities*. University Park, PA: Penn State Press.

Kedar, Benjamin Z. (1982) 'The Patriarch Eraclius', in B.Z. Kedar, H.E. Mayer and R.C. Smail (eds) *Outremer: Studies in the History of the Crusading Kingdom of Jerusalem Presented to Joshua Prawer*. Jerusalem: Israel Academy of Sciences, pp. 177–204.

Kedar, Benjamin Z. (1983) 'Gerard of Nazareth, a Neglected Twelfth-Century Writer of the Latin East', *Dumbarton Oaks Papers*, 37: 55–77.

Kedar, Benjamin Z. (1984) *Crusade and Mission: European Approaches toward the Muslims*. Princeton, NJ: Princeton University Press.

Kedar, Benjamin Z. (1990) 'The Subjected Muslims of the Frankish Levant', in James Powell (ed.) *Muslims under Latin Rule*. Princeton, NJ: Princeton University Press, pp. 135–74.

Kedar, Benjamin Z. (1992) 'The Battle of Hattin Revisited', in Benjamin Kedar (ed.) *The Horns of Hattin*. Aldershot: Ashgate and Yad Izhak Ben Zvi, pp. 190–207.

Kedar, Benjamin Z. (1997) 'L'appel de Clermont vu de Jérusalem', in *Le concile de Clermont de 1095 et l'appel à la croisade*. Rome: Ecole française de Rome, pp. 287–94.

Kedar, Benjamin Z. (1998) 'The "Tractatus de locis et statu sancta terre Ierosolimitane" ', in J. France and W. Zajac (eds) *The Crusades and their Sources: Studies Presented to Bernard Hamilton*. Aldershot: Ashgate, pp. 113–33.

Keen, Maurice (1984) *Chivalry*. New Haven, CT: Yale University Press.

Kennan, Elizabeth (1971) 'Innocent III and the First Political Crusade', *Traditio*, 27: 231–51.

Köhler, M.A. (1991) *Allianzen und Verträge zwischen frankischen und islamischen Herrschern in Voderen Orient*. Berlin: Gruyter.

Krautheimer, Richard (1942) 'Introduction to an Iconography of Medieval Architecture', *Journal of the Warburg and Courtauld Institutes*, 5: 1–33.

Kühnel, Beata (1977) 'Crusader Sculpture at the Ascension Church on the Mount of Olives in Jerusalem', *Gesta*, 16: 41–50.

Kühnel, Gustav (1988) *Wall Painting in the Latin Kingdom of Jerusalem*. Berlin: Verlag.

La Monte, J.L. (1932) 'To what extent was the Byzantine emperor the suzerain of the Latin Crusading States?' *Byzantion*, 7: 253–64.

Lambert, Malcolm (1998) *The Cathars*. Oxford: Blackwell.

Leopold, Anthony (2000) *How to Recover the Holy Land. The Crusade Proposals of the Late Thirteenth and Early Fourteenth Centuries*. Aldershot: Ashgate.

Lilie, Ralph-Joahhnes (1993) *Byzantium and the Crusader States*. Oxford: Oxford University Press.

Linehan, Peter (1992) *History and the Historians of Medieval Spain*. Oxford: Clarendon Press.

Livermore, Harold (1990) 'The Conquest of Lisbon and its Author', *Portuguese Studies*, 6: 1–16.

Lock, Peter (1995) *The Franks in the Aegean 1204–1500*. London: Longman.

Loud, Graham (2000) *The Age of Robert Guiscard. Southern Italy and the Norman Conquest*. Harlow: Longman.

Lunt, W.E. (1917) 'A Papal Tenth Levied in the British Isles from 1274 to 1280', *English Historical Review*, 32. 49–89.

Luttrell, Anthony (1996) 'The Earliest Templars', in M. Balard (ed.) *Autour de la première croisade*. Paris: Sorbonne, pp. 193–202.

Luttrell, Anthony (1997) 'The Earliest Hospitallers', in Benjamin Z. Kedar, J.S.C. Riley-Smith and Rudolf Hiestand (eds) *Montjoie: Studies in Crusade History in Honour of Hans Eberhard Mayer*. Aldershot: Ashgate, pp. 37–54.

Lyman, Thomas (1992) 'The Counts of Toulouse, the Reformed Canons and the Holy Sepulchre', in Benjamin Kedar (ed.) *The Horns of Hattin*. Aldershot: Ashgate and Yad Izhak Ben Zvi, pp. 63–80.

Lyons, Malcolm C. and Jackson, David (1982) *Saladin: The Politics of the Holy War*. Cambridge: Cambridge University Press.

McGinn, Bernard (1978) 'Iter sancti Sepulchre: the Piety of the First Crusaders', in Bede Lackner and Kenneth Philip, *Essays in Medieval Civilization. The Walter Prescott Webb Memorial Lectures 12*. Austin, TX: University of Texas, pp. 33–72.

McKay, Angus (1977) *Spain in the Middle Ages: From Frontier to Empire, 1000–1500*. Basingstoke: Macmillan.

Madelin, L. (1916) 'La Syrie franque', *Revue des deux mondes*, 38: 314–58.

Madelin, L. (1918) *L'expansion française; de la Syrie au Rhin*. Paris.

Magdalino, Paul (1993) *The Empire of Manuel I Komnenos*. Cambridge: Cambridge University Press.

Maier, Christoph (1996) *Preaching the Crusades. Mendicant Friars and the Cross in the Thirteenth Century*. Cambridge: CUP.

Maier, Christoph (2000) *Crusade Propaganda and Ideology: Model Sermons for the Preaching of the Cross*. Cambridge: Cambridge University Press.

Mayer, Hans E. (1972) *The Crusades*. Trans. J. Gillingham. Oxford: Oxford University Press. (2nd edn 1988.)

Mayer, Hans E. (1977) *Bistümer, Klöster und Stifte im Königreich Jerusalem*. Stuttgart: MGH.

Mayer, Hans E. (1978) 'Latins, Muslims and Greeks in the Latin Kingdom of Jerusalem', *History*, 63: 175–92.

Mayer, Hans E. (1982) 'The Concordat of Nablus', *Journal of Ecclesiastical History*, 33: 531–43.

Mayer, Hans E. (1990) 'The Wheel of Fortune: Seigneurial Vicissitudes under Kings Fulk and Baldwin III of Jerusalem', *Speculum*, 65: 860–77.

Mayer, Hans E. (1992) 'The Beginnings of King Amalric of Jerusalem', in Benjamin Z. Kedar (ed.) *The Horns of Hattin. Proceedings of the Second Conference of the Society for the Study of the Crusades and the Latin East, July 1987*. Jerusalem: Yad Izhak Ben Zvi, pp. 121–35.

Menédez Pidal, Ramon (1956) *La Espana del Cid*. Madrid: Espasa calpa.

Moore, R.I. (1987) *The Formation of a Persecuting Society*. Oxford: Blackwell.

Moore, R.I. (2000) *The First European Revolution, c.970–1215*. Oxford: Blackwell.

Morony, M.G. (1990) 'The Age of Conversions: A Reassessment', in M. Gervers and R.J. Bikhazi (eds), *Conversion and Continuity. Indigenous Christian Communities in Islamic Lands: Eighth to Eighteenth Centuries*. Toronto: Pontifical Institute for Medieval Studies, pp. 135–50.

Morris, Colin (1984) 'Policy and Visions: the Case of the Holy Lance at Antioch', in John Gillingham and J.C. Holt (eds) *War and Government in the Middle Ages: Essays in Honour of J.O. Prestwich*. Woodbridge: Boydell and Brewer, pp. 33–45.

Morris, Colin (1996) 'Peter the Hermit and the Chroniclers', in J. Phillips (ed.) *The First Crusade. Origins and Impact*. Manchester: Manchester University Press, pp. 21–34.

Morris, Colin (1997) 'Bringing the Holy Sepulchre to the West: S. Stefano, Bologna, from the Fifth to the Twentieth Century, in R.N. Swanson (ed.) *The Church Retrospective*. Woodbridge: Boydell and Brewer, pp. 31–59.

Morris, Colin (2000) 'Memorials of the Holy Places and Blessings from the East: Devotion to Jerusalem before the Crusades', in R.N. Swanson (ed.) *The Holy Land, Holy Lands and Christian History*. Woodbridge: Boydell and Brewer, pp. 90–109.

Mouriki, Doula (1985/6) 'Thirteenth-century Icon Painting in Cyprus', *The Griffon*, 1–2: 9–112.

Muldoon, James (1979) *Popes, Lawyers and Infidels*. Liverpool: Liverpool University Press.

Murray, Alan V. (1990) 'The Title of Godfrey of Bouillon as Ruler of Jerusalem', *Collegium Medievale*, 3: 163–78.

Murray, Alan V. (1992) 'The Army of Godfrey of Bouillon 1096–99: Structure and Dynamics of an Army on the First Crusade', *Revue Belge de philologie et d'histoire*, 70: 301–29.

Murray, Alan V. (1994) 'Baldwin II and his Nobles: Baronial Factionalism and Discontent in the Kingdom of Jerusalem, 1118–24', *Nottingham Medieval Studies*, 38: 60–85.

Murray, Alan V. (1995) 'Ethnic Identity in the Crusader States', in S. Forde, L. Johnson and A.V. Murray (eds) *Concepts of National Identity in the Middle Ages*. Leeds: University of Leeds, pp. 59–73.

Murray, Alan V. (2000) *The Crusader Kingdom of Jerusalem. A Dynastic History 1099–1125*. Oxford: Unit for Prosopographical Research.

Nicholson, Helen (2001) *The Knights Hospitaller*. Woodbridge: Boydell and Brewer.

Nicholson, R.L. (1958) 'The Growth of the Latin States, 1118–1144', in
K. Setton (gen. ed.) *A History of the Crusades. I: The First Hundred
Years*. ed. Marshall W. Baldwin. Philadelphia: University of
Pennsylvania Press, pp. 410–48.

Nicol, Donald (1988) *Byzantium and Venice*. Cambridge: Cambridge
University Press.

O'Callaghan, Joseph (2003) *Reconquest and Crusade in Medieval Spain*.
Philadelphia: University of Pennsylvania Press.

Oikonomides, Nicholas (1976) 'La decomposition de l'empire byzantin a
la veille de 1204 et les origins de l'empire de Nicée a propos de la
Partitio Romaniae', *Actes du XVe Congrès international d'études
Byzantines*. Athens: I, 3–28.

Ostrogorsky, George (1969) *History of the Byzantine State*. Revised
edition. New Brunswick, NJ: Rutgers University Press.

Pahlitzsch, Johannes (2001) *Graeci und Suriani im Palästina der
Kreuzfahrerzeit. Beiträge und Quellen zur Geschichte des griechisch-
orthodoxen Patriarchats von Jerusalem*. Berlin: Duncker and
Humblot.

Palmer, Andrew (1992) 'The History of the Syrian Orthodox in
Jerusalem, II. Queen Melisende and the Jacobite Estates',
Oriens Christianus, 76: 74–94.

Payne, Robert (1984) *The Dream and the Tomb. A History of the
Crusades*. New York: Stein and Day.

Panagopoulos, Beata (1979) *Cistercian and Mendicant Monasteries in
Medieval Greece*. Chicago.

Phillips, Jonathan (1996) *Defenders of the Holy Land*. Oxford:
Clarendon Press.

Phillips, Jonathan (1997) 'St Bernard, the Low Countries, and the Lisbon
Letter of the Second Crusade', *Journal of Ecclesiastical History*, 48:
485–97.

Phillips, Jonathan (2001) 'Papacy, Empire and the Second Crusade', in
J. Phillips and M. Hoch (eds) *The Second Crusade: Scope and
Consequences*. Manchester: Manchester University Press,
pp. 15–31.

Powell, James M. (1986) *Anatomy of a Crusade, 1213–21*. Philadelphia:
University of Pennsylvania Press.

Powicke, F.M. and C.R. Cheny (eds) (1981) *Councils and Synods, with other Documents relating to the English Church*. Vol 2. Oxford: Clarendon Press.

Powicke, Maurice (1962) *The Thirteenth Century*. Oxford: OUP.

Prawer, Joshua (1972) *The Latin Kingdom of Jerusalem: European Colonialism in the Middle Ages*. London: Weidenfeld and Nicolson.

Prawer, Joshua (1980) *Crusader Institutions*. Oxford: Clarendon Press.

Prawer, Joshua (1985) 'Social Classes in the Crusader States: The "Minorities" ', in K. Setton (gen. ed.) *History of the Crusades. V: The Impact of the Crusades on the Near East*, ed. Norman P. Zacour and Harry W. Hazard. Madison: University of Wisconsin Press.

Prawer, Joshua (1988) *A History of the Jews in the Latin Kingdom of Jerusalem*. Oxford: Oxford University Press.

Price, Richard (2000) 'The Holy Land in Old Russian Culture', in R.N. Swanson (ed.) *The Holy Land, Holy Lands and Christian History*. Woodbridge: Boydell and Brewer, pp. 250–62.

Queller, Donald and Madden, Thomas (1997) *The Fourth Crusade: The Conquest of Constantinople 1201–4*, 2nd edition. Philadelphia: University of Pennsylvania Press.

Rabbat, Nassa (1995) 'The Ideological Significance of the Dar al-'Aal in the Medieval Islamic Orient', *International Journal of Middle East Studies*, 3–28.

Reilly, Bernard (1992) *The Contest of Christian and Muslim Spain 1031–1157*. Cambridge, MA: Blackwell.

Reuter, Timothy (2001) 'The "non-crusade" of 1149–50', in Jonathan Phillips and Martin Hoch (eds) *The Second Crusade. Scope and Consequences*. Manchester: MUP, pp. 150–63.

Rey, E.G. (1866) *Essai sur la domination française en Syrie durant le moyen âge*. Paris.

Richard, Jean (1949) 'Le début des relations entre la papauté et les Mongols de Perse', *Journal Asiatique*, 237: 291–7.

Richard, Jean (1989) 'Aux origins d'un grand lignage: des Paladii à Renaud de Châtillon', in *Receuil de melanges offerts a Karl Ferdinand Werner à l'occasion de son 65e anniversaire*. Maulévrier: Hérault, pp. 409–18.

Richard, Jean (1992) *Saint Louis, Crusader King of France*. Trans. Jean Birrell. Cambridge: Cambridge University Press.

Richard, Jean (1999) *The Crusades*. Cambridge: Cambridge University Press.

Riley-Smith, J.S.C. (1967) *The Knights of St John in Jerusalem and Cyprus, c.1050–1310*. London: Macmillan.

Riley-Smith, J.S.C. (1973a) *The Feudal Nobility and the Kingdom of Jerusalem, 1174–1277*. London: Macmillan.

Riley-Smith, J.S.C. (1973b) 'Government in Latin Syria and the Commercial Privileges of Foreign Merchants', in D. Baker (ed.) *Relations between East and West in the Middle Ages*. Edinburgh: Edinburgh University Press, pp. 109–32.

Riley-Smith, J.S.C. (1977) *What Were the Crusades?* Basingstoke: Macmillan.

Riley-Smith, J.S.C. (1979) 'The Title of Godfrey of Bouillon', *Bulletin of the Institute of Historical Research*, 52: 83–6.

Riley-Smith, J.S.C. (1980) 'An Approach to Crusading Ethics', *Reading Medieval Studies*, 6: 177–92.

Riley-Smith, J.S.C. (1986) *The First Crusade and the Idea of Crusading*. London: Athlone.

Riley-Smith, J.S.C. (1987) *The Crusades: A Short History*. London: Athlone.

Riley-Smith, J.S.C. (1992) 'Family Traditions and Participation in the Second Crusade', in M. Gervers (ed.) *The Second Crusade and the Cistercians*. New York: St Martin's Press, pp. 101–8.

Riley-Smith, J.S.C. (1995), ed. *The Oxford Illustrated History* of the Crusades. Oxford: OUP.

Riley-Smith, J.S.C. (1997) *The First Crusaders*. Cambridge: Cambridge University Press.

Riley-Smith, J.S.C. (2002) *What were the Crusades?*, 2nd edition. Basingstoke: Palgrave Macmillan.

Riley-Smith, J.S.C. and Riley-Smith, Louise (1981) *The Crusades: Idea and Reality*. London: Edward Arnold.

Robinson, I.S. (1973) 'Gregory VII and the Soldiers of Christ', *History*, 58: 169–92.

Robinson, I.S. (1999) *Henry IV of Germany, 1056–1106*. Cambridge: Cambridge University Press.

Rosenwein, B.H. (1989) *To Be the Neighbor of St Peter: The Social Meaning of Cluny's Property 909–1049*. Ithaca, NY: Cornell University Press.

Rowe, John G. (1992) 'The origins of the Second Crusade: Pope Eugenius III, Bernard of Clairvaux and Louis VII of France' in M. Gervers (ed.) *The Second Crusade and the Cistercians*. New York: St Martin's Press, pp. 79–90.

Runciman, Steven (1951–4) *A History of the Crusades*. 3 vols: vol 1, 1951; vol 2, 1952; vol 3, 1954. Cambridge: Cambridge University Press.

Runciman, Steven (1958) 'The Pilgrimages to Palestine before 1095', in K. Setton (gen. ed.) *History of the Crusades. I: The First Hundred Years*, ed. Marshall W. Baldwin. Philadelphia: University of Pennsylvania Press, pp. 68–78.

Runciman, Steven (1958) *The Sicilian Vespers: A History of the Mediterranean World in the Later Thirteenth Century*. Cambridge: Cambridge University Press.

Russell, Frederick H. (1975) *The Just War in the Middle Ages*. Cambridge: Cambridge University Press.

Sayers, Jane (1994) *Innocent III: Leader of Europe, 1198–1216*. London: Longman.

Schein, Sylvia (1979) '*Gesta Dei per Mongolos*. The Genesis of a Non-Event', *English Historical Review*, 94: 805–19.

Schein, Sylvia (1991) *Fidelis Crucis: The Papacy, the West and the Holy Land 1274–1314*. Oxford: Oxford University Press.

Shepard, Jonathan (1996) 'Cross-purposes: Alexius Comnenus and the First Crusade', in J. Phillips (ed.) *The First Crusade: Origins and Impact*. Manchester: Manchester University Press, pp. 107–29.

Siberry, Elizabeth (1983) 'Missionaries and Crusaders, 1095–1274: Opponents or Allies?', *Studies in Church History*, 20: 103–10.

Siberry, Elizabeth (1985) *Criticism of Crusading 1095–1274*. Oxford: Oxford University Press.

Sivan, Emmanuel (1968) *L'Islam et la crosiade: idéologie et propaganda dans les reactions musulmanes aux croisades*. Paris: Adrien Maisonneuve.

Smail, R.C. (1956) *Crusading Warfare 1097–1193*. Cambridge: Cambridge University Press.

Smail, R.C. (1969) 'Latin Syria and the West, 1149–87', *Transactions of the Royal Historical Society*, 19: 1–20.

Smail, R.C. (1977) 'The International Status of the Latin Kingdom of Jerusalem, 1150–92', in P.M. Holt (ed.) *The Eastern Mediterranean Lands in the Period of the Crusades*. Warminster: Aris and Phillips, pp. 23–43.

Smail, R.C. (1982) 'The Predicaments of Guy of Lusignan', in B.Z. Kedar, H.E. Mayer and R.C. Smail (eds) *Outremer: Studies in the History of the Crusading Kingdom of Jerusalem Presented to Joshua Prawer*. Jerusalem: Israel Academy of Sciences, pp. 159–76.

Somerville, Robert (1972) *The Councils of Urban II. Vol 1: Decreta Claromontensia*. Amsterdam: Hakkert.

Southern, Richard William (1970) *Western Society and the Church in the Middle Ages*. Harmondsworth: Penguin.

Stern, H. (1936) 'Les representations des conciles dans l'Eglise de la Nativité à Bethléem', *Byzantion*, 11: 101–52.

Stern, H. (1938) 'Les representations des conciles dans l'Eglise de la Nativité à Bethléem. 2ème partie. Les inscriptions' *Byzantion*, 13: 415–49.

Strayer, Joseph R. (1962) 'The Political Crusades of the Thirteenth Century', in K. Setton (gen. ed.) *A History of the Crusades. II: The Later Crusades 1189–1311*, ed. Robert L. Wolff and Harry W. Hazard. Philadelphia: University of Pennsylvania Press, pp. 343–75.

Sumption, Jonathan (1975) *Pilgrimage. An Image of Medieval Religion*. London: Faber.

Sumption, Jonathan (1978) *The Albigensian Crusade*. London: Faber.

Tabbaa, Yasser (1986) 'Monuments with a Message: Propagation of Jihad under Nur al-Din', in V.P. Goss (ed.) *The Meeting of Two Worlds: Cultural Exchange Between East and West during the Period of the Crusades*. Kalamazoo: Medieval Studies Institute, pp. 223–40.

Thorau, Peter (1992) *The Lion of Egypt: Sultan Baybars I and the Near East in the Thirteenth Century*. Trans. by P.M. Holt. London: Longman.

Throop, Palmer (1940) *Criticism of the Crusade: A Study of Public Opinion and Crusade Propaganda*. Amsterdam: Swets and Zeitlinger.

Tibble, Stephen (1989) *Monarchy and Lordships in the Latin Kingdom of Jerusalem*. Oxford: Clarendon Press.

Treadgold, Warren (1997) *A History of the Byzantine State and Society*. Stanford, CA: University of California Press.

Tyerman, Christopher (1985) 'The Holy Land and the Crusades of the Thirteenth and Fourteenth Centuries', in P.W. Edbury (ed.) *Crusade and Settlement*. Cardiff: University of Wales Press, pp. 105–12.

Tyerman, Christopher (1988) *England and the Crusades, 1095–1588*. Chicago, IL: University of Chicago Press.

Tyerman, Christopher (1998) *The Invention of the Crusades*. Basingstoke: Macmillan.

Van Cleve, Thomas (1962) 'The Fifth Crusade', in K. Setton (gen. ed.) *A History of the Crusades. II: The Later Crusades 1189–1311*, ed. Robert L. Wolff and Harry W. Hazard. Philadelphia: University of Pennsylvania Press, pp. 377–428.

Van Moolenbroek, Jaap (1987) 'Signs in the heavens in Groningen and Friesland in 1214: Oliver of Cologne and Crusading Propaganda', *Journal of Medieval History*, 13: 251–72.

Vasiliev, A.A. (1937) 'The Opening Stages of the Anglo-Saxon Immigration to Byzantium in the Eleventh Century', *Annales de l'institut Kondakov* 9: 39–41.

Waley, Daniel (1961) *The Papal State in the Thirteenth Century*. London: Macmillan.

Weiler, Bjorn (2000) 'Gregory IX, Frederick II and the liberation of the Holy Land, 1230–9'.

Weiler, Bjorn (2003) 'The "Negotium Terrae Sanctae" in the Political Discourse of Latin Christendom, 1215–1311', *International History Review*, 25: 1–36.

Weitzmann, Kurt (1963) 'Thirteenth-century Crusader Icons on Mt Sinai', *Art Bulletin*, 45: 179–203.

Weitzmann, Kurt (1966) 'Icon Painting in the Crusader Kingdom', *Dumbarton Oaks Papers*, 20: 49–83.

Wessley, Stephen (2000) 'The Role of the Holy Land for the Early Followers of Joachim of Fiore', in R.N. Swanson (ed.) *The Holy Land, Holy Lands and Christian History*. Woodbridge: Boydell and Brewer, pp. 181–91.

White, Stephen (1988) *Custom, kinship and Gifts to Saints: the 'Laudatio Parentum' in Western France 1050–1150*. Chapel Hill, NC: University of North Carolina Press.

Index

Volume II

A NECESSARY EVIL?

Slavery and the Debate Over the Constitution

Edited by

JOHN P. KAMINSKI

Published for
The Center for the Study of the
American Constitution

MADISON HOUSE

Madison 1995

Kaminski, Ed.
A Necessary Evil?
Slavery and the Debate Over the Constitution

Copyright © 1995 by Madison House Publishers, Inc.
All rights reserved.

Printed in the United States of America

LIBRARY OF CONGRESS CATALOGING-IN-PUBLICATION DATA

A necessary evil? slavery and the debate over the Constitution /
edited by John P. Kaminski
p.   cm. — (Constitutional heritage series ; v. 2)
"Published for the Center for the Study of the American Constitution."
Includes bibliographical references and index.
ISBN 0-945612-16-8 (cloth). — ISBN 0-945612-33-8 (paper)
1. Slavery—Law and legislation—United States—History.
I. Kaminski, John P.   II. University of Wisconsin—Madison.
Center for the Study of the American Constitution.   III. Series.
KF4545.S5N43
326'.0973—DC20   95-9711
CIP

Volume II in the
CONSTITUTIONAL HERITAGE SERIES
ISSN 0895-9633

Published for
The Center for the Study of the American Constitution
by
MADSION HOUSE PUBLISHERS, INC.
P.O. Box 3100, Madison, Wisconsin 53704

FIRST EDITION

Contents

For Jan

Preface

THE PERIOD OF THE AMERICAN REVOLU-
tion has attracted relatively little in-depth attention from scholars interested
in the institution of slavery. Only a handful of books and articles concentrate
on this period—most scholars saunter through the last quarter of the eighteenth century with their eyes focused on the antebellum period when slavery
and abolitionism took center stage in national politics.

During the last quarter century, while I have been editing the public
debates and private papers dealing with the ratification of the Constitution, I
have found that slavery was a crucial issue for the Founders. By collecting and
publishing here in one volume the complete documentation on the issue of
slavery during the debate over the ratification of the Constitution, it is hoped
that readers will gain a deeper appreciation of the founding generation—slave-
holders, opponents of slavery, and slaves—as it grappled with its seemingly
insoluble moral, social, and economic dilemma.

To understand more completely the attitude of the Revolutionary generation toward slavery, documents have been included in this volume from
the beginning of the war for independence through the abolition of the foreign slave trade. The debate in the first federal Congress over slavery indicates
the official governmental stance on the institution under the new Constitution. Although Congress would be allowed to prohibit the foreign slave trade
in 1808, slavery itself was not to be a national issue—it was to fall within the
jurisdiction of state governments. Emancipation would continue to move for-

ward gradually in the Northern and Middle states; slaveholding, however, would thrive and expand in the South.

A number of scholars have offered their advice and assistance. Brent Tarter of the State Library of Virginia sent me copies of manuscript petitions submitted by Methodists to the legislature in 1785 seeking an endorsement of emancipation in Virginia. Mary A. Y. Gallagher of *The Papers of Robert Morris* reminded me of James Madison's quandary with his slave Billey. Helen E. Veit, Kenneth R. Bowling, and Charlene Bangs Bickford of the *Documentary History of the First Federal Congress* shared with me information on the congressional debates over the ten-dollar tax on slave imports and the anti-slavery petitions. The complete text of these debates has now been published in their volumes. My colleague and friend Richard Leffler did his usual excellent job of critiquing the entire manuscript. Professor William L. Van Deburg of the Afro-American Studies program at the University of Wisconsin-Madison offered sound advice and insightful comments on the manuscript. Gregory M. Britton, director of Madison House, provided his encouragement and assistance, along with critical editorial judgments throughout the publication process. Finally, my wife Janice, to whom this book is dedicated, has been her usual patient self in sharing me with the eighteenth century.

Introduction

AMERICAN SOCIETY WAS ALMOST TWO centuries old by the time—in 1776—when most Americans decided that their liberty could best be preserved outside of the British Empire. With a stirring expression of their sentiments they declared:

> We hold these truths to be self-evident, that all men are created equal, that they are endowed by their Creator with certain unalienable Rights, that among these are Life, Liberty and the pursuit of Happiness.

Noble principles indeed. Eleven years later, disgruntled with their governments—state and confederation—Americans issued a second statement of principles. This one, in the form of a new federal constitution, proclaimed:

> We, the People of the United States, in order to form a more perfect union, establish justice, insure domestic tranquility, provide for the common defence, promote the general welfare, and secure the blessings of liberty to ourselves and our posterity, do ordain and establish this Constitution for the United States of America.

Something dramatic had happened to the character of the American people in the intervening decade between the signing of the Declaration of Independence and the promulgation of the Constitution. The principles for which Americans were willing to die—*freedom*, *equality*, and *unalienable rights*—had given way to the Constitution's call for *justice*, *tranquility*, *defense*, *general welfare*, and *liberty*. Americans qualified their earlier expression of universal equality by applying it only to certain groups of people. They also wrote a constitution that strongly protected personal property. In the eighteenth century that meant condoning, sanctioning, and even rewarding the institution of slavery.

This book looks at American attitudes toward a system that held an entire race in bondage during this formative period of the new nation. It asks the question of how Americans could enter their Revolutionary struggle for independence with an empathy toward black slaves deprived of their liberties, only to have many Americans change that empathy into an institutionalized justification of slavery and a blatant racism that stigmatized freed blacks. A majority of Americans in 1776 favored the closing of the African slave trade and, at least philosophically, the idea of a general emancipation of slaves; but a decade later the overall attitude of the country had changed. In the South, slavery became more strongly entrenched than ever, and the "positive good" thesis of slavery sprang to life. In the North, a small minority's intense fervor for emancipation grew steadily alongside a tolerance for the continued existence of slavery in the South and a persistent, mean-spirited racism in the North.

By the early 1820s, it was clear that the Revolution had brought a radical democratizing change to America, marked by vastly expanded political, social, and economic liberty for the white population. In the North, while there were remnants of slavery, it was essentially ended. But in the South, it was clear that the vast extension of liberty was not going to apply to slaves. As slavery became again a moral and political issue, men like Jefferson were frightened by the implications of a crisis based on the unresolved conflict between the liberty of the Revolution and the continuation of slavery. In the wake of the Missouri Compromise, Jefferson expressed his fear for the future. "I regret that I am now to die in the belief, that the useless sacrifice of themselves by the generation of 1776, to acquire self-government and happiness to their country, is to be thrown away by the unwise and unworthy passions of their sons, and that my only consolation is to be, that I live not to weep over it." Many others would shed tears as the conflict over slavery ended in bloody civil war.

Arguably the best way to understand the minds of America's founding generation is to let them speak for themselves. These people thought deeply about issues of race and slavery, and the contradictions of their situation were apparent to many of them. They almost all hated slavery for what it did to blacks as well as what it did to themselves, their children, and their country. They saw how the institution of slavery made petty tyrants out of masters, while at the same time masters struggled in a life and death conflict with Great Britain to throw off the imperial yoke of tyranny. The writings of this founding generation document this struggle of conscience and their ultimate failure to eradicate slavery.

The documents presented here are arranged to give the reader the most direct access to these attitudes. The first chapter lays out the foundations of the discussions of slavery in the nascent republic and focuses on the ever present rhetoric of freedom. Chapter 2 describes the Constitutional Convention debate over the continuation of the African slave trade, the apportionment of slaves for representation and taxation, and the late inclusion of a provision to return runaway slaves to their masters. The third, fourth, and fifth chapters focus on attitudes in New England, the Middle States, and the South on slavery during the debate over the ratification of the new federal Constitution. After ratification, the first federal Congress took up the issue of slavery and the foreign slave trade as racism grew in both the North and the South. This is the subject of the sixth chapter. The last chapter looks specifically at three slaveholding founders—Washington, Jefferson, and Madison—and how their personal failures to endorse a realistic plan for emancipation contributed to the nation's tragic inability to resolve its most serious paradox.

A Note on the Text

The texts of the documents used in this volume have been transcribed verbatim from the original sources. No punctuation has been added and misspellings have not been corrected except for obvious typographical errors. Because of space limitations, excerpts have often been printed.

1

Laying Slavery's
Foundations

B Y THE TIME OF THE REVOLUTION, *America had developed into a diversity of colonies ranging in age from the "Old Dominion" of Virginia established for over 150 years to Georgia which was barely forty years old. Within each colony antagonisms persisted between the more established eastern coastal regions, the maturing hinterlands, and the expanding backcountry. The majority of every colony was of English ancestry, but significant minorities of other Europeans existed everywhere. All of the colonies had slave populations. Although slaves were most numerous and important to the Southern colonies, Northern merchants also prospered from shipping the slave-produced staples of the South. Northern merchants, particularly those in Rhode Island and Massachusetts, also dominated the foreign slave trade. Thus, by the time of the Revolution, slavery was an integral component of American society. The economy, especially the burgeoning Southern agricultural economy and the thriving commerce of the North, benefited from this repressive source of labor.*

In October 1774 the First Continental Congress proposed a Continental Association which called for an economic boycott of British shipping to convince Parliament to soften its imperial policies. Among its provisions the Association prohibited the importation of African slaves. Within a year all of the states adopted the Association, although Georgia objected to the slave-trade provision. A year-and-a-half later, the Second Continental Congress opened all American ports to foreign commerce in direct violation of the British navigation acts. Slave importations, however, remained prohibited. Although future Americans would allude to the humanitarianism of the Continental Association, in reality the measure was first and foremost an economic weapon against British commerce.

When the war ended and Americans achieved independence, some Northern merchants and Southern planters immediately revived the African slave trade. In reaction to this, abolition societies sprang up in Pennsylvania, New York, and New England, with Quakers in the vanguard. Quakers, despised by many for their pacifism during the war, extended their humanitarianism to the plight of slaves and freedmen. These early abolitionists focused on the slave trade because of its odious nature. Free men, women, and children were taken from their families in Africa, incarcerated and transported across the Atlantic in pestilential slave ships on which many died of disease. They were exposed to the tyrannies of the crews and then sold on the slave markets of North America and the West Indies. In addition to these atrocities, the slave trade could be attacked more easily than the institution of slavery itself because no slaveowner would lose personal property if the trade were abolished.

In October 1783, Quakers from Pennsylvania to Virginia petitioned Congress attacking the renewal of the African trade and asking Congress "to discourage and prevent so obvious an Evil." A congressional committee reported that Congress should recommend to the states that they "enact such laws as to their wisdom may appear best calculated" to prohibit the slave trade as outlined in the Continental Association. Congress rejected even this mild measure. Quakers again petitioned Congress in January 1785, but with no success.

Between 1783 and 1787 various states—Maryland, Rhode Island, New York, New Jersey, Delaware and South Carolina—passed laws prohibiting the slave trade or tightening existing laws, while North Carolina laid a prohibitory duty on slave importations. But this state-by-state action was slow, limited, and reversible. Continental action was sought. In 1785 Congress proposed an amendment to the Articles of Confederation giving it the power to regulate commerce. Most of the states granted Congress this power, but Georgia and South Carolina refused to allow Congress power over the slave trade. Like most of the other states, Georgia and South Carolina also adopted the Continental imposts of 1781 and 1783 which authorized Congress to levy a tariff on imports, but both states refused to allow a federal tariff on imported slaves.

THOMAS JEFFERSON
A Summary View of the Rights of British America, July 1774

Originally drafted as instructions for Virginia's delegates to the First Continental Congress, Thomas Jefferson's "Summary View" was tabled and milder instructions were adopted. Without his knowledge, friends of Jefferson published his radical instructions as a twenty-three-page pamphlet in Williamsburg. Within a year, the pamphlet was reprinted in Philadelphia and twice in London. Consequently, when Jefferson arrived in Philadelphia as a delegate to the Second Continental Congress, he had already established a reputation as a radical opponent of British policy and as an eloquent writer.

Among the objections listed in the pamphlet, Jefferson criticized King George III for vetoing colonial legislation that would have prohibited the African slave trade.

Jefferson felt that the king yielded to pressure from British commercial interests who were profiting from the slave trade. He meted out no censure to Virginians—or any other Americans—for their role in the African slave trade. Interestingly, Jefferson alluded to the fact that slaves would eventually be enfranchised. First, however, the slave trade had to be prohibited and then the slaves emancipated.

The criticism against the king expressed here was repeated by Jefferson in his draft of the Declaration of Independence, only to be deleted by Congress.

That we next proceed to consider the conduct of his majesty, as holding the executive powers of the laws of these states, and mark out his deviations from the line of duty: By the constitution of Great Britain, as well as of the several American states, his majesty possesses the power of refusing to pass into a law any bill which has already passed the other two branches of legislature. His majesty, however, and his ancestors, conscious of the impropriety of opposing their single opinion to the united wisdom of two houses of parliament, while their proceedings were unbiassed by interested principles, for several ages past have modestly declined the exercise of this power in that part of his empire called Great Britain. But by change of circumstances, other principles than those of justice simply have obtained an influence on their determinations; the addition of new, and sometimes opposite interests. It is now, therefore, the great office of his majesty, to resume the exercise of his negative power, and to prevent the passage of laws by any one legislature of the empire, which might bear injuriously on the rights and interests of another. Yet this will not excuse the wanton exercise of this power which we have seen his majesty practise on the laws of the American legislatures. For the most trifling reasons, and sometimes for no conceivable reason at all, his majesty has rejected laws of the most salutary tendency. The abolition of domestic slavery is the great object of desire in those colonies, where it was unhappily introduced in their infant state. But previous to the enfranchisement of the slaves we have, it is necessary to exclude all further importations from Africa; yet our repeated attempts to effect this by prohibitions, and by imposing duties which might amount to a prohibition, have been hitherto defeated by his majesty's negative: Thus preferring the immediate advantages of a few British corsairs to the lasting interests of the American states, and to the rights of human nature, deeply wounded by this infamous practice. Nay, the single interposition of an interested individual against a law was scarcely ever known to fail of success, though in the opposite scale were placed the interests of a whole country. That this is so shameful an abuse of a power trusted with his majesty for other purposes, as if not reformed, would call for some legal restrictions.

The Continental Congresses

THE CONTINENTAL ASSOCIATION
October 20, 1774[1]

The First Continental Congress met from September 5 to October 26, 1774. The most important action taken by the delegates (other than calling a second congress) was the passage of the Continental Association, which committed the colonies to resume the economic war against Parliament begun with the non-importation and non-consumption policies of 1768–69. Among other things, Congress agreed that the colonies would not import or purchase slaves from Africa after December 1, 1774. This prohibition on the slave trade, reconfirmed by the Second Continental Congress, was primarily economic rather than a recognition of the evils of slavery. Although petitioned to continue its prohibition, Congress took no action to maintain its official ban; and as the war neared an end in late 1782 and early 1783, the African slave trade resumed.

We will neither import nor purchase, any slave imported after the first day of December next; after which time, we will wholly discontinue the slave trade, and will neither be concerned in it ourselves, nor will we hire our vessels, nor sell our commodities or manufactures to those who are concerned in it.

A PROPOSAL TO FREE THE SLAVES
Fredericksburg, Va., June 9, 1775[2]

Most Americans, in the South as well as in the North, saw and admitted the inconsistencies in their fight for liberty and equality in their struggle against Great Britain, while permitting slavery to exist at home. The search for a viable plan of emancipation continued until the passage of the Thirteenth Amendment in 1865.

This 1775 proposal offered an innovative way in which America could escape from its dilemma. Owners would be compensated, slaves would assist in the war effort, and after victory the slaves would be resettled in Canada apart from white American society. Three problems persisted: what would happen if American attempts to capture Canada failed; what kind of society would develop in Canada if blacks were settled among the existing white population; and how could white Americans in areas with large slave populations be convinced of the propriety and the safety of arming black men? Although many blacks served alongside whites in the state militias and in the Continental army and navy, only Rhode Island formed an all black regiment.

---

[1]Worthington C. Ford et al., eds, *Journals of the Continental Congress, 1774–1789* . . . (34 vols., Washington, D.C., 1904–1937), I, 77.

[2]Adams Papers, Massachusetts Historical Society. The author of this letter addressed to John Adams is unknown.

. . . To proclaim instant Freedom to all [indentured] Servants that will join in the Defence of America, is a Measure to be handled with great Delicacy, as so great, so immediate a Sacrafice of Property, may possibly draw off many of the Americans themselves from the common Cause.

But is not such a Measure absolutely necessary? And might not a proper Equivalent be made to the Masters, out of the Large Sums of Money which at all Events must be struck, in the present Emergency?

If America should neglect to do this, will not Great Britain engage these Servants to espouse her Interest, by proclaiming Freedom to them, without giving any Equivalent to the Masters? To give Freedom to the Slaves is a more dangerous, but equally necessary Measure.

Is it not incompatible with the glorious Struggle America is making for her own Liberty, to hold in absolute Slavery a Number of Wretches, who will be urged by Despair on one Side, and the most flattering Promises on the other, to become the most inveterate Enemies to their present Masters?

If the Inhabitants of Quebec should assist Great Britain, would not true Wisdom dictate to the other Colonies, to lead their Slaves to the Conquest of that Country, and to bestow that and Liberty upon them as a Reward for their Bravery and Fidelity?

Might not a considerable Quit Rent reserved upon their Lands, and a moderate Tax upon their Labours, stipulated beforehand, in a Course of Years, sink the Money struck, and refund to the Colonies the Price of the Slaves now paid to their Masters for their Freedom?

SAMUEL HOPKINS TO THOMAS CUSHING
Newport, R.I., December 29, 1775[3]

Much honored Sir: The degree of acquaintance I have with you, through your indulgence; and your known candour, condescention and goodness, encourages me to address you on an affair, which, in my view, is very interesting, and calls for the particular attention of the honorable members of the Continental Congress.

They have indeed manifested much wisdom and benevolence in advising to a total stop of the slave trade, and leading the united American Colonies to resolve not to buy any more slaves, imported from Africa. This has rejoiced the hearts of many benevolent, pious persons, who have been long convinced of the unrighteousness and cruelty of that trade, by which so many Hundreds of thousands are enslaved. And have we not reason to think this

[3] Adams Papers, Massachusetts Historical Society. Hopkins was the pastor of the First Congregational Church in Newport. Since 1776 he had published several antislavery items. Cushing had been speaker of the Massachusetts colonial House of Representatives almost continuously from 1766 to 1774. At this time he was a delegate to Congress. From 1780 until his death in 1788 he served as lieutenant governor.

has been one means of obtaining the remarkable, and almost miraculous protection and success, which heaven has hitherto granted to the united Colonies, in their opposition to unrighteousness and tyranny, and struggle for *liberty*?

But if the slave trade be altogether unjust, is it not equally unjust to hold those in slavery, who by this trade have been reduced to this unhappy state? Have they not a right to their liberty, which has been thus violently, and altogether without right, taken from them? Have they not reason to complain of any one who withholds it from them? Do not the cries of these oppressed poor reach to the heavens? Will not God require it at the hands of those who refuse to let them go out free? If practising or promoting the slave trade be inconsistent with what takes place among us, in our struggle for liberty, is not retaining the slaves in bondage, whom by this trade we have in our power, equally inconsistent? And is there not, consequently, an inconsistence in resolving against the former, and yet continuing the latter?

And if the righteous and infinitely good Governor of the world, has given testimony of his approbation of our resolving to put a stop to the slave trade, by doing such wonders in our favor; have we not reason to fear he will take his protection from us, and give us up to the power of oppression and tyranny, when he sees we stop short of what might be reasonably expected; and continue the practice of that which we ourselves have, implicitly at least, condemned, by refusing to let the oppressed go free, and *to break every yoke*?

Does not the conduct of Lord Dunmore, and the ministerialists, in taking the advantage of the slavery practised among us, and encouraging all slaves to join them, by promising them liberty,[4] point out the best, if not the only way to defeat them in this, viz. granting freedom to them ourselves, so as no longer to use our neighbour's service without wages, but give them for their labours what is equal and just?

And suffer me further to query, Whether something might not be done to send the light of the gospel to these nations in Africa, who have been injured so much by the slave trade? Would not this have a most direct tendency to put a stop to that unrighteousness; and be the best compensation we can make them? At the same time it will be an attempt to promote the most important interest, the *kingdom of Christ*, in obedience to his command, "Go, teach all nations."

[4]On November 7, 1775, Lord Dunmore, governor of Virginia, issued a proclamation offering freedom to any slave who abandoned his rebel master and joined the king's military forces.

DECLARATION OF INDEPENDENCE: DELETED CLAUSES
June 28, 1776

Thomas Jefferson had several reasons for including a paragraph on slavery in the Declaration of Independence. The primary motive in adopting a declaration was to justify to the world America's secession from the British Empire. America's struggle for its own rights against British imperial domination seemed to be compromised by its preservation of a widespread system of enslavement of an entire race of people. To counter this inconsistency, Jefferson blamed George III with vetoing colonial attempts to prohibit the African slave trade (a criticism first raised by Jefferson in his Summary View of the Rights of British America) *and with inciting slaves to rise up and murder their masters. Many delegates to Congress, however, objected to the clause. According to Jefferson's recollection, it "was struck out in complaisance to South Carolina and Georgia, who had never attempted to restrain the importation of slaves, and who on the contrary still wished to continue it." Northerners also "felt a little tender under those censures; for tho' their people have very few slaves themselves yet they had been pretty considerable carriers of them to others." Sensing the strong opposition, Jefferson reluctantly acquiesced in the deletion.*

He has waged cruel war against human nature itself, violating its most sacred rights of life and liberty in the persons of a distant people who never offended him, captivating & carrying them into slavery in another hemisphere, or to incur miserable death in their transportation thither. This piratical warfare, the opprobrium of INFIDEL powers, is the warfare of the CHRISTIAN king of Great Britain. Determined to keep open a market where MEN should be bought & sold, he has prostituted his negative for suppressing every legislative attempt to prohibit or to restrain this execrable commerce. And that this assemblage of horrors might want no fact of distinguished die, he is now exciting those very people to rise in arms among us, and to purchase that liberty of which he has deprived them by murdering the people on whom he also obtruded them: thus paying off former crimes committed against the LIBERTIES of one people, with crimes which he urges them to commit against the LIVES of another.

Revolutionary Declarations of Rights

At the time of independence, statements of the equality of man and the natural rights of man filled newspapers, pamphlets and broadsides. These concepts, said by Jefferson to be "in the air," were also incorporated into the Declaration of Independence and several state declarations of rights. Despite these lofty statements, almost one-fifth of all Americans in 1776 were black slaves. The rhetoric of emancipation burned brightly on the eve of the Revolution but faded (in some places more rapidly than in others) after independence was achieved.

VIRGINIA DECLARATION OF RIGHTS
June 12, 1776

Article 1. That all men are by nature equally free and independent, and have certain inherent rights, of which, when they enter into a state of society, they cannot, by any compact, deprive or divest their posterity; namely, the enjoyment of life and liberty, with the means of acquiring and possessing property, and pursuing and obtaining happiness and safety.

THE DECLARATION OF INDEPENDENCE
July 4, 1776

We hold these truths to be self evident; that all men are created equal; that they are endowed by their Creator with certain unalienable rights; that among these are life, liberty and the pursuit of happiness.

DELAWARE DECLARATION OF RIGHTS AND CONSTITUTION
September 11, 1776

Declaration of Rights, Article 10. That every member of society hath a right to be protected in the enjoyment of life, liberty, and property. . . .

Constitution, Article 26. No person hereafter imported into this state from Africa ought to be held in slavery under any pretence whatever, and no negro, indian, or mulatto slave, ought to be brought into this state for sale from any part of the world.

PENNSYLVANIA DECLARATION OF RIGHTS
September 28, 1776

Article 1. That all men are born equally free and independent, and have certain natural, inherent, and unalienable rights, amongst which are, the enjoying and defending life and liberty, acquiring, possessing, and protecting property, and pursuing and obtaining happiness and safety.

VERMONT DECLARATION OF RIGHTS
July 8, 1777

Chapter 1. That all men are born equally free and independent, and have certain natural, inherent and unalienable rights, amongst which are the enjoying and defending life and liberty; acquiring, possessing and protecting property, and pursuing and obtaining happiness and safety. Therefore, no male person, born in this country, or brought from over sea, ought to be holden by law to serve any person as a servant, slave or apprentice, after he arrives to the age of twenty-one years, nor female in like manner, after she arrives to the age of eighteen years, unless they are bound by their own consent after they arrive to such age, or bound by law for the payment of debts, damages, fines, costs, or the like.

MASSACHUSETTS DECLARATION OF RIGHTS
June 16, 1780

Article I. All men are born free and equal, and have certain natural, essential, and unalienable rights; among which may be reckoned the right of enjoying and defending their lives and liberties; that of acquiring, possessing, and protecting property; in fine, that of seeking and obtaining their safety and happiness.

NEW HAMPSHIRE BILL OF RIGHTS
June 3, 1784

Article 1. All Men are born equally free and independent; therefore, all government of right originates from the people, is founded in consent, and instituted for the general good.

Article 2. All men have certain natural, essential, and inherent rights; among which are—the enjoying and defending life and liberty—acquiring, possessing and protecting property—and in a word, of seeking and obtaining happiness.

JONATHAN DICKINSON SERGEANT
Plan to Free the Slaves, August 1776

The need to raise militia and an army inspired various plans to offer freedom to any slave within a certain age who was willing to enlist. If slaves were not allowed to enlist, they presented an obvious danger when many of the white males were away serving in the militia or in the Continental service. Slaves might combine together in insurrections, individually attack their master's families and property, run away, or enlist in the service of the British. To encourage them to enlist their slaves, some state legislatures offered masters bounties or exemptions from bearing arms for themselves or their sons. New York offered freedom to any slave who served in the military for three years. Although many blacks served in the army or militia, Rhode Island was the only state that actually had a black regiment created by an act of the legislature in February 1778.

Jonathan Dickinson Sergeant, a former member of Congress from New Jersey who had recently resigned to help draft a state constitution, sent his plan for creating a black battalion to John Adams on August 13, 1776. Four days later, Adams responded: "Your Negro Battalion will never do. S. Carolina would run out of their Wits at the least Hint of such a Measure."

At this Time of general Danger, when every one is anxiously considering by what Means our Liberties may be preserved, I hope to be at least forgiven, if I attempt to suggest a Hint which, perhaps, by wiser Heads, may be improved to publick Advantage.

The Calling out our Militia in such Numbers for the Defence of our Country is attended with this Difficulty among others, that the Slaves left at home excite an Alarm for the Safety of their Families; an Alarm which, on such Occasions, is industriously increased by designing Men, who make it their Business to obstruct every Measure which is taken for the publick Good.

I would therefore desire that it may be considered whether a Method might not be devised for employing those Slaves as Soldiers in the publick Service.

Suppose the Congress to enlist under proper Officers a Number of Slaves within a certain Age sufficient to form a Battalion, paying their Masters according to a certain Rate (say fifty Pounds a piece) and as a farther Compensation for their additional Value let the Master be exempted from bearing Arms. Many Slaves would willingly enlist and I suppose a great many Masters would be glad to purchase an Exemption from bearing Arms upon these Terms.

Let every one of these Slaves become free as soon as by Stoppages from his Pay or otherwise he can reimburse the Money advanced for his Purchase and as a Security to the Publick let the Survivors be answerable for the Deficiencies of such as may die in the Service. This will not be heavier upon the Survivor than if each Individual was bound to make good the full Amount of his real Value.

Let these People, during the Time of their Redemption, be on their good Behaviour. Let every great Offence or gross Misconduct be punished by reducing them back to Slavery.

Other Regulations may be found necessary. I shall only add that if Peace should be restored before these people had redeemed themselves, they might be set to labour on some publick Works until they had made Satisfaction. Or also possibly it might be as well, instead of the Plan of their redeeming themselves by Stoppages, to enlist them at Once for 7 or 10 Years at 30 shillings a Month, instead of 50 shillings.

There are two or three Objections to this Scheme which deserve to be considered.

1. It may be said that these People will want Courage. Slaves generally are Cowards: but set Liberty before their Eyes as the Reward of their Valour and I believe we should find them sufficiently brave. Neither the Hue of their Complexion nor the Blood of Africk have any Connection with Cowardice. It is their Condition as Slaves that stifles every noble Exertion. Change their Conditions and You will change their Tempers. If any one has further doubts upon this subject, let them consider the free Negroes of Jamaica who purchased their Freedom by Arms, or the Case of the brave Caribbs.

2. The Danger of putting Arms into such Hands may be objected. This can only be obviated by restricting their Numbers, so as not to suffer them to bear any large Proportion to the whites. When at length they had wrought out their own Freedom they would have the same interest with the Rest of the Community in quelling Insurrections.

3. Some may be narrow enough to enquire what is to become of those People when they are free and discharged? I answer, let them have Land, let them form a Settlement of Blacks if they will. There is Room enough on this Continent for them and us too.

If this Experiment should be thought worth trying and should answer any valuable Purpose I shall rejoice to have furnished these Hints; if otherwise I am content.

PRINCE HALL AND OTHER AFRICAN AMERICANS
A Petition to the Massachusetts Legislature, January 13, 1777[5]

Born of an English father and a mother of mixed race, Prince Hall was freed from slavery in 1770. In this petition, Hall and seven other blacks ask the Massachusetts legislature to grant them the same rights that the colonists were fighting for in their conflict with Great Britain. The principles of the American Revolution were known to the slave population, which recognized the contradiction between these principles and the institution of slavery. Since 1773, Massachusetts slaves had petitioned the legislature but with no success. On one occasion, violence narrowly was avoided

[5]Massachusetts Archives, Boston.

*when the frustration of angry slaves was channeled into a petition. The legislature
read Hall's petition on March 18, 1777, and drafted a bill "for preventing the Practice of holding Persons in Slavery." The bill was referred to the Confederation Congress
where it died.*

To the Honorable Counsel & House of Representatives for the State of
Massachusitte Bay in General Court assembled, Jan. 13, 1777.
    The petition of A Great Number of Blackes detained in a State of sla-
very in the Bowels of a free & Christian Country Humbly shuwith that your
Petitioners apprehend that they have in Common with all other men a Natural and Unaliable Right to that freedom which the Great Parent of the Unavers
hath Bestowed equalley on all menkind and which they have Never forfuted
by any Compact or agreement whatever—but that thay were Unjustly Dragged
by the hand of cruel Power from their Derest friends and sum of them Even
torn from the Embraces of their tender Parents—from A popolous, Pleasant
and plentiful contry and in violation of Laws of Nature and of Nations and in
defiance of all the tender feelings of humanity Brought hear Either to Be sold
Like Beasts of Burthen & Like them Condemnd to Slavery for Life—Among
A People Professing the mild Religion of Jesus A people Not Insensible of
the Secrets of Rationable Being Nor without spirit to Resent the unjust
endeavours of others to Reduce them to a state of Bondage and Subjection.
Your honouer Need not to be informed that A Life of Slavery Like that of
your petitioners Deprived of Every social privilege of Every thing Requiset
to Render Life Tolable is far worse then Nonexistance.
    In imitation of the Lawdable Example of the Good People of these States
your petitioners have Long and Patiently waited the Event of petition after
petition By them presented to the Legislative Body of this state and cannot
but with Grief Reflect that their Sucess hath ben but too similar. They Cannot but express their Astonishment that It has Never Bin Consirdered that
Every Principle from which Amarica has Acted in the Course of their unhappy Deficultes with Great Briton Pleads Stronger than A thousand
arguments in favowrs of your petitioners. They therfor humbly Beseech your
honours to give this petition its due weight & consideration and cause an act
of the Legislatur to be past Wherby they may Be Restored to the Enjoyments
of that which is the Naturel Right of all men—and their Children who were
Born in this Land of Liberty may not be heald as Slaves after they arive at the
age of Twenty one years so may the Inhabitance of thes States No longer
chargeable with the inconsistancey of acting themselves the part which they
condem and oppose in others Be prospered in their present Glorious struggle
for Liberty and have those Blessing to them, &c.

PENNSYLVANIA ACT FOR THE GRADUAL ABOLITION OF SLAVERY
March 1, 1780

The contradiction between slavery and the ideals of the American revolution-aries was apparent to whites as well as blacks. Pennsylvania, because of its large Quaker population, became the first state to pass a law calling for the abolition of slavery. Although wealthy and strong in numbers, Quakers were unable actively to pursue their antislavery policies because they were despised for their pacifism during the war. Quakers, however, worked behind the scenes in favor of emancipation. Ma-jor concessions were needed in the act—prolonging the period when all slaves would be free—before enough votes could be obtained for passage. The preamble of the act was written by Thomas Paine, then serving as the clerk of the Pennsylvania assembly.

An ACT for the gradual abolition of Slavery.

When we contemplate our abhorrence of that condition to which the arms and tyranny of Great-Britain were exerted to reduce us; when we look back on the variety of dangers to which we have been exposed, and how mi-raculously our wants in many instances have been supplied, and our deliverances wrought, when even hope and human fortitude have become unequal to the conflict; we are unavoidably led to a serious and grateful sense of the manifold blessings which we have undeservedly received from the hand of that Being from whom every good and perfect gift cometh. Impressed with these ideas, we conceive that it is our duty, and we rejoice that it is in our power, to extend a portion of that freedom to others, which hath been ex-tended to us; and a release from that state of thraldom, to which we ourselves were tyrannically doomed, and from which we have now every prospect of being delivered. It is not for us to inquire, why, in the creation of mankind, the inhabitants of the several parts of the earth were distinguished by a differ-ence in feature or complexion. It is sufficient to know, that all are the work of an Almighty Hand. We find in the distribution of the human species, that the most fertile, as well as the most barren parts of the earth, are inhabited by men of complexion different from ours, and from each other, from whence we may reasonably, as well as religiously infer, that He who placed them in their various situations, hath extended equally His care and protection to all, and that it becometh not us to counteract His mercies. We esteem it a pecu-liar blessing granted to us, that we are enabled this day, to add one more step to universal civilization, by removing as much as possible, the sorrows of those who have lived in undeserved bondage, and from which, by the assumed au-thority of the Kings of Britain, no effectual legal relief, could be obtained. Weaned by a long course of experience, from those narrow prejudices and partialities we had imbibed, we find our hearts enlarged with kindness and benevolence, towards men of all conditions and nations; and we conceive ourselves at this particular period extraordinarily called upon, by the bless-ings which we have received, to manifest the sincerity of our profession, and to give a substantial proof of our gratitude.

And whereas the condition of those persons who have heretofore been denominated Negroe and Mulatto slaves, has been attended with circumstances, which not only deprived them of the common blessings that they were by nature entitled to, but has cast them into the deepest afflictions, by an unnatural separation and sale of husband and wife from each other, and from their children; an injury the greatness of which, can only be conceived, by supposing, that we were in the same unhappy case. In justice therefore, to persons so unhappily circumstanced, and who, having no prospect before them, whereon they may rest their sorrows and their hopes, have no reasonable inducement, to render that service to society, which they otherwise might; and also, in grateful commemoration of our own happy deliverance, from that state of unconditional submission, to which we were doomed by the tyranny of Britain.

Be it enacted, and it is hereby enacted, by the Representatives of the Freemen of the Commonwealth of Pennsylvania, in General Assembly met, and by the authority of the same That all persons, as well Negroes and Mulattos as others, who shall be born within this State, from and after the passing of this Act, shall not be deemed and considered as servants for life or slaves; and that all servitude for life, or slavery of children, in consequence of the slavery of their mothers, in the case of all children born within this State, from and after the passing of this Act as aforesaid, shall be, and hereby is utterly taken away, extinguished and for ever abolished.

Provided always, and be it further enacted by the authority aforesaid, That every Negroe and Mulatto child born within this State, after the passing of this Act as aforesaid, who would, in case this Act had not been made, have been born a servant for years, or life or a slave, shall be deemed to be and shall be by virtue of this Act, the servant of such person or his or her assigns, who would in such case have been intitled to the service of such child, until such child shall attain unto the age of twenty eight years, in the manner and on the conditions whereon servants bound by indenture for four years, are or may be retained and holden; and shall be liable to like correction and punishment, and intitled to like relief in case he or she be evily treated by his or her master or mistress, and to like freedom dues[6] and other privileges as servants bound by indenture for four years, are or may be intitled, unless the person to whom the service of any such child shall belong, shall abandon his or her claim to the same, in which case the Overseers of the Poor of the city, township or district respectively, where such child shall be so abandoned, shall by indenture bind out every child so abandoned, as an apprentice for a time not exceeding the age herein before limited, for the service of such children. . . .

---

[6] The customary payment to an indentured servant at the end of the indenture. This often included two suits of clothing, a cash payment, an allotment of food, and sometimes land.

Virginia Manumission Law
May 1782

Despite the wish of some slaveholders to free their slaves, private, voluntary manumission in Virginia had been illegal since 1723. Individual slaves could be freed only by special act of the legislature. With the revolutionary fervor in favor of liberty and strong lobbying efforts of Quakers, the Virginia legislature passed a manumission law in 1782 allowing slaveholders (while alive or in their wills) to free their slaves. Many slaveowners freed thousands of slaves under this law. For many other slaveowners, however, the law only increased their guilt, knowing that they now could legally free their slaves but for various reasons refused—the state no longer stood as a buffer between slaveowners and their consciences.

Some slaveowners opposed the manumission act because of the unsettling effect of having free blacks in the community. In 1785 in response to the Methodist petitions to the legislature in support of general emancipation, slaveowners mounted a strenuous counter-petition campaign to safeguard their slaves and to repeal the manumission act of 1782, which they argued had "pernicious Effects." Later newly-freed slaves were required by law to leave the state within a year of their emancipation. As the abolition sentiment waned and as the fear of free blacks increased, the legislature repealed the manumission law, and it once again became illegal for slaveowners to free their slaves.

An Act to authorize the manumission of slaves.

I. WHEREAS application hath been made to this present general asembly, that those persons who are disposed to emancipate their slaves may be empowered so to do, and the same hath been judged expedient under certain restrictions: *Be it therefore enacted,* That it shall hereafter be lawful for any person, by his or her last will and testament, or by any other instrument in writing, under his or her hand and seal, attested and proved in the county court by two witnesses, or acknowledged by the party in the court of the county where he or she resides, to emancipate and set free, his or her slaves, or any of them, who shall thereupon be entirely and fully discharged from the performance of any contract entered into during servitude, and enjoy as full freedom as if they had been particularly named and freed by this act. . . .

The Resumption of the Slave Trade

As the Revolutionary War drew to a close, the slave trade was resumed. Because the Southern states had not enacted laws to prohibit it, the slave trade could be resumed without new legislation. Much of the trade was undertaken by Northern merchants in Northern-owned ships.

DAVID RAMSAY TO BENJAMIN RUSH
Charleston, August 22 and September 9, 1783[7]

[August 22] . . . I sincerely regret with you that the Slave trade will be resumed. Had a law been necessary to open this trade I would have opposed it; but, there has never been any debate on the subject. We are now at liberty to trade where we please under that general sanction. This infamous traffic will be resumed without any thing being said on the subject. As to stepping out & volunteering to combat the practice, no good can be expected from that quarter. The Don Quixot who would attempt it would have the consolation of having done his duty to balance the calumny & public odium he would be the subject of without the least prospect of his being able to put a stop to the inhuman traffic. If a law was necessary to legalize the trade, opposition might be made with hope of doing something, but the contrary is contended & it will be resumed without one word being said on the subject by authority.[8]

[September 9] . . . The genius of our people is entirely turned from war to commerce. Schemes of business & partnerships for extending commerce are daily forming. The infamous African trade which you so justly abhor will among other branches of business be renewed. I am informed by a Gentleman lately from Philadelphia that the guilt of this is not confined to one State; for he says that sundry vessels have been lately fitted out in your city on the same business with a view to sell their slaves either in the West Indies or the Southern States. I greatly fear that the mad notions entertained of the separate States sovereignty will forever prevent the interference of the Continental

---

[7]Rush Manuscripts, Library Company of Philadelphia. Printed in *David Ramsay, 1749–1815, Selections from His Writings*, edited by Robert L. Brunhouse in the *Transactions of the American Philosophical Society*, 55 (Philadelphia, 1965), 76. Ramsay, a Charleston, S.C., physician and historian, had studied medicine with Rush in Philadelphia. Ramsay represented Charleston in the state House of Representatives from 1776 to 1790, and was a delegate to Congress from 1782 to 1785.

[8]In February and September 1783, bills were introduced into the South Carolina legislature to prohibit the slave trade, but they failed to pass; instead, the legislature levied a duty of £3 sterling on each imported slave.

Sovereignty on this subject. I sincerely lament the renewal of this trade, I shall for my own part have no concern with it.

NEW YORK *INDEPENDENT JOURNAL*, JUNE 2, 1787

What, observes a correspondent, must be the feelings of those unfortunate Americans, when they shall be informed that the very hands from which they so anxiously expect relief, have ever since the peace, been employed in equipping vessels, armed with the instruments of death, and loaded, among their other articles, with the *badges* and *insignias* of slavery, to deprive the poor, unhappy Africans of their liberty? Blush O ye *Bostonians* and *Pennsylvanians** for your degenerate sons, who have thus embarked in this monstrous and abominable trade and traffic! Blush also for yourselves, because you have long beheld this wicked practice, without making the least effort to prevent it. There is no difference between *permitting* it to be done in your harbours and the actual *doing* it yourselves. The equipment of vessels, in your own ports, by your own subjects, for the open and avowed purpose of enslaving their fellow men, bears ample proof of your sanction and approbation; and the stigma will be fixed on you as a people, not on the individuals who are immediately engaged in the business for pecuniary motives.

*It is almost needless to mention, that vessels for the coast of Africa, have been no where fitted out in America, since the treaty of peace, but at the ports of Boston and Philadelphia.

MASSACHUSETTS OUTLAWS SLAVERY, 1783
Commonwealth of Massachusetts v. Nathaniel Jennison[9]

In 1783 Nathaniel Jennison was arrested for beating his runaway slave Quock Walker. Jennison defended himself on the grounds that Walker was his slave, but Walker argued that the Massachusetts Declaration of Rights of 1780 made slavery unconstitutional. Chief Justice William Cushing speaking for the Massachusetts Supreme Court accepted Walker's interpretation, and slavery was declared illegal and unconstitutional in Massachusetts. Despite this decision slavery persisted in Massachusetts for several more years and only gradually disappeared until no slaves were reported in the 1790 federal census.

[9]Albert B. Hart, ed., *Commonwealth History of Massachusetts* (5 vols., New York, 1927–1930), IV, 37–38.

As to the doctrine of slavery and the right of Christians to hold Africans in perpetual servitude, and sell and treat them as we do our horses and cattle, that (it is true) has been heretofore countenanced by the Province Laws formerly, but nowhere is it expressly enacted or established. It has been a usage—a usage which took its origin from the practice of some of the European nations, and the regulations of British government respecting the then Colonies, for the benefit of trade and wealth. But whatever sentiments have formerly prevailed in this particular or slid in upon us by the example of others, a different idea has taken place with the people of America, more favorable to the natural rights of mankind, and to that natural, innate desire of Liberty, which with Heaven (without regard to color, complexion, or shape of noses—features) has inspired all the human race. And upon this ground our Constitution of Government, by which the people of this Commonwealth have solemnly bound themselves, sets out with declaring that all men are born free and equal—and that every subject is entitled to liberty, and to have it guarded by the laws, as well as life and property—and in short is totally repugnant to the idea of being born slaves. This being the case, I think the idea of slavery is inconsistent with our own conduct and Constitution; and there can be no such things as perpetual servitude of a rational creature, unless his liberty is forfeited by some criminal conduct or given up by personal consent or contract. . . .

Verdict Guilty.

The Origins of the Three-Fifths Clause

THE CONFEDERATION CONGRESS AND
THE POPULATION AMENDMENT OF 1783

The controversy over the apportionment of federal taxes divided the Second Continental Congress as it debated the draft of Articles of Confederation in 1777. The initial draft, written primarily by John Dickinson of Pennsylvania, called for general expenses to be apportioned among the states on the basis of total population, excluding Indians not paying taxes. Southern delegates vehemently argued that only the white population should be counted in apportioning federal expenses among the states and threatened that if slaves were counted, there would be no confederation. Unable to compromise on a population basis for sharing expenses, Congress provided in the Articles that expenses be shared on the basis of the estimated value of all lands granted to or surveyed for individuals, including the improvements made on that land.

Congress soon saw the futility of apportioning expenses on the basis of estimates of land values when those estimates were to be made by the state governments themselves. States vied with one another to see which could come up with lower estimates, thus reducing their shares of the federal expenses.

In February 1783 Congress appointed a special committee to consider its financial powers. On March 6, the committee recommended a federal tariff for twenty-five years and a new system apportioning federal expenses using total population (excluding slaves under a certain age) as the basis. Northern delegates favored the plan; southern delegates opposed it. Soon both sides agreed to eliminate the age provision concerning slaves, and a committee recommended that "two blacks be rated as equal to one freeman." Heated debate ensued as northerners moved for a 4-to-3 ratio while southerners favored a 2-to-1 or even a 4-to-1 ratio. As a compromise, James Madison proposed a 5-to-3 ratio which was accepted, then rejected, and finally accepted again "without opposition."

Within three years, eleven state legislatures adopted the amendment. Only New Hampshire and Rhode Island rejected it; but, because of the requirement for unanimity to adopt amendments to the Articles of Confederation, the population amendment failed. In 1786, however, Congress used population figures (including three-fifths of slaves) in apportioning the quotas of federal expenses among the states. This formula was adopted by the Constitutional Convention in apportioning representation in the House of Representatives and in levying direct taxes among the states. Frequent references were made during the debate over the ratification of the Constitution to the population amendment as the will of the people because it had been adopted by eleven states.

REPORT OF THE NORTH CAROLINA CONGRESSIONAL DELEGATES
Philadelphia, March 24, 1783[10]

. . . We have been attempting with much pains to fix on some mode by which the quota of the several States might be determined according to the 8th Article of the Confederation, i.e. according to the value of located Lands & their improvements. The Rule is good and plain but the question is extremely difficult. How shall the value be fixed? Let the appropriated Lands and their improvements be valued by the Inhabitants of the respective States and we have great reason to believe, from proofs before us, that the valuation would be unequal and unjust, for instance, the average value of lands as they are now rated for the purpose of taxation in the State of Virginia is one-third higher than the value of Lands as they are rated in Pennsylvania though it is certain that the Lands in Pennsylvania are at an average worth one-third more than the Lands in Virginia. If such valuation should be made in fixing the continental Quota, Pennsylvania when compared with Virginia would not pay quite half the sum she ought to pay. We have many other Arguments which either prove the different frauds or the diversity of opinions respecting the value of Lands which prevail in different States. It is presumed that the valuation would be more uniform and just if it was made by a Set of Commis-

---

[10]To Governor Alexander Martin, Paul H. Smith et al., eds., *Letters of Delegates to Congress, 1774–1789* (21 vols. to date, Washington, D.C., 1976–), 20:90–91.

sioners who should view all the lands and buildings in the United States. But there is reason to believe that such process, like estates entailed, would be perpetual and it would be an even chance which would come first, the fixing the quotas or the day of Judgment. The eastern States, who consider the valuation Scheme as impracticable, talk much of fixing the quotas according to the number of Inhabitants, making considerable allowance for slaves. Some of them propose to exclude all Slaves under 16 Years, which would be rating two slaves for one free man. We presume that the Southern States would meet them upon this ground or even upon ground somewhat lower for the sake of preventing Jealousies, a Contention and delay but we fear that if an attempt should be made to alter or amend the mode of fixing the quota, those very men would again talk of a Slave being equal to a white man.

DEBATES AND PROCEEDINGS IN CONGRESS
March 6 to April 18, 1783[11]

March 6

The committee on Revenue made a report which was ordered printed for each member, and to be taken up on Monday next.

March 7

Printed copies of the Report above-mentioned were delivered to each member, as follows, viz. . . .

(11) That as a more convenient and certain rule of ascertaining the proportions to be supplied by the States respectively to the common Treasury, the following alteration in the articles of confederation and perpetual union between these States, be and the same is hereby, agreed to in Congress, & the several States are advised to authorize their respective delegates to subscribe and ratify the same, as part of the said instrument of Union, in the words following, to wit.

(12) "So much of the 8th of the Articles of Confederation & perpetual Union between the thirteen States of America as is contained in the words following to wit 'All charges of war &c (to the end of the paragraph)'—is hereby revoked and made void, and in place thereof, it is declared and Concluded, the same having been agreed to in a Congress of the United States, that all charges of war, and all other expences that shall be incurred for the common defence or general welfare and allowed by the U.S. in Congress assembled shall be defrayed out of a common treasury, which shall be sup-

---

[11]Taken from the *Journals of the Continental Congress*, XIV, 259–61; XXV, 921, 922, 948–49, 952, 962.

plied by the several States in proportion to the number of inhabitants of every age, sex & condition, except Indians not paying taxes in each State; which number shall be triennially taken & transmitted to the U.S. in Congress assembled, in such mode as they shall direct and appoint; provided always that in such numeration no persons shall be included who are bound to servitude for life, according to the laws of the State to which they belong, other than such as may be between the ages of _____."

March 27

Theodorick Bland (Va.) opposed it: said that the value of land was the best rule, and that at any rate no change should be attempted untill its practicability should be tried.

James Madison (Va.) thought the value of land could never be justly or satisfactorily obtained; that it would ever be a source of contentions among the States, and that as a repetition of the valuation would be within the course of the 25 years, it would unless exchanged for a more simple rule mar the whole plan.

Nathaniel Gorham (Mass.) was in favor of the paragraph. He represented in strong terms the inequality & clamors produced by valuations of land in the State of Massachusetts & the probability of the evils being increased among the States themselves which were less tied together & more likely to be jealous of each other.

Hugh Williamson (N.C.) was in favor of the paragraph.

James Wilson (Pa.) was strenuous in favor of it, said he was in Congress when the Articles of Confederation directing a value of land was agreed to, that it was the effect of the impossibility of compromising the different ideas of the Eastern and Southern States as to the value of Slaves compared with the Whites, the alternative in question

Abraham Clark (N.J.) was in favor of it. He said that he was also in Congress when this article was decided; that the Southern States would have agreed to numbers, in preference to the value of land if one-half their Slaves only should be included; but that the Eastern States would not concur in that proportion.

It was agreed on all sides that, instead of fixing the proportion by ages, as the report proposed, it would be best to fix the proportion in absolute numbers. With this view & that the blank might be filled up, the clause was recommitted.

March 28

The Committee last mentioned reported that two blacks be rated as equal to one freeman.

Oliver Wolcott (Conn.) was for rating them as 4 to 3.

Daniel Carrol (Md.) as 4 to 1.

Hugh Williamson (N.C.) said he was principled against slavery; and that he thought slaves an incumbrance to Society instead of increasing its ability to pay taxes.

Stephen Higginson (Mass.) as 4 to 3.

John Rutledge (S.C.) said, for the sake of the object he would agree to rate Slaves as 2 to 1, but he sincerely thought 3 to 1 would be a juster proportion.

Samuel Holten (Mass.) as 4 to 3.

Samuel Osgood (Mass.) said he could not go beyond 4 to 3.

On a question for rating them as 3 to 2 the votes were N.H. ay, Mass. no, R.I. divided, Conn. ay, N.J. ay, Penn. ay, Del. ay, Md. no, Va. no, N.C. no, S.C. no (5 to 5, with 1 divided).

The paragraph was then postponed by general consent, some wishing for further time to deliberate on it; but it appearing to be the general opinion that no compromise would be agreed to.

After some further discussions on the report in which the necessity of some simple and practicable rule of apportionment came fully into view, Mr. Madison said that in order to give a proof of the sincerity of his professions of liberality, he would propose that Slaves should be rated as 5 to 3. Mr. Rutledge seconded the motion.

James Wilson (Pa.) said he would sacrifice his opinion to this compromise.

Arthur Lee (Va.) was against changing the rule, but gave it as his opinion that two slaves were not equal to one freeman.

On the question for 5 to 3 it passed in the affirmative N.H. ay, Mass. divided, R.I. no, Conn. no, N.J. ay, Penn. ay, Md. ay, Va. ay, N.C. ay, S.C. ay (7 to 2, with 1 divided).

A motion was then made by Mr. Bland, seconded by Mr. Lee to strike out the clause so amended and on the question "shall it stand" it passed in the negative; N.H. ay, Mass. no, R.I. no, Conn. no, N.J. ay, Penn. ay, Del. no, Md. ay, Va. ay, N.C. ay, S.C. no (6 to 5); so the clause was struck out.

The arguments used by those who were for rating slaves high were, that the expence of feeding & cloathing them was as far below that incident to freemen as their industry & ingenuity were below those of freemen; and that the warm climate within which the States having slaves lay, compared with the rigorous climate & inferior fertility of the others, ought to have great weight in the case & that the exports of the former States were greater than of the latter. On the other side it was said that Slaves were not put to labour as young as the children of laboring families—that, having no interest in their labor, they did as little as possible, & omitted every exertion of thought requisite to facilitate & expedite it; that if the exports of the States having slaves exceeded those of the others, their imports were in proportion, slaves being employed wholly in agriculture, not in manufactures; & that in fact the balance of trade formerly was much more against the Southern States than the others.

April 1

Congress resumed the Report on Revenue &c. Alexander Hamilton who had been absent when the last question was taken for substituting numbers in place of the value of land, moved to reconsider that vote. He was seconded by Mr. Osgood. Those who voted differently from their former votes were influenced by the conviction of the necessity of the change and despair on both sides of a more favorable rate of the slaves. The rate of 3/5 was agreed to without opposition.

April 18

The plan of Revenue was then passed as it had been amended.

"That as a more convenient and certain rule of ascertaining the proportions to be supplied by the states respectively to the common treasury, the following alteration in the Articles of Confederation and perpetual union, between these states be, and the same is hereby agreed to in Congress; and the several states are advised to authorise their respective delegates to subscribe and ratify the same as part of the said instrument of union, in the words following, to wit:

"So much of the 8th of the Articles of Confederation and perpetual union, between the thirteen states of America, as is contained in the words following, to wit:

"'All charges of war and all other expences that shall be incurred for the common defence or general welfare, and allowed by the United States in Congress assembled, shall be defrayed out of a common treasury, which shall be supplied by the several states in proportion to the value of all land within each State granted to or surveyed for any person, as such land and the buildings and improvements thereon shall be estimated according to such mode as the United States in Congress assembled shall, from time to time, direct and appoint,' is hereby revoked and made void; and in place thereof it is declared and concluded, the same having been agreed to in a Congress of the United States, that 'all charges of war and all other expences that have been or shall be incurred for the common defence or general welfare, and allowed by the United States in Congress assembled, except so far as shall be otherwise provided for, shall be supplied by the several states in proportion to the whole number of white and other free citizens and inhabitants, of every age, sex and condition, including those bound to servitude for a term of years, and three-fifths of all other persons not comprehended in the foregoing description, except Indians, not paying taxes, in each State; which number shall be triennially taken and transmitted to the United States in Congress assembled, in such mode as they shall direct and appoint.'"

Lafayette's Emancipation Plan

On October 17, 1784, James Madison wrote Thomas Jefferson describing his impression of the Marquis de Lafayette. "The time I have lately passed with the M. has given me a pretty thorough insight into his character. With great natural frankness of temper he unites much address; with very considerable talents. A strong thirst of praise and popularity. In his politics he says his three hobby-horses are the alliance between France and the United States, the union of the latter and the manumission of the slaves. The two former are the dearer to him as they are connected with his personal glory. The last does him real honor as it is a proof of his humanity."

Lafayette sought George Washington's cooperation in pursuing an experiment to free some slaves and to establish them as tenants on a plantation. If successful, this plan would show that freedmen could lead productive lives and not become a pestilence on society. Washington's participation in this scheme would also encourage other Southern slaveowners to follow his example. Although Washington failed to participate in such a project, Lafayette went ahead with his experiment on a plantation he purchased on Cayenne, an island in the French West Indies.

The Philadelphia reformer Benjamin Rush proposed a plan similar to Lafayette's in 1794. Rush donated 5,200 acres to the Pennsylvania Abolition Society to be parceled out among free blacks in a new settlement to be named Benezet.[12]

Other Europeans who assisted America during the Revolution also felt compassion for the slaves. For example, Thaddeus Kosciusko of Poland on his last departure from the United States in 1798, left written authorization with Thomas Jefferson "appropriating, after his death, all the property he had in our public funds, the price of his military services here, to the education and emancipation of as many of the children of bondage in this country as it should be adequate to."[13]

MARQUIS DE LAFAYETTE TO GEORGE WASHINGTON
Cadiz, February 5, 1783[14]

Now, my dear General, that You are Going to Enjoy Some Ease and Quiet, Permit me to Propose a plan to You Which Might Become Greatly Beneficial to the Black Part of Mankind. Let us Unite in Purchasing a Small Estate Where We May try the Experiment to free the Negroes, and Use them only as tenants. Such an Example as Yours Might Render it a General Practice, and if We Succeed in America, I Will Chearfully Devote a part of My time to Render the Method fascionable in the West Indies. If it Be a Wild

[12] Rush to the President of the Pennsylvania Abolition Society, 1794, Rush Manuscripts, Library Company of Philadelphia.

[13] Thomas Jefferson to M. Jullien, Monticello, July 23, 1818.

[14] Stanley J. Idzerda et al., eds., *Lafayette in the Age of the American Revolution* (5 vols., Ithaca, N.Y., 1977–1983), v, 91–92.

Scheme, I Had Rather Be Mad that Way, than to Be thought Wise in the other tack.

GEORGE WASHINGTON TO MARQUIS DE LAFAYETTE
Newburgh, N.Y., April 5, 1783[15]

The scheme, my dear Marqs. which you propose as a precedent, to encourage the emancipation of the black people of this Country from that state of Bondage in which they are held, is a striking evidence of the benevolence of your Heart. I shall be happy to join you in so laudable a work; but will defer going into a detail of the business, till I have the pleasure of seeing you.

MARQUIS DE LAFAYETTE TO HENRY KNOX
Chavaniac, Auvergne, June 12, 1785[16]

I Confidentially intrust to You, my dear Sir, that I am about purchasing a fine plantation in a french Colony, to make the experiment for Enfranchising Our Negro Brethren, god grant it may Be propagated!

MARQUIS DE LAFAYETTE TO GEORGE WASHINGTON
Sarguemines on the French Frontier, July 14, 1785[17]

You remember an idea which I imparted to you three years ago. I am going to try it in the French colony of Cayenne, but will write more fully on the subject in my other letters.

MARQUIS DE LAFAYETTE TO GEORGE WASHINGTON
Paris, February 6, 1786[18]

Another secret I intrust to you, my dear General, is that I have purchased for a hundred and twenty five thousand French livres a plantation in the Colony of Cayenne and am going to free my Negroes in order to make that experiment which you know is my hobby horse.

---

[15]Ibid., 119.
[16]Ibid., 330.
[17]Louis Gottschalk, ed., *The Letters of Lafayette to Washington, 1777–1799* (New York, 1944), 301.
[18]Ibid., 309.

GEORGE WASHINGTON TO MARQUIS DE LAFAYETTE
Mount Vernon, May 10, 1786[19]

The benevolence of your heart my Dr. Marqs. is so conspicuous upon all occasions, that I never wonder at any fresh proofs of it; but your late purchase of an estate in the colony of Cayenne, with a view of emancipating the slaves on it, is a generous and noble proof of your humanity. Would to God a like spirit would diffuse itself generally into the minds of the people of this country; but I despair of seeing it. Some petitions were presented to the Assembly, at its last Session, for the abolition of slavery, but they could scarcely obtain a reading. To set them afloat at once would, I really believe, be productive of much inconvenience and mischief; but by degrees it certainly might, and assuredly ought to be effected; and that too by Legislative authority.

QUAKERS PETITION THE CONFEDERATION CONGRESS
October 4, 1783

As the war with Great Britain ended and the African slave trade resumed, Quakers mounted campaigns on both the state and federal levels to prohibit the importation of slaves. Over 500 Quakers signed a petition to Congress reminding the delegates of their "solemn declarations often repeated in favour of universal liberty" and asking the delegates to prohibit the importation of slaves.

On October 6, 1783, a delegation of four Quakers led by Anthony Benezet carried the petition to Congress then meeting in Princeton, New Jersey. Congress read the petition on October 8 and referred it to a committee. In January the committee avoided direct congressional interference in the slave trade as it recommended that the state legislatures "enact such laws as to their wisdom may appear best calculated to" end the foreign slave trade. Divided along sectional lines, Congress was unable to adopt even this limited proposal which would have transferred the burden of anti-slave importation to the states.

Being through the favour of Divine Providence met as usual at this season in our annual Assembly to promote the cause of Piety and Virtue, We find with great satisfaction our well meant endeavours for the relief of an oppressed part of our fellow Men have been so far blessed, that those of them who have been held in bondage by Members of our Religious Society are generally restored to freedom, their natural and just right.

[19]John C. Fitzpatrick, ed., *The Writings of George Washington* . . . (39 vols., Washington, D.C., 1931–1944), xxviii, 424.

Commiserating the afflicted state into which the Inhabitants of Africa are very deeply involved by many Professors of the mild and benign doctrines of the Gospel, and affected with a sincere concern for the essential Good of our Country, We conceive it our indispensible duty to revive the lamentable grievance of that oppressed people in your view as an interesting subject evidently claiming the serious attention of those who are entrusted with the powers of Government, as Guardians of the common rights of Mankind and advocates for liberty.

We have long beheld with sorrow the complicated evils produced by an unrighteous commerce which subjects many thousands of the human species to the deplorable State of Slavery.

The Restoration of Peace and restraint to the effusion of human Blood we are persuaded excite in the minds of many of all Christian denominations gratitude and thankfulness to the all wise Controuler of human events; but we have grounds to fear, that some forgetful of the days of Distress are prompted from avaricious motives to renew the iniquitous trade for Slaves to the African Coasts, contrary to every humane and righteous consideration, and in opposition to the solemn declarations often repeated in favour of universal liberty, thereby increasing the too general torrent of Corruption and licentiousness, and laying a foundation for future Calamities.

We therefore earnestly sollicit your Christian interposition to discourage and prevent so obvious an Evil, in such manner as under the influence of Divine Wisdom you shall see meet—

Signed in and on behalf of our Yearly Meeting held in Philadelphia for Pennsylvania, New Jersey, and Delaware, and the Western parts of Maryland and Virginia dated the fourth day of the tenth Month 1783

RHODE ISLAND GRADUAL ABOLITION LAW
February 1784

Despite Rhode Island's libertarian tradition, Rhode Island slave traders were the principal American carriers of slaves from the African coast to the New World throughout the colonial era. Typically Rhode Island slavers would send rum and other goods to Africa to be traded for slaves. These human cargoes were usually traded in the West Indies for money, sugar, and molasses, which would be transported back to New England to be made into rum. Some slaves were sold in the Southern states while others were brought back to Rhode Island to serve their bondage as domestic servants or as field hands on plantations or smaller farms. According to the most reliable sources, over 100,000 human beings were stolen away from Africa and brought to the New World by Rhode Island slavers between 1709 and 1807. The slave trade was a Rhode Island staple and an integral component of the New England economy.

Beginning in 1719 Rhode Island's influential Quaker population began its public criticism of the slave trade. Anglicans and Congregationalists, particularly the

Reverend Samuel Hopkins, joined the crusade to ban the obnoxious trade. Not until the Revolutionary movement, however, did anti-slavery forces gain success. In 1774 the Providence town meeting refused to accept six slaves the town inherited from an intestate master, proclaiming "that it is unbecoming the character of freemen to enslave the said negroes and they do hereby give up all claim of right or property in them." The town then, at the urging of Quaker Moses Brown and Stephen Hopkins, petitioned the colonial assembly that "Whereas, the inhabitants of America are engaged in the preservation of their rights and liberties; and as personal liberty is an essential part of the rights of mankind, the deputies of the town are directed to use their endeavors to obtain an act of the General Assembly, prohibiting the importation of negro slaves into this colony; and that all negroes born in the colony should be free, after attaining to a certain age." In June 1774 the legislature prohibited the importation of slaves into the colony but failed to take any action on general emancipation.

During the Revolution, the anti-slavery movement in Rhode Island gained momentum. To augment its Continental recruits, Rhode Island raised a black regiment—the only such force during the Revolution—which fought with distinction. In October 1779 the legislature prohibited the sale of Rhode Island slaves outside of the state without their consent because such action would tend to "aggravate the Condition of Slavery, which this General Assembly is disposed rather to alleviate, till some favorable Occasion may offer for its total Abolition." In December 1783 Quakers submitted a petition to the legislature advocating emancipation. A committee brought in a bill which was submitted to the towns for their consideration. The assembly defeated the bill on its initial vote, but the bill passed after amendments were added including allowing Rhode Islanders to participate in the slave trade outside of the state. The act sanctioned voluntary manumission and provided that all children born to slave mothers after March 1, 1784, should be free. In the 1790 federal census Rhode Island counted 948 slaves.

An ACT authorizing the manumission of Negroes, Mulattos and others, and for the gradual abolition of slavery.

Whereas all men are entitled to life, liberty and the pursuit of happiness, and the holding mankind in a state of slavery, as private property, which has gradually obtained by unrestrained custom and the permission of the laws, is repugnant to this principle, and subversive of the happiness of mankind, the great end of all civil government:

Be it therefore enacted by this General Assembly, and by the authority thereof it is enacted, That no person or persons, whether Negroes, Mulattos or others, who shall be born within the limits of this State, on or after the first day of March, A.D. 1784, shall be deemed or considered as servants for life, or slaves; and that all servitude for life, or slavery of children, to be born as aforesaid, in consequence of the condition of their mothers, be and the same is hereby taken away, extinguished and forever abolished.

And whereas humanity requires, that children declared free as aforesaid remain with their mothers a convenient time from and after their birth; to enable therefore those who claim the services of such mothers to maintain and support such children in a becoming manner, *It is further enacted by the authority aforesaid,* That such support and maintenance be at the expence of the respective towns where those reside and are settled: *Provided however,* That the respective Town-Councils may bind out such children as apprentices, or otherwise provide for their support and maintenance, at any time after they arrive to the age of one year, and before they arrive to their respective ages of twenty-one, if males, and eighteen, if females.

And whereas it is the earnest desire of this Assembly, that such children be educated in the principles of morality and religion, and instructed in reading, writing and arithmetic: *Be it further enacted by the authority aforesaid,* That due and adequate satisfaction be made as aforesaid for such education and instruction. And for ascertaining the allowance for such support, maintenance, education and instruction, the respective Town-Councils are hereby required to adjust and settle the accounts in this behalf from time to time, as the same shall be exhibited to them: Which settlement so made shall be final; and the respective towns by virtue thereof shall become liable to pay the sums therein specified and allowed.

And be it further enacted by the authority aforesaid, That all persons held in servitude or slavery, who shall be hereafter emancipated by those who claim them, shall be supported as other paupers, and not at the separate expence of the claimants, if they become chargeable; provided they shall be between the ages of twenty and forty years, and are of sound body and mind; which shall be judged of and determined by the Town-Councils aforesaid.

An Englishman's Perspective

Throughout the Revolution, Richard Price defended Americans in their struggle with Parliament. Price, a British clergyman and writer on theology, morals, finance, and politics, continued his praise of America after the war in his pamphlet Observations on the Importance of the American Revolution, and the Means of Making It a Benefit to the World. *Published in England in 1784, the pamphlet was reprinted eight times in the United States by the end of 1786. Price praised the American government as "equitable" and "liberal," predicted that America would become the refuge of the oppressed of the world, and encouraged Americans to complete the fight for liberty by freeing their slaves. Most Americans admired and appreciated Price, but his thoughts on slavery alienated many slaveowners who were not ready for a general emancipation.*

RICHARD PRICE
Observations on the Importance of the American Revolution, 1784

Of the NEGRO TRADE and SLAVERY.

The Negro Trade cannot be censured in language too severe. It is a traffick which, as it has hitherto been carried on, is shocking to humanity, cruel, wicked, and diabolical. I am happy to find that the united states are entering into measures for discountenancing it, and for abolishing the odious slavery which it has introduced. 'Till they have done this, it will not appear they deserve the liberty for which they have been contending. For it is self evident, that if there are any men whom they have a right to hold in slavery, there may be others who have had a right to hold them in slavery.—I am sensible, however, that this is a work which they cannot accomplish at once.

The emancipation of the Negroes must, I suppose, be left in some measure to be the effect of time and of manners. But nothing can excuse the United States, if it is not done with as much speed and at the same time with as much effect as their particular circumstances and situation will allow. I rejoice that on this occasion I can recommend to them the example of my own country.—In *Britain*, a *Negro* becomes a *freeman* the moment he sets his foot on *British* ground.

RICHARD PRICE TO JOHN JAY
Newington Green, near London, July 9, 1785[20]

I directed to you in autumn last some copies of my pamphlet on the American Revolution.

This was an effort of my zeal to promote, according to the best of my judgment, the improvement and happiness of mankind in general and of the United States in particular. The recommendations in it of measures to abolish gradually the Negro-trade and Slavery and to prevent too great an inequality of property have I find offended some of the leading men in South Carolina; and I have been assured from thence that such measures will never be encouraged there. Should a like disposition prevail in many of the other States, it will appear that the people who have struggled so bravely against being enslaved themselves are ready enough to enslave others; the event which had raised my hopes of seeing a better state of human affairs will prove only an introduction to a new scene of aristocratical tyranny and human debasement; and the friends of liberty and virtue in Europe will be sadly disappointed and mortified.

---

[20]Henry P. Johnston, ed., *The Correspondence and Public Papers of John Jay* (4 vols., New York, 1891), III, 159. John Jay of New York was the Confederation's Secretary for Foreign Affairs.

NEW YORK'S ATTEMPT TO ABOLISH SLAVERY
Objections of the Council of Revision to the
Gradual Abolition Bill, March 21, 1785

Approximately 21,000 slaves lived in New York and were owned by about 8,500 families. About 6,700 families owned between one and three slaves; only 700 families owned more than six slaves. Large numbers of slaves were owned by the manor lords and by merchants in Albany and New York City. Federalist leaders such as Chancellor Robert R. Livingston and Philip Schuyler (Alexander Hamilton's father-in-law) owned many slaves; and slaveholding was also popular among the largely Dutch Hudson River Valley yeoman farmers.

Spurred by the efforts of the newly formed New York Abolition Society, the legislature in March 1785 adopted a bill for the gradual emancipation of slaves. According to the state constitution, all bills passed by the legislature had to be submitted to the Council of Revision composed of the governor, the chancellor, and the three justices of the state supreme court. The Council had a veto power that could be overridden by a two-thirds vote of each house.

The Council objected to the gradual emancipation bill because of its provisions disfranchising free blacks and persons of color. Such disfranchisement would create a divided, unstable society. The Council returned its objections to the state senate, which on March 23 overrode the Council's objections by a vote of 15 to 4. Three days later the House reconsidered the bill and by a vote of 23 to 17 failed (by four votes) to override the Council's objections.

In April the legislature with Council approval passed a bill prohibiting the sale of slaves within New York, providing for the voluntary manumission of slaves, and safeguarding the right to trial by jury in capital cases for all slaves "according to the Course of the Common Law." Three years later the legislature prohibited the importation of slaves and the purchase of slaves for export out of the state. An attempt to pass another gradual emancipation act failed in 1790. It was not until 1796 that New York enacted a gradual abolition act.

Present His Excellency Governor Clinton, The Honorable Mr. Chancellor Livingston, The Honorable Mr. Justice Hobart.

Mr. Chancellor Livingston to whom was committed the Bill entitled An Act for the gradual abolition of slavery within this State." Reported certain objections being again read and considered.—

The Council object against the said Bill becoming a Law of this state.—

1st. Because the last clause of the Bill enacts that no Negro, Mulatto, or Mustee shall have a legal vote in any case whatsoever which implicatively excludes Persons of this description from all share in the Legislature and from those offices in which a Vote may be necessary as well as from the important privilege of electing those by whom they are to be governed. The Bill having in other instances placed the children that shall be born of Slaves in the rank of Citizens, agreeable both to the spirit and letter of the Constitution [of

New York] they are as such entitled to all the Priviledges of Citizens nor can they be deprived of these essential Rights without shocking those Principles of equal Liberty which every page in that Constitution labours to enforce.—

2ndly. Because it holds up a Doctrine which is repugnant to the principle on which the United States justify their separation from Great Britain and either enacts what is wrong or supposes that those may rightfully be charged with the Burden of Government who have no representative share in imposing them.—

3rdly. Because this Class of disenfranchised and discontented Citizens who at some future period may be both numerous and wealthy may under the direction of ambitious and factious Leaders become dangerous to the state and effect the ruin of a Constitution whose benefits they are not permitted to enjoy.—

4thly. Because the Creation of an Order of Citizens who are to have no Legislative or Representative share in the Government, necessarily lays the foundation of an Aristocracy of the most dangerous and malignant kind rendering Power permanent and Hereditary in the Hands of those Persons who deduce their Origin through White Ancestors only, tho' these at some future period should not amount to a fiftieth part of the people. That this is not a chimerical supposition will be apparent to those who reflect that the term Mustee is indefinite. That the desire of power will induce those who possess it to exclude competitors by extending it as far as possible. That supposing it to extend to the seventeenth generation, every man will have the Blood of many more than 200,000 Ancestors running in his Veins and that if any of these should have been coloured his Posterity will by the operation of this Law be disfranchised so that if only one thousandth part of the Black Inhabitants now in this State should intermarry with the white their Posterity will amount to so many Millions that it will be difficult to suppose a fiftieth part of the People born within this State two hundred years hence who may be entitled to share in the benefits which our Excellent Constitution intended to secure to every Free Inhabitant of the State.—

5thly. Because the last clause of the Bill being General deprives those Black, Mulatto, and Mustee Citizens who have heretofore been entitled to a Vote of this essential priviledge and under the Idea of Political expediency without their having been charged with any offence, disfranchises them in direct violation of the established rules of justice against the Letter and spirit of the Constitution and tends to support a Doctrine which is inconsistent with the most obvious principles of Government, that the Legislature may arbitrarily dispose of the dearest Rights of their Constituents.—

Ordered that a Copy of the said Bill with a Copy of the preceding objections thereto, signed by his Excellency the Governor be delivered to the Honorable the Senate by Mr. Chancellor Livingston.

The Attempt to Abolish Slavery in Virginia

Many individuals and groups used the American Revolution and its philosophical underpinnings to argue for the general emancipation of slaves. At their 1780 conference, Methodists declared "that slavery is contrary to the laws of God, man, and nature, and hurtful to society, contrary to the dictates of conscience and pure religion, and doing that which we would not others do to us and ours." Four years later at their Christmas Conference, Methodists condemned slavery as "contrary to the golden law of God on which hang all the law and the prophets and the unalienable rights of mankind, as well as every principle of the revolution."

The high-water mark of abolitionism in Virginia occurred in the spring of 1785 when Methodist ministers travelled around the state collecting signatures on anti-slavery petitions to the state legislature. The petition, drafted by Bishop Thomas Coke, aroused opposition, as many slaveowners feared that it threatened their property. Counter-petitions circulated throughout the state justifying slavery, invoking the principles of the Revolution, and calling for the repeal of the manumission law of 1782.

While the anti-slavery petitions circulated, Bishop Coke and Bishop Francis Asbury visited George Washington to enlist his support. Washington told the ministers that he held similar sentiments and that he had told the state leaders his attitude toward slavery. The General, however, refused to sign a petition, but promised to send his sentiments to the legislature if it considered the petitions. (The following year Washington wrote to Lafayette endorsing a gradual, compensated emancipation effected by legislative authority.)

Read in the House of Delegates in early November, all of the anti-slavery petitions were rejected without dissent. The counter-petitions fared better. On December 14, 1785, Speaker Benjamin Harrison broke a 51-to-51 tie vote in favor of appointing a committee to bring in a bill for the repeal of the manumission act of 1782. On December 24 delegates defeated the bill, and on January 17, 1786, an amendment to the manumission bill also was defeated. No other serious effort at emancipation would arise in Virginia.

METHODISTS PETITION AGAINST SLAVERY
November 8, 1785[21]

To the Honourable the General Assembly of the State of Virginia; The Petition of the underwritten Electors of the said State.

Humbly Sheweth: That your Petitioners are clearly and fully persuaded that Liberty is the Birthright of Mankind, the right of every rational Creature

[21]Legislative Petitions, Frederick County, Received November 8, 1785, Virginia State Library.

without exception, who has not forfeited that right to the laws of his Country: That the Body of Negroes in this State have been robbed of that right without any such Forfeiture, and therefore ought in Justice to have their right restored: That the Glorious and ever Memorable Revolution can be Justified on no other principles, but what do plead with greater force for the Emancipation of our Slaves; in proportion as the Oppression exercised over them exceeds the Oppression formerly exercised by Great Britain over these States. That the Argument, "They were Prisoners of War, when they were Originally purchased" is utterly invalid, for no right of Conquest can Justly subject any Man to perpetual Slavery, much less his posterity: That the Riches & Strength of every Country consists in the number of its Inhabitants who are Interested in the support of its Government; and therefore to bind the Vast Body of Negroes to the State by the powerful ties of Interest will be the highest Policy. That the Argument drawn from the difference of Hair, Features and Colour, are so beneath the Man of Sense, much more the Christian, that we would insult the Honourable Assembly by enlarging upon them.—That the fear of the Enormities which the Negroes may commit, will be groundless, at least if the Emancipation be gradual, as the Activity of the Magistrates and the provision of Houses of Correction where Occasion may require, will easily Suppress the gross, flagrant, Idleness either of Whites or Blacks. But above all, that deep Debasement of Spirit, which is the necessary Consequence of Slavery, incapacitates the human Mind (except in a few instances) for the Reception of the Noble and enlarged principles of the Gospel; and therefore to encourage or allow of it, we apprehend to be most opposite to that Catholic Spirit of Christianity, which desires the Establishment of the Kingdom of Christ over all the World, and produces in the Conduct every Action consonant to that Desire. That of Consequence, Justice, Mercy and Truth, every Virtue that can Adorn the Man or the Christian, the Interest of Religion, the honour & real Interest of the State, and the Welfare of Mankind do unanswerably, uncontroulably plead for the Removal of this grand Abomination; And therefore that we humbly entreat the Honourable the Assembly, as their Superior Wisdom may dictate to them, to pursue the most Prudential, but effectual Method for the immediate or Gradual Exterpation of Slavery: And your Petitioners, as in Duty bound shall ever pray &c.

PETITION IN FAVOR OF SLAVERY
November 10, 1785[22]

To the honorable the General Assembly of Virginia, The Remonstrance and Petition of the free Inhabitants of the County of Pittsylvania.

Gentlemen, When the British Parliament usurped a Right to dispose of our property without our Consent, we dissolved the Union with our Parent State, and established a Constitution and form of Government of our own, that our property might be secure in future; in Order to effect this we risked our Lives and Fortunes, and waited through Seas of Blood. Divine Providence smiled on our Enterprize, & Crowned it with Success, and our Rights of Liberty and Property are now as well secured to us as they can be by any human Constitution or form of Government,

But notwithstanding this, we understand an Attempt is now afoot to dispossess us of a very important part of our property. An Attempt made by the Enemies of our Country, Tools of the British Administration, and supported by certain deluded Men among us, To WREST FROM US OUR SLAVES by an Act of the Legislature for a General Emancipation of them. They have the Address indeed to cover their Design with the Veil of Piety and Liberallity of Sentiment. But it is unsupported by the Word of God, and Productive of Ruin to this State.

It is unsupported by the Sacred Scriptures. Under the Old Testament Dispensation, Slavery was permitted by the Deity himself. For thus it is recorded, Leviticus Chap. 25 Verses 44, 45, 46. Both they, Bond Men and Bond Maids which thou shall have, shall be of the Heathen that are round about you; of them Shall ye buy Bond Men and Bond Maids.—Moreover, of the Children of the Strangers, that do sojourn among you, of them shall ye buy, and of their families that are with you, which they beget in your Land, and they shall be your Possession; and ye shall take them as an Inheritance for your Children after you, to Inherit them for a Possession; they shall be your Bond Men forever."—This Permission to Possess and Inherit Bond-men we have Reason to believe, was continued through all the Revolutions of the Jewish Government down to the Advent of our Lord. And we do not find that either he or his Apostles abridged it. The Freedom which the Followers of Jesus were taught to expect, was a Freedom from the Bondage of Sin & Satan and from the Dominion of their Lusts and Passions, But as to their OUTWARD CONDITION whatever that was before they embraced Christianity, whether BOND or FREE, it remained the same afterwards. This St. Paul hath expressly told us (1 Cor. Chap. 7 Ver. 20) where he is speaking directly to

[22]Legislative Petitions, Pittsylvania County, Received November 10, 1785, Virginia State Library.

this very Point; "Let every Man Abide in the same Calling wherein he is called"; and at Ver 24 "Let every Man wherein he is called, therein abide with God." Thus it is evident, that the said Design is unsupported by the Divine Word.

It is also ruinous to the State. For it involves in it, and is productive of Want, Poverty, Distress and Ruin to the free Citizens;—Neglect, Famine, & Death to the helpless black Infant and superannuated Parent; the Horrors of all the Rape, Murders, Roberies, and Outrages which a vast Multitude of unprincipled, unpropertied, vindictive and Remorseless Banditti are capable of perpetrating;—inevitable Bankruptcy to the Revenue, & Consequently Breach of public Faith, & Loss of Credit with foreign Nations;—and lastly, sure and final Ruin to this now free and flourishing Country.

WE therefore your Remonstrants and Petitioners do Solemnly adjure and humbly pray you, that you will Discountenance and utterly reject every Motion and Proposal for emancipating our Slaves;—that as the Act lately made, empowering the Owners of Slaves to liberate them, has been, and is still, in Part, productive of many of the above pernicious Effects, you will immediately and totally repeal it—and that as many of the Slaves liberated by the said Act have been Guilty of Thefts and Outrages, Insolences & Violences Destructive to the Peace, Safety, & Happiness of Society, you will make effectual Provision for the due Government of them in Future.

And your Remonstrants & Petitioners will ever pray, &c. &c.

James Madison to George Washington
Richmond, November 11, 1785[23]

. . . The pulse of the House of Delegates was felt on Thursday with regard to a general manumission by a petition presented on that subject. It was rejected without dissent but not without an avowed patronage of its principles by sundry respectable members. A motion was made to throw it under the table, which was treated with as much indignation on one side, as the petition itself was on the other. There are several petitions before the House against any step towards freeing the slaves, and even praying for a repeal of the law which licenses particular manumissions.

---

[23]Washington Papers, Library of Congress.

John Brown to Moses Brown
A Slave Trader's Rationalization, Providence, November 27, 1786[24]

Moses and John Brown were two of five brothers born to a prominent Provi-
dence, R.I., commercial family. John had a natural ability to adjust to changing
political and economic circumstances and soon became known as the "Providence Co-
lossus." He and his brothers regularly participated in the slave trade before the
Revolution. In 1773, however, several months after the death of his wife, Moses
Brown converted from being a Baptist to a Quaker. The rest of his life was devoted to
the amelioration of the condition of slaves and freedmen. John Brown underwent no
such conversion and continued the slave trade.

In 1784 John Brown represented the town of Providence in the assembly. De-
spite instructions from the town meeting to vote for the gradual emancipation bill,
Brown led the opposition to the measure. Because of this violation of his instructions,
he was not reelected to legislature later in the year. He was reelected, however, in
1786. When the legislature prohibited the slave trade, John Brown refocused his
commercial attention on the newly opening trade with the Far East.

Your Esteemed Favour was this Day at Diner handed me by your Son. I
have not yet had time to peruse the Treetice you was so kind as to Accompany
with the Letter but will do it Soone, by begining it this Evening. You mention
that you had heard, as Last Evening, I had it in Contemplation the Sending
an Other Vessell to Affrica in the Slave Trade. Its true that I have not onley
had an Other Voyage in Contemplation but have been prepairing the Cargo
for this 4 Months past & this Day before I Received yours began taking in the
Ballis having before Shipt the Captin & Mate. I have no Doubt of your
Sinsearity, in Your Exurtions to Discorage the Slave Trade, and did I Consider it as You do I would by no means be Concernd in it, but from the best
Information I can Git & that has beene from Grait Numbers the Slaves are
possitively better off, that is brought from the Coast than those who are Left
behind or then those would be was they not brought away—More aspetially
those who are Caried Among the French, as I propose this Vessill as Well as
the One Allredy Gone will Land them on High spanolia, where all Accounts
Agree they are better Treeted then in Aney part of the English West Indies.
This Trade has beene permitted by the Supreame Govenour of all things for
time Immemmoriel and whenever I am Convinced, as you are, that its Rong
in the Sight of God, I will Immediately Deasist, but while its not only allowd
by the Supreame Govenour of all States but by all the Nations of Europe, and
perhaps the most of Any by that Very Government which has beene & is Still
So Much Esteemed & Incoraged by those Very people who Appears the Most
Active in Writing and Clammering Against the Trade I cannot thinke this
State ought to Decline the Trade. I fulley Agree with You that no past Proffit

[24]Peck Collection, Rhode Island Historical Society.

that I have purtooke in the trade Can be My Inducement having Lost & that Very Graitly in allmost Every Voyage to Guiney I have beene Concernd in but You are Sencible much property has beene Acquired by this Trade from Newport. I Lately heard Severil of their principle people Say that the Merchants of Newport Very Scarsly Ever Cleard any property in Aney other Trade & that all the Estates that had Ever beene Acquired in that Town had beene Got in the Guiney Trades. It may be as you Suppose Determind by that power which presides over all Events that no Inhabitant of this Town Shall Ever Prosper in the Slave Trade. You Mention that the Melasses Trade may be Caried on without Extending it to the Guiney Trade. I agree it may be in a Limmited Degree. The Newfoundlands as well as all other British ports you are Sencible is Shet up from this Commerce which Used to be Very Grait.

I owe an Enormus Sum of Money in Europe & am Striving in Every Trade Which Appears Lawfull & Right to me, to pay as Much of the Debt as possable Dureing My Life time as I Wish Most Ardently to Leave My Famely Less Invoulved In Debt then is Now the Case. You are Sencible I have Tryd the Tobacco Trade, have a Ship Now Gone to Verginnia for a Load, have Tryd the Fishery having fifteen Schooners in the Buissiness, but have not yet beene So Fortunate as to Lessen the Debt. Should the Brigg I am now Fitting to Guiney & the Snow Captain Cooke who Saild in July Last the Ship Captain Sheldon now Gone to Verginnia & bound to France together with the Fisherman, which I propose Shall all Sail Very early in the Spring in order to Make up a Cargo of the best & Earlyest Spring Fish for Billbo [Bilbao, Spain]; I Say Should all these Succeede but Tollorably Well I hope to Discharge a Very Large proportion of My Debts, and I do Assure You I have not nor never had one Feeling in my Mind but that the Guiney Trade, or the Slave Trade as you More Explissetly Call it, was and is as Just & Right as Aney Trade I am or Ever Was Concernd in & Vastly More So then to Send a Vessill to Jamaica with a Two Fold or Double Intention nay Three Fold ContredICTION to the professions of Some Owners of Vessills. First, if the Slave Trade is Rong why will the Men Who thinkes So Incorage it by Sending for the produce Raised & made by those Slaves So Rongfully & unritchously Imported from their own Native Country. Are they Not doing as bad as tho they was to Undertake to Incorage A Theif by purchasing his Goods tho they knew they was Stole from an Honest Man?

Secondly, to thus Incorage the Trade of Carying Slaves to the West Indies have they Not Coverd their property with Fals papers & by this Means Indeverd to Disseive the people they was bound Among, by bringing their produce to this Countery Directly Conterary to their Laws?

& Thirdly, to Intreduce the produce of the English Islands In to this State Under Different Cullers & Different papers their others being Conseal'd, & all this Done by Owners & a Captin who are So Consheus bound as to use None but the plain Langwige & the better to Cover their Wickedness they Appear Among the Foremust to write & Taulke against the Guiney Trade & The Lord deliver me from Such Wolves in Sheeps Cloathing.

If you Incline to have a Law passed in this State to Stop the Guiney Trade, I once told you & I now Repeete it that I Shall be happey to Step out of the Seet Whenever it Can be So much better Fild as it will be when you Accept. I am Fulley Sencible that you Can Searve the Town & State at Large all better as a Legislater than I Can and wish with all My hart you woud once more Concent to Searve the publick in that Way. Our Different Sentiments Respecting the Guiney Trade will Weigh Nothing in My Mind against your taking my place in the House.

I will not Detain you aney Longer. I am Exceeding Sorry to Differ So much from you in this Buissiness, but I have Charrity for You that you thinke you are doing Gods Service when you go According to the Lite of Your Contience. I only wish for the Same Charritable Disposistion towards me that will I go According to the Dicketates of my Contious I may have mercey Extended to me tho I am not Endowd with that Devinc Light to See the Guiney Trade with the Same Eyes as you do. I Respect you as a Brother & a Friend. I Respect your Children and Sincearly wish they mought be Indulged to be More Furmillior with mine. I am Shure my Childron has a perticular Regard for Yours—

THE NORTHWEST ORDINANCE
July 13, 1787

Between 1784 and 1787, Congress considered various ordinances for the sale and government of the western lands ceded to it by the states. An ordinance drafted in 1784 by Thomas Jefferson provided for the prohibition of slavery in Congress' territory north and west of the Ohio River, but the ordinance never received approval from Congress. In 1785 Congress passed an ordinance for the surveying and sale of the Northwest Territory and two years later Congress adopted an ordinance creating the governmental structure for the territory. The Northwest Ordinance provided for two stages: a territorial stage and full-fledged statehood on a par with the original thirteen states. An abbreviated bill of rights was incorporated into the ordinance, and slavery was prohibited.

Southern delegates to Congress supported this prohibition of slavery for economic, political, and demographic reasons. They did not want a competing plantation economy to develop in the North. Furthermore, by prohibiting slavery from the Northwest Territory, a tacit understanding existed that slavery was permissable in the territory southwest of the Ohio River. Finally, migration patterns indicated that most settlers moving to the Northwest Territory were emigrating from New England, and thus were predisposed to opposing slavery. These settlers would probably build anti-slavery provisions into their state constitutions.

As a counterbalance to the prohibition of slavery, Congress included a fugitive-slave clause in the Northwest Ordinance. This clause served as a model for the Constitutional Convention, which was meeting simultaneously in Philadelphia.

Article the Sixth. There shall be neither slavery nor involuntary servitude in the said territory otherwise than in punishment of crimes whereof the party shall have been duly convicted: provided always that any person escaping into the same from whom labor or service is lawfully claimed in any one of the original states, such fugitive may be lawfully reclaimed and conveyed to the person claiming his, or her labor, or service as aforesaid.

2

The Constitutional
Convention and Slavery

T HROUGHOUT THE LATE SPRING AND
*summer of 1787, delegates from twelve American states met in Philadelphia and
drafted a new federal constitution. The old constitution—the Articles of Confederation—was acknowledged to be defective, and amendments to it were necessary to meet
the Union's needs. Repeated attempts to revise the Articles had failed. Beginning in
February 1781, several amendments were sent to the states for their approval but
none received the necessary unanimous ratification of the state legislatures. In 1786
seven amendments strengthening Congress were prepared but never received congressional endorsement because of the sectional explosion that occurred when Northern
states agreed to forgo the free navigation of the Mississippi River (an extremely important commercial avenue for the South) in exchange for a commercial treaty with
Spain (an important venture for the North). This dispute brought to the fore the
implicit sectionalism that had always divided the American colonies and the states
during the Revolution. From this time on, most national issues would be seen through
a sectional lens.*

*When the Constitutional Convention assembled in Philadelphia to revise the
Articles of Confederation, a sense of urgency gripped the delegates. This might be the
last opportunity for Americans to determine for themselves the kind of government
they wished to establish. The principles Americans had fought for during the Revolution seemed to be in serious jeopardy. The delegates had to succeed in reorganizing
government or risk splitting the Union into separate confederacies, or the possibility
of a restoration of monarchy and the downfall of the republic.*

Almost immediately the Convention delegates agreed to abandon the Articles of Confederation and create a new constitution. Never did the delegates consider eradicating slavery. The Revolutionary rhetoric of freedom and equality had been left behind; Americans in general and the delegates to the Convention in particular wanted a united, well-ordered, and prosperous society in which private property—including slave property—would be secure.

Throughout the intense debates in the Convention, sectionalism had a divisive effect. Seldom did delegates forget that the states were united primarily in name. In reality the country was economically, politically, socially, and morally divided. Virginia delegate James Madison suggested that the major division in the country was not between large and small states but between slave and non-slave states. (Although all of the states except Massachusetts had slave populations, the states from Pennsylvania northward had far fewer slaves than the Southern states, and slavery was far less important to the economies of the Northern states.) Madison felt that the interests of both slave states and non-slave states needed to be protected. He suggested that the representation in one house of Congress be based on each state's free population only and the representation in the other house be based on each state's total population including all of the slaves. This proposal never received serious consideration.

On June 2, 1787, the Pennsylvania Abolition Society petitioned the Convention for an end to the slave trade. The petition was entrusted to Tench Coxe, one of the society's secretaries, who delivered it to Pennsylvania Convention delegate Benjamin Franklin, president of the Pennsylvania Abolition Society. Coxe strongly advised Franklin not to submit the petition. Franklin agreed "that the memorial, in the beginning of the deliberations of the convention, might alarm some of the Southern states, and thereby defeat the wishes of the enemies of the African trade."[1]

Despite the sectionalism that divided the country, the impulse for union was so great that the Convention delegates reconciled the fundamental differences in American society through numerous compromises. Although never specifically mentioning the word "slavery" in the text of the Constitution, many of these compromises protected the interests of slaveowners in particular and the Southern states in general. In the end, the delegates accepted a constitution that everyone agreed was imperfect. The inadequacies of the document, however, merely reflected the fundamental differences that divided American society.

Four specific provisions in the new Constitution dealt directly with slavery, while many other provisions did so indirectly. The four specific provisions on slavery were: (1) three-fifths of the slave population was to be considered in apportioning direct taxes and representation in the U.S. House of Representatives; (2) the foreign slave trade could not be prohibited before 1808 and a tax levied on imported slaves could not exceed ten dollars per slave; (3) runaway slaves had to be returned to their masters "on demand" and could not be emancipated; and (4) no amendment to the Constitution prohibiting the slave trade could be adopted before 1808.

---

[1]For the text of the petition, see Chapter 4, *Pennsylvania Gazette*, March 5, 1788.

A partial list of constitutional provisions that indirectly affected slavery in-cluded (1) authorizing Congress to call forth the militia to help suppress domestic insurrections (including slave uprisings); (2) prohibitions on both the federal and state governments from levying export duties, thereby guaranteeing that the products of a slave economy (tobacco, indigo, rice, etc.) would not be taxed; (3) providing for the indirect election of the president through electors based on representation in Congress, which, because of the three-fifths clause, inflated the influence of the white Southern vote; (4) requiring a three-fourths approval of the states to adopt amendments to the Constitution, thus giving the South a veto power over all potential amendments; and (5) limiting the privileges and immunities clause to "citizens," thus denying these protections to slaves and in some cases to free blacks.

The Three-Fifths Clause

The organization of Congress was the first and one of the most difficult problems faced by the Constitutional Convention. The large states (Virginia, Massachusetts, and Pennsylvania) wanted a bicameral Congress with representation based on either population or wealth. The smaller states wanted to retain the equal representation of the states either in a strengthened unicameral Congress as under the Articles of Confederation or in a bicameral Congress. If population was to be the method of apportioning representation among the states, the status of slaves would have to be determined. The North did not want slaves to be counted in apportioning representatives; the South wanted all of the slaves to be counted. This impasse was finally resolved with a compromise. Three-fifths of the slaves would be counted in apportioning both representation and direct taxation.

The fraction three-fifths was known as the federal ratio. In April 1783 the Confederation Congress had proposed an amendment to the Articles of Confederation that would have changed the method of apportioning federal expenses among the states. The method established in the Articles of Confederation called for expenses to be apportioned among the states based on the value of land. Population would serve as the new standard of apportioning expenses among the states.

During the debate in Congress over this amendment in 1783, the Northern delegates argued that all of the slaves should be counted, while the Southern delegates maintained that slaves, as property, ought not to be counted. A compromise decided that three-fifths of the slaves should be counted. Because this amendment was adopted by only eleven states and not by the required unanimous approval of all of the states, the amendment was not formally adopted. (Rhode Island and New Hampshire did not ratify it.) Congress, however, used the population ratio in apportioning its requisition on the states for funds in 1786. (For the origins of the three-fifths clause, see the debate over the population amendment to the Articles of Confederation in Chapter 1.)

ARTICLE I, SECTION 2. Representatives and direct Taxes shall be apportioned among the several States which may be included within this Union, according to their respective Numbers, which shall be determined by adding to the whole Number of free Persons, including those bound to Service for a Term of Years, and excluding Indians not taxed, three fifths of all other persons.

MAY 29, 1787

Edmund Randolph (Va.): Resolved therefore that the rights of suffrage in the National Legislature ought to be proportioned to the Quotas of contribution, or to the number of free inhabitants, as the one or the other rule may seem best in different cases.

MAY 30

James Madison (Va.) observing that the words "or to the number of free inhabitants," might occasion debates which would divert the Committee from the general question whether the principle of representation should be changed, moved that they might be struck out.

Rufus King (Mass.) observed that the quotas of contributions which would alone remain as the measure of representation, would not answer, because waiving every other view of the matter, the revenue might hereafter be so collected by the general Government that the sums respectively drawn from the States would not appear; and would besides be continually varying.

Mr. Madison admitted the propriety of the observation, and that some better rule ought to be found.

Alexander Hamilton (N.Y.) moved to alter the resolution so as to read "that the rights of suffrage in the national Legislature ought to be proportioned to the number of free inhabitants. *Richard Dobbs Spaight* (N.C.) seconded the motion.

It was then moved that the Resolution be postponed, which was agreed to.

Mr. Randolph and Mr. Madison then moved the following resolution— "that the rights of suffrage in the national Legislature ought to be proportioned."

It was moved and seconded to amend it by adding "and not according to the present system"[2]—which was agreed to.

[2]Under the Articles of Confederation each state had one vote in Congress.

It was then moved and seconded to alter the resolution so as to read "that the rights of suffrage in the national Legislature ought not to be according to the present system."

It was then moved and seconded to postpone the Resolution moved by Mr. Randolph and Mr. Madison, which being agreed to:

Mr. Madison, moved, in order to get over the difficulties, the following resolution—"that the equality of suffrage established by the articles of Confederation ought not to prevail in the national Legislature, and that an equitable ratio of representation ought to be substituted." This was seconded by *Gouverneur Morris* (Pa.), and being generally relished, would have been agreed to: when,

George Read (Del.) moved that the whole clause relating to the point of Representation be postponed; reminding the Committee that the deputies from Delaware were restrained by their commission from assenting to any change of the rule of suffrage, and in case such a change should be fixed on, it might become their duty to retire from the Convention.

By several it was observed that no just construction of the Act of Delaware, could require or justify a secession of her deputies, even if the resolution were to be carried through the House as well as the Committee. It was finally agreed however that the clause should be postponed: it being understood that in the event the proposed change of representation would certainly be agreed to, no objection or difficulty being started from any other quarter than from Delaware.

The motion of Mr. Read to postpone being agreed to.

JUNE 9

William Paterson (N.J.) moves that the Committee resume the clause relating to the rule of suffrage in the National Legislature.

[A long debate followed over whether representation of the states in Congress should remain equal or become proportional.]

JUNE 11

Roger Sherman (Conn.) proposed that the proportion of suffrage in the first branch should be according to the respective numbers of free inhabitants; and that in the second branch or Senate, each State should have one vote and no more. He said as the States would remain possessed of certain individual rights, each State ought to be able to protect itself: otherwise a few large States will rule the rest. The House of Lords in England he observed had certain particular rights under the Constitution, and hence they have an equal vote with the House of Commons that they may be able to defend their rights.

John Rutledge (S.C.) proposed that the proportion of suffrage in the first branch should be according to the quotas of contribution. The justice of this rule he said could not be contested.

Pierce Butler (S.C.) urged the same idea: adding that money was power; and that the States ought to have weight in the Government in proportion to their wealth.

Mr. King and James Wilson (Pa.) in order to bring the question to a point moved "that the right of suffrage in the first branch of the national Legislature ought not to be according to the rule established in the articles of Confederation, but according to some equitable ratio of representation. . . ."

On the question for agreeing to Mr. King's and Mr. Wilson's motion it passed in the affirmative Massachusetts ay, Connecticut ay, New York no, New Jersey no, Pennsylvania ay, Delaware no, Maryland divided, Virginia ay, North Carolina ay, South Carolina ay, Georgia ay (7 to 3, with 1 divided).

It was then moved by *Mr. Rutledge* seconded by *Mr. Butler* to add to the words "equitable ratio of representation" at the end of the motion just agreed to, the words "according to the quotas of contribution." On motion of *Mr. Wilson* seconded by *Charles Pinckney* (S.C.), this was postponed; in order to add, after the words "equitable ratio of representation" the words following "in proportion to the whole number of white and other free Citizens and inhabitants of every age, sex and condition including those bound to servitude for a term of years and three fifths of all other persons not comprehended in the foregoing description, except Indians not paying taxes, in each State," this being the rule in the Act of Congress agreed to by eleven States, for apportioning quotas of revenue on the States, and requiring a Census only every 5–7, or 10 years.

Elbridge Gerry (Mass.) thought property not the rule of representation. Why then should the blacks, who were property in the South, be in the rule of representation more than the Cattle and horses of the North.

On the question,—Massachusetts, Connecticut, New York, Pennsylvania, Maryland, Virginia, North Carolina, South Carolina, and Georgia were in the affirmative; New Jersey and Delaware in the negative (9 to 2).

Mr. Sherman moved that a question be taken whether each State shall have one vote in the second branch. Every thing he said depended on this. The smaller States would never agree to the plan on any other principle than an equality of suffrage in this branch. *Oliver Ellsworth* (Conn.) seconded the motion. On the question for allowing each State one vote in the second branch.

Massachusetts no, Connecticut ay, New York ay, New Jersey ay, Pennsylvania no, Delaware ay, Maryland ay, Virginia no, North Carolina no, South Carolina no, Georgia no (5 to 6).

JUNE 30

Mr. Madison contended that the States were divided into different interests not by their difference of size, but by other circumstances; the most material of which resulted partly from climate, but principally from the effects of their having or not having slaves. These two causes concurred in forming the great division of interests in the United States. It did not lie between the large & small States: it lay between the Northern & Southern, and if any defensive power were necessary, it ought to be mutually given to these two interests. He was so strongly impressed with this important truth that he had been casting about in his mind for some expedient that would answer the purpose. The one which had occurred was that instead of proportioning the votes of the States in both branches [of Congress], to their respective numbers of inhabitants computing the slaves in the ratio of 5 to 3, they should be represented in one branch according to the number of free inhabitants only; and in the other according to the whole number counting the slaves as if free. By this arrangement the Southern Scale would have the advantage in one House, and the Northern in the other. He had been restrained from proposing this expedient by two considerations; one was his unwillingness to urge any diversity of interests on an occasion when it is but too apt to arise of itself—the other was the inequality of powers that must be vested in the two branches, and which would destroy the equilibrium of interests.

[*Between June 11 and July 2 the heated debate over representation continued, centering on the issue of continuing a system of equal state representation or adopting a new system of proportional representation of the states based upon population or wealth. On July 2 the Convention appointed a grand committee (one delegate from each state) to consider the representation impasse. The committee reported on July 5, recommending that in the first branch of the legislature (i.e., the House of Representatives) each state then in the Union be allowed one representative for every 40,000 inhabitants with three-fifths of the slaves being counted among the inhabitants. Debate then shifted to the numbers of representatives allotted to each state. On July 6 the Convention appointed a committee of five to consider the apportionment of representatives among the states in the first House of Representatives. The committee (consisting of three northerners—Gouverneur Morris, Nathaniel Gorham and Rufus King—and two Southerners—Edmund Randolph and John Rutledge) reported on July 9 that the first House of Representatives should consist of 56 members apportioned accordingly: New Hampshire 2, Massachusetts 7, Rhode Island 1, Connecticut 4, New York 5, New Jersey 3, Pennsylvania 8, Delaware 1, Maryland 4, Virginia 9, North Carolina 5, South Carolina 5, and Georgia 2. Dividing the country at Delaware, the North received thirty-one representatives to the South's twenty-five. Southerners did not like the division but were pleased that the ratio for apportioning representatives still included three-fifths of the slaves. Northerners liked the initial apportionment of the House of Representatives but some of them worried about the long-term prospects if slaves were part of the ratio. Some Northerners also worried about the reaction of their constituents to allowing representation for three-fifths of the slaves.]*

JULY 9

Mr. Randolph disliked the report of the Committee but had been unwilling to object to it. He was apprehensive that as the number was not to be changed till the National Legislature should please, a pretext would never be wanting to postpone alterations [i.e., reapportionment], and keep the power in the hands of those possessed of it. He was in favor of the commitment to a member from each State.

Mr. Paterson considered the proposed estimate for the future according to the Combined rule of numbers and wealth, as too vague. For this reason New Jersey was against it. He could regard negro slaves in no light but as property. They are no free agents, have no personal liberty, no faculty of acquiring property, but on the contrary are themselves property, and like other property entirely at the will of the Master. Has a man in Virginia a number of votes in proportion to the number of his slaves? And if Negroes are not represented in the States to which they belong, why should they be represented in the General Government? What is the true principle of Representation? It is an expedient by which an assembly of certain individuals chosen by the people is substituted in place of the inconvenient meeting of the people themselves. If such a meeting of the people was actually to take place, would the slaves vote? They would not. Why then should they be represented. He was also against such an indirect encouragement of the slave trade; observing that Congress in their act relating to the change of the 8 article: of Confederation had been ashamed to use the term "slaves" and had substituted a description.

Mr. Madison reminded Mr. Paterson that his doctrine of Representation which was in its principle the genuine one, must forever silence the pretensions of the small States to an equality of votes with the large ones. They ought to vote in the same proportion in which their citizens would do, if the people of all the States were collectively met. He suggested as a proper ground of compromise, that in the first branch the States should be represented according to their number of free inhabitants; and in the second which had for one of its primary objects the guardianship of property, according to the whole number, including slaves.

Mr. Butler urged warmly the justice and necessity of regarding wealth in the apportionment of Representation.

Mr. King had always expected that as the Southern States are the richest, they would not league themselves with the Northern unless some respect were paid to their superior wealth. If the latter expect those preferential distinctions in Commerce and other advantages which they will derive from the connection they must not expect to receive them without allowing some advantages in return. Eleven out of 13 of the States had agreed to consider Slaves in the apportionment of taxation; and taxation and Representation ought to go together.

On the question for committing the first paragraph of the Report to a member from each State.

Massachusetts ay, Connecticut ay, New York no, New Jersey ay, Pennsylvania ay, Delaware ay, Maryland ay, Virginia ay, North Carolina ay, South Carolina no, Georgia ay (9 to 2).

[On July 10 the committee reported a new apportionment for the first House of Representatives. The 65 representatives would be divided accordingly: New Hampshire 3, Massachusetts 8, Rhode Island 1, Connecticut 5, New York 6, New Jersey 4, Pennsylvania 8, Delaware 1, Maryland 6, Virginia 10, North Carolina 5, South Carolina 5, Georgia 3. The North again dominated this time by an increased margin of 36 to 29. South Carolina delegates John Rutledge and Charles Cotesworth Pinckney moved for the reduction in New Hampshire's representation from three to two, thus reestablishing the difference in representation between the North and South to six representatives. Massachusetts delegate Rufus King defended New Hampshire's allotment of three representatives and suggested that the Convention had reached a pivotal point.]

Mr. King. He believed [northerners] to be very desirous of uniting with their Southern brethren, but did not think it prudent to rely so far on that disposition as to subject them to any gross inequality. He was fully convinced that the question concerning a difference of interests did not lie where it had hitherto been discussed, between the great and small States; but between the Southern and Eastern. For this reason he had been ready to yield something in the proportion of representatives for the security of the Southern. No principle would justify the giving them a majority. They were brought near as an equality as was possible. He was not averse to giving them a still greater security, but did not see how it could be done.

Charles Cotesworth Pinckney (S.C.). The Report before it was committed was more favorable to the Southern States than as it now stands. If they are to form so considerable a minority, and the regulation of trade is to be given to the General Government, they will be nothing more than overseers for the Northern States. He did not expect the Southern States to be raised to a majority of representatives, but wished them to have something like an equality. At present by the alterations of the Committee in favor of the Northern States they are removed farther from it than they were before. One member indeed had been added to Virginia which he was glad of as he considered her as a Southern State. He was glad also that the members of Georgia were increased.

Hugh Williamson (N.C.) was not for reducing New Hampshire from 3 to 2, but for reducing some others. The Southern Interest must be extremely endangered by the present arrangement. The Northern States are to have a majority in the first instance and the means of perpetuating it.

General Pinckney urged the reduction, dwelt on the superior wealth of the Southern States, and insisted on its having its due weight in the Government.

Gouverneur Morris regretted the turn of the debate. The States he found had many Representatives on the floor. Few he feared were to be deemed the Representatives of America. He thought the Southern States have by the report more than their share of representation. Property ought to have its weight, but not all the weight. If the Southern States are to supply money, the Northern States are to spill their blood. Besides, the probable Revenue to be expected from the Southern States has been greatly overrated. He was against reducing New Hampshire.

Mr. Randolph was opposed to a reduction of New Hampshire, not because she had a full title to three members: but because it was in his contemplation first to make it the duty instead of leaving it to the discretion of the Legislature to regulate the representation by a periodical census. Secondly to require more than a bare majority of votes in the Legislature in certain cases, and particularly in commercial cases.

On the question for reducing New Hampshire from 3 to 2 Representatives it passed in the negative (8 to 2).

July 11

Mr. Butler and General Pinckney insisted that blacks be included in the rule of Representation, *equally* with the Whites: and for that purpose moved that the words "three-fifths" be struck out.

Mr. Gerry thought that 3/5 of them was to say the least the full proportion that could be admitted.

Nathaniel Gorham (Mass.). This ratio was fixed by Congress as a rule of taxation. Then it was urged by the Delegates representing the States having slaves that the blacks were still more inferior to freemen. At present when the ratio of representation is to be established, we are assured that they are equal to freemen. The arguments on the former occasion had convinced him that 3/5 was pretty near the just proportion and he should vote according to the same opinion now.

Mr. Butler insisted that the labour of a slave in South Carolina was as productive and valuable as that of a freeman in Massachusetts, that as wealth was the great means of defence and utility to the Nation they were equally valuable to it with freemen; and that consequently an equal representation ought to be allowed for them in a Government which was instituted principally for the protection of property, and was itself to be supported by property.

George Mason (Va.) could not agree to the motion, notwithstanding it was favorable to Virginia because he thought it unjust. It was certain that the slaves were valuable, as they raised the value of land, increased the exports and imports, and of course the revenue, would supply the means of feeding and supporting an army, and might in cases of emergency become themselves soldiers. As in these important respects they were useful to the community at large, they ought not to be excluded from the estimate of Representation. He could not however regard them as equal to freemen and could not vote for them as such. He added as worthy of remark, that the Southern States have this peculiar species of property, over and above the other species of property common to all the States.

Mr. Williamson reminded Mr. Gorham that if the Southern States contended for the inferiority of blacks to whites when taxation was in view, the Eastern States on the same occasion contended for their equality. He did not however either then or now, concur in either extreme, but approved of the ratio of 3/5.

On Mr. Butler's motion for considering blacks as equal to Whites in the apportionment of Representatives.

Massachusetts no, Connecticut no, New York not on the floor, New Jersey no, Pennsylvania no, Delaware ay, Maryland no, Virginia no, North Carolina no, South Carolina ay, Georgia ay (7 to 3). . . .

The next clause as to 3/5 of the negroes being considered.

Mr. King being much opposed to fixing numbers as the rule of representation, was particularly so on account of the blacks. He thought the admission of them along with Whites at all, would excite great discontents among the States having no slaves. He had never said as to any particular point that he would in no event acquiesce in and support it; but he would say that if in any case such a declaration was to be made by him, it would be in this. He remarked that in the temporary allotment of Representatives made by the Committee, the Southern States had received more than the number of their white and three fifths of their black inhabitants entitled them to.

Mr. Gorham supported the propriety of establishing numbers as the rule. . . . He was aware that there might be some weight in what had fallen from his colleague, as to the umbrage which might be taken by the people of the Eastern States. But he recollected that when the proposition of Congress changing the 8th article: of Confederation was before the Legislature of Massachusetts the only difficulty then was to satisfy them that the negroes ought not to have been counted equally with the whites instead of being counted in the ratio of three fifths only.

Mr. Wilson did not well see on what principle the admission of blacks in the proportion of three fifths could be explained. Are they admitted as Citizens? Then why are they not admitted on an equality with White Citizens? Are they admitted as property? Then why is not other property admitted into the computation? These were difficulties however which he thought must be overruled by the necessity of compromise. He had some apprehensions also from the tendency of the blending of the blacks with the whites, to give disgust to the people of Pennsylvania as had been intimated by his Colleague

[Gouverneor Morris]. But he differed from him in thinking numbers of inhabitants so incorrect a measure of wealth.

Gouverneur Morris was compelled to declare himself reduced to the dilemma of doing injustice to the Southern States or to human nature, and he must therefore do it to the former. For he could never agree to give such encouragement to the slave trade as would be given by allowing them a representation for their negroes, and he did not believe those States would ever confederate on terms that would deprive them of that trade.

On the Question for agreeing to include 3/5 of the blacks.

Massachusetts no, Connecticut ay, New Jersey no, Pennsylvania no, Delaware no, Maryland no, Virginia ay, North Carolina ay, South Carolina no, Georgia ay (4 to 6).

JULY 12

Gouverneur Morris moved to add to the clause empowering the Legislature to vary the Representation according to the principles of wealth and number of inhabitants a "proviso that taxation shall be in proportion to Representation."

Mr. Butler contended again that Representation should be according to the full number of inhabitants including all the blacks; admitting the justice of Mr. Gouverneur Morris's motion.

William R. Davie (N.C.) said it was high time now to speak out. He saw that it was meant by some gentlemen to deprive the Southern States of any share of Representation for their blacks. He was sure that North Carolina would never confederate on any terms that did not rate them at least as 3/5. If the Eastern States meant therefore to exclude them altogether the business was at an end.

William Samuel Johnson (Conn.) thought that wealth and population were the true, equitable rules of representation; but he conceived that these two principles resolved themselves into one; population being the best measure of wealth. He concluded therefore that the number of people ought to be established as the rule, and that all descriptions including blacks *equally* with the whites, ought to fall within the computation. As various opinions had been expressed on the subject, he would move that a Committee might be appointed to take them into consideration and report thereon.

Gouverneur Morris. It had been said that it is high time to speak out, as one member, he would candidly do so. He came here to form a compact for the good of America. He was ready to do so with all the States. He hoped and believed that all would enter into such a Compact. If they would not he was ready to join with any States that would. But as the Compact was to be voluntary, it is in vain for the Eastern States to insist on what the Southern States will never agree to. It is equally vain for the latter to require what the other States can never admit; and he verily believed the people of Pennsylvania will never agree to a representation of Negroes. What can be desired by these States more than has been already proposed; that the Legislature shall from time to time regulate Representation according to population and wealth.

General Pinckney desired that the rule of wealth should be ascertained and not left to the pleasure of the Legislature; and that property in slaves should not be exposed to danger under a Government instituted for the protection of property.

The first clause in the Report of the first Grand Committee was postponed.

Mr. Ellsworth. In order to carry into effect the principle established, moved to add to the last clause adopted by the House the words following "and that the rule of contribution by direct taxation for the support of the Government of the United States shall be the number of white inhabitants, and three fifths of every other description in the several States, until some other rule that shall more accurately ascertain the wealth of the several States can be devised and adopted by the Legislature."

Mr. Butler seconded the motion in order that it might be committed.

Mr. Randolph was not satisfied with the motion. The danger will be revived that the ingenuity of the Legislature may evade or pervert the rule so as to perpetuate the power where it shall be lodged in the first instance. He proposed in lieu of Mr. Ellsworth's motion "that in order to ascertain the alterations in Representation that may be required from time to time by changes in the relative circumstances of the States, a census shall be taken within two years from the first meeting of the General Legislature of the United States, and once within the term of every ____ years afterwards, if all the inhabitants in the manner and according to the ratio recommended by Congress in their resolution of the 18th day of April 1783 [rating blacks at 3/5 of their number]; and, that the Legislature of the U.S. shall arrange the Representation accordingly,"—He urged strenuously that express security ought to be provided for including slaves in the ratio of Representation. He lamented that such a species of property existed. But as it did exist the holders of it would require this security. It was perceived that the design was entertained by some excluding slaves altogether; the Legislature therefore ought not to be left at liberty.

Mr. Ellsworth withdraws his motion and seconds that of Mr. Randolph.

Mr. Wilson observed that less umbrage would perhaps be taken against an admission of the slaves into the Rule of representation, if it should be so expressed as to make them indirectly only an ingredient in the rule, by saying that they should enter into the rule of taxation: and as representation was to be according to taxation, the end would be equally attained. He accordingly moved and was seconded so to alter the last clause adopted by the House, that together with the amendment proposed the whole should read as follows— "provided always that the representation ought to be proportioned according to direct taxation, and in order to ascertain the alterations in the direct taxation which may be required from time to time by the changes in the relative circumstances of the States. Resolved that a census be taken within two years from the first meeting of the Legislature of the United States, and once within the term of every ____ years afterwards of all the inhabitants of the U.S. in the manner and according to the ratio recommended by Congress in their Resolution of the eighteenth day of April 1783; and that the Legislature of the U.S. shall proportion the direct taxation accordingly."

Mr. King. Although this amendment varies the aspect somewhat, he had still two powerful objects against tying down the Legislature to the rule of numbers, first they were at this time an uncertain index of the relative wealth of the States. Secondly if they were a just index at this time it can not be supposed always to continue so. He was far from wishing to retain any unjust advantage whatever in one part of the Republic. If justice was not the basis of the connection it could not be of long duration. He must be shortsighted indeed who does not foresee that whenever the Southern States shall be more numerous than the Northern, they can and will hold a language that will awe them into justice. If they threaten to separate now in case injury shall be done them, will their threats be less urgent or effectual when force shall back their demands. Even in the intervening period, there will be no point of time at which they will not be able to say, do us justice or we will separate. He urged the necessity of placing confidence to a certain degree in every Government and did not conceive that the proposed confidence as to a periodical readjustment, of the representation exceeded that degree.

Charles Pinckney moved to amend Mr. Randolph's motion so as to make "blacks equal to the whites in the ratio of representation." This he urged was nothing more than justice. The blacks are the labourers, the peasants of the Southern States: they are as productive of pecuniary resources as those of the Northern States. They add equally to the wealth, and considering money as the sinew of war, to the strength of the nation. It will also be politic with regard to the Northern States, as taxation is to keep pace with Representation.

On Mr. Pinckney's motion for rating blacks as equal to Whites instead of as 3/5—

Massachusetts no, Connecticut no, New Jersey no, Pennsylvania no, Delaware no, Maryland no, Virginia no, North Carolina no, South Carolina ay, Georgia ay (2 to 9).

On the question on the whole proposition; as proportioning representation to direct taxation and both to the white and 3/5 of the black inhabitants, and requiring a Census within six years—and within every ten years afterwards.

Massachusetts divided, Connecticut ay, New Jersey no, Pennsylvania ay, Delaware no, Maryland ay, Virginia ay, North Carolina ay, South Carolina divided, Georgia ay (6 to 2 with 2 divided).

The Foreign Slave Trade

ARTICLE I, SECTION 9. The Migration or Importation of such Persons as any of the States now existing shall think proper to admit, shall not be prohibited by the Congress prior to the Year one thousand eight hundred and eight, but a Tax or duty may be imposed on such Importation, not exceeding ten dollars for each Person.

ARTICLE V. . . . Provided that no Amendment which may be made prior to the Year One thousand eight hundred and eight shall in any Manner affect the first and fourth Clauses in the Ninth Section of the first Article. . . .

JULY 23

Elbridge Gerry moved that the proceedings of the Convention for the establishment of a National Government (except the part relating to the Executive), be referred to a Committee to prepare and report a Constitution conformable thereto.

Charles Cotesworth Pinckney reminded the Convention that if the Committee should fail to insert some security to the Southern States against an emancipation of slaves, and taxes on exports, he should be bound by duty to his State to vote against their Report.

The appointment of a Committee as moved by Mr. Gerry was Agreed to nem. con.[3]

AUGUST 6

John Rutledge delivered in the Report of the Committee of Detail as follows: a printed copy being at the same time furnished to each member. . . .

Article VII, section 4. No tax or duty shall be laid by the Legislature [i.e., Congress] on articles exported from any State; nor on the migration or importation of such persons as the several States shall think proper to admit, nor shall such migration or importation be prohibited.

AUGUST 8

Rufus King. . . . The admission of slaves [into the formula for apportioning representation] was a most grating circumstance to his mind, and he believed would be so to a great part of the people of America. He had not

---

[3] *Nemine contradicente* (no one contradicting).

made a strenuous opposition to it heretofore because he hoped that this concession would have produced a readiness which had not been manifested, to strengthen the General Government and to mark a full confidence in it. The Report under consideration had by the tenor of it, put an end to all those hopes. In two great points the hands of the Legislature were absolutely tied. The importation of slaves could not be prohibited—exports could not be taxed. Is this reasonable? What are the great objects of the General System? First defence against foreign invasion. Secondly against internal sedition. Shall all the States then be bound to defend each; and shall each be at liberty to introduce a weakness which will render defence more difficult? Shall one part of the U.S. be bound to defend another part, and that other part be at liberty not only to increase its own danger, but to withhold the compensation for the burden? If slaves are to be imported shall not the exports produced by their labor, supply a revenue the better to enable the General Government to defend their masters?—there was so much inequality and unreasonableness in all this, that the people of the Northern States could never be reconciled to it. No candid man could undertake to justify it to them. He had hoped that some accommodation would have taken place on this subject; that at least a time would have been limited for the importation of slaves. He never could agree to let them be imported without limitation and then be represented in the National Legislature. Indeed he could so little persuade himself of the rectitude of such a practice, that he was not sure he could assent to it under any circumstances. At all events, either slaves should not be represented, or exports should be taxable.

Roger Sherman regarded the slave trade as iniquitous; but the point of representation having been settled after much difficulty and deliberation, he did not think himself bound to make opposition; especially as the present article as amended did not preclude any arrangement whatever on that point in another place of the Report.

James Madison objected to 1 for every 40,000 inhabitants as a perpetual rule. The future increase of population if the Union should be permanent, will render the number of Representatives excessive.

Mr. Sherman and Mr. Madison moved to insert the words "not exceeding" before the words "1 for every 40,000," which was agreed to nem. con.

Gouverneur Morris moved to insert "free" before the word inhabitants. Much he said would depend on this point He never would concur in upholding domestic slavery. It was a nefarious institution. It was the curse of heaven on the States where it prevailed. Compare the free regions of the Middle States, where a rich and noble cultivation marks the prosperity and happiness of the people, with the misery and poverty which overspread the barren wastes of Virginia, Maryland and the other States having slaves. Travel through the whole Continent and you behold the prospect continually varying with the appearance and disappearance of slavery. The moment you leave the Eastern States and enter New York, the effects of the institution become visible, passing through the Jerseys and entering Pennsylvania every criterion of superior improvement witnesses the change. Proceed southwardly and every step you take through the great region of slaves presents a desert increasing, with the increasing proportion of these wretched beings. Upon what principle is it that the slaves shall be computed in the representation? Are they men? Then make them Citizens and let them vote. Are they property? Why then is no other property included? The Houses in this city [Philadelphia] are worth more than all the wretched slaves which cover the rice swamps of South Carolina. The admission of slaves into the Representation when fairly explained comes to this: that the inhabitants of Georgia and South Carolina who goes to the Coast of Africa, and in defiance of the most sacred laws of humanity tears away his fellow creatures from their dearest connections and damns them to the most cruel bondage, shall have more votes in a Government instituted for the protection of the rights of mankind, than the Citizen of Pennsylvania or New Jersey who views with a laudable horror, so nefarious a practice. He would add that Domestic slavery is the most prominent feature in the aristocratic countenance of the proposed Constitution. The vassalage of the poor has ever been the favorite offspring of Aristocracy. And What is the proposed compensation to the Northern States for a sacrifice of every principle of right, of every impulse of humanity. They are to bind themselves to march their militia for the defence of the Southern States; for their defence against those very slaves of whom they complain. They must supply vessels and seamen in case of foreign Attack. The Legislature will have indefinite power to tax them by excises, and duties on imports: both of which will fall heavier on them than on the Southern inhabitants; for the bohea tea used by a Northern freeman, will pay more tax than the whole consumption of the miserable slave, which consists of nothing more than his physical subsistence and the rag that covers his nakedness. On the other side the Southern States are not to be restrained from importing fresh supplies of wretched Africans, at once to increase the danger of attack, and the difficulty of defence; nay they are to be encouraged to it by an assurance of having their votes in the National Government increased in proportion, and are at the same time to have their exports and their slaves exempt from all contributions for the public service. Let it not be said that direct taxation is to be proportioned to representation. It is idle to suppose that the General Government can stretch its hand directly into the pockets of the people scattered over so vast a Country. They can only do it through the medium of exports, imports, and excises. For what then are all these sacrifices to be made? He would sooner submit himself to a tax for paying for all the negroes in the United States, than saddle posterity with such a Constitution.

Jonathan Dayton (N.J.) seconded the motion. He did it he said that his sentiments on the subject might appear whatever might be the fate of the amendment.

Mr. Sherman did not regard the admission of Negroes into the ratio of representation, as liable to such insuperable objections. It was the freemen of the Southern States who were in fact to be represented according to the taxes paid by them, and the Negroes are only included in the Estimate of the taxes. This was his idea of the matter.

Charles Pinckney considered the fisheries and the Western frontier as more burdensome to the U.S. than the slaves. He thought this could be demonstrated if the occasion were a proper one.

James Wilson thought the motion premature. An agreement to the clause would be no bar to the object of it.

On the Question on the motion to insert "free" before "inhabitants,"

New Hampshire no, Massachusetts no, Connecticut no, New Jersey ay, Pennsylvania no, Delaware no, Maryland no, Virginia no, North Carolina no, South Carolina no, Georgia no (1 to 10).

AUGUST 16

Luther Martin (Md.) proposed to vary the Section 4, article VII so as to allow a prohibition or tax on the importation of slaves. In the first place as five slaves are to be counted as 3 free men in the apportionment of Representatives; such a clause would leave an encouragement to this traffic. In the second place slaves weakened one part of the Union which the other parts were bound to protect: the privilege of importing them was therefore unreasonable. And in the third place it was inconsistent with the principle of the revolution and dishonorable to the American character to have such a feature in the Constitution.

Mr. Rutledge did not see how the importation of slaves could be encouraged by this Section. He was not apprehensive of insurrections and would readily exempt the other States from the obligation to protect the Southern against themselves.—Religion and humanity had nothing to do with this question. Interest alone is the governing principles with nations. The true question at present is whether the Southern States shall or shall not be parties to the Union. If the Northern States consult their interest, they will not oppose the increase of Slaves which will increase the commodities of which they will become the carriers.

Oliver Ellsworth was for leaving the clause as it stands. Let every State import what it pleases. The morality or wisdom of slavery are considerations belonging to the States themselves. What enriches a part enriches the whole, and the States are the best judges of their particular interest. The old confederation had not meddled with this point, and he did not see any greater necessity for bringing it within the policy of the new one.

Charles Pinckney. South Carolina can never receive the plan if it prohibits the slave trade. In every proposed extension of the powers of the Congress, that State has expressly and watchfully excepted that of meddling with the importation of negroes. If the States be all left at liberty on this subject, South Carolina may perhaps by degrees do of herself what is wished, as Virginia and Maryland have already done.

AUGUST 22

Mr. Sherman was for leaving the clause as it stands. He disapproved of the slave trade; yet as the States were now possessed of the right to import slaves, as the public good did not require it to be taken from them, and as it was expedient to have as few objections as possible to the proposed scheme of Government, he thought it best to leave the matter as we find it. He observed that the abolition of Slavery seemed to be going on in the U.S. and that the good sense of the several States would probably by degrees complete it. He urged on the Convention the necessity of despatching its business.

George Mason. This infernal traffic originated in the avarice of British Merchants. The British Government constantly checked the attempts of Virginia to put a stop to it. The present question concerns not the importing States alone but the whole Union. The evil of having slaves was experienced during the late war. Had slaves been treated as they might have been by the Enemy, they would have proved dangerous instruments in their hands. But their folly dealt by the slaves, as it did by the Tories. He mentioned the dangerous insurrections of the slaves in Greece and Sicily; and the instructions given by Cromwell to the Commissioners sent to Virginia, to arm the servants and slaves, in case other means of obtaining its submission should fail. Maryland and Virginia he said had already prohibited the importation of slaves expressly. North Carolina had done the same in substance. All this would be in vain if South Carolina and Georgia be at liberty to import. The Western people are already calling out for slaves for their new lands, and will fill that Country with slaves if they can be got through South Carolina and Georgia.

Slavery discourages arts and manufactures. The poor despise labor when performed by slaves. They prevent the immigration of Whites, who really enrich and strengthen a country. They produce the most pernicious effect on manners. Every master of slaves is born a petty tyrant. They bring the judgment of heaven on a Country. As nations can not be rewarded or punished in the next world they must be in this. By an inevitable chain of causes and effects providence punishes national sins, by national calamities. He lamented that some of our Eastern brethren had from a lust of gain embarked in this nefarious traffic. As to the States being in possession of the Right to import, this was the case with many other rights, now to be properly given up. He held it essential in every point of view that the General Government should have power to prevent the increase of slavery.

Mr. Ellsworth. As he had never owned a slave could not judge of the effect of slavery on character: He said however that if it was to be considered in a moral light we ought to go farther and free those already in the Country.—As slaves also multiply so fast in Virginia and Maryland that it is cheaper to raise than import them, whilst in the sickly rice swamps foreign supplies are necessary, if we go no farther than is urged, we shall be unjust towards South Carolina and Georgia. Let us not intermeddle. As population increases poor laborers will be so plenty as to render slaves useless. Slavery in time will not be a speck in our Country. Provision is already made in Connecticut for abolishing it. And the abolition has already taken place in Massachusetts. As to the danger of insurrections from foreign influence, that will become a motive to kind treatment of the slaves.

Charles Pinckney. If slavery be wrong, it is justified by the example of all of the world. He cited the case of Greece, Rome and other ancient States; the sanction given by France, England, Holland and other modern States. In all ages one-half of mankind have been slaves. If the Southern States were let alone they will probably of themselves stop importations. He would himself as a Citizen of South Carolina vote for it. An attempt to take away the right as proposed will produce serious objections to the Constitution which he wished to see adopted.

General Pinckney declared it to be his firm opinion that if himself and all his colleagues were to sign the Constitution and use their personal influence, it would be of no avail towards obtaining the assent of their Constituents. South Carolina and Georgia cannot do without slaves. As to Virginia she will gain by stopping the importations. Her slaves will rise in value, and she has more than she wants. It would be unequal to require South Carolina and Georgia to confederate on such unequal terms. He said the Royal assent before the Revolution had never been refused to South Carolina as to Virginia. He contended that the importation of slaves would be for the interest of the whole Union. The more slaves, the more produce to employ the carrying trade; the more consumption also, and the more of this, the more of revenue for the common treasury. He admitted it to be reasonable that slaves should be dutied like other imports, but should consider a rejection of the clause as an exclusion of South Carolina from the Union.

Abraham Baldwin (Ga.) had conceived national objects alone to be before the Convention, not such as like the present were of a local nature. Georgia was decided on this point. That State has always hitherto supposed a General Government to be the pursuit of the central States who wished to have a vortex for every thing—that her distance would preclude her from equal advantage—and that she could not prudently purchase it by yielding national powers. From this it might be understood in what light she would view an attempt to abridge one of her favorite prerogatives. If left to herself, she may probably put a stop to the evil. As one ground for this conjecture, he took notice of the section of ____ which he said was a respectable class of people, who carried their ethics beyond the mere *equality of men*, extending their humanity to the claims of the whole animal creation.

Mr. Wilson observed that if South Carolina and Georgia were themselves disposed to get rid of the importation of slaves in a short time as had been suggested, they would never refuse to Unite because the importation might be prohibited. As the Section now stands all articles imported are to be taxed. Slaves alone are exempt. This is in fact a bounty on that article.

Mr. Gerry thought we had nothing to do with the conduct of the States as to Slaves, but ought to be careful not to give any sanction to it.

John Dickinson (Del.) considered it as inadmissible on every principle of honor and safety that the importation of slaves should be authorized to the States by the Constitution. The true question was whether the national happiness would be promoted or impeded by the importation, and this question ought to be left to the National Government not to the States particularly interested. If England and France permit slavery, slaves are at the same time excluded from both those Kingdoms. Greece and Rome were made unhappy by their slaves. He could not believe that the Southern States would refuse to confederate on the account apprehended; especially by the General Government.

Hugh Williamson stated the law of North Carolina on the subject, to wit that it did not directly prohibit the importation of slaves. It imposed a duty of £5 on each slave imported from Africa. £10 on each from elsewhere, and £50 on each from a State licensing manumission. He thought the Southern States could not be members of the Union if the clause should be rejected, and that it was wrong to force any thing down, not absolutely necessary, and which any State must disagree to.

Mr. King thought the subject should be considered in a political light only. If two States will not agree to the Constitution as stated on one side, he could affirm with equal belief on the other, that great equal opposition would be experienced from the other States. He remarked on the exemption of slaves from duty whilst every other import was subjected to it, as an inequality that could not fail to strike the commercial sagacity of the Northern and Middle States.

John Langdon (N.H.) was strenuous for giving the power to the General Government. He could not with a good conscience leave it with the States who could then go on with the traffic, without being restrained by the opinions here given that they will themselves cease to import slaves.

General Pinckney thought himself bound to declare candidly that he did not think South Carolina would stop her importations of slaves in any short time, but only stop them occasionally as she now does. He moved to commit the clause that slaves might be made liable to an equal tax with other imports which he thought right and which would remove one difficulty that had been started.

Mr. Rutledge. If the Convention thinks that North Carolina, South Carolina and Georgia will ever agree to the plan, unless their right to import slaves be untouched, the expectation is vain. The people of those states will never be such fools as to give up so important an interest. He was strenuous against striking out the Section, and seconded the motion of General Pinckney for a commitment.

Gouverneur Morris wished the whole subject to be committed including the clauses relating to taxes on exports and to a navigation act. These things may form a bargain among the Northern and Southern States.

Pierce Butler declared that he never would agree to the power of taxing exports.

Mr. Sherman said it was better to let the Southern States import slaves than to part with them, if they made that a sine qua non. He was opposed to a tax on slaves imported as making the matter worse, because it implied they were *property*. He acknowledged that if the power of prohibiting the importation should be given to the General Government that it would be exercised. He thought it would be its duty to exercise the power.

George Read was for the commitment provided the clause concerning taxes should also be committed.

Mr. Sherman observed that that clause had been agreed to and therefore could not be committed.

Edmund Randolph was for committing in order that some middle ground might, if possible, be found. He could never agree to the clause as it stands. He would sooner risk the constitution. He dwelt on the dilemma to which the Convention was exposed. By agreeing to the clause, it would revolt the Quakers, the Methodists, and many others in the States having no slaves. On the other hand, two States might be lost to the Union. Let us then, he said, try the chance of a commitment.

On the question for committing the remaining part of Section 4 and 5 of article: 7. New Hampshire no, Massachusetts absent, Connecticut ay, New Jersey ay, Pennsylvania no, Delaware no, Maryland ay, Virginia ay, North Carolina ay, South Carolina ay, Georgia ay (7 to 3).

AUGUST 24

William Livingston (N.J.), from the Committee of Eleven, to whom were referred the two remaining clauses of the 4th Section and the 5 and 6 Section of the 7th article [of the report of the Committee of Detail] delivered in the following Report:

"Strike out so much of the 4th Section as was referred to the Committee and insert—'The migration or importation of such persons as the several States now existing shall think proper to admit, shall not be prohibited by the Legislature prior to the year 1800, but a tax or duty may be imposed on such migration or importation at a rate not exceeding the average of the duties laid on imports. . . .'"

AUGUST 25

The Report of the Committee of eleven being taken up.

General Pinckney moved to strike out the words "the year 1800" as the year limiting the importation of slaves, and to insert the words "the year 1808."

Nathaniel Gorham seconded the motion.

Mr. Madison. Twenty years will produce all the mischief that can be apprehended from the liberty to import slaves. So long a term will be more dishonorable to the American character than to say nothing about it in the Constitution.

On the motion; which passed in the affirmative.

New Hampshire ay, Massachusetts ay, Connecticut ay, New Jersey no, Pennsylvania no, Delaware no, Maryland ay, Virginia no, North Carolina ay, South Carolina ay, Georgia ay (7 to 4).

Gouverneur Morris was for making the clause read at once, "the importation of slaves into North Carolina, South Carolina and Georgia shall not be prohibited etc." This he said would be most fair and would avoid the ambiguity by which, under the power with regard to naturalization, the liberty reserved to the States might be defeated. He wished it to be known also that this part of the Constitution was a compliance with those States. If the change of language however should be objected to by the members from those States, he should not urge it.

Colonel Mason was not against using the term "slaves" but against naming North Carolina, South Carolina and Georgia, lest it should give offence to the people of those States

Mr. Sherman liked a description better than the terms proposed, which had been declined by the old Congress and were not pleasing to some people.

George Clymer (Pa.) concurred with Mr. Sherman.

Mr. Williamson said that both in opinion and practice he was, against slavery; but thought it more in favor of humanity, from a view of all circumstances, to let in South Carolina and Georgia on those terms, than to exclude them from the Union.

Gouverneur Morris withdrew his motion.

Mr. Dickinson wished the clause to be confined to the States which had not themselves prohibited the importation of slaves, and for that purpose moved to amend the clause so as to read "The importation of slaves into such of the States as shall permit the same shall not be prohibited by the Legislature of the United States until the year 1808"—which was disagreed to nem. con.

The first part of the report was then agreed to, amended as follows.

"The migration or importation of such persons as the several States now existing shall think proper to admit, shall not be prohibited by the Legislature prior to the year 1808."

New Hampshire, Massachusetts, Connecticut, Maryland, North Carolina, South Carolina, and Georgia ay; New Jersey, Pennsylvania, Delaware, and Virginia no (7 to 4).

Mr. Baldwin in order to restrain and more explicitly define "the average duty" moved to strike out of the second part the words "average of the duties laid on imports" and insert "common impost on articles not enumerated" which was agreed to nem. con.

Mr. Sherman was against this second part, as acknowledging men to be property, by taxing them as such under the character of slaves.

Mr. King and Mr. Langdon considered this as the price of the first part.

General Pinckney admitted that it was so.

Colonel Mason. Not to tax, will be equivalent to a bounty on the importation of slaves.

Mr. Gorham thought that Mr. Sherman should consider the duty, not as implying that slaves are property, but as a discouragement to the importation of them.

Gouverneur Morris remarked that as the clause now stands it implies that the Legislature may tax freemen imported.

Mr. Sherman in answer to Mr. Gorham observed that the smallness of the duty showed revenue to be the object, not the discouragement of the importation.

Mr. Madison thought it wrong to admit in the Constitution the idea that there could be property in men. The reason of duties did not hold, as slaves are not like merchandise, consumed, etc.

Colonel Mason (in answer to Gouverneur Morris) the provision as it stands was necessary for the case of Convicts in order to prevent the introduction of them.

It was finally agreed nem. con. to make the clause read "but a tax or duty may be imposed on such importation not exceeding ten dollars for each person," and then the second part as amended was agreed to.

SEPTEMBER 10

[When the Convention finished its consideration of the method of amending the Constitution, John Rutledge proposed an additional clause prohibiting an amendment altering the prohibition of Congress' ban on the foreign slave trade before 1808 or setting a limit on the amount of tax to be laid on slave importations.]

Mr. Rutledge said he never could agree to give a power by which the articles relating to slaves might be altered by the States not interested in that property and prejudiced against it. In order to obviate this objection, these words were added to the proposition: "provided that no amendments which may be made prior to the year 1808, shall in any manner affect the 4 and 5 sections of the VII article."

On the vote, New Hampshire divided, Massachusetts ay, Connecticut ay, New Jersey ay, Pennsylvania ay, Delaware no, Maryland ay, Virginia ay, North Carolina ay, South Carolina ay, Georgia ay (9 to 1, with 1 divided).

The Fugitive Slave Clause

Article IV, section 2. No Person held to Service or Labour in one State, under the Laws thereof, escaping into another, shall, in Consequence of any Law or Regulation therein, be discharged from such Service or Labour, but shall be delivered up on Claim of the Party to whom such Service or Labour may be due.

August 28

[Report of the Committee of Detail considered on the extradition of criminals from one state to another state.]

Pierce Butler and *Charles Pinckney* moved "to require fugitive slaves and servants to be delivered up like criminals."

James Wilson. This would oblige the Executive of the State to do it at the public expence.

Roger Sherman saw no more propriety in the public seizing and surrendering a slave or servant, than a horse.

Mr. Butler withdrew his proposition in order that some particular provision might be made apart from this article.

August 29

Mr. Butler moved to insert after article XV [in the Committee of Detail report as amended on 28 August] "If any person bound to service or labor in any of the United States shall escape into another State, he or she shall not be discharged from such service or labor, in consequence of any regulations subsisting in the State to which they escape, but shall be delivered up to the person justly claiming their service or labor," which was agreed to nem. con.

3

New England Debates Slavery and the Constitution

NEW ENGLAND DELEGATES to the *Constitutional Convention readily accepted constitutional provisions that safeguarded slavery and enhanced the position of the Southern states in the new federal Union. In exchange for these concessions, the delegates from the southernmost states made commercial concessions to New England. New England delegates divorced questions about slavery from morality. This willingness of New England delegates to tolerate, condone, and even reward slaveholding was not shared by many of their constituents during the debate over the ratification of the Constitution.*

Perhaps the most conscientious opponents of the Constitution in New England were Quakers. Most influential in Rhode Island and Massachusetts, Quakers strongly objected to the slave provisions of the Constitution. William Rotch of Nantucket felt that the "very foundation [of the Constitution] was on Slavery & Blood." Not only did Quakers deplore the barbarities of the African slave trade, which the Constitution allowed to continue until 1808, but they particularly were distressed by the loss of Massachusetts as an asylum for runaway slaves. Because of the Constitution's fugitive-slave clause, it seemed that runaway slaves would no longer be safe even in Massachusetts where the state supreme court had ruled that slavery was illegal.

The fugitive-slave clause had serious effects even before the Constitution was ratified. Cato, a forty-year-old runaway slave, had lived and worked in Nantucket for two years after his escape from his master John Slocum in Newport, Rhode Island. After reading the fugitive-slave clause in the proposed Constitution, Slocum traveled to Nantucket and threatened Cato with legal action once the Constitution went into effect. Fearing lifelong slavery and unwilling to leave friends, family, and his job for

Canada, Cato agreed to go back into slavery for one year. Many other runaways felt compelled to agree "with their masters to Serve a Certain time and then take manumissions." Such a provision, according to Moses Brown, "Wounded the Cause of Liberty & the Rights of Men."

Other New England Antifederalists, particularly those in Massachusetts, objected to the three-fifths clause, which allotted representation in the House of Representatives and apportioned direct taxation on the basis of population, counting three-fifths of the slaves in these computations. Slaves were property—not agents of free will—and as such should not be considered in representation, "which is the corner stone of a free government." The Constitution was a covenant that would degrade freemen to the level of slaves; and, because Massachusetts had no slaves and other New England states had few, the massive numbers of slaves in the Southern states would disproportionately weigh against the influence of New England in the U.S. House of Representatives. Furthermore, since Federalists argued that direct taxation would not be necessary except in crises, Southern states faced no imminent tax burden under the three-fifths clause. Thus New England would "receive no kind of benefit" from the three-fifths clause. In fact New Englanders "shall have committed ourselves to the mercy of the states having slaves, without any consideration whatever. Indeed, should direct taxes be necessary, shall we not by increasing the representation of those States, put it in their power to prevent the levying such taxes, and thus defeat our own purposes? Certainly we shall, and having given up a substantial and essential right [in representation], shall in lieu of it, have a mere visionary advantage" in levying direct taxes.

New England Federalists countered the attack on the three-fifths clause by suggesting that slaves were not to be represented nor were slaveowners themselves to receive extra representation. "The representation is given to the State," which then by a provision of the Constitution is required to allow the electors of the most numerous branch of the state legislature to vote for representatives. Population was merely a convenient method for determining the relative strength of the states. Eleven states had previously ratified an amendment to the Articles of Confederation that would apportion federal expenses based on population counting three-fifths of the slaves. According to Federalists, "it was the language of America," and as such the Confederation Congress implemented this change in its 1786 requisition on the states.

Perhaps the most emotional debate in New England over slavery and the Constitution focused on the provision that prohibited Congress from closing the African slave trade before 1808. Federalists pointed out that, despite the twenty-year restriction on Congress, the states themselves faced no limitation whatsoever. All previously enacted state prohibitions against the African slave trade remained in force and new state prohibitions could be enacted at any time. Federalists argued that the Constitution was a considerable improvement over the Articles of Confederation, which gave Congress no authority to restrict or prohibit the African slave trade; the new Constitution provided "a dawn of hope for the final abolition of the horrid Traffick." In fact, said Federalists, the slave-trade clause was "a great Point gained of the southern states." Some Federalists went so far as to predict that this provision would naturally lead to the gradual and eventual total abolition of slavery. "The friends to liberty and humanity, may look forward with satisfaction to the period, when slavery shall not exist in the United States; while the enlightened patriot will approve of the system, which renders its abolition gradual."

Opponents of the Constitution could not see any justification for two more decades "of that most cursed of all trades." In fact, they said, the Constitution did not require the closing of the foreign slave trade in 1808. "It is wholly optional with the Congress." If the Constitutional Convention could not prohibit such an abominable business, why would Congress do so when the slave states had further strengthened their position in Congress after twenty more years of importing slaves, while increased slave populations were represented in Congress under the three-fifths clause? It would have been "much more glorious" for America if the slave trade had been abolished at "the first moment the constitution should be established."

New England Antifederalists vehemently proclaimed the "doctrine of imputed guilt in civil society." "Whether we go ourselves to Africa to procure slaves, or employ others to do it for us, or purchase them at any rate of others, it matters not a whit. It is an old saying and a true one, 'The partaker is as bad as the thief.'" "We become consenters to and partakers in the sin and guilt of this abominable traffic." New Englanders should not feel responsible for righting the evils of the entire world, but they ought not by their actions to prolong the slave trade a day longer than absolutely necessary. "We do not esteem ourselves under any necessity to go to Spain or Italy to suppress the Inquisition of those countries, nor of making a journey to the Carolinas to abolish the detestable custom of enslaving the Africans: but, sir, we will not lend the aid of our ratification to this cruel and inhuman merchandise, not even for a day." There was a qualitative distinction between tolerating existing iniquities and becoming the guarantors of such evils. In framing a constitution, men must take responsibility for their actions. "Can we who have fought so hard for Liberty give our consent to have it taken away from others?" Quaker minister James Neal in the Massachusetts ratifying convention went so far as to suggest that free Americans should suffer the same fate as those we enslave—"we should lose our Liberties as the Africans lost theirs." The only solution was to expunge the entire provision or "we ought not to vote to give life to a constitution, which at its first breath will be branded with eternal infamy, by having a stamp of slavery and oppression upon it."

Federalists disagreed with the doctrine of imputed guilt. Such a doctrine would force Massachusetts, for instance, to disconnect itself from every state in the Union because all, except for Massachusetts, still allowed slavery. Some states had passed gradual emancipation acts; others had not. "Each State is sovereign and independent to a certain degree, and they have a right, and will regulate their own internal affairs, as to themselves appears proper; and shall we refuse to eat, or to drink, or to be united, with those who do not think, or act, just as we do, surely not. We are not in this case partakers of other men's sins, for in nothing do we voluntarily encourage the slavery of our fellow men."

Federalists readily acknowledged the imperfection of the Constitution. Concessions had to be granted to assure the permanence of the Union and the stability and

well being of the economy. After the ratification and implementation of the Constitu-
tion, it would be time to remedy society's other evils, including slavery. As much as
New England Federalists may have detested slavery, they desired Union more. If the
continuation of the slave trade for twenty years—and the continuation of slavery for
an indefinite time—was the price they had to pay for an invigorated Union under the
Constitution, so be it.

MOSES BROWN TO JAMES PEMBERTON
Providence, October 17, 1787[1]

Thy favour of the 19th Ulo. came duly to hand and the Contents observ'd.—The subject of Slavery now calls my particular attention, and it is with Sattisfaction I observe what thou mentions of the care and attention of friends [i.e., Quakers] & others in England to revive before Parliament the Iniquity & Impolacy of the slave Trade, but it is with feelings very reverse I observe in the Proceedings of the Federal Convention Two Articles which according to the Construction of friends here Millitate against Our Testimony in Support of Liberty, or against Slavery.—It is with Reluctance I say any thing against the doings of that respectable Body, but observing in the publick papers that friends in your City [Philadelphia] with others are United in Approving the Constitution, and apprehending under Our Construction of it, friends will not be Clear without bearing Testimony against those parts which give Countenance to if not directly Incourage Slavery, I thought best to Write thee on the subject.—It appears Necessary that friends as far as possible be United on such parts as Effects their Testimony. I may therefore without further Appology for Touching on a subject which is so Intimately Connected with publick affairs, inform thee how we are Affected, that if we misconceive the Intention we may by being better Informed be released from Our present Uneasiness. I say Our, as there is no friend I have Converes'd with on the subject but what has been disagreeably affected. On my reading the Doings of the Convention, the 3d paragraph of the 2d Secn of the 4th Article Sensibly affected me with an apprehension that it was designd to Distroy the Present Assylum of the Massachusets from being as a City of Refuge for the poor Blacks, many of whom had resorted there on Acct of their Constitution or Bill of rights declaring in the first Article "That all men are born free & Equal &c," and there being no Laws in that State to support slavery, the Negroes on Entering that state are as free as they are on Entering into Great Brittain and the southern people have not been able by Applycation to the

---

[1]Pemberton Papers, Historical Society of Pennsylvania. In 1773 Moses Brown converted from being a Baptist to a Quaker and spent the rest of his life ameliorating the condition of slaves and freedmen. Pemberton, a wealthy Quaker Philaldelphia merchant, was vice president of the Pennsylvania Society for Promoting the Abolition of Slavery.

Governour, Judges or other Authority to Recover those they had held as Slaves who chose to Stay there. I have Considerd it a great favour to that people, and an Opening in Wisdom in that State, for the Exaltation of Truth, Testimony over the Oppressors of the African race; the Strikeing at which in that dark covert way it seems to be in Struck me with great Disapprobation and a fear that Light & Truth was not so prevalent in that Body as I had hoped for, indeed I thought it an Indignity, or a violation of Right, accompanyd with Insult on the great Principle of that first Article of the Massachusets [constitution] which had been rattified by the Declaration of Congress and other States, and Contrary to the Divine Law Express'd in Deuterony 23d, 15V 16 from which Grenville Sharp[2] Drew an Argument correspondent with the Law of the Land, which I think conclusive on the point, to have an Article in the Constitution of these States so Repugnant to the Principles of Liberty, Truth & Righteousness Afflicted me, when those respectable Characters compare this Article under Our Construction with their own Declaration in the Preamble Expressive of the Design & End of the Confederacy Viz "to Establish Justice Secure the Blessings of Liberty to Ourselves & our posterity" they may Easily See the Contrariety if not inconsistancy. But the Poor Devoted Africans seems in the 9th. Secn. of the 1st Article, tho the subjects of Unrighteous Revenue to be left out of the power of Congress to Consider them as Men & so Entituled to Liberty and their protection, nor yet are they yielded by the States to Congress as Commerce, but Left to the Averice or Oppression of the Subjects of any State, with the concurrance of the Convention, the Usage of Importing them being so far Acknoledged by them as a right, that the Constitution is not to Admit of being mended in that particular, by which Countenance or Establishment of Slavery for 21 years, the Incouragement of a Reformation is obstructed and the states may fall back from their present Light into great Darkness on this Subject, and the Recovery from this Gross Evil, for which this Land Mourns be long Obstructed.—We having no member from this State at the convention,[3] I have not heard how this subject was Treated, whether it was slid over as a matter of Little Consequence or Insisted on by the southern members against the Wish and Inclination of the Middle & Northern Ones, perhaps full information how the subject was Treated and the Article are intended to be Understood might remove some of Our feelings on this Ocation. It seems to Exhibit a poor Example of Confidence in Congress the Southern states being not wiling to Leave the

---

[2]Granville Sharp was an English supporter of American rights before and during the Revolution. He strenuously worked for the abolition of the slave trade and the emancipation of slaves. Through lawsuits he was able in 1772 to win recognition for the principle that as soon as a slave set foot on English soil he was free. In 1787 he helped found the English society for the abolition of slavery.

[3]Rhode Island was the only state that refused to send delegates to the Constitutional Convention.

Commerce in men under their Controul and Regulation as well as Other Matters, had this been done and nothing more said about it. Nor no infringment on the Constitution of the Massachusets in this Respect I should have been very Easy to have Trusted Congress with it, and as it affects the Essential Rights of Men, those States or men that could not be prevailed on to have Confided in Congress, Congress nor the other States, methinks can have little reason to Confide in them, had the period of 21 years been fixed for Abolishing Slavery as some Writers your way seems to represent, it would have been doing something, but that will be then to do with probably Less Ability to Effect it, if it be not now held up to be a sore Evil as it now stands and if possible to be Amended. When I consider us in New England calld upon once a Quarter and, to answer Conscientiously that we bear a faithful Testimony against Slavery, I cannot see how we can approve those Articles. I was & am very Sorry we have this renew'd Tryal for to me it is a pretty Close one to be Considerd as Oppos'd to that Constitution which is intended as a Reformation of the Govnt of these states, as I am Sensible there is Need of it. I know not that I should Object to any Other part, 'tho I think I can see wherein Friends may be Affected, and it behoves us to Act Wisely in this matter, to bear Our Testimony faithfully wherein that is Affected, Trusting in the protection of Divine Providence more than in this or Any Change of Government, Remembering Stephen Crisps Saying "Take heed of that part in you which Trusts & Relies upon any Sort of the men of this World"[4]— Inclosd I send a late publication on the African Trade Written by a presbeterian minister S. H. in Newport,[5] and the Testimony of the Baptists at a late assosiation.[6] As I expect it will be agreable to friends here to hear from friends with you, on this subject I am free thou should Shew this to Such Discreet friends for their Advice, as thou mayst think propper.

PROVIDENCE *UNITED STATES CHRONICLE*, OCTOBER 18, 1787[7]

It is with Satisfaction we communicate the 12th Article of the Minutes of the Baptist, or Warren Association, at their yearly Meeting, held at Chelmsford, last Month, composed of 45 Churches, to our Readers, as follows:—

[4]An English Quaker preacher, Crisp made this statement in *An Epistle to Friends Concerning the Present and Succeeding Times*, first printed in London in 1666 and reprinted in Philadelphia in 1780.

[5]Samuel Hopkins' "Crito" was published in the *Providence Gazette* on October 6 and 13. It described the slave trade as "a *national sin*." Brown distributed fifty copies of Crito to members of the Rhode Island legislature which on October 31 debated and enacted a law prohibiting the slave trade. Hopkins had tried to get Crito published in the *Newport Herald* in August, but the printer refused, fearing that the harsh criticism of Newport slave traders might offend customers.

[6]See the report of the Warren Baptist Association immediately below.

[7]Reprinted in the Boston *American Herald* on October 22.

"Notwithstanding the great Expense of Blood and Treasure during the last War, to ward off Slavery from ourselves, we are informed, That in various Parts of this Country, many have recurred to the horrid Practice of sending our Shipping to Africa, to bring from thence the Natives, and to sell them as Slaves in the West-Indies: And as *Man-stealing* is a capital Crime by the Laws of God—See Deut. vii. I. Tim. i 10.—We, therefore, earnestly desire all our Brethren, to guard against giving the least Countenance to that Heaven-daring Wickedness."

SAMUEL HOPKINS TO MOSES BROWN
Newport, October 22, 1787[8]

I am hurt by the doings of the convention respecting the *Slave Trade*. It is as you suppose. They have carefully secured the practice of it in these States for 20 years, and prevented any Asylum for slaves during that term, unless every individual State, should suppress this trade. They have taken it out of the hands of Congress. We cannot determine that the major part of the delegates were pleased with this. Some of the southern delegates no doubt, insisted upon it that the introduction of slaves should be secured, and obstinately refused to consent to any constitution, which did not secure it. The others therefore consented, rather than have no constitution, or one in which the delegates should not be unanimous. I fear that is an *Achan*, which will bring a curse, so that we cannot prosper.[9] At the same time it appears to me that if this constitution be not adopted by the States, as it now stands, we shall have none, and nothing but anarchy and confusion can be expected.—I must leave it with the Supreme Ruler of the universe, who will do right, and knows what to do with these States, to answer his own infinitely wise purposes; and will vindicate the oppressed, and break the arm of the oppressor in his own way and time; and cause the wrath of man to praise Him.

It has been objected by some of the ministers against prefering a memorial to the General Assembly respecting the Slave trade; That the present ruling part in the Assembly, have appeared to be so destitute of all principles of justice, or regard to it; and have acted such an iniquitous part, that there is an impropriety in applying to *them* for justice; especially for the ministers of the Gospel to do it, whom they hold in the highest contempt, and would embrace any opportunity to pour contempt upon them, which we should give them by laying such a petition before them. This prevents any thing of that kind being done at present.

---

[8]Moses Brown Papers, Rhode Island Historical Society. Hopkins was pastor of the First Congregational Church in Newport. He had published several essays attacking slavery and the slave trade.

[9]Achan's actions brought the wrath of God upon the people of Israel (Joshua 7).

WILLIAM ROTCH, SR., TO MOSES BROWN
Nantucket, November 8, 1787[10]

Thou queries how friends can be active in establishing the new form of Government, which so much favours Slavery; alass in this point I must refer thee to some advocate for it, as to my own part my heart has been often pained since the publication of the doings of the Convention; and much disappointed I am as I had entertaind some hope that so many wise men, would have form'd some System of Government, founded on equity & justice, that thereby it might have acquir'd some strength and energy, and that it might be on such a basis that we as a Society might lend our aid in establishing it so far as it tended to peace and morality; but we may say in truth that the wisdom of man (as man) can or shall not work the Righteousness of God; and whatever high encomiums are given to it (the Constitution) it is evident to me it is founded on *Slavery* and that is on *Blood*, because I understand, some of the Southern members utterly refused doing any thing unless this horid part was admitted, which occasions me to say its very foundation was on Slavery & Blood, as that I suppose was the corner stone; there are many parts which meets my approbation, as it so nearly approaches a mix'd Monarchy, wch. I think a good exchange for an Ideal Republican; but the valuable parts are all marred to a great degree in my view, and often brings me into a deep consideration of the consequence that must eventually attend; can we expect additional judgments will not vissit our land when the people have given their own late declarations the lie in so bare faced a manner, some complaint has been made, that no Bill of Rights was prefixed to their doings; but how was it possible that it could be done, for what are those Rights, except the very Rights in part, and the most valuable part which they have declard they would not protect; and I much fear it will be taken for an implicit encouragement, to pursue the trade, though I sincerely wish the Northern States may manifest a better principle of justice than the united Wisdom of the Continent has exampled them in, & as thou observes it breaks up this Assylum of liberty (the Massachusets) thus my dear Friend thou may see I can by no means alleviate thy mind in this respect, except in being united with thee as a fellow sufferer in this great cause of oppression, interceding when a little strength is afforded that the Father of Mercies may in his own time open the hearts of the people and incline them to true justice and judgment, and grant that the bonds of the oppressed may be loosed, and indeed notwithstanding, I fear that for a time, an encouragement may be the consequence of that article; yet as the work is on the wheels I fully believe it is the determination of Heaven that Slavery shall be abolished, though it may be through some sore judgments; I am much affected at this time with this Subject from a recent Instance, though it may

[10]Austin Collection, Moses Brown School, Brown University. Rotch, a Quaker, was a wealthy whale merchant in Nantucket, Mass.

be view'd by some as of the mildest kind, yet the depravity of the heart that is guilty of the fact sinks deep in my mind; Jno Slocum of Newport[11] has lately been here and demanded Cato as his Servant, Cato left Slocum and made this place his Assylum for liberty, was hir'd by my Son S Rodman[12] with whom he had lived near two years, and on considering, that if the new Constitution should be adopted (which was hardly doubted) he would then be fully within his power, where he had nothing to expect, but perpetual Slavery, in this circumstance he was advisd to compromise the matter with Slocum, which he did & agreed to serve him a year & then to be free, a manumission on these conditions was obtaind, but Slocums proposal was to lodge it in the hands of one of his own children, but this base intimation was rejected by Catoes friends, & S Rodman has it in possession; Cato is a very honest orderly man set out yesterday on his new years Slavery, he is I believe near forty years old, and altho it is but for a year, yet the darkness of that heart that requires it, gives me so bad an opinion of this man, that I think in this enlightend age, he ought not to lay claim to any great degree of Christianity.—I doubt I shall tire thee, I will therefore come to thy proposition, my heart is warm'd toward those poor blacks and I feel sometimes willing to spend and be Spent if I could contribute to their enlargement; I intend for Bedford if well in a week or ten days, when its not improbable I may meet thee somewhere, but perhaps it may be in Newport, as I hardly feel clear of J Slocum without either writing or seeing him, the later I would prefer if [I] could have thy company.

Moses Brown to James Thornton, Sr.
Providence, November 13, 1787[13]

I shall . . . touch on An Other [subject] where we seem to be happily United; that is the Stoping of the African Slave Trade the Effect of Our Applycation to the Assembly of this state having been Sattisfactory I Inclose thee a Copy of Our Address & the Act Obtaind thereupon[14] which I had struck off for my friend, doubtless thou hast heard by the friend from your way of the Applycation to the Massachusets allso. On my getting home from Our Assembly the 1st. of this mo. I sent off a Copy of the Act with a Letter to their Committee, no friend on the Committee being ready to accompany me

[11]Probably John Slocum, a Newport merchant, whose household included three blacks in 1774 and one in 1790.

[12]Samuel Rodman, a Quaker, was Rotch's business partner and son-in-law.

[13]Quaker Collection, Haverford College, Haverford, Pa. Thornton, of Byberry, Pa., had emigrated to America from England in 1750. He was a farmer and a highly esteemed Quaker minister.

[14]Stimulated by a Quaker petition, the legislature in October 1787 passed a law prohibiting any Rhode Islander from engaging in the slave trade.

or I should have gone down to Boston again on the subject, I am in hopes of hearing dayly of their doing something to the like purpose, I may now mention my desire you may be favourd to bring about an Act similar in Pennsylvania Seeing there is now no hopes from Congress, the present not being Competent and the new if it Takes place have Bard that Door of hope for 21 years & I fear from that Concession much Longer, indeed their doings on this subject aspecialy the 3d paragraph of the 2d sectn. of the 4th Article appears Calculated on purpose 'tho, Plauseably Coverd, to distroy the present Effect of the 1st Article of the Massachusets Bill of Rights by which all Negroes when in that Jurisdiction are Declared free, as well and on the same ground as in England and no Law there can support a Claimmer in Carrying One Out of that Assylum or City of Refuge which it has been to many, many Others have agreed with their masters to Serve a Certain time and then take manumisions, by means of this, their Retreat from the Injuries of Slavery, but alas instead of Extending Humanity and good Will to that People the Convention has, I think very Unhapily Wounded the Cause of Liberty & the rights of Men, the Justice of Such an Assylum is supported by the Divine Law Deut 23. 15–16 which Grenvill Sharp has Adopted in an Argument proving the Right of protection & of Protectors to slaves who Escape from their masters, which I could Wish had been laid before that Assembly as it appears to me Conclusive on the point, tho I must Confess till I saw it I had some doubts whether it was best for friends to protect them 'tho this doubt arose from a fear of Blame and so hurting Our Testimony, not from the Right they had to Take their Liberty when Ever they Could, I mentiond this matter in a Letter sometime since to James Pemberton & Queryd how We could Unite in the present federal system, & answer Our Query "Whether we bare a faithfull Testimony against slavery" Once a Quarter I Wish for the Cause of Humanity, Justice, Liberty & Religion that this Usurpation Over the Massachusets Constitution

[The remainder of the letter is missing.]

WILLIAM SYMMES, JR., TO PETER OSGOOD, JR.
Andover, Mass., November 15, 1787[15]

According to my promise I sit down to sketch out my reasons for objecting against the Federal Constitution. The essay will doubtless be imperfect; but I design it for your perusal only, & I can safely rely upon your goodness for all necessary allowances,—

[15]Willis Papers, Maine Historical Society. Symmes, a twenty-seven-year-old Harvard graduate, was a lawyer. He had been elected to the Massachusetts ratifying Convention as an Antifederalist, but became convinced that the Constitution should be adopted. He voted to ratify the Constitution and "In so doing, I stand acquitted to my own conscience, I hope and trust I shall to my constituents." Osgood, a merchant, represented the town of Andover in the Massachusetts House of Representatives.

I will consider the objectionable passages in course as they occur in the System, as well for your convenience as my own.

1.—The apportionment of taxes.

It appears to me that this will operate unequally against the northern States. Let us suppose that two fifths of the slaves in the five southern states amount at least to 150,000 persons. What reason can be given why, if taxes must be proportioned by population only, this should be rejected?—The profits of their labour are nothing? I deny the fact; for I believe that every negro that cultivates the valuable staples, Tobacco, Wheat, Rice, Indigo, &c. raises a greater profit to his master than any white can raise from his labour here.— What then?—That the southern Nabobs squander it all in Luxuries, & so the States there are made if anything, poorer?—Very good—The Convention then have patronized Luxury, & taxed Industry & Oeconomy. But three fifths include all the working slaves. Neither will this answer; for the northern States are taxed as much for an infant or a decrepit old man, as fore a vigorous youth.

How then shall we be taxed? I say not in proportion to actual wealth at present, but in proportion to a State's advantages for acquiring wealth. The soil & climate of Virginia are better than those of this State—The staples of Virginia are in high demand—Its Rivers the finest in the world. How rich might Virginia be!—But Virginia is not rich—What then?—Shall a man need no better excuse from taxes than Idleness? He will hardly pay his *private* debts so easily.—

Taxes must certainly grow out of the ground. What then is more evident than that the best land & the best produce (supposing the advantages of Commerce to be equal) should pay or (if you please, produce) the most?— And are not our long winters in which we consume the labours of the summer, to be considered? No-But yes, I beg pardon—they *are* considered—We pay the more.—

LANDHOLDER VI (OLIVER ELLSWORTH)
Connecticut Courant, December 10, 1787[16]

[Answer to George Mason's Objections]: *The general legislature is restrained from prohibiting the further importation of slaves for twenty odd years*—

But every state legislature may restrain its own subjects; but if they should not, shall we refuse to confederate with them? their consciences are there

---

[16]This essay was also printed in the Hartford *American Mercury* on December 10 and reprinted twenty-one times within two months from New Hampshire to South Carolina. Ellsworth had been a delegate to the Constitutional Convention. He voted to adopt the Constitution in the Connecticut ratifying convention and was elected one of the state's first U.S. senators. He served as Chief Justice of the U.S. from 1796 to 1800. For George Mason's objections, see Chapter 5.

own, tho' their wealth and strength are blended with ours. Mr. Mason has himself about three hundred slaves, and lives in Virginia, where it is found by prudent management they can breed and raise slaves faster than they want them for their own use, and could supply the deficiency in Georgia and South-Carolina; and perhaps Col. Mason may suppose it more humane to breed than import slaves—those imported having been bred and born free, may not so tamely bear slavery as those born slaves, and from their infancy inured to it; but his objections are not on the side of freedom, nor in compassion to the human race who are slaves, but that *such importation render the United States weaker, more vulnerable, and less capable of defence.* To this I readily agree, and all good men wish the entire abolition of slavery, as soon as it can take place with safety to the public, and for the lasting good of the present wretched race of slaves. The only possible step that could be taken towards it by the convention was to fix a period after which they should not be imported.

CORNELIUS
Springfield *Hampshire Chronicle*, December 18, 1787

In the case of direct taxation, the rule of apportionment among the several States, I take to be very unequal, and in its operation, will prove exceedingly injurious to the Eastern States. These States, compared with the Southern, have always abounded in people more than in wealth; and from the nature of their climate and soil, will forever continue to do the same. Yet, by this rule of apportionment, a great allowance is made in favour of the Southern States: Three free persons, including those bound to service for a term of years, being reckoned equal to five slaves. In the Eastern States there are but few slaves; in Massachusetts, there are none. There are in that State, a large number of negroes; and in apportioning the taxes, three of these negroes are to be reckoned equal to five in the Southern States.

**THOMAS B. WAIT TO GEORGE THATCHER
Portland, Maine, January 8, 1788**[17]

There is a certain darkness, duplicity and studied ambiguity of expression runing thro' the whole Constitution which renders a Bill of Rights peculiarly necessary.—As it now stands but very few individuals do, or ever will understand it.—Consequently, Congress will be its own *interpreter*—The article respecting taxation and representation is neither more or less than a

---

[17]Thatcher Papers, Chamberlain Collection, Boston Public Library. Wait was the owner and printer of the Portland *Cumberland Gazette*, Maine's only newspaper. Thatcher, a Biddeford, Maine, lawyer, was a delegate to Congress.

puzling Cap; and you, my friend, had the pleasure of *wearing* it, at my office, an hour or two—and then pulled it off, *just as wise* as when you put it on.— But you will now perhaps tell me that you can explain it entirely to my satisfaction—possibly you can; but that may not happen completely to satisfy Congress—if it should not, why they will put a different one,—one that may not satisfy *either you or me*—But Some persons have *guessed* the meaning to be this—that *taxation and representation should be in proportion to all the freemen and slaves in each state-counting five of the latter to three of the former*—If these were the ideas of the Convention, what a strange collection of words do we find in the Constitution to express them!—Who, in the name of God, but the *majority* of that honl. body, would ever have tho't of expressing like ideas in like words!—But bad as may be the *mode* of *expression*, the *ideas*, in my opinion, are worse—

By this *interpretation* the article in question is an egregious imposition on the northern states—Tell me, if you can, why a southern *negro*, in his present debased condition, is any more intitled to representation, than a northern *Bullock?*—Both are mere pieces of property—and nothing more! The latter is equally a *free agent* with the former.—

SALEM MERCURY, JANUARY 8, 1788

The Hon. Richard Henry Lee, Esq. of Virginia, who has written so much about the danger of losing our LIBERTY by the adoption of the New Constitution, is the master of several hundred slaves.

MARK ANTONY
Boston *Independent Chronicle*, January 10, 1788[18]

"Here under leave of BRUTUS, *and the rest,* (For Brutus is
an honourable man; So are they all, all honourable men,)
Come I to speak."*————Julius Caesar.

Mess'rs, Adams and Nourse: Among the various artifices of those who are opposed to the federal constitution or to any efficient plan of government, none is more natural, and perhaps none more successful, than to excite a jealousy between the inhabitants of the several States. Upon this plan the suggestions of Brutus, whose speculations have appeared in your paper, appear frequently to be founded. I particularly allude, at present, to his third

---

[18] Mark Antony is a response to Brutus III, *New York Journal*, November 15, which was reprinted in the Boston *Independent Chronicle* on December 13. For the text of Brutus III, see Chapter 4 (below).

number, in which he objects to the mode of representation proposed in the new constitution, expatiating largely upon an idea, at which the feelings of Freemen must reluct, that the system of slavery in the southern States, is patronized and encouraged by the proposed mode of representation. Such misrepresentations ought surely to be exposed. With many it is needless; but those who have perceived the futility of his observations, will excuse me for the sake of those who have not.

The equal voice of unequal States in Congress, is a well grounded objection of long standing, to the present Confederation.[19] A new system therefore, which should not remedy this defect, would most surely be rejected by the larger States; while the minor States would feel alarmed at the reform, unless they were secured, by a provision in some measure conformable to the spirit of the confederation. In this dilemma, originating from opposite interests, human wisdom, perhaps, could not devise a happier expedient than the new frame of government proposes. A House of Representatives chosen by the States, according to their respective numbers, gives weight to the larger States, in some measure, proportioned to their magnitude; while the small States, are secured from the danger or the apprehension of being overborn by their more powerful confederates, by an equal voice in the Senate.

In this new construction of a House of Representatives, the number, which each State shall send, becomes an interesting question. The general answer is easy, that it ought to be in proportion to the supplies furnished to the public chest. This equitable rule has become a political maxim, which *Brutus* himself enforces when it suits his convenience. The next enquiry is, by what rule taxes shall be proportioned, which when ascertained, is also the rule of representation. The mode prescribed in the confederation, has long been esteemed injudicious and impracticable. It has not hitherto governed the Continental requisition; and if executed and observed, would probably prove unsatisfactory and unjust. Those who are acquainted with State *Valuations*, will concur with the above conclusions, especially when they calculate its operation among thirteen different powers. Some other rule of apportionment became indispensible. The number of inhabitants in each State, has obtained the preference to any other system: And for the combined advantages of simplicity, certainty, facility and equity, none probably can be found more eligible. Here a difficulty arises, with respect to the slaves in the southern States, nor would the difficulty be lessened probably if they were infranchised. Five of them are computed to be equal to three freemen. Their comparative value cannot be demonstrated; but it is acknowledged that they are not equal to free persons, in an estimation of this nature; and the ratio established, being the result of compromise, the presumption is in favour of its propriety.

[19]Under the Articles of Confederation, each state was to be represented in Congress by a minimum of two and a maximum of seven delegates. Despite the size of the delegation, however, each state had only one vote.

This connected system of representation and taxation is thus expressed in the proposed Constitution: "Representatives and direct taxes shall be proportioned among the several States, which may be included within this Union, according to their respective numbers, which shall be determined by adding to the whole number of free persons, including those bound to service for a term of years, and excluding Indians not taxed, three-fifths of all other persons." Few sentences I believe of equal import, can be produced, so perspicuous and comprehensive. But the critical *Brutus* is offended with its construction, and, after suggesting its ambiguity, proceeds to give us an abridged sense of it in his own superior style. "What a strange and unnecessary accumulation of words (says he) are here used, to conceal from the public eye, what might have been expressed in the following concise manner—*Representatives are to be proportioned among the States respectively, according to the number of freemen and slaves inhabiting them, counting five slaves for three freemen?*" The charge of studied concealment, which *Brutus* so illiberally suggests, is applicable only to himself. It frequently happens that precision is lost in conciseness; but *Brutus* has sacrificed the truth. The careful reader will observe, that the article under consideration proportions representatives and taxes according to numbers. But the pretended abridgment fabricated by *Brutus*, mentions Representatives only. The difference is material. In the constitution, it is a fair and equitable establishment: As represented by *Brutus*, one essential branch is omitted, upon which its consistency depended, and being thus mutilated it has been exposed to his objections. This artifice has in some measure succeeded to his wishes, for some have been misled by his suggestions. But, my countrymen, he deceives and abuses you. For what has *Brutus* attacked? Is it the alteration of the rule of apportioning taxes, from a valuation of property to numbers? No.—His reason dictated, and probably his own experience in public affairs, demonstrated the propriety of the measure. Does he disapprove of the ratio between free persons and other persons, in this great political estimate? Upon this matter he is silent. Does he condemn the proportion of Representatives to taxes? No.—In the same performance, when cavilling against the Senate he observes, "on every principle of equity and propriety, representation in a government, should be in exact proportion to the number of aids, afforded by the persons represented—The same principle applies to States, as to individuals in this respect; and if *Brutus* had been honest, his abilities would have saved him from inconsistency. The number of persons, whatever their condition or degree, being reckoned to any State, to render it chargeable in the public contributions, in the same proportion ought that State to be represented. As the slaves are not personally chargeable with taxes, so they are not concerned in representation. But says *Brutus*, "Why is the number of members in the assembly, to be encreased on their account? Is it because in some of the States, a considerable part of the property of the inhabitants, consists in a number of their fellow men, who are held in bondage, in defiance of every idea of benevolence, justice, and religion, and contrary to all the principles of liberty, which have been publickly avowed, in the late glorious revolution? If this be a just ground of representation, the horses in some of the States, and the oxen in others, ought to be represented: for a great share of property in some of them consists in these animals." &c. Is the man really misled, or does he only attempt to mislead others, and to avail himself of our strong disapprobation of slavery? The practice of slavery among our confederates ought to be regreted by us, but it is evidently beyond our controul. Do we in fact countenance or give encouragement to it, by consenting to this rule of apportionment, more than we should by concurring with another? Suppose, for instance, Representatives and direct taxes were to be apportioned by a valuation, instead of numbers, and thirty thousand pounds property, should give one representative, instead of thirty thousand persons. If Virginia exceeded Massachusetts in the valuation, thirty thousand pounds, as it would bear its additional proportion of the public burdens, it would be entitled to an additional representation. For greater convenience and certainty, the rule of apportionment is changed from a valuation to numbers. Shall not the slaves be reckoned? The objection of the northern States has hitherto been, that they were not to be estimated nearer at par with the free. Virginia, we will suppose, is found to contain thirty thousand persons more than Massachusetts, reckoning slaves in the ratio proposed by the constitution. We all agree it ought to be proportionably chargeable in the federal contributions; shall it not have its proportionate representation? Having granted it in the former instance, with what pretence of equity or propriety can we deny it in the latter? And is slavery any more promoted or affected in one case than in the other? The number of persons, and of slaves, necessarily, among the rest, is fixed upon only as a criterion to determine each State's proportion in the public contributions, to which representation ought to be adequate. The sophistry of *Brutus* wholly arises from this circumstance; for if the proportion was determined by any other criterion, the States holding slaves would probably have as large a representation, as under the mode proposed. If they had not, it would be because their taxes were less, which could only evince, probably that the criterion was not so certain or equitable as the one under consideration. *Brutus* has mentioned *horses* and *oxen*. If the number of those animals was the rule of apportionment of taxes, upon the principle above established, and which *Brutus* concedes, the representation of the states would in fact be according to the number of horses or oxen, found in them respectively; and it might then be said they were represented or that they increased the representation, in the same sense as *Brutus* suggests it respecting the slaves. In reality they have no concern in the representation, any further than they are used, with other persons, in a certain proportion, to determine the States proportion of taxes, from *which*, representation originating, as the effect from the cause, is therefore determined by the same rule. The representation is given to the State, and the Representatives are to be chosen, *by the electors of the most numerous branch of the State legislature*, according to the second article of the proposed federal constitution. In these elections the slaves have no part: and here we may feel a degree of regret, that in any quarter of the United States, such a proportion of our fellow creatures, should be deprived of a share of political and civil liberty. To this only do the objections of Brutus, and his warm declamations apply: for, whatever may be their intended operation, upon an entire view of the true sense of the article in question, part of which he has artfully suppressed they evidently do not affect the proposed constitution.

The acts of power, which some of the States see fit to exercise with respect to their internal concerns, may be repugnant to our notions of justice; but shall we therefore refuse to confederate with them? *Brutus* himself surely, could not have this in contemplation. Does *Brutus* wish the slaves emancipated! It is a dictate of humanity, and we need no stimulous to join with him most cordially. But even in this laudible pursuit, we ought to temper the feelings of humanity with political wisdom. Great numbers of slaves becoming citizens, might be burdensome and dangerous to the Public. These inconveniences ought to be regarded. *M. Montesquieu*, whom *Brutus* quotes, and whom we all revere, after mentioning the embarrassment of the Roman Senate, in this respect, sometimes limiting, at other times facilitating the infranchisement of slaves, with great modesty observes, "much less can I determine what ought to be the regulations of a good republic, in an affair of this kind; this depends on too many circumstances." Of this he is certain, that "their condition should be more favoured in the civil, than in the political State"— As federalists, and I think as patriots, we ought to agree with him. This subject doubtless engaged the attention of the late respectable Convention. But, in the immensity of their object, it was not their province to establish those minute provisions, which properly belong partly to federal, partly to State Legislation. They probably went as far as policy would warrant, or practicability allow. The friends to liberty and humanity, may look forward with satisfaction to the period, when slavery shall not exist in the United States; while the enlightened patriot will approve of the system, which renders its abolition gradual.

To return to *Brutus*, from whom I have in some measure wandered. I have endeavoured to expose the fallacy and futility of his objections, to a very important article of the proposed Constitution; whether his mistakes were willful or designed, let the impartial determine. Certain it is, that under a pretence of abbreviating the article, he has given a false and imperfect representation of it, and under that representation, has pointed a number of objections, calculated to engage the feelings of the people, but which do not apply to the article as it stands in the Constitution. The zeal of Brutus may have led him into error, and that zeal may be honest: But his apparent ability prevents me from supposing him inconscious of the fallacy of his own observations. He might consider that the number of those who read, is greater than of those who examine; and that the feelings of the people might be so engaged, as to mislead their judgment. If he was influenced by those considerations to urge conscious fallacies upon the public mind, the investigation of truth is not his object; his patriotism is pretension; his zeal suspicious, and as he writes with design, we ought to read with caution.

Massachusetts Ratification Convention Debates
January 17 to February 4, 1788[20]

January 17

[Article I, section 2, paragraph 3 under discussion.]

Rufus King rose to explain it. There has, says he, been much misconception of this section. It is a principle of this constitution, that representation and taxation should go hand in hand. This paragraph states, that the numbers of free persons shall be determined, by adding to the whole number of free persons, including those bound to service for a term of years, and excluding Indians not taxed, three fifths of all other persons. These persons are the slaves. By this rule is representation and taxation to be apportioned. And it was adopted, because it was the language of all America. According to the confederation, ratified in 1781, the sums for the general welfare and defence, should be apportioned according to the surveyed lands, and improvements thereon, in the several states. But that it hath never been in the power of Congress to follow that rule; the returns from the several states being so very imperfect.

{*Mr. King*: The principle on which this paragraph is founded is, that taxation and representation should go hand in hand. By the Confederation, the apportionment is upon surveyed land, the buildings and improvements. The rule could never be assessed. A new rule has been proposed by Congress, similar to the present rule, which has been adopted by eleven States—all but New Hampshire and Rhode Island.

William Widgery objects to the rule, as apprentices are not freemen, but blunders about it.

Francis Shurtliff: His difficulty is, our negroes are free, but those of other States are not. But the number of representatives first chosen.

General Samuel Thompson: The rule is unequal; as we have more children than the luxurious inhabitants of the southern States. Congress will have no impost or excise, but lay the whole tax on polls. We live longer than they live. We live to one hundred; they to forty.

John Taylor: If eleven States have agreed to the rule of polls, twelve have agreed to alter the Confederation. So the agreement of eleven States is no reason. . . .

---

[20]The primary text is taken from the Convention debates printed in the *Massachusetts Centinel*. The text within braces { } is taken from Theophilus Parsons' notes of the debates, printed in *Debates and Proceedings in the Convention of the Commonwealth of Massachusetts Held in the Year 1788* . . . (Boston, 1856), 285–320.

William Widgery wants to know whether all white infants are free persons? If they are, we are over-taxed.}

Mr. Widgery asked, if a boy of six years of age, was to be considered as a free person?

{*Rufus King*: All persons born free are counted among free persons, to which three-fifths of all persons born or imported slaves, make the census.

William Widgery: If Mr. King is right, then we shall pay one-quarter of the debt.

Nathaniel Gorham: Mr. Wedgery is totally in the wrong. It will lessen our old proportion nearly one-seventh. As eleven States have agreed to this rule, among which was Massachusetts, it is a rule most likely to be adopted. As to representation mentioned—

Colonel Abraham Fuller: The arguments against the representation are groundless. As the rule of proportion is by numbers, five slaves to three freemen is but equal, for slaves are but chattels.

Francis Dana: The old rule of apportionment by lands was against this State. Our lands are worth more by the acre. Lands cultivated by slaves are not worth as much as lands cultivated by freemen. Slaves are their masters' moneys, and at their risk, and it would be unjust to tax a slave as much as a freeman. If we think there should be a difference, the only question would be, what difference. The States have agreed, in Convention, on materials, which we have not. The southern States have not half the value of buildings we have, arising from the climate and manner of living.

Francis Shurtliff wants to know whether five smart negro slaves are to be equal to three of our children?}

Rufus King, in answer, said, all persons born free, were to be considered as freemen; and to make the idea of taxation by numbers more intelligible, said, that five Negro children of South-Carolina, are to pay as much tax as the three governours of New-Hampshire, Massachusetts, and Connecticut.

Francis Dana: In reply to the remark of some gentlemen, that the southern states were favoured in this mode of apportionment, by having 5 of their negroes set against 3 persons in the eastern, the honourable judge observed, that the negroes of the southern states, work no longer than when the eye of the driver is on them. Can, asked he, that land flourish like this, which is cultivated by the hands of freemen? And are not *three* of these independent freemen of more real advantage to a state, than *five* of those poor slaves? As a friend to equal taxation, he rejoiced that an opportunity was presented in this Constitution, to change this unjust mode of apportionment: Indeed, concluded he, from a survey of every part of the Constitution, I think it the best that the wisdom of men could suggest.

Samuel Nasson remarked on the statement of the honourable Mr. King, by saying that the honourable gentleman should have gone further, and shewn us the other side of the question. It is a good rule that works both ways—and the gentleman should also have told us, that three of our infants in the cradle, are to be rated as high as five of the working negroes of Virginia. Mr. N.

adverted to a statement of Mr. *King*, who had said, that five Negro children of S. Carolina were equally rateable as three governours of New-England, and wished, he said, the Hon. Gentleman had considered this question upon the other side—as it would then appear that this State will pay as great a tax for three children in the cradle, as any of the southern States will for five hearty working Negro men. He hoped, he said, while we were making a new government, we should make it better than the old one: for if we had made a bad bargain before, as had been hinted, it was a reason why we should make a better one now.

{*Mr. Nasson* thinks both sides should be stated. Mr. King says five of their infant slaves are equal to three of our governors; but three of our infants are equal to five of their healthy strong slaves. Besides though our climates make us build houses, yet we have to work all summer for winter. Also, the representation is unequal between us and New Hampshire; also, our negroes are all free, and theirs are slaves.}

Benjamin Randall begged leave to answer a remark of the Hon. Mr. Dana, which he thought reflected on the barrenness of the southern states. He spoke from his own personal knowledge, he said, and he could say, that the land in general in those states was preferable to any he ever saw.

{*Mr. Randall*: Lands in the southern States are as good as ours; if not better. It produces every thing. Mr. Dana is mistaken; but as to the slaves, he is about right. The laboring part of the free men in the southern States can live upon two days' work, as easily as we can upon six. They can work all winter, we cannot.}

Francis Dana rose to set the gentleman right, he said it was not the *quality* of the lands, but the *manner* of tilling it, that he alluded to.

January 18

The 3d paragraph of the 2d section of article I. still under consideration.

Tristram Dalton opened the conversation with some remarks on Mr. Randal's positive assertions of the fertility of the southern states—who said from his own observation, and from accounts he had seen, which were better, he could say that the gentleman's remark was not perfectly accurate—the Hon. Gentleman shewed why it was not so, by stating the inconsiderable product of the land; which, though it might in part be owing to the faithlessness and ignorance of the slaves who cultivate it, he said, was in a greater measure owing to the want of *heart* in the soil.

Benjamin Randall: I rise to make an observation on the suggestion of the Hon. Gentleman from Newbury. I have, sir, travelled into the southern states, and should be glad to compare our knowledge on the subject together. In Carolina, Mr. President, if they dont get more than 20 or 30 bushels of corn from an acre, they think it a small crop. On the low lands they sometimes get 40. I hope, sir, these great men of eloquence and learning will not try to *make* arguments to make this Constitution go down, right or wrong. An old saying, sir, is, that *a good thing don't need praising;* but, sir, it takes the best men in the state to gloss this Constitution, which they say is the best that human wisdom can invent. In praise of it, we hear the Rev. Clergy, the Judges of the Supreme Court, and the ablest Lawyers, exerting their utmost abilities.—Now, sir, suppose all this artillery turned the other way, and these great men would speak half as much against it, we might complete our business, and go home in 48 hours. Let us, sir, consider we are acting for the *people*, and for ages unborn; let us deal fairly and above board. Every one comes here to discharge his duty to his constituents, and I hope none will be biassed by the best orators; because we are not acting for ourselves: I think Congress ought to have power, such as is for the good of the nation, but what it is, let a more able man than I tell us.

    Thomas Dawes said, he was sorry to hear so many objections raised against the paragraph under consideration. He thought them wholly unfounded; that the black inhabitants of the southern states must be considered either as slaves, and as so much property, or in the character of so many free men; if the former, why should they not be wholly represented? Our *own* State laws and Constitution would lead us to consider those blacks as *free men*, and so indeed would our own ideas of natural justice: If then, they are free men, they might form an equal basis for representation as though they were all white inhabitants. In either view, therefore, he could not see that the Northern States would suffer, but directly to the contrary. He thought, however, that gentlemen would do well to connect the passage in dispute with another article in the Constitution, that permits Congress, in the year 1808, wholly to prohibit the importation of slaves, and in the mean time to impose a duty of ten dollars a head on such blacks as should be imported before that period. Besides, by the new Constitution, every particular state is left to its own option totally to prohibit the introduction of slaves into its own territories. What could the Convention do more? The members of the Southern States, like ourselves, have *their* prejudices. It would not do to abolish slavery, by an act of Congress, in a moment, and so destroy what our Southern brethren consider as property. But we may say, that although slavery is not smitten by an apoplexy, yet it has received a mortal wound and will die of a consumption.

    {*Mr. Dawes*: Though slaves are reckoned five equal to three now, yet in a few years slavery must be abolished, and in the mean time, slaves may be taxed on importation, sixty shillings per head. Slavery will not die of an apoplexy, but of a consumption.

    Benjamin Randall: Sorry to hear it said that after 1808 negroes would be free. If a southern man heard it, he would call us pumpkins.

    William Widgery objects to Mr. Dana's description of the southern States. Their land is better than ours.

Francis Dana says he never compared the value of the eastern or southern lands; he compared only the mode of cultivation.

William Widgery says if this rule is for an equal poll tax, he has no objection. But for a rule of apportionment, it is unjust, southern land being better than ours. In Virginia, one thousand acres has forty-eight polls; in Massachusetts, a family of six, to fifty acres, makes one hundred and twenty polls to the one thousand acres. In legislation, one southern man with sixty slaves, will have as much influence as thirty-seven freemen in the eastern States.

Caleb Strong: This mode of census is not new. Our General Court have considered it, and the General Court have agreed. The southern States have their inconveniences; none but negroes can work there; the buildings are worth nothing.

James Neal, of Kittery, spoke against the slave trade. We shall all suffer for joining with them, when they allow the slave trade.

George Cabot asks the gentleman from Sharon [Benjamin Randall], whether, in his five hundred miles travel, he saw five thousand people who live as well as five thousand people of the lowest sort here. As to the slave trade, the southern States have the slave trade, and are sovereign States. this Constitution is the best way to get rid of it.

Benjamin Randall says he believes he has, but is not certain. If they do not, it is their own fault.

Samuel Nasson: Southern States are not poor.}

January 25

[Article I, section 9 under discussion.]

{*Tristram Dalton*: In favor of the first paragraph because we gain a right in time to abolish the slave trade.}

James Neal from Kittery went over the ground of objection to this section on the idea, that the slave trade was allowed to be continued for 20 years. His profession, he said, obliged him to bear witness against any thing that should favour the making merchandize of the bodies of men; and unless his objection was removed, he could not put his hand to the constitution. Other gentlemen said, in addition to this idea, that there was not even a provision that the negroes ever shall be free.

{*Daniel Cooley* asks, whether negro slaves, emigrating into this State, will not be considered as a poll, to increase our ratio of taxes?

Rev. Isaac Backus answers Mr. Neal, and shows we have now gained a check which we had not before, and hopes in time we shall stop the slave trade.

William Bodman says, those born slaves in the southern States may still continue slaves.}

General Samuel Thompson exclaimed: Shall it be said, that after we have established our own independence and freedom, we make slaves of others.

Oh! Washington, what a name has he had! How he has immortalized him-self!—but he holds those in slavery who have as good a right to be free as he has—He is still for self; and in my opinion, his character has sunk 50 per cent.

{*General Thompson*: If the southern States would not give up the right of slavery, then we should not join with them—Washington's character fell fifty per cent. by keeping slaves—it is all a contrivance, and Washington at the head—our delegates over-powered by Washington and others.}

On the other side gentlemen said, that the step taken in this article, towards the abolition of slavery, was one of the beauties of the Constitution. They observed that in the confederation there was no provision whatever for its ever being abolished; but this constitution provides, that Congress may, after 20 years, totally annihilate the slave trade; and that all the states, except two, have passed laws to this effect, it might reasonably be expected, that it would then be done—in the interim, all the states were at liberty to prohibit it.

{*William Jones, of Bristol*, objects to Article V., because we can't amend this section for twenty years }

January 16

The debate on the 9th section still continued desultory—and consisted of similar objections, and answers thereto, as had before been used.—Both sides deprecated the slave-trade in the most pointed terms—on one side it was pathetically lamented, by Mr. Nason, Major Lusk, Mr. Neal, and others, that this Constitution provided for the continuation of the slave trade for 20 years.—On the other, the Hon. Judge Dana, Mr. Adams, and others rejoiced that a door was now to be opened, for the annihilation of this odious, abhorent practice, in a certain time.

January 30

General William Heath: By my indisposition, and absence, I have lost several important opportunities; I have lost the opportunity of expressing my sentiments with a candid freedom, on some of the paragraphs of the system, which have lain heavy on my mind. I have lost the opportunity of expressing my warm approbation on some of the paragraphs. I have lost the opportunity of asking some questions for my own information, touching some of the para-graphs, and which naturally occurred, as the system unfolded. I have lost the opportunity of hearing those judicious, enlightening, and convincing argu-ments, which have been advanced during the investigation of the system,—this is my misfortune, and I must bear it. The paragraph respecting the migration or importation of such persons, as any of the States now existing shall think proper to admit, &c. is one of those considered during my absence, and I have heard nothing on the subject, save what has been mentioned this morning, but I think the gentlemen who have spoken, have carried the matter rather too far on both sides,—I apprehend that it is not in our power to do any thing for, or against, those who are in slavery in the southern States. No gentleman within these walls detests every idea of slavery more than I do: It is generally detested by the people of this Commonwealth,—and I ardently hope that the time will soon come, when our brethren in the southern States will view it as we do, and put a stop to it, but to this we have no right to compel them. Two questions naturally arise if we ratify the Constitution, shall we do any thing by our act to hold the blacks in slavery—or shall we become partakers of other men's sins. I think neither of them: Each State is sovereign and independent to a certain degree, and they have a right, and will regulate their own internal affairs, as to themselves appears proper; and shall we refuse to eat, or to drink, or to be united, with those who do not think, or act, just as we do, surely not. We are not in this case partakers of other men's sins, for in nothing do we voluntarily encourage the slavery of our fellow men, a restriction is laid on the federal government, which could not be avoided and a union take place: The federal Convention went as far as they could, the migration or importation, &c. is confined to the States now *existing only*, new States cannot claim it. Congress by their ordinance for erecting new States, some time since, declared, that the new States shall be republican, and that there shall be no slavery in them.[21] But whether those in slavery in the southern States, will be emancipated after the year 1808, I do not pretend to determine, I rather doubt it.

The conversation on the Constitution by paragraphs being ended,

Theophilus Parsons moved, that the Convention do assent to and ratify this Constitution.

James Neal rose and said, that as the Constitution at large was now under consideration, he would just remark, that the article which respected the Africans was the one which lay on his mind—and unless his objections to that were removed, it must, how much soever he likes the other parts of the Constitution, be a sufficient reason for him to give his negative to it.

February 1

Samuel Nasson: Representation and taxation to be apportioned according to numbers. This, sir, I am opposed to; it is unequal. I will shew an instance in point—We know for certainty, that in the town of Brooklyn, persons are better able to pay their taxes, than in the parts I represent: Suppose the tax is laid on polls: Why the people of the former place will pay their tax ten times

[21]See the Northwest Ordinance, adopted by Congress, July 13, 1787 (Chapter 1).

as easy, as the latter—thus helping that part of the community, which stands in the least need of help: On this footing the poor pay as much as the rich: And in this a way is laid, that five slaves shall be rated no more than three children. Let gentlemen consider this—a farmer takes three small orphans, on charity, to bring up—they are bound to him—when they arrive at 21 years of age, he gives each of them a couple suits of clothes, a cow, and two or three young cattle—we are rated as much for these, as a farmer in Virginia is for five slaves, whom he holds for life—they and their posterity—the male and the she ones too.

February 4

Thomas Lusk . . . entered largely into the consideration of the 9th section and in the most pathetick and feeling manner, described the miseries of the poor natives of Africa, who are kidnapped, and sold for slaves—with the brightest colours, he painted their happiness and ease on their native shores; and contrasted them with their wretched, miserable, and unhappy condition in a state of slavery.

Rev. Isaac Backus: I have said very little in this honourable Convention; but I now beg leave to offer a few thoughts upon some points in the Constitution proposed to us.—Much hath been said, about the importation of slaves into this country. I believe that, according to my capacity, no man abhors that wicked practice more than I do, and would gladly make use of all lawful means, towards the abolishing of slavery in all parts of the land.—But let us consider where we are, and what we are doing. In the articles of confederation, no provision was made to hinder the importation of slaves into any of these States; but a door is now opened, hereafter to do it; and each State is at liberty now to abolish slavery as soon as they please. And let us remember our former connection with Great-Britain, from whom many in our land think we ought not to have revolted: How did they carry on the slave-trade! I know that the Bishop of Gloucester, in an annual sermon in London, in February, 1766, endeavoured to justify their tyrannical claims of power over us, by casting the reproach of the slave-trade upon the Americans. But at the close of the war, the Bishop of Chester, in an annual sermon, in February, 1783, ingenuously owned, that their nation is the most deeply involved in the guilt of that trade, of any nation in the world; and also, that they have treated their slaves in the West-Indies, worse than the French or Spaniards have done theirs.—Thus slavery grows more and more odious through the world;—and, as an honourable gentleman [Thomas Dawes] said some days ago, "Though we cannot say, that slavery is struck with an apoplexy, yet we may hope it will die with a consumption." And a main source, sir, of that iniquity, hath been an abuse of the covenant of circumcision, which gave the seed of Abraham to destroy the inhabitants of Canaan, and to take their houses, vineyards, and all their estates as their own; and also to buy and hold others as servants. And as christian privileges are much greater than those of the Hebrews were, many have imagined that they had a right to seize upon the lands of the heathen, and to destroy or enslave them as far as they could extend their power. And from thence the mystery of iniquity carried many into the practice of making *merchandise of slaves and souls of men*. But all ought to remember, that when God promised the land of Canaan to Abraham and his seed, he let him know that they were not to take possession of that land, until the *iniquity of the Amorites was full*; and then they did it under the immediate direction of heaven; and they were as real executors of the judgment of God upon those heathens, as any person ever was an executor of a criminal justly condemned. And in doing it they were not allowed to invade the lands of the Edomites, who sprang from Esau, who was not only of the seed of Abraham, but was born at the same birth with Israel; and yet they were not of that church. Neither were Israel allowed to invade the lands of the Moabites, or of the children of Ammon, who were of the seed of Lot. And no officer in Israel had any legislative power, but such as were immediately inspired. Even David, the man after God's own heart, had no legislative power, but only as he was inspired from above; and he is expressly called a *Prophet* in the New Testament. And we are to remember that Abraham and his seed, for four hundred years, had no warrant to admit any strangers into that church, but by buying of him as a servant, with money. And it was a great privilege to be bought, and adopted into a religious family for seven years, and then to have their freedom. And that covenant was expressly repealed in various parts of the New-Testament; and particularly in the first epistle to the Corinthians, wherein it is said, *Ye are bought with a price; therefore glorify God in your body, and in your spirit, which are God's. And again, circumcission is nothing, and uncircumcission is nothing, but the keeping of the commandments of God. Ye are bought with a price, be not ye the servants of men*. Thus the gospel sets all men upon a level; very contrary to the declaration of an honourable gentleman in this house "That the Bible was contrived for the advantage of a particular order of men."

The Republican Federalist V (James Warren)
Massachusetts Centinel, January 19, 1788[22]

To the Members *of the* Convention *of* Massachusetts.

Let us now proceed to the provision in the system for a representation of the people, which is the *corner stone* of a free government. The Constitution provides, art. 1st, sect. 2, "that representatives and direct taxes shall be apportioned among the several States, which may be included within this

---

[22]Warren, a Plymouth merchant-farmer, had been one of the leaders in the Revolutionary movement. At this time he was Speaker of the Massachusetts House of Representatives.

union, according to their respective numbers, which shall be determined by adding to the whole number of free persons, including those bound to service for a term of years, and excluding Indians not taxed, three fifths of all other persons." "Representatives" then are to be "apportioned among the several States, according to their respective numbers," and *five* slaves, in composing those numbers, are to be classed with *three* freemen—By which rule, fifty thousand slaves, having neither *liberty or property*, will have a representative in that branch of the legislature—to which more especially will be committed, the *protection of the liberties*, and *disposal of all the property* of the freemen of the Union—for thus stands the new Constitution. Should it be said, that not *the slaves* but their *masters* are to send a representative, the answer is plain—If the *slaves* have a *right* to be represented, they are *on a footing* with *freemen*, *three* of *whom* can then have no more than an equal right of representation with *three slaves*, and these when qualified by property, may elect or be elected represen tatives, *which is not the case*: But if they have not a right to be represented, their masters can have no right derived from their *slaves*, for *these* cannot transfer to others what they have not themselves. Mr. Locke, in treating of political or civil societies, chap. 7, sect. 85, says, that men "being in the state of slavery, not capable of any property, cannot, in that state, be considered as any part of civil society, the chief end whereof, is the preservation of property." If slaves, then, are no part of civil society, there can be no more reason in admitting them, than there would be in admitting the *beasts* of the field, or *trees* of the forest, to be classed with *free electors*. What covenant are the freemen of Massachusetts about to ratify? A covenant that will degrade them to the *level of slaves*, and give to the States who have as many blacks as whites, *eight* representatives, *for the same number of freemen* as will enable this State to elect *five*—Is this an *equal*, a *safe*, or a *righteous* plan of government? indeed it is *not*. But if to encrease these objections, it should be urged, "that representation being regulated by the same rule as taxation, and taxation being regulated by a rule intended to ascertain the relative property of the States, representation will then be regulated by the principles of property." This *answer* would be the only one that could be made, for representation, according to the new Constitution is to be regulated, either by *numbers* or *property*.

Let us now inquire of those who take this ground, what right they have to put a construction on the constitution, which is repugnant to the express terms of the Constitution itself? This provides, "that representatives shall be apportioned among the several States, *according to their respective numbers*." Not a word of *property* is mentioned, but the word "numbers" is repeatedly expressed—Admitting however that property was intended by the Constitution as the rule of representation, does this *mend the matter*? it will be but a short time, after the adoption of the new constitution, before the State *legislatures*, and *establishments in general* will be so *burthensome* and *useless* as to make the people desirous of being rid of them, for they will not be able to support them. The *State appointment* of Representatives will then cease, but *the principle of representation according to property*, will undoubtedly be retained, and before *it is established* it is necessary to consider whether it is a just *one*, for if *once* it is *adopted* it will *not be easily altered*.—According to this principle, a man worth £50,000 is to have as many votes for representatives in the new Congress, as *one thousand men*, worth £50 each: And *sixty such nabobs* may send *two* representatives, while *sixty thousand* freemen having £50 *each* can only send the same number. Does not this establish in the representative branch of the new Congress a principle of aristocracy, with a vengeance? The Constitutions of the several States, admit of no such principle, neither can any freeman with safety thus surrender not only the intire disposition of their property, but also, the controul of their liberties and lives to a few opulent citizens. Should it be said that the rule of federal taxation, being advantageous to the State, it should be content with the same rule for representation. The answer is plain, the rule gives no advantage, but is supposed to be advantageous to Massachusetts, and to be an accommodation very beneficial to the southern States: But admiting this State will be benefited by the rule, is it disposed to sell its birthright, the right of an equal representation in the federal councils *for so small a consideration?* Would this State give up that right to any State that would pay our whole proportion of *direct* and *indirect taxes?* Shall we relinquish some of the most essential rights of government, which are our only security for every thing dear to us, to avoid our proportion of the publick expense? shall we give up all we have, for a small part of it? This if agreed to, would be no great evidence of our wisdom or foresight. But it is not probable, in the opinion of some of the ablest advocates for the new system, that *direct taxes* will ever be levied on the States, and if not, the provision for levying such taxes will be *nugatory*: We shall receive no kind of benefit from it, and shall have committed ourselves to the mercy of the states having slaves, *without any consideration whatever*. Indeed, should direct taxes be necessary, shall we not by increasing the representation of those States, put it in their power to prevent the levying such taxes, and thus defeat our own purposes? Certainly we shall, and having given up a substantial and *essential* right, shall in lieu of it, have a mere *visionary advantage*. Upon the whole then, it must be evident, that we might as well have committed ourselves to the parliament of Great-Britain, under the idea of a *virtual representation* as in this manner resign ourselves to the federal government.

THE YEOMANRY OF MASSACHUSETTS
Massachusetts Gazette, January 25, 1788

Another thing they tell us, that the constitution must be good, from the characters which composed the Convention that framed it. It is graced with the names of a Washington and a Franklin. Illustrious names, we allow— worthy characters in civil society. Yet we cannot suppose them, to be infallible guides, neither yet that a man must necessarily incur guilt to himself merely by dissenting from them in opinion.

We cannot think the noble general, has the same ideas with ourselves, with regard to the rules of right and wrong. We cannot think, he acts a very consistent part, or did through the whole of the contest with Great-Britain; who, notwithstanding he wielded the sword in defence of American liberty, yet at the same time was, and is to this day, living upon the labours of several hundreds of miserable Africans, as freeborn as himself; and some of them very likely, descended from parents who, in point of property and dignity in their own country, might cope with any man in America. We do not conceive we are to be overborne by the weight of any names, however revered.

SAMUEL HOPKINS TO LEVI HART
Newport, January 29, 1788[23]

I received yours of the 12th inst. Yesterday. I thank you for your exertions with regard to the *slave trade*. I should have been glad to be informed, whether what was reported to Mr. Brown be true, viz. that they are going into this trade at Middletown and Norwich. I hear they threatened to carry it on here and at Providence yet; but question whether they will do it, as they will expose themselves so much by it.

The new constitution, you observe, guarantees this trade for twenty years. I fear, if it be adopted, this will prove an Achan in our camp. How does it appear in the sight of Heaven, and of all good men, well informed, that *these States*, who have been fighting for liberty, and consider themselves as the highest and most noble example of zeal for it, cannot agree in any political constitution, unless it indulge and authorize them to inslave their fellow men! — I think if this constitution be not adopted, as it is, without any alteration, we shall have none, and shall be in a state of anarchy and probably of civil war: Therefore, I wish to have it adopted: but still, as I said, *I fear*—And perhaps civil war will not be avoided, if it be adopted. Ah! These unclean spirits, like frogs—They, like the furies of the poets, are spreading discord, and exciting men to contention and war, wherever they go: And they can spoil the best constitution that can be formed. When Congress shall be formed on the new plan, these frogs will be there; for they go forth to *the kings of the earth* in the first place. They will turn the members of that august body into devils, so far as they are permitted to influence them. Have they not already got possession of most of the men who will or can be chosen and appointed to a place in that assembly? I suppose that even good christians, are not out of the reach of influence from these *frogs*. "Blessed is he that watcheth, and keepeth his garments."

[23]Misc. Mss., S. Hopkins folder, New-York Historical Society. Hart was pastor of the Second or North Congregational parish in Preston, Conn. In 1775 he published an attack on the slave trade entitled *Liberty Described and Recommended*. . . . Hart refused election to the Connecticut convention even though he supported the Constitution.

ADELOS
Northampton *Hampshire Gazette*, February 6, 1788

Mr. Printer: In the present political contest, respecting the establishment of the Federal Constitution, I have been rather a silent spectator—I
have heard and read much—said little—The writers on both sides the question, have shewn spirit and learning. I confess the advocates for it have
manifested a spirit (some of them at least) rather unbecoming, and somewhat
bordering upon persecution: this I hope, however, will not be justified by
people in general; but that every one may be allowed to speak his sentiments,
and have fair play at disquisition. It is a matter of importance and requires
sobriety. In this view of the matter, I beg leave to offer a few remarks for
public consideration. I am not about at this time to impugn every part and
parcel of the new frame of government; but if I may be allowed the natural,
inherent, the grand birthright of all the human race, I will very freely and
boldly oppose one article of it;—page 12, beginning of section 9. which however strange it may appear, has been but little noticed. It permits, in express
terms, of that most cursed of all trades, the African slave-trade. I must confess
it will be very wonderful to me, if the Massachusettensians (above all people
in the world) should hold up their hands to give efficacy to a constitution
which admits of slavery; and not only so, but Congress is expressly restricted
from making any provision against it, for the term of twenty-one years, let
the mischiefs resulting therefrom be ever so great. It is strange, I say, if Massachusetts should give countenance to this, after establishing a constitution of
their own, fronted with these words, "*All men are born free and equal*;" and in
consequence of which have emancipated many wretched Africans, and delivered them from masters more sordid to them (many of them) than they were
to the brutal herd. I cannot see but the first moment we adopt the Federal
Constitution as it stands, we rase our own to the very foundation. We allow
that freedom and equity are the natural rights of every man born into the
world; but if we vote this, we vote to take away those rights, and to sport
ourselves with the liberties of mankind. I wish to know how one man came by
his right to the service of another, without his free consent, and a proper
recompense when required? Whether we go ourselves to Africa to procure
slaves, or employ others to do it for us, or purchase them at any rate of others,
it matters not a whit. It is an old saying and a true one, "The partaker is as bad
as the thief." It is well known that this trade is carried on by violence and
rapine; nay, murder is not, I presume, out of the question. Who gave mankind a right thus to play the devil with one another? We reprobate the conduct
of the Algerines; their conduct truly is highly reprehensible;—they enslave
the Americans,[24]—the Americans enslave the Africans; which is worst? Six of
one and a half a dozen of the other. Congress may make laws to punish pira-

---

[24]After the Revolution, American commerce lost the protection of the British
navy. In the Mediterranean Sea the Barbary states preyed on American merchantmen.

cies and felonies committed on the high seas; but yet we may go to Africa, and lay waste and destroy what we please; captivate thousands of free born men, without the least provocation—bring them to America and doom them to perpetual bondage, and all with impunity: Congress are not to be allowed to prevent it. The thought is truly shocking, and nature shudders at the recollection!

Flimsy indeed, is the argument of the Connecticut Landholder, in support of the Constitution, that "slaves are so numerous in the southern states, should an emancipation take place, they will be undone,"—truly wretched enough! So then, if by fraud and violence, I have got the possession of my neighbour's estate, reduced him to misery and slavery, the laws may not restore it to him, the rightful owner again, lest I should be undone. Too weak even for idiotcy itself. I think upon the whole the article ought to be expunged; or that we ought not to vote to give life to a constitution, which at its first breath will be branded with eternal infamy, by having a stamp of slavery and oppression upon it.

A FRIEND TO THE RIGHTS OF THE PEOPLE: ANTI-FŒDERALIST I
Exeter *Freeman's Oracle*, February 8, 1788

To the Inhabitants of NEW-HAMPSHIRE

The grand topick of the day is the New Constitution, much has been said for, much has been said against it by able writers.—On one side, it is warmly asserted, that the liberties of the people, are sufficiently secure, as it now stands—On the other it is urged with equal vehemence, they are not, amendments must be made—a Bill of Rights prefixed, or we are undone; so that it is very difficult, for common people to know what is right, any thing that may serve to throw light upon the subject, may be very useful at this juncture. Both sides, it appears to me, so far as I have had opportunity of reading, have kept the Constitution too much out of view. There seems to be a necessity of a more particular and impartial examination of the thing itself, which is the bone of so much contention.—It is therefore of vast importance that the Constitution should be well considered, and carefully examined, for upon that and nothing else, rationally explained, according to the common usage of words, must we found our judgment of its goodness or badness;— My design therefore is to quote, and make some remarks upon some of the most capital propositions. . . .

On October 1, 1783, Moroccan pirates seized the ship *Betsey* and held the crew hostage for almost two years. Two other ships—the *Maria* and the *Dauphin*—were captured and their crews imprisoned in Algiers. Because the U.S. was unwilling and unable to pay ransom or build a navy that could force their release, the American sailors lanquished in prison in Algiers and some were sold into slavery.

Remark 5. Upon the slave trade, sect. 9. The migration, or importation of such persons, as any of the states shall think proper to admit, shall not be prohibited by Congress prior to the year eighteen hundred and eight—By the importation of the persons above-mentioned is doubtless meant the Guinea trade, by which thousands and millions of poor negroes have been wrested from their native country, their friends and all that is dear to them, and brought into a state of the most abject slavery and wretchedness—By the above article, this cruel and barbarous practice is not to be prohibited by Congress for twenty years to come, and even then, it is not said, it shall cease—Here is a permission granted, for the enslaving and making miserable our fellow men, totally contrary to all the principles of reason, justice, benevolence and humanity, and all the kind and compassionate dictates of the Christian Religion. Can we then hold up our hands for a Constitution that licenses this bloody practice? Can we who have fought so hard for Liberty give our consent to have it taken away from others? May the powers above forbid.

JEREMY BELKNAP TO BENJAMIN RUSH
Boston, February 12, 1788[25]

It gives me great Pleasure to hear of the intended effort of your Society to obtain a Law prohibiting the African Trade. Rhode-Island, bad as they are in some respects, have set us a good Example in *this* Instance, they have by Law prohibited under a severe Penalty the buying and selling slaves in foreign Parts & to render Conviction easy the Evidence of one Seaman belonging to the Vessell is sufficient for a Condemnation—

In our late Convention something was said by way of objection to the Constitution because "it *established*" (as the speakers said) "the importation of Slaves for 20 years" Several of the antifoederal Party urged this, but none more violently than a certain Quaker Preacher [James Neal] who went so far as to predict that the same measure should be meted to us—i.e. that we should lose our Liberties as the Africans lost theirs—He was answered very ably by Mr. [Theophilus] Parsons who construed that article into a dawn of hope for the final abolition of the horrid Traffick & spoke of it as a great Point gained of the southern states. However the Quaker remained inflexible & as I know him to be a Man of influence in the Circle of *Friends* at the Eastward I suppose he will prejudice the minds of a great Part of that fraternity against the Constitution. The reason of my mentioning this to you is to desire you to inform me whether among the Quakers of Pennslva: any such Construction is put on that article wch respects the Migration or Importation of Foreigners—I think there must be some Men of Sense among them who cannot be so

[25]Rush Papers, Library Company of Philadelphia. Rush replied to Belknap on February 28, 1788 (Chapter 4).

prejudiced, but I wish to be made certain of it & I think I shall be able to make a good use of the Information.

JOSHUA ATHERTON
Speech in the New Hampshire Ratifying Convention,
ca. February 13, 1788[26]

I can not be of the opinion of the honorable gentleman who last spoke, that this paragraph is either so unjust, or so inoffensive, as they seem to imagine, or that the objections to it are so totally void of foundation. The idea that strikes those that are opposed to this clause, so disagreeably and so forcibly, is, hereby, it is conceived (if we ratify this constitution) that we become *consenters to* and *partakers in* the sin and guilt of this abominable traffic, at least for a certain period, without any positive stipulation that it shall even then be brought to an end. We do not behold in that valuable acquisition, so much boasted of by the honorable member from Portsmouth, "*that an end is then to be put to slavery.*"[27] Congress may be as much or more puzzled to put a stop to it then than we are now. The clause has not secured its abolition.

We do not think we are under any obligation to perform works of supercrogation in the reformation of mankind; we do not esteem ourselves under any necessity to go to Spain or Italy to suppress the Inquisition of those countries, nor of making a journey to the Carolinas to abolish the detestable custom of enslaving the Africans: but, sir, we will not lend the aid of our ratification to this cruel and inhuman merchandise, not even for a day.

There is a great distinction in not taking part in the most barbarous violation of the sacred laws of God and humanity, and our becoming guarantees for its exercise for a term of years. Yes, sir, it is our full purpose to wash our hands clean of it, and, however unconcerned spectators we may remain of such predatory infractions of the laws of our nation, however unfeeling we may subscribe to the ratification of manstealing, with all its baneful consequences, yet I can not but believe, in justice to human nature, that if we reverse the consideration, and bring this claimed power somewhat nearer to our own doors, we shall form a more equitable opinion of its claim to this ratification.

Let us figure to ourselves a company of these manstealers, well equipped for the enterprise, landing on our coast. They seize or carry off the whole or a part of the town of Exeter. Parents are taken and children left, or possibly they may be so fortunate as to have a whole family taken and carried off to-

[26]Transcribed from Daniel F. Secomb, *History of the Town of Amherst, Hillsborough County, New Hampshire* . . . .(Concord, N.H., 1883), 230–31.

[27]Probably John Langdon, one of New Hampshire's delegates to the Constitutional Convention and one of three Portsmouth delegates to the state ratifying convention.

gether by these relentless robbers. What must be their feelings in the hands of their new and arbitrary masters! Dragged at once from every thing they held dear to them, stripped of every comfort of life, like beasts of prey, they are hurried on a loathsome and distressing voyage to the coast of Africa, or some other quarter of the globe where the greatest price may waft them, and here, if any thing can be added to their miseries, comes on the heart-breaking scene—a parent sold to one, a son to another, and a daughter to a third; brother is cleft from brother, sister from sister, and parents from their darling offspring. Broken with every distress that human nature can feel, and bedewed with tears of anguish, they are dragged into the last stage of depression and slavery, never, never to behold the faces of one another again. The scene is too affecting; I have not fortitude to pursue the subject.

JEREMY BELKNAP TO BENJAMIN RUSH
Boston, April 7, 1788[28]

By your very obliging Letter I had the pleasure to be assured of the friendly disposition of the Quakers in Phila to the federalists and Constitution, the contrary of which has been confidently asserted here, & what will not the virulence of party zeal lead [—] to assert? I have taken some pains to let some of my Friends of that persuasion know the Truth for I found that even they had been deceived into a belief that all was wrong & that the Constitution was *intended* to establish Slavery for twenty years at least.

CONSIDER ARMS, MALACHI MAYNARD, AND SAMUEL FIELD
Northampton *Hampshire Gazette*, April 9 and 16, 1788

After a month of heated debates, the Massachusetts Convention narrowly ratified the Constitution on February 6 by a vote of 187 to 168. Immediately after the vote and over the next few days, several delegates who had voted against ratification publically stated that they would accept the will of the majority and support the Constitution. News of the acquiescence of these Antifederalists was disseminated rapidly throughout America and was applauded by Federalists.

Three Hampshire County delegates who had voted against ratification, however, continued to oppose the Constitution, and they decided to publish their reasons of dissent in the Hampshire Gazette, *one of the county's two newspapers. These three delegates, who apparently did not speak in the Convention debates, were Samuel Field of Deerfield and Consider Arms and Malachi Maynard of Conway, all of whom had been figures of controversy before. During the Revolution, Field and Arms were Loyalists; while Arms and Maynard were active in Shays's Rebellion, during the winter of 1786–87. After the Shaysites were defeated, Arms and Maynard were*

[28]Rush Papers, Library Company of Philadelphia.

among the 790 insurgents who took the oath of allegiance to Massachusetts. They were two of the twenty-nine insurgents in the Massachusetts Convention, all but one of whom voted against ratification.

Sensitive to their controversial pasts, Field, Arms, and Maynard concluded their reasons by promising not "to be disturbers of the peace," but "to be subject to 'the powers that be'" if the Constitution were adopted and put into effect. No newspapers other than the Hampshire Gazette *printed their dissent.*

Samuel Field, a graduate of Yale College, was variously a divinity student, lawyer, merchant, farmer, and poet. As a Sandemanian in religion, he remained loyal to the king, but stayed neutral during the fighting. He represented Deerfield in the state legislature in 1773, 1774, and 1791. Consider Arms, a farmer and large landholder, was elected to the Massachusetts Provincial Congress in 1774 but did not attend. During the war, he was a Loyalist. He represented Conway in the legislature in 1788. Malachi Maynard, a farmer, was Conway town treasurer and assessor throughout most of the 1780s and 1790s.

On April 23, a writer using the pseudonym Philanthrop answered Field, Arms, and Maynard in the Hampshire Gazette. *He could not understand why they published their dissent because "Their objections have adorned the news papers for several months past, and often received answers, which to many persons appeared satisfactory." Philanthrop challenged their objections to the power of Congress to regulate federal elections, the power of Congress over the military and finances, the three-fifths clause, and, most particularly, the slave-trade clause. He also defended the Massachusetts Convention against charges that "unfair methods were practised" in order to produce a majority in favor of the Constitution and that the Convention supported the slave trade. For a reply to Philanthrop's comments concerning the slave trade, see* Phileleutherus, Hampshire Gazette, *May 21.*

MR. PRINTER, We the Subscribers being of the number, who did not assent to the ratification of the Federal Constitution, under consideration in the late State Convention, held at Boston, to which we were called by the suffrages of the corporations to which we respectively belong—beg leave, through the channel of your paper, to lay before the public in general, and our constituents in particular, the reasons of our dissent, and the principles which governed us in our decision of this important question.

Fully convinced, ever since the late revolution, of the necessity of a firm, energetic government, we should have rejoiced in an opportunity to have given our assent to such an one; and should in the present case, most cordially have done it, could we at the same time [have] been happy to have seen the liberties of the people and the rights of mankind properly guarded and secured. We conceive that the very notion of government carries along with it the idea of justice and equity, and that the whole design of instituting government in the world, was to preserve men's properties from rapine, and their bodies from violence and bloodshed.

These propositions being established, we conceive must of necessity produce the following consequence, viz. That every constitution or system, which does not quadrate with this original design, is not government, but in fact a subversion of it.

Having premised thus much, we proceed to mention some things in this constitution, to which we object, and to enter into an enquiry, whether, and how far they coincide with those simple and original notions of government beforementioned.

In the first place—as direct taxes are to be apportioned according to the numbers in each state, and as Massachusetts has none in it but what are declared freemen, so the whole, blacks as well as whites, must be numbered; this must therefore operate against us, as two fifths of the slaves in the southern states are to be left out of the numeration; consequently, three Massachusetts infants will increase the tax equal to five sturdy, full grown negroes of theirs, who work every day in the week for their masters, saving the Sabbath, upon which they are allowed to get something for their own support. We can see no justice in this way of apportioning taxes; neither can we see any good reason why this was consented to on the part of our delegates. . . .

But we pass on to another thing, which (aside from every other consideration) was, and still is an insuperable objection in the way of our assent. This we find in the 9th section under the head of restrictions upon Congress, viz. "The migration or importation of such persons as any of the states now existing shall think proper to admit, shall not be prohibited by the Congress, prior to the year one thousand eight hundred and eight," &c. It was not controverted in the Convention, but owned that this provision was made purely that the southern states might not be deprived of their profits arising from that most *nefarious trade* of enslaving the Africans. The hon. Mr. King himself, who was an assistant in forming this constitution, in discoursing upon the slave trade, in the late Convention at Boston, was pleased to design it by this epithet, *nefarious*, which carries with it the idea of something peculiarly wicked and abominable: and indeed we think it deserving of this and every odious epithet which our language affords, descriptive of the iniquity of it. This being the case, we were naturally led to enquire why we should establish a constitution, which gives licence to a measure of this sort—How is it possible we could do it consistent with our ideas of government? consistent with the principles and documents we endeavour to inculcate upon others? It is a standing law in the kingdom of Heaven, "Do unto others as ye would have others do unto you." This is the royal law—this we often hear inculcated upon others. But had we given our affirmative voice in this case, could we have claimed to ourselves that consistent line of conduct, which marks the path of every honest man? Should we not rather have been guilty of a contumelious repugnancy, to what we profess to believe is equitable and just? Let us for once bring the matter home to ourselves, and summons up our own feelings upon the occasion, and hear the simple sober verdict of our own hearts, were we in the place of those unhappy Africans—this is the test, the proper *touch-stone* by which to try the matter before us. Where is the man, who under the influence of sober dispassionate reasoning, and not void of natural affection, can lay his hand upon his heart and say, I am willing my sons and my daughters should be torn from me and doomed to perpetual slavery? We presume that man is not to be found amongst us: And yet we think the consequence is fairly drawn, that this is what every man ought to be able to say, who voted for this constitution. But we dare say this will never be the case here, so long as the country has power to repel force by force. Notwithstanding this we will practise this upon those who are destitute of the power of repulsion: from whence we conclude it is not the tincture of a skin, or any disparity of features that are necessarily connected with slavery, and possibly may therefore fall to the lot of some who voted it, to have the same measure measured unto them which they have measured unto others. If we could once make it our own case, we should soon discover what distress & anxiety, what poignant feelings it would produce in our own breasts, to have our infants torn from the bosoms of their tender mothers—indeed our children of all ages, from infancy to manhood, arrested from us by a banditti of lawless ruffians, in defiance of all the laws of humanity, and carried to a country far distant, without any hopes of their return—attended likewise with the cutting reflection, that they were likely to undergo all those indignities, those miseries, which are the usual concomitants of slavery. Indeed when we consider the depredations committed in Africa, the cruelties exercised towards the poor captivated inhabitants of that country on their passage to this—crowded by droves into the holds of ships, suffering what might naturally be expected would result from scanty provisions, and inelastic infectious air, and after their arrival, drove like brutes from market to market, *branded* on their naked *bodies* with *hot irons*, with the initial letters of their masters names—fed upon the entrails of beasts like swine in the slaughter-yard of a butcher; and many other barbarities, of which we have documents well authenticated: then put to the hardest of labour, and to perform the vilest of drudges—their master (or rather *usurpers*) by far less kind and benevolent to them, than to their horses and their hounds. We say, when we consider these things (the recollection of which gives us pain) conscience applauds the decision we have made, and we feel that satisfaction which arises from acting agreeable to its dictates. When we hear those barbarities pled for—When we see them voted for, (as in the late Convention at Boston) when we see them practised by those who denominate themselves Christians, we are presented with something truely *heterogeneous*—something *monstrous* indeed! Can we suppose this line of conduct keeps pace with the rule of right? Do such practices coincide with the plain and simple ideas of government beforementioned? By no means. We could wish it might be kept in mind, that the very notion of government is to protect men in the enjoyment of those privileges to which they have a natural, therefore an indefeasible right; and not to be made an engine of rapine, robbery and murder. This is but establishing inequity, by law founded on usurpation. Establishing this constitution is, in our opinion, establishing the most ignominious kind of theft, man-stealing, and so heinous and agrivated was this crime considered, by one who cannot err, that under the Jewish theocracy it was punished with death. Indeed what can shew men scarcely more hardened, than being guilty of this crime? for there is *nothing else* they will stick at in order to perpetrate this.

The question therefore—Why should we vote for the establishment of this system? recoils upon us armed with treple force—force which sets at defiance, the whole power of sophistry, employed for the defence of those, who by a "cursed thirst for gold," are prompted on to actions, which cast an indelible stain upon the character of the human species—actions at which certain quadrupeds, were they possessed of Organs for the purpose, would discover a BLUSH.

But we were told by an honourable gentleman who was one of the framers of this Constitution, that the two southernmost states, absolutely refused to confederate at all, except they might be gratified in this article. What then? Was this an argument sufficient to induce us to give energy to this article, thus fraught with iniquity? By no means. But we were informed by that gentleman, further that those two states pled, that they had lost much of their property during the late war. Their slaves being either taken from them by the British troops, or they themselves taking the liberty of absconding from them, and therefore they must import more, in order to make up their losses. To this we say they lost no property, because they never had any in them, however much money they might have paid for them. For we look upon it, every man is the sole proprietor of his own liberty, and no one but himself hath a right to convey it unless by some crime adequate to the punishment, it should be made forfeit, and so by that means becomes the property of government: But this is by no means the case in the present instance. And we cannot suppose a vendee, can acquire property in any thing, which at the time of purchase, he knew the vendor had no right to convey. This is an acknowledgment, we are constrained to make as a tribute due to justice and equity. But suppose they had lost real property; so have we; and indeed where is the man, but will tell us he has been a great loser by means of the war? And shall we from thence argue that we have a right to make inroads upon another nation, pilfer and rob them, in order to compensate ourselves for the losses we have sustained by means of a war, in which they had been utterly neutral? Truly upon this plan of reasoning it is lawful thus to do, and had we voted the constitution as it stands, we must have given countenance to conduct equally criminal, and more so, if possible. Such arguments as the above seem to be calculated and designed for idiotcy. We however acknowledge, we think them rather an affront, even upon that.

The hon. Gentleman above named, was asked the question—What would be the consequence, suppose one or two states, upon any principle, should refuse confederating? His answer was—"The consequence is plain and easy—they would be compelled to it; not by force of arms; but all commerce with them would be interdicted; their property would be seized in every port they should enter, and by law made forfeit: and this line of conduct would soon reduce them to order." This method of procedure perhaps no one would be disposed to reprehend; and if eleven, or even nine states were agreed, could they not, ought they not to take this method, rather than to make a compact with them, by which they give countenance, nay even bind themselves (as the case may be) to aid and assist them in sporting with the liberties of others, and accumulating to themselves fortunes, by making thousands of their fellow creatures miserable. To animadvert upon the British manoevres at that time, would not fall within the compass of our present design. But that the Africans had a right to depart, we must assert, and are able to prove it from the highest authority perhaps that this Commonwealth does or ever did afford. In a printed pamphlet, published in Boston in the year 1772, said to be the report of a Committee, and unanimously voted by said town, and ordered to be sent to the several towns in the state for their consideration. In said pamphlet we find the following *axiom*, which we will quote verbatim,—page 2d—"All men have a right to remain in a state of nature as long as they please, and in case of intolerable oppression, civil or religious, to leave the society they belong to, and enter into another."[29] If it can by any kind of reasoning be made to appear, that this authority is not pertinently adduced in the case before us, then we think it can by the same reasoning be investigated, that black is white and white is black—that oppression and freedom are exactly similar, and benevolence and malignity synonymous terms.

The advocates for the constitution seemed to suppose, that this restriction being laid upon Congress only for a term of time, is the "fair dawning of liberty." That "it was a glorious acquisition towards the final abolition of slavery." But how much more glorious would the acquisition have been, was such abolition to take place the first moment the constitution should be established. If we had said that after the expiration of a certain term the practice should cease, it would have appeared with a better grace; but this is not the case, for even after that, it is wholly optional with the Congress, whether they abolish it or not. And by that time we presume the enslaving the Africans will be accounted by far less an inconsiderable affair than it is at present: therefore conclude from good reasons, that the "*nefarious practice*" will be continued and increased as the inhabitants of the country shall be found to increase.

[29]This Boston pamphlet, probably written by Samuel Adams, was entitled *The Votes and Proceedings of the Freeholders and Other Inhabitants of the Town of Boston, in Town Meeting Assembled, According to Law. . . .* The quoted material is from the first part of the pamphlet: "a State of the *Rights* of the Colonists and of this Province in particular." The Boston town meeting approved the pamphlet, which was based upon a report of a committee of correspondence, and ordered 600 copies distributed throughout Massachusetts.

This practice of enslaving mankind is in direct opposition to a fundamental maxim of truth, on which our state constitution is founded, viz. "All
men are born free and equal." This is our motto. We have said it—we cannot
go back. Indeed no man can justify himself in enslaving another, unless he can
produce a commission under the broad seal of Heaven, purporting a licence
therefor from him who created all men, and can therefore dispose of them at
his pleasure.

We would not be thought to detract from the character of any person,
but to us it is somewhat nearly paradoxical, that some of our leading characters in the law department (especially in the western counties) after having (to
their honour be it spoken) exerted themselves to promote, and finally to effect the emancipation of slaves, should now turn directly about, and exhibit to
the world principles diametrically opposite thereto: that they should now appear such strenuous advocates for the establishment of that diabolical trade of
importing the Africans.[30] But said some, it is not we who do it—and compared it to entering into an alliance with another nation, for some particular
purpose; but we think this by no means a parallel. We are one nation, forming
a constitution for the whole, and suppose the states are under obligation,
whenever this constitution shall be established, reciprocally to aid each other
in defence and support of every thing to which they are entitled thereby, right
or wrong. Perhaps we may never be called upon to take up arms for the defence of the southern states, in prosecuting this abominable traffick.

It is true at present there is not much danger to be apprehended, and for
this plain reason are those innocent Africans (as to us) pitched upon to drag
out their lives in misery and chains. Such is their local situation—their unpol-
ished manners—their inexperience in the art of war, that those invaders of
the rights of mankind know they can, at present, perpetrate those enormities
with impunity. But let us suppose for once, a thing which is by no means
impossible, viz. that those Africans should rise superior to all their local and
other disadvantages, and attempt to avenge themselves for the wrongs done
them? Or suppose some potent nation should interfere in their behalf, as
France in the cause of America, must we not rise and resist them? Would not

[30]The reference is probably to Massachusetts Convention members Theodore
Sedgwick of Stockbridge and Caleb Strong of Northampton. In 1781 Sedgwick represented and won the freedom of Elizabeth Freeman, a slave known as "Mumbet."
Two years later Sedgwick was a member of a committee of the state House of Representatives which was asked to draw up a bill declaring that slavery had never been
legal in Massachusetts. The bill passed the House, but failed in the Senate. Strong
was then involved in two of three cases regarding Quock Walker between 1781 and
1783. Walker had run away from his master and was hired as a servant by another
man. Walker's master pursued, caught, and beat him badly. As a result of these cases,
Walker was declared free and received damages for the beating. Chief Justice William
Cushing, one of the judges in the third case, in which neither Sedgwick nor Strong
took part, declared in his charge to the jury that under Article I of the state Declaration of Rights slavery was unconstitutional. (For the Walker case, see Chapter 1.)

the Congress immediately call forth the whole force of the country, if needed, to oppose them, and so attempt more closely to rivet their manacles upon them, and in that way perpetuate the miseries of those unhappy people? This we think the natural consequence which will flow from the establishment of this constitution, and that it is not a forced, but a very liberal construction of it. It was said that "the adoption of this Constitution, would be ominous of much good, and betoken the smiles of Heaven upon the country." But we view the matter in a very different light; we think this lurch for unjust gains, this lust for slavery, portentous of much evil in America, for the cry of inno- cent blood, which hath been shed in carrying on this execrable commerce, hath undoubtedly reached to the Heavens, to which that cry is always di- rected, and will draw down upon them vengeance adequate to the enormity of the crime. To what other cause, than a full conviction, of the moral evil in this practice, together with some fearful forebodings of punishment there- fore arising in the minds of the Congress in the year 1774, can it be imputed, that drew from them at that time, (at least an implied) confession of guilt, and a solemn, explicit promise of reformation? This is a fact, but lest it should be disputed, we think it most safe for ourselves to lay before our readers, an extract from a certain pamphlet, entitled "Extracts from the votes and pro- ceedings of the American Continental Congress, held at Philadelphia, on the 5th of September, 1774, &c." In the 22d page of this same pamphlet, we find the following paragraph, viz. "Second. That we will neither import, nor pur- chase any slave imported, after the first day of December next; after which time we will wholly discontinue the slave-trade, and will neither be concerned in it ourselves, nor will we hire our vessels nor sell our commodities or manu- factures to those who are concerned in it." The inconsistency of opposing slavery, which they thought designed for themselves, and by clandestine means, procuring others to enslave at the same time—it is very natural to suppose would stare them in the face, and at all times guard them against breaking their resolution. Hence it appears to us unaccountably strange, that any per- son who signed the above resolve, should sign the federal constitution. For do they not hold up to view principles diametrically opposite? Can we sup- pose that what was morally evil in the year 1774, has become in the year 1788, morally good? Or shall we change evil into good and good into evil, as often as we find it will serve a turn? We cannot but say the conduct of those who associated in the year 1774 in the manner above, and now appear advocates for this new constitution, is highly inconsistent, although we find such con- duct has the celebrated names of a *Washington* and an *Adams* to grace it. And this may serve as a reason why we could not be wrought upon by another argument, which was made use of in the Convention in favour of the consti- tution, viz. *the weight of names*—a solid argument with some people who belonged to the Convention, and would have induced them to comply with measures of almost any kind. It was urged that the gentlemen who composed the federal Convention, were men of the greatest abilities, integrity and eru- dition, and had been the greatest contenders for freedom. We suppose it to be true, and that they have exemplified it, by the manner, in which they have earnestly dogmatized for liberty—But notwithstanding we could not view this argument, as advancing any where towards infallibility—because long before we entered upon the business of the Convention, we were by some means or other possessed with a notion (and we think from good authority) that "*great men are not always wise.*" And to be sure the weight of a name adduced to give efficacy to a measure where liberty is in dispute, cannot be so likely to have its intended effect, when the person designed by that name, at the same time he is brandishing his sword, in the behalf of freedom for himself—is likewise tyranizing over two or three hundred miserable Africans, as free born as himself. . . .

Notwithstanding what has been said, we would not have it understood, that we mean to be disturbers of the peace, should the states receive the constitution; but on the contrary, declare it our intention, as we think it our duty, to be subject to "the powers that be," wherever our lot may be cast.

PHILANTHROP
Northampton *Hampshire Gazette*, April 23, 1788

There is another objection which the gentlemen declare is insuperable, that Congress may not prohibit the importation of slaves prior to the year 1808; and that the states will be obliged to aid each other in defence and support of every thing, to which they are intitled by the constitution; but are we not already under the most solemn engagements to aid each other in defence and support of every thing, to which we are respectively intitled by the confederation; and by the confederation Congress would never have the power to prohibit the slave trade, the several states retained the right of doing as they pleased in that respect.

The idea that by civil connections, we become partakers of each others sins, I believe is of late date. While this country was connected with Great Britain, we heard nothing from the gentlemen to prove, that the connections ought to be broken because that nation promoted the slave-trade, more than any other people under Heaven. Before the late war, several of the colonies wished to abolish that trade, and passed laws for the purpose, which were negatived by the king; but I presume the gentlemen never urged that as an argument to prove, that we were absolved from our allegiance; if the principle is just, we cannot again be connected with Great Britain for the reasons abovementioned; nor with France, for they have adopted the same policy. Although the European nations, being replenished with white inhabitants, forbid the importation and slavery of negroes at home, yet all of them which have colonies or islands in America, encourage the slave trade. Nor can we be connected even with Connecticut or New-York, for the laws of those states justify slavery; our hand must therefore be against every man, and every man's hand against us.

A few years since, slavery was countenanced by the laws of this state, and many persons imported and held negroes in servitude: but were other persons therefore guilty, who had no concern in the practice, and ever expressed their disapprobation of it. If at that time a man in the state owned a negro, or was guilty of any other *nefarious* practice, and this doctrine of imputed guilt in civil society is just, the gentlemen were bound in conscience to drive the man out of the state, or leave it themselves.

From the printed debates in the Convention it appears, that the advocates for the constitution declared themselves as averse to slavery, as those who opposed it. Notwithstanding which, the gentlemen suggest that the late Convention at Boston voted for the barbarities exercised upon the negroes: and that establishing the constitution, is establishing the most ignominious kind of theft. A very small regard to decency, would have prevented these and several other gross and unmanly reflections, on the major part of the assembly, which is acknowledged to have been the fullest representation of the people ever known in the state. Should the gentlemen write further on the subject, I hope they will do it with less acrimony; and if their Christian charity, of which they obtain the credit of a full share, compels them to bless the negroes, I hope it will restrain them from cursing the Convention.

PHILELEUTHEROS
Northampton *Hampshire Gazette*, May 21 and June 4, 1788

Quid miseros toties in aperta Pericula Cives, projius?
—Virgil

I have somewhat to say in regard to a piece under the signature of *Philanthrop*, inserted in the *Hampshire Gazette* of April 23, intended as an answer to the three gentlemen, who have lately published their reasons of dissent from the Federal Constitution—Herein I shall study brevity. . . .

He says, "he hopes if charity obliges them to bless the negroes, it will not lead them to curse the Convention,"—this is evidently intended as a sarcasm upon the gentlemen, for vindicating the injured rights of the Africans. Why is it thus? Is it because their colour is not so fashionable in this country as in Africa? Or that their features are not supposed to be so beautifully arranged as those of the Americans, and therefore not to be treated with that humanity which it is generally thought belongs to white people? Or does Philanthrop suppose, with some scoffers at sacred history, that those Africans did not proceed from the loins of the same parent with himself, and so on that score to be treated like brutes? He seems to charge them with cursing the Convention, only because they said that body voted for those barbarities which are exercised towards the Africans. Does this amount to a curse? can rehearsing a plain matter of fact, concerning any man or body of men, by good logic, be construed into a curse? Did not the Convention vote, that the Congress should not have it in their power to prohibit the slave-trade until the year eighteen hundred and eight? Were they not sensible, therefore, that this trade would be carried on, with all those barbarities annexed to it? Did they not very generally confess that it was a most wicked piece of business? Where then is the great crime in saying such barbarities were voted for? If it is so, let Philanthrop point out the criminality of it, and no doubt he will obtain credence. If he make charge he ought certainly to adduce some proof to support it. He must be possessed of some entirely new constructed spy-glass, else he never could have spied out curses where there were none. As little ground had he to dream, from any which dropped from the gentlemen, that they had a desire to return back to Great-Britain, as he had to tax them with cursing the Convention. Two constitutions were offered to the people of Massachusetts before one could obtain. Must we infer from thence, that those who opposed the first, did it with a view of becoming subject to Great-Britain! Let him answer, if he please.

I wish, seeing Philanthrop thought fit to take the matter up, that he had made use of less false reasoning—I will give an instance of it. He says, "While under Great-Britain, we heard nothing from the gentlemen, because that nation more than any under heaven, carried on that trade:" and "that whenever a neighbour had imported a slave, their duty was either to drive him out of the country or to quit it themselves." It would not, one would think, require a degree of discernment far beyond mediocrity, to discover the fallacy of this argument. Will the gentleman pretend there is no difference between a man's being born under a government, in the framing of which, consequently he could have no hand and in making one himself? If a man in a state of nature assists in forming a constitution, which gives countenance to iniquity and all kinds of cruelty, does he not by his own act incur guilt to his conscience? But if he is born under such a government, where is the guilt? Neither hath he any right to take upon himself the office of an executioner, and so undertake to punish those who are guilty, as Philanthrop would suppose—all he can have any right to is to remonstrate and exhort. He indeed may share in the calamities which may fall upon the community in consequence thereof; but he has this consolation, that those calamities are not derived through his means, and therefore he cannot share in the guilt.

When a man is called to establish a frame of government, it ought to be such an one as his conscience will justify, or he must give his voice against it. I must say the whole Philanthrop has said on this subject is fallacious, from the foundation to the top-stone. His first position seems calculated to deceive the people, whether designedly or not, I do not say—the purport of which is evidently this, that by civil connections people cannot become partakers of each others sins. What is the import of "civil connections?" Let them import what they will, it is evident Philanthrop would have us think, that in establishing this constitution, we have no more connection in the wickedness which the southern states may commit, with our hearty and free consent and engagements to support them in, than we should be with any one nation under heaven in their inequities, provided we had connected with them for the sake of their trade, &c. But these are not parallel—I wish Philanthrop had ingenuity enough to confess it. It is idle for him to pretend, that establishing this constitution is no more than barely establishing certain civil connections with the southern states; was this the case I would agree with him. I am sensible America may enter into a treaty and connect with Spain for certain purposes, and still not become partakers in the iniquity of the inquisition. But if America and Spain were about to form themselves into one politic national body, and America should in that case pledge to Spain their lives and fortunes for the support of the inquisition, (cousin german,* to the slave-trade) I should certainly suppose, upon good grounds, they would become partakers of the sin, connected with the inquisition. But lest I should be too lengthy, I shall wave what I have to say further upon Philanthrop's piece, until another opportunity, only adding my wish, that if he writes again, his sentiments may better agree with his signature.

[June 4] I have but a word to say at this time to Philanthrop—It would seem by his discourse that he would not be thought a friend to the cruelties exercised in carrying on the Slave-Trade—but yet rather than lose his darling constitution, he would have that trade carried on—Here I am reminded of some conversation to which I was lately a witness. A gentleman lately conversing upon this matter said he disliked the Slave-Trade but he would not have the Constitution set aside on that account—he was asked the question whether, if he had a son a slave in Georgia, he would not overset the Constitution if he could? His reply was "Indeed I would"—This I doubt not is the case with Philanthrop, and every other advocate of the Constitution—I could wish him to reconcile this with humanity—perhaps Philanthrop wishes to reap the benefits arising from the Slave-Trade and let the southern States bear the iniquity—I once heard of a certain man who sent a horse to market by his neighbour—the horse was sold and the returns made, which were so much more than the owner expected or thought he was worth, he had doubt upon his mind whether he ought to receive the whole proceeds—he called a council of neighbours to determine the matter—the result upon the whole was, seeing the horse was sold and the pay made, he who sent him to market should keep the amount for which he was sold—and he who sold him, should bear all the sin—this is the only way I can conceive of, for Philanthrop to get out—and a pretty miserable *get out* too, if I rightly judge of the matter—If we cannot connect with the southern states without giving countenance to blood and carnage, and all kinds of fraud and injustice, I say let them go—Often do we see unhappy convicts led forth to execution for only taking the property of another—The Congress by the new Constitution are to make laws for the

*Rather twin brother.

punishment of piracy and murder upon the high seas—But all this seems only
to respect white people—the Africans may be pirated, hacked and tortured,
and all with impunity. The killing a negroe in the southern states is no more
accounted of, than the killing a dog—this is a fact.

> "Is there not? sure there is, "some chosen curse,
> Some hidden thunder in the store of heaven.
> Red with uncommon wrath, to blast the man,
> Who sport with lives and drain out human blood."

MASSACHUSETTS CENTINEL
June 14, 1788

Mr. Russell: If, for no other consideration than that it opens a door for
the abolition of the Slave Trade, in America, in a given number of years—the
new proposed Constitution for the United States is incomparably preferable
to the old one, in which no provision is made either for the suppression or
circumscription of this wicked trade—and must therefore meet the wishes,
and derive the support of every friend to humanity, and the common rights of
mankind.

It cannot be without a blush, that a European, a friend to liberty, and
the American revolution, in perusing the newspapers of the southern States,
sees page after page, advertisements for the sale, exchange, &c. of men—of
men, too, who are not guilty of any crime—and whose misery is caused by the
barbarity and *avarice* of men, who dare to call themselves Christian. Well may
infidels deny the faith, while the professed followers of the cross are guilty of
such crimes. Well may they boast the *purity* of worship.

"Slavery," says a celebrated writer, "in whatever mode it exists, will ever
be considered as unlawful, except, perhaps, where it is inflicted as a punishment for some enormous crime. That a man should be treated in the same
manner as a beast, or a piece of household furniture, and bought and sold,
and entirely subjected to the will of another, whose equal he is by nature, and
all this for no crime on his part, but merely because he is of a certain colour,
born in a certain country, or descended of certain parents—is absolutely inconsistent with every idea of justice and humanity.

"We are taught by reason and religion to consider every man as our
brother, and to regard him with the same degree of affection, with which we
regard ourselves. Not to confine our benevolence to those of our own colour,
country, or kindred, but to extend it unto all who are endowed with the same
common nature. No impassible mountains, no unnavigable oceans, no inhos-
pitable deserts are to be considered as boundaries, to intercept the force and
authority of this principle. Like the sun, its power and influence ought to
extend to the uttermost parts of the earth, and prove a connecting band of
union with all our fellow creatures. But slavery is utterly irreconcilable with

such a principle, being directly subversive of the dearest and most obvious rights of human nature, and perfectly repugnant to the plainest principles of Christianity.

"Is the pecuniary interest of a few individuals a consideration more to be attended to than the happiness of whole nations? Are millions of the human race to be doomed to a life of unmerited misery—to wear the galling chains of slavery, tyranny and oppression, in order to fill the bags of avarice, or answer the demands of extravagance, of a few particular persons? If robbery and theft were totally suppressed, the finances of the thief and the highwayman would no doubt be materially affected: But are those practices, therefore, to be tolerated and encouraged? Is the robber, therefore to be permitted to provide the necessaries of life, by robbing his neighbour of his purse on the highway? If not, why is the slave-merchant to be permitted to acquire the luxuries and superfluities of life, by robbing thousands of his fellow-creatures of their liberty, on the coast of Africa? What an inconsistent creature is man? How often are his boasted powers of reason led captive by prejudice and custom? How absurd is it, in the case before us, that one person shall be allowed to rise to wealth and affluence, and another condemned to lose his life, for actions, which have the same motive and the same tendency, merely because they are circumstantially different; and yet how must our astonishment encrease, when we consider, that the difference is such as adds greatly to the absurdity; and that the action, which exposes a man to infamy and death, wants only greater aggravation of guilt, and more extensive and pernicious effects, to be the means of advancing him to riches and honour!!

SIMEON BALDWIN'S ORATION
New Haven, July 4, 1788[31]

The best system of government cannot insure freedom, riches, and national respect, without the vigilance, the industry and the virtuous exertions of the people. The labours of the patriot and the friend of humanity are not yet completed. It is their task to remove those blemishes which have hitherto sullied the glory of these States. We may feed our vanity with the pompous recital of noble atchievements—we may pride ourselves in the excellency of our government—we may boast of the anticipated glories of the western continent:—But virtue will mourn that injustice and ingratitude have, in too many instances, had the countenance of law—Humanity will mourn that an odious slavery, cruel in itself, degrading to the dignity of man, and shocking to human nature, is tolerated, and in many instances practised with barbarian cruelty.—Yes, even in this land of boasted freedom, this asylum for the oppressed, that inhuman practice has lost its horrors by the sanction of custom.

[31]Baldwin was a New Haven lawyer.

To remedy this evil will be a work of time.—God be thanked it is already begun. Most of the northern & middle states have made salutary provision by law for the future emancipation of this unfortunate race of men, and it does honour to the candour and philanthropy of the southern states, that they consented to that liberal clause in our new constitution evidently calculated to abolish a slavery upon which they calculated their riches. It is the duty of every friend to his country to lead his fellow citizens to rational reflections upon these interesting subjects to abolish as much as possible the vices peculiar to us as a nation and as individuals, and to disseminate still farther those principles of wisdom and virtue which form the pillars of republican government.

PROVIDENCE *UNITED STATES CHRONICLE*
July 17, 1788

Mr. Wheeler: A Number of black Inhabitants of Providence, pleased with the Prospect of a Stop being put to the Trade to Africa in our Fellow-Creatures, by the Adoption of the Federal Constitution, met on the 4th Instant, in Celebration of that happy Event—and after Dining on the Product of their own Industry, drank the following Toasts—which you are desired to publish.

1. The Nine States that have adopted the Federal Constitution.
2. May the Natives of Africa enjoy their natural Privileges unmolested.
3. May the Freedom of our unfortunate Countrymen (who are wearing the Chains of Bondage in different Parts of the World) be restored to them.
4. May the Event we this Day celebrate enable our Employers to pay us in hard Cash for our Labour.[32]
5. The Merchants and others who take the Lead in recommending Restoration of Equity and Peace.
6. His Excellency General Washington.
7. The Humane Society of Philadelphia.
8. Hon. John Brown, Esq.[33]
9. May Unity prevail throughout all Nations.

[32]A critical reference to the radical fiscal policies of the Country Party which included state-issued paper currency that depreciated greatly and then was used to pay the state's large wartime debt, which had accumulated in the hands of wealthy speculators who dominated the Mercantile Party.

[33]The toast to Brown, a former Providence assemblyman, is inexplicable since he reentered the foreign slave trade after the Revolution. For Brown's letter justifying the slave trade in 1786, see Chapter 1.

4

The Middle States
Debate Slavery
and the Constitution

D URING THE MEETING OF THE CON-
stitutional Convention, delegates from the Middle states opposed slavery with
vehemence and moral righteousness rarely displayed by New Englanders. But during
the public debate over ratification of the Constitution, this opposition to slavery was
muted. Without an asylum for runaway slaves such as Massachusetts, little attention
centered on the fugitive-slave clause in the Middle states. Rather a gentlemanly,
high-toned debate focused on the political consequences of the three-fifths clause and a
slightly more emotional debate addressed the clause prohibiting Congress from closing
the African slave trade before 1808.

Federalists from the Middle states were unjustifiably optimistic that under the
Constitution slavery would perish. They placed great confidence in the effect that the
prohibition of the foreign slave trade would have on slavery itself. Federalists con-
gratulated themselves that in twenty-one years the foreign slave trade would be
prohibited and America in less than 180 years from its founding would "possess a
degree of liberality and humanity which has been unknown during so many centuries, and which is yet unattained in so many parts of the globe." "The abolition of
slavery," Federalists suggested, "is put within the reach of the federal government."
"The foundation for banishing slavery out of this country" was laid. By the time a
newborn slave reached adulthood, "the supreme power of the United States shall
abolish slavery altogether," and, in the meantime, a tax or duty could be laid on slave
importations which will operate "as a partial prohibition." Furthermore, all newly
formed states "will be under the control of Congress in this particular; and slaves will
never be introduced amongst them." When they considered the situation in the South,

Federalists felt that Northerners "will find more reason to rejoice that the power should be given at all, than to regret that its exercise should be postponed for twenty years." James Wilson, defending the slave trade clause in the Pennsylvania convention, declared that "If there was no other lovely feature in the Constitution, but this one, it would diffuse a beauty over its whole countenance. Yet the lapse of a few years and Congress will have power to exterminate slavery from within our borders." "Instead of finding fault with what has been gained," Pennsylvania Chief Justice Thomas McKean was "happy to see a disposition in the United States to do so much." Steady progress toward gradual abolition was the goal; any attempt to speed up the process might jeopardize the emancipation movement.

Antifederalists attacked the unlikely optimism of their opponents. Congress, they argued, merely had the option to prohibit the foreign slave trade in 1808, but it would have absolutely no power to abolish the institution of slavery itself. In fact, Northerners in 1808 would be equally as willing "to ease the southern states, as it is prettily called, — as at present." Regard for the American character would continue to fall throughout the world while the nation continued a commerce "that can only be conducted upon rivers of human tears and blood." A government built on a foundation of "the flagrant evil of Slavery" could only expect to fail. According to Antifederalists, Union with the Southern states would cost too high a price.

JAMES PEMBERTON TO JOHN PEMBERTON
Philadelphia, September 20, 1787[1]

The Expectation of our Politicians has been much turned towards the determination of this Convention [i.e., the Constitutional Convention], the members of which being under an injunction of Secrecy, their proceedings have been kept very close; how they will now relish the Plan, time will make manifest, but the late [Confederation] Congress had become so very low in general estimation, a Change with enlarged powers, & a proper balance seemed to be absolutely necessary, but yet, unless there is an increase of Virtue among the People, all the efforts of human wisdom, & policy will avail little to promote their real happiness, and welfare—I have given thee these outlines of the new plan of a Foederal Government, with a view to mention, that we entertained a hope, that it's establishment would have been more conspicuous on the principles of Equity & moral Justice by a Provision against the iniquitous Slave trade, but the influence of the Southern Governmts has diverted them from that very important Object, so far as to obtain a prohibition

---

[1]Pemberton Papers, Historical Society of Pennsylvania. James Pemberton and his brother John were leaders of the Quaker community and the antislavery movement in Pennsylvania. In 1787 James Pemberton was a vice president of the Pennsylvania Society for Promoting the Abolition of Slavery. John Pemberton was in Scotland on a preaching tour.

against the Congress medling therewith for 21 years, as appears by the ninth Section of the first Article of the Plan—which Says—viz.

"The migration, or importation of such Persons as any of the States now existing shall think proper to admit, shall not be prohibited by Congress prior to the year 1808, but a tax or duty may be imposed on such importation, not exceeding ten dollars for each person" which is further defended by a fifth Article, which after liberty given for the mode of proposing future amendments to this intended Constitution, sets forth a Proviso, "that no amendment which may be made prior to the year 1808 shall in any manner affect the first and fourth clauses in the ninth Section of the first Article."

PENNSYLVANIA GAZETTE, SEPTEMBER 26, 1787[2]

It is remarkable, that while the foederal government lessens the power of the *states*, it increases the privileges of *individuals*. It holds out additional security for liberty, property and life, in no less than *five* different articles, which have no place in any one of the state constitutions. It moreover provides an effectual check to the African trade, in the course of one and twenty years. How honorable to America—to have been the first Christian power that has borne a testimony against a practice, that is alike disgraceful to religion, and repugnant to the true interests and happiness of Society.

DR. BENJAMIN RUSH TO DR. JOHN COAKLEY LETTSOM
Philadelphia, September 28, 1787[3]

To the influence of Pennsylvania chiefly is to be ascribed the prevalence of sentiments favorable to African liberty in every part of the United States. You will see a proof of their operation in the new constitution of the United States. In the year one thousand eight hundred and eight there will be an end of the African trade in America. No mention was made of *negroes* or *slaves* in this constitution, only because it was thought the very words would contaminate the glorious fabric of American liberty and government. Thus you see the cloud which a few years ago was no larger than a man's hand, has descended in plentiful dews and at last cover'd every part of our land. . . .

[2]Reprinted twenty-seven times from Vermont to Virginia by October 18.
[3]Rush Papers, Library Company of Philadelphia. Rush of Philadelphia and Lettsom of London were prolific writers on medical subjects and on a wide range of reforms.

A CITIZEN OF AMERICA (NOAH WEBSTER)
An Examination into the Leading Principles of the Federal Constitution, Philadelphia, October 17, 1787[4]

But, say the enemies of slavery, negroes may be imported for twenty-one years. This exception is addressed to the quakers; and a very pitiful exception it is.

The truth is, Congress cannot prohibit the importation of slaves, during that period; but the laws against the importation into particular states, stand unrepealed. An immediate abolition of slavery would bring ruin upon the whites, and misery upon the blacks, in the southern states. The constitution has therefore wisely left each state to pursue its own measures, with respect to this article of legislation, during the period of twenty-one years.

ROBERT WALN TO RICHARD WALN
Philadelphia, October 3, 1787[5]

Thee wou'd not censure the Convention so severely if thee knew every circumstance respecting the Negroes—by far the greatest part wished to abolish that trade entirely, & a resolution was offer'd for that purpose but the Southern Delegates positively refus'd their consent & threatened to retire if the motion was not withdrawn—and as their absence would have broken up the House, it was thought best to withdraw it, and introduce another (which was carried) & which will put it in the power of Congress at the end of 21 years to put a total stop to that iniquitous traffic—and as each state is still at liberty to enact such laws for the abolition of slavery as they may think proper, the Convention cannot be charg'd with holding out any encouragement to it.

AN AMERICAN CITIZEN (TENCH COXE)
On the Federal Government, No. 4, October 21, 1787[6]

In considering the respective powers of the President, the Senate and the House of representatives, under the foederal constitution, we have seen *a*

---

[4]Noah Webster, a native of Connecticut, moved to Philadelphia in 1786. In 1785 he wrote a pamphlet advocating a stronger central government. In mid-September 1787 Philadelphia Federalists asked him to write a pamphlet defending the newly proposed Constitution. The excerpt printed here is taken from this pamphlet, which circulated throughout the country.

[5]Autograph Collection of the Historical Society of Pennsylvania. Robert Waln and his cousin Richard were wealthy Philadelphia merchants.

[6]This final essay in a four-part series, first appeared in a four-page Federalist anthology printed by Hall and Sellers of the *Pennsylvania Gazette*. On October 24 the essay was printed in the *Pennsylvania Gazette* and in the Philadelphia *Independent Gaz-*

part of the wholesome precautions, which are contained in the new system. Let us examine what *further securities for the safety and happiness of the people* are contained in the general stipulations and provisions.

. . . the importation of slaves from any foreign country is, by a clear implication, held up to the world as equally inconsistent with the dispositions and the duties of the people of America. A solid foundation is laid for exploding the principles of negro slavery, in which many good men of all parties in Pennsylvania, and throughout the union, have already concurred. The *temporary* reservation of any particular matter must ever be deemed an admission that it should be done away. This appears to have been well understood. In addition to the arguments drawn from liberty, justice and religion, opinions against this practice, *founded in sound policy*, have no doubt been urged. Regard was necessarily paid to the peculiar situation of our southern fellow-citizens; but they, on the other hand, have not been insensible of *the delicate situation of our national character on this subject*.

TIMOTHY MEANWELL
Philadelphia *Independent Gazetteer*, October 29, 1787

Friend Oswald:[7] As I sometimes (though not very often) read newspapers, and when I do read any, I generally give thine the preference; because if there is any dispute a-going, thee generally has both sides of the question. I have also read the new constitution which is offered to us, and I am very sorry to inform thee that I don't altogether like it. . . .

I also heard a story going that the importation of slaves was allowed for twenty-one years, I thought now I certainly had in my power to catch them in telling untruths, I took up the constitution once more, and went to searching again, (for I was sure my friend _____ [Benjamin Franklin], whose character I very much respect, would never attempt to encourage or connive at slavery, he who is famed throughout the world as the champion of liberty, nor friend _____ [Jared Ingersoll], who is one of the members of our society for the abolition of slavery, would never agree to so inhuman a traffic as that of carrying on a trade in the human species) but to my mortification, I found that this assertion was too true; for in the 9th section of the new constitution, this traffic is allowed: however, I thought this part would never do at any rate, and I was in hopes that some well disposed people would petition and have this

etteer. It was reprinted in nine other newspapers from Massachusetts to South Carolina by December 10. Tench Coxe, a secretary of the Pennsylvania Abolition society, was a Philadelphia merchant and one of the most active Federalist essayists during the ratification debate.

[7]Eleazer Oswald, a highly partisan Antifederalist, was the owner and printer of the *Independent Gazetteer*.

article erased and abolished as a disgrace to the annals of America—But methinks I wont be too censorious but examine further, perhaps I shall find some method by which this clause may be evaded or repealed, but to my mortification, the further I went the worse I liked it—I had been told that there was a clause reserving a right to amend the constitution.—Ah thinks I, here is a hole in which the importation of slaves will be thrust out of the constitution; I pushed on in search of the clause, I found it, but what was my surprise when I found it, for in the 5th article, I find that there are two clauses which cannot in the new constitution be repealed till after the year 1808, and perhaps never will after that time, one of which, is that of allowing the importation of slaves for 21 years. . . .

I am afraid friend Oswald, that I have trespassed too much upon thy patience; upon the whole, I have I think, now sufficiently considered the constitution to inform thee that I dont like it, and there is so many things in it incompatible with the known principles of Friends, that I think they will deviate much from their profession if they have any thing to do with it, or give any aid or assistance in establishing of it, for if they do assist in establishing of it, remember, they have lent their aid in abolishing of the liberty of conscience; in encouraging and establishing the importation of slaves for 21 years—they also give their assent to the raising and keeping up of a standing army, all of which are totally incompatible with the principles of Friends, and I hope they will steer clear of having any hand in the establishing of these several facts. I hope they will adhere to their good old rules of neither setting up nor pulling down governments, that is to say, of neither setting up their new government, nor of pulling down the good old constitution of Pennsylvania, which has secured and protected them in so many civil and religious privileges. I have made free to write thus much to thee at present, which thou art at liberty to communicate, if thou thinkest proper. The time is short wherein the liberty of the press may be preserved. Before it is too late and becomes shackled and restrained, I beg leave to communicate my sentiments, though perhaps at this time it may in some measure be dangerous, but the constitution of Pennsylvania protects me, I have a right to enjoy that protection, which is secured to me by the 12th section of the bill of rights.—"That the people have a right to freedom of speech, and of writing and publishing their sentiments; therefore the freedom of the press ought not to be restrained."

If any thing else occurs to me, I shall make free to write to thee upon the subject; and as I always hold myself open to conviction, if I have not formed a right idea of the matter, or if I have taken it up wrong, I will thank any friend to set me right. I am with the greatest esteem and respect, Thy assured and well-wishing friend.

Spank Town, 10th month 20th day, 1787

PLAIN TRUTH TO TIMOTHY MEANWELL
Philadelphia *Independent Gazetteer*, October 30, 1787

Friend Timothy: Thou hast abused thy name and the sect to which thou pretendest to belong, by telling things that are false; hence I conclude that thou dost not mean well.

Thou sayest that "the importation of slaves is allowed by the new federal constitution for twenty-one years," and thence thou hast falsely insinuated as if friend _____ [Franklin] and friend _____ [Ingersoll] had given their sanction to this unchristian practice. This is the same mode of arguing, that deistical and profane writers adopt to oppose the dictates of Jesus Christ; they take their own construction of some particular phrase, and then applying it to a foreign subject, they think they prove, that our Redeemer was inconsistent with himself; But an enlightened Christian need only look into the book, and the cheat appears evident. Thus it is with thy uncandid assertion. Let us look into the constitution—"The migration or importation of such persons as any of the states now existing shall think proper to admit, shall not be prohibited by the congress prior to the year 1808."—Would not every candid reader conclude from this, that in twenty one years such importations *may be prohibited*; and would he not bless God, that in this new country, we should, in less than 150 years, possess a degree of liberality and humanity, which has been unknown during so many centuries, and which is yet unattained in so many parts of the globe.

What alteration does the new constitution make in the present system adopted by many of the states relative to slavery? NONE contrary to that system; but in favor of it, has taken a power of checking this abominable importation, by laying duties on it. The constitution says, by implication, to such states,—"well done ye good and faithful servants, continue your endeavors to compleat the glorious work—our assistance is not very far distant; for, ere the child now born, shall arrive to an age of manhood, the supreme power of the United States shall abolish slavery altogether, and in the mean time they will oppose it as much as they can."

I fear, Timothy, that thy disturbed spirit has led thee to make these remarks, before thou hadst read the letter written by the President of the Convention.[8]

"The Constitution (saith he) which we now present, is the result of a spirit of amity, and of that mutual deference and concession which the peculiarity of our political situation rendered indispensable." Now is it not the duty of every candid objector to consider this before he makes his remarks? The Convention in fact tells every opposer, "Friend we have considered thy objection, but points of small magnitude must give way to general good; if

---

[8]George Washington to the President of Congress, Philadelphia, September 17, 1787, National Archives.

thy objection had been insisted on, as thou insistest, we should have made no constitution at all, for we never could have agreed."

As to the idea that the new constitution is disagreeable to our society, who ever saw friends prefer anarchy, confusion, a bloodshed to the blessings of good government? and what else but anarchy, confusion, and bloodshed can be expected from a refusal of this constitution, and the consequent dissolution of the union?

I wish thy reformation, and am thy friend.

Timothy Meanwell
Philadelphia *Independent Gazetteer*, November 3, 1787

Friend Oswald: When I wrote to thee last, it was with a wish that if I had imbibed a wrong idea of our proposed new constitution that some friend or other would endeavour to set me right: I have seen the publication of a person under the *pirated* signature of *Plain Truth*; I call it *pirated*, because the signature is affixed to that which is not *true*, in which performance he pretends to remove some of the objections which had taken hold of my mind relative to the new constitution; instead of answering or removing any of those objections, he has the more firmly fixed them with me, he has not removed a single one, neither has he touched upon any part of my letter, only that which alludes to the importation of slaves, and instead of his removing of this objection, he has confirmed it, and has inserted a part of the new constitution, which he says will prove *my* assertion, to be false; the constitution runs thus, "The migration or importation of persons, as any of the states now existing shall think proper to admit, *shall not be prohibited by the* Congress, prior to the year 1808;" mark these words, "*they shall not be prohibited, &c.*" this confirms the very objection which I had to the clause allowing of the importation of slaves; he goes on further, and says, "Would not every candid reader conclude from this, that in one and twenty years, such importations *may be prohibited, &c.*" here he does not entertain the least shadow of a prohibition taking place before 21 years, and after that time he says, a prohibition *may take place*; if he meant in the least degree to have removed my objections, he ought not to have relied upon *maybe's*; such flimsey arguments will not in the least tend to conviction with me, I must have more solid reasoning than such pitiful *may-be's* as these: he will I trust, excuse me if I am a little severe in my reply, after being treated so cavalierly by him. He goes on further and says, "What alteration does the present constitution make in the present system adopted by many of the states relative to slavery;" he draws the answer himself, which is not the answer of *Plain Truth*, and says "None contrary to that system:" this is false, because, if the new constitution is adopted, it will in a great measure contravene the act of assembly of Pennsylvania for the gradual abolition of slavery, which act expressly forbids the further importation of slaves; the new constitution says they *shall not be prohibited* for 21 years, and goes no further, and does not say that a prohibition shall take place after that period: this is left to be determined by the *great Congress* hereafter to be chosen, who may or may not abolish this inhuman traffic after 21 years, as they in their great goodness and unbounded wisdom may think proper—This new constitution by implication says thus—Ill done ye bad and faithless servants, continue your endeavours to complete the inglorious work which you have begun, our assistance is not far distant, for ere the child now born shall arrive to the age of manhood, the supreme power of the United States, after having established slavery for 21 years, they will entirely and irrevocably fix it altogether, and in the mean time the new constitution will encourage it all that it can.

I fear thou hast the signature of *Plain Truth*, that thy disturbed spirit has led thee to make these remarks, before thou hast examined the new constitution.

I observe by a part of thy performance, that thou either are tolerably well versed in the scriptures, or at least thou would have the public to entertain that opinion of thee. There is some matters in the scriptures which I sometimes have heard mentioned, and am at a loss to find answers for; I have no doubt but that from thy acquaintance with that old-fashioned and too much neglected book, thou canst answer them; I will give thee one of them, if thou canst answer thou wilt do me a kindness, *Pray who was David's grandfather's nurse?* this answer I expect thou wilt give as a result of thy own information, and not seek it from any divine; I shall then perhaps, entertain a better opinion of thy knowledge in the scriptures than I do at present, some other queries may be offered to thee as they occur.

I must, before I conclude, beg leave to inform thee, that I had a very great suspicion that thou want not what thou wouldst have the world to believe thou art, that is, that thou art not that *venerable old dame* designated by the title of *Plain Truth*, but that thou hadst *pirated* her name, to affix to thy publication, having had some considerable acquaintance with this *venerable lady*. I made it my business to wait upon *her*, in hopes that if I had imbibed a wrong notion of the new constitution *she* would set me right. After making some enquiry for *her*, (for thou must understand, *she* is so often attacked and abused *she* is obliged frequently to remove *her* quarters) I found her inhabiting of a very neat small house up a little ally out of doors tied to a stump; after passing of the usual compliments, I told *her* that I had seen a publication under *her* signature in the Independent Gazetteer, of the 30th October, which I did not altogether understand, and I waited upon *her* for *her* explanation of the matter. *She* immediately denied having wrote or published any thing. I answered that I had the paper in my pocket and would shew it to *her*; *she* took it and perused it, and told me that it was none of *her* performance, and that *she* totally disclaimed every iota contained in that publication as *her* performance; the *venerable lady* seemed very much enraged and out of humour, thinking that any person should thus so much traduce *her* as to publish *untruths* under *her* signature, and told me *she* wished *she* could detect the impostor, and requested that I would have the advertisement herewith sent,* (which *she* handed me) immediately published that the impostor might be detected. I told *her* I would take special care to have it done. *She* told me *she* took it very kind in my waiting on *her* on this occasion, and requested immediate information if any more impositions should appear in future, when after the usual salutations we parted.

All the other objections which I have made to the new constitution I presume are acceded to, as none of them have been contradicted.

I fear friend Oswald I have trespassed too much upon thy good nature, but having been informed that thou wast very obliging, made me trouble thee thus long, which I hope thou wilt excuse; and should the same person who has signed himself *Plain Truth* appear again unless he adduces better arguments than his last, I shall not take up the time of thy readers, as too much of thy valuable paper will be lost by answering of him, I choose not to be throwing of *pearls before swine*, but only hope that he may not enjoy the first fruits of this new constitution, especially to have that part retorted on him that relates to slavery, by his being sold as a slave to the Dey of Algiers or to Botany Bay, that he may not there know the blessings of liberty, and be debarred from the enjoyment of it, perhaps he may then repent that he had not paid a more minute attention to this new constitution, and wish that he had given the establishment of it all the opposition in his power, however if he has such contracted notions of liberty, I leave him to *Satan to buffet with*. I am, With the greatest respect, Thy assured Friend.

Spank Town, October 31st, 1787.

*Stop Thief! Stop Thief!

Whereas a certain person under the signature of *Plain Truth* published in the Independent Gazetteer of the 30th October, has made free to pirate my name and palm it upon the public attached to a performance, which I totally disavow and hereby disclaim every iota of. I also hereby offer a reward adequate to the trouble to any person who shall detect the impostor and deliver him to me.

Given under my hand and seal at my palace at a little house up an alley out of doors tied to a stump, this 31st day of October, 1787.

AN OFFICER OF THE LATE CONTINENTAL ARMY (WILLIAM FINDLEY)
Philadelphia *Independent Gazetteer*, November 6, 1787[9]

The objections that have been made to the new constitution, are these: . . .
20. The importation of slaves is not to be prohibited until the year 1808, and SLAVERY will probably resume its empire in Pennsylvania.

PLAIN TRUTH
Philadelphia *Independent Gazetteer*, November 7, 1787[10]

Friend Oswald: Thy correspondent, *Timothy Meanwell*, is a weak man. I would in charity hope, that the very unworthy motive which appears to actuate him, is rather the consequence of imbecility in his brain than corruptness in his heart: and I am the more inclined to think of him in this compassionate manner, as I find his invention fails, even his scurrility, which is a feeble imitation of some of his silly predecessors. How often alas! has the public been nauseated by the paltry witticisms against my signature!

I shall not trouble thee with any more personal observations on this poor man; but it may not be amiss to shew how falsely and how weakly he has quoted that part of the new constitution which is supposed to relate to the slave trade. The constitution saith, that, this importation "shall not be prohibited by the Congress prior to the year 1808."—This is thus represented by Timothy; "mark these words (saith he) *shall not be prohibited*," and thence he draws this wise conclusion, that, the new government (notwithstanding this importation is already prohibited in at least 10 of the states) hath "*established slavery for* 21 *years.*" Suppose this federal compact had said that *Congress* should not prohibit theft; would that abolish the state laws and establish theft in Pennsylvania?—If friend Timothy were in that case to reason in such a manner, and to act accordingly, I fear he would discover his error, under the discipline of the wheel-barrow.[11] . . .

---

[9]Reprinted as a broadside, in a pamphlet, in the November issue of the Philadelphia *American Museum*, and in eight newspapers from Pennsylvania to Massachusetts by January 9, 1788.

[10]Except for the first two paragraphs, this item was reprinted twelve times by February 5 from Vermont to Virginia.

[11]A reference to Pennsylvania's public-works system in which convicts sometimes were chained to a wheelbarrow.

CENTINEL III (SAMUEL BRYAN)
Philadelphia *Independent Gazetteer,* November 8, 1787[12]

Section the 9th begins thus,—"The migration or importation of such persons, as any of the states, now existing, shall think proper to admit, shall not be prohibited by Congress, prior to the year 1808, but a duty or tax may be imposed on such importation not exceeding ten dollars for each person." And by the fifth article this restraint is not to be removed by any future convention. We are told that the objects of this article, are slaves, and that it is inserted to secure to the southern states, the right of introducing negroes for twenty-one years to come, against the declared sense of the other states to put an end to an odious traffic in the human species; which is especially scandalous and inconsistent in a people, who have asserted their own liberty by the sword, and which dangerously enfeebles the districts, wherein the laborers are bondmen. The words dark and ambiguous; such as no plain man of common sense would have used, are evidently chosen to conceal from Europe, that in this enlightened country, the practice of slavery has its advocates among men in the highest stations. When it is recollected that no poll tax can be imposed on *five* negroes, above what *three* whites shall be charged; when it is considered, that the impost on the consumption of Carolina field negroes, must be trifling, and the excise, nothing, it is plain that the proportion of contributions, which can be expected from the southern states under the new constitution, will be very unequal, and yet they are to be allowed to enfeeble themselves by the further importation of negroes till the year 1808. Has not the concurrence of the five southern states (in the convention) to the new system, been purchased too dearly by the rest, who have undertaken to make good their deficiencies of revenue, occasioned by their willful incapacity, without an equivalent?

PLAIN TRUTH
Philadelphia *Independent Gazetteer,* November 10, 1787[13]

[Answers objections raised by An Officer of the Late Continental Army.]
"20. The importation of slaves is not to be prohibited until the year 1808, and SLAVERY will probably resume its empire in Pennsylvania."
20. This is fully answered in my letter to Timothy, but it may not be amiss to repeat that Congress will have no power to meddle in the business

---

[12]Reprinted in the *Pennsylvania Herald,* November 10; Philadelphia *Freeman's Journal,* November 14; *New York Journal,* November 20; Providence *United States Chronicle,* January 3; Boston *American Herald,* January 7, 1788; and in a New York pamphlet anthology.
[13]Reprinted in the November issue of the Philadelphia *American Museum* and in the *Carlisle Gazette,* November 28.

'till 1808. All that can be said against this offending clause is, that we may have no alteration in this respect for 21 years to come, but 21 years is fixed as a period when we may be better, and in the mean time we cannot be worse than we are now.

BRUTUS III (MELANCTON SMITH)
New York Journal, November 15, 1787[14]

The words are "representatives and direct taxes, shall be apportioned among the several states, which may be included in this union, according to their respective numbers, which shall be determined by adding to the whole number of free persons, including those bound to service for a term of years, and excluding Indians not taxed, three fifths of all other persons."—What a strange and unnecessary accumulation of words are here used to conceal from the public eye, what might have been expressed in the following concise manner. Representatives are to be proportioned among the states respectively, according to the number of freemen and slaves inhabiting them, counting five slaves for three free men.

"In a free state," says the celebrated Montesquieu, "every man, who is supposed to be a free agent, ought to be concerned in his own government, therefore the legislature should reside in the whole body of the people, or their representatives." But it has never been alleged that those who are not free agents, can, upon any rational principle, have any thing to do in government, either by themselves or others. If they have no share in government, why is the number of members in the assembly, to be increased on their account? Is it because in some of the states, a considerable part of the property of the inhabitants consists in a number of their fellow men, who are held in bondage, in defiance of every idea of benevolence, justice, and religion, and contrary to all the principles of liberty, which have been publickly avowed in the late glorious revolution? If this be a just ground for representation, the horses in some of the states, and the oxen in others, ought to be represented— for a great share of property in some of them, consists in these animals; and they have as much controul over their own actions, as these poor unhappy creatures, who are intended to be described in the above recited clause, by the words, "all other persons." By this mode of apportionment, the representatives of the different parts of the union, will be extremely unequal; in some of the southern states, the slaves are nearly equal in number to the free men; and

[14]Part of a series of sixteen Antifederalist essays, this essay was reprinted in the Philadelphia *Freeman's Journal*, November 21; Philadelphia *Independent Gazetteer*, November 23; Boston *Independent Chronicle*, December 13. For a response, see Mark Antony, Boston *Independent Chronicle*, January 10, 1788, in Chapter 3. Melancton Smith, a New York City merchant, was a leading advisor of Governor George Clinton. He was the Antifederalist "manager" of the N.Y. ratifying convention, but advocated ratification after the Constitution had been adopted by ten states.

for all these slaves, they will be entitled to a proportionate share in the legis-lature—this will give them an unreasonable weight in the government, which can derive no additional strength, protection, nor defence from the slaves, but the contrary. Why then should they be represented? What adds to the evil is, that these states are to be permitted to continue the inhuman traffic of importing slaves, until the year 1808—and for every cargo of these unhappy people, which unfeeling, unprincipled, barbarous, and avaricious wretches, may tear from their country, friends and tender connections, and bring into those states, they are to be rewarded by having an increase of members in the general assembly.

James Pemberton to Moses Brown
Philadelphia, November 16, 1787[15]

I lately recd. thy acceptable letter of 17th ulto. by which I perceive that thy mind has been exercised in like manner with many others of thy brethren in these parts, on account of the present stirrings among the people in their political pursuits, and that the members of our religious Society may be pre-served in a conduct consistent with our profession, to promote which, some caution was verbally given in our late yearly meeting, it being the united sense of the Solid & Judicious among friends, that our union, and safety depended on our quietude, & forbearance to intermix with the people in their political consultations, and debates on the present occasion; the like cautionary advice has been repeated in our quarterly meeting, and if the Representatives from the other quarters, and other concerned friends who attended the yearly meet-ing perform their duty, care will be taken to revive, and diffuse the same in the quarterly meetings now coming on, and transmit it to the monthly meet-ings, as also to their members individually as occasion offers, the weighty part of friends being much united in Judgment on the subject, but we are numer-ous, and there are many among us weak & unstable who stand in need of suitable counsel on occasions of this kind, as on others, in which our testi-mony is concerned; that there is ground to apprehend, that divers were precipitately drawn in to sign petitions to the Assembly towards the close of their session in the ninth month last to promote a speedy Election of Del-egates for the State Convention,[16] which proceeded from inattention, and I believe many have been Since convinced of the impropriety of their conduct in that matter, from whence the people who are active in these concerns may

[15]Moses Brown Papers, Rhode Island Historical Society. For Brown's letter to Pemberton of October 17, 1787, see Chapter 3.
[16]In mid-September 1787 Federalists mounted a strenuous petition campaign in the City and County of Philadelphia, requesting that the assembly speedily call a state ratifying convention.

have taken occasion to represent the Judgment of the Society being favorable to their cause, but an Election of Delegates for the proposed State Convention has since been held in this City, and other parts, and I do not find, that our members have intermeddled any way, except a few inexperienced young men, and others who are resolute to follow their own wills without due consideration, and run with the multitude at all hazards—

Altho' it is most consistent & safe for us to avoid an active part in the business now in agitation, yet we can but observe those things which are exceptionable in the plan of Government recommended by the late General Convention, and that there are several parts which may affect Civil & religious liberty, at the same time Charity leads me to conclude that they have done the best they could under the circumstances attending their deliberations, and Some of the Delegates appologize for its imperfections particularly in respect to that part which appears to give countenance to the Slave trade for twenty one years, tho' the construction they put on those Sections is, that they only limit the power of the Foederal Legislature, and are not intended to restrain the Legislatures of the respective States from enacting such laws, or Supplements to laws already in force, as they shall judge expedient for the prohibition of the trade, or the abolition of Slavery within their own jurisdiction, and some of our Lawyers have given their opinion to the same purport; There was a desire prevailed in the Convention to subvert the enormous traffic, which the Representatives from So Carolina, & the Adjacent States being aware of, vigorously opposed, and is Said to be the Sole cause of this very inconsistent part of their System professed to be founded on liberal principles, and is given out among other reasons by the Virginia Delegates who declined Signing, for their dissent; However should the plan be adopted, which seems not to be improbable; it will be requisite for the Advocates for the Enslaved Negroes to consider, whether consistent with their laudable desire for their emancipation, and the Suppression of the iniquitous commerce to Africa for Slaves, they ought not firmly to remonstrate against those very exceptionable parts of a Constitution said to be intended to hold up a Standard of impartial Liberty, and I hope friends here, and others will not be inattentive to a Subject of such weighty importance. . . .

The Essay on the Slave trade thou Sent me I have delivered to one of our news-printers for republication, and observe it is inserted in a paper of this day; The Act lately passed by your Assembly to prevent the Slave Trade does them credit, but I fear it is not sufficiently explicit to prevent evasion; There is an Intention of applying to the assembly of this Government for a Similar law, instances having lately occurred to make it expedient.

A COUNTRYMAN FROM DUTCHESS COUNTY I (HUGH HUGHES)
New York Journal, November 21, 1787[17]

As you have several Times intimated a Wish to know my Sentiments, relative to the conduct of the late Convention, as well as of the Constitution, which they have offered to the Consideration of the People, I shall freely, as often as convenient, communicate whatever occurs to me on the Subject, as most worthy of Observation, if not already publicly discussed. When the latter is the Case, perhaps I may drop a Sentiment concerning the Propriety, or Impropriety, of the Discussion, etc. But all this, my Friend, will in a great Measure, depend on your reciprocating; for I am too phlegmatic to write, unless answered.

In the first Place then, most unfeignedly do I wish, and that for the Sake of Humanity, that the Convention never had existed; and, for the Sake of our old illustrious Commander in chief, I wish, as they have departed from their Instructed, that they had offered a Constitution more worthy of so great a Character. But, as he has acted entirely in a Ministerial Capacity, so I wish to consider him, whenever I am obliged to mention his venerable Name, or allude to it. Not that I think any Name, however great, can justify Injustice, or make Slavery more eligible than Freedom, and beg to be so understood.

Yet, when I consider the original Confederation, and Constitutions of the States which compose the Union, as well as the Resolutions of several of the States, for calling a Convention to *amend* the Confederation, which it admits, but not *a new one*, I am greatly at a Loss to account for the surprizing Conduct of so many wise Men, as must have composed that honorable Body. In fact, I do not know, at present, whether it can be accounted for; unless it be by supposing a Predetermination of a Majority of the Members to reject their Instructions, and all authority under which they acted.

If this be the Case, the Transition to prostrating every Thing that stood in their Way, though ever so serviceable or sacred to others, was natural and easy.—However, I do not even wish to think so unfavorably of the Majority; but rather, that several of them, were, by different Means, insidiously drawn into the Measures of the more artful and designing Members, who have long envied the great Body of the People, in the United States, the Liberties which they enjoy.—And, as a Proof of their being Enemies to the Rights of Mankind, permit me to refer you to the first Clause, of the 9th Section, of the first Article of the new Constitution, which is framed to deprive Millions of the human Species of their natural Rights, and, perhaps, as many more of their Lives in procuring others! That Clause, you will immediately perceive, has

---

[17]This first of six essays was written by Hugh Hughes, a Dutchess County landholder who had served as Continental deputy quartermaster general during the war. He was part of the inner circle of Antifederalist publicists that regularly supplied material for the New York City press.

been purposely so contrived for reviving that wicked and inhuman Trade to Africa.—That Trade in Blood, and every Vice, of which the Avarice, Pride, Insolence and Cruelty, of Man is capable! A Trade, which, if ever permitted, will entail eternal Infamy on the United States, and all that they have ever said or done in Defence of Freedom.—Will it not be said, that the greatest Sticklers for Liberty, are its worst Enemies?—For these Gentlemen, no doubt, mean to treat the United States, if they adopt the new Constitution, as they have some of their Colleagues; that is, make Cloaks of them, to cover their Wickedness.

At the Moment it is adopted by the States, in its present Form, that Moment the *external* Turpitude of it is transferred to the Adopters; and the Framers of it will immediately say, it was called for by the People, of whom they were but the Servants, and, that the Adoption is a Proof of the Assertion.

. . . as I have often told you, such is the unfortunate Lot of Humanity, that there are a Thousand brilliant Characters, to one that is always consistent, and, of this, Dr. Franklin, and Mr. John Dickinson, are two recent Examples among the Many. The Doctor is at the Head of a humane Institution for promoting the Emancipation of Slaves, or abolishing Slavery; yet lends his Assistance to frame a Constitution which evidently has a Tendency not only to enslave all those whom it ought to protect; but avowedly encourages the enslaving of those over whom it can have no Manner of Right, to exercise the least shadow of Authority.

Mr. Dickinson, a few Years before the Revolution, publicly impeached the Doctor's Conduct for offering to attempt a Change in the chartered Privileges of Pennsylvania, and now joins him in destroying a far superior Constitution, yes, thirteen far superior Constitutions, and opening a Trade which is a Disgrace to Humanity! Will not such Conduct leave these Gentlemen Monuments of much departed Fame? As I have several of their Publications by me, which, I imagine you never saw, I purpose in my Next, to let them speak for themselves, if you have no Objection.

ALGERNON SIDNEY
Philadelphia *Independent Gazetteer,* November 21, 1787

It is stipulated, most wickedly stipulated, that there shall be a power to import Negroes for twenty-one years, and that this power shall be irrevocable. It is said that this was done with a design to ease the southern states. There will in all probability be the same wish to ease the southern states, as it is prettily called, at the expiration of twenty-one years as at present. The infernal custom therefore of selling Negroes, if we adopt this new constitution, may not only prevail forever among the people of the southern states, but may be entered into by all the states of the union. For it will be contended that if one state has a right to import Negroes, another should not be debarred from the privilege; and the continental court, which is to be paramount to all others, if it is inclined to act consistently, must allow the validity of the plea.

Cato V (George Clinton)
New York Journal, November 22, 1787[18]

. . . that the slave trade, is to all intents and purposes permanently established; and a slavish capitation, or poll-tax, may at any time be levied—these are some of the many evils that will attend the adoption of this government.

A Countryman from Dutchess County II (Hugh Hughes)
New York Journal, November 23, 1787

In the Conclusion of my First . . . I promised that Mr. Dickinson, or the famous Author of the Farmer's Letters,[19] and Doctor Franklin should speak for themselves; I now offer you as a Specimen of the Farmer's Rhetoric, the second Paragraph of his first Letter, which appears thus—"From my Infancy I was taught to love Humanity and Liberty. Enquiry and Experience have since confirmed my Reverence for the Lessons then given me, by convincing me more fully of their Truth and Excellence. Benevolence towards Mankind excites Wishes for their Welfare, and such Wishes endear the Means of Fulfilling them. Those can be found in Liberty alone, and therefore her sacred Cause ought to be espoused by every Man, on every occasion, to the utmost of his Power. As a Charitable, but poor, Person does not withhold his Mite, because he can not relieve all the Distresses of the Miserable; so let not any honest Man suppress his Sentiments concerning Freedom, however small their Influence is likely to be. Perhaps he may touch some Wheel that will have a greater Effect than he expects." What gracious Sentiments, and how sweetly expressed!—But what are Sentiments, or the tenderest Expressions, when not accompanied by corresponding Actions? They certainly render the Author a greater Object of our Pity, if not of Contempt.—How is it possible to reconcile the first Clause of the 9th Section, in the first Article of the new Constitution, with such universal Benevolence to all Mankind?

Will this Gentleman say, that the Africans do not come within the Description of "Mankind?" If he should, will he be believed?—Besides, he seems to have run counter to a generally received Maxim in educating the rational

---

[18]Reprinted in the New York *Daily Advertiser*, November 24, and the *Albany Gazette*, December 6. George Clinton had been the only governor of New York since 1777.
    [19]John Dickinson's famous "Letters from a Farmer in Pennsylvania" (1767).

as well as the irrational Creation; as he acknowledges, that he was early instructed in Virtue, which, now, in advanced Life, he seems either to have forgotten or stifled?

Had Cornwallis, Rawdon, Arnold, or any of the British, Marauding, Butchers, signed such a Clause, there would have been a Consistency; but, for the benevolent Author of the Farmer's Letters, which every where seem to breathe the pure Spirit of Liberty and Humanity, to lend his once venerated Name, for promoting that which the Framers of the Clause were either ashamed or afraid, openly, to avow, exceeds Credulity itself, were it not for occular Demonstration.

Is this the Way by which we are to demonstrate our Gratitude to Providence, for his divine Interposition in our Favor, when oppressed by Great Britain?—Who could have imagined, that Men lately professing the highest Sense of Justice and the Liberties of Mankind, could so soon and easily be brought to give a Sanction to the greatest Injustice and Violation of those very Liberties? Strange Inconsistency and painful Reflection!—And the more so, when it is considered, that not only Individuals in Europe, as well as in each of these states; but that several of the nations of Europe have, for some years before the revolution, been endeavouring to put a Stop to a Trade, which was a Disgrace to the very Name of Christianity itself. Nay, that Numbers among those whom we so lately considered as Enemies to Liberty, are now using every Means in their Power to abolish Slavery! Will not a contrary Conduct of the States tarnish the Lustre of the American revolution, by violating the Law of Nations, and entailing endless Servitude on Millions of the human Race, and their unborn Posterity? Can any Person, who is not deeply interested in enslaving this Country, believe, that the Contrivers of such a diabolical Scheme had any Regard for the most sacred Rights of human Nature?

It really seems to have been, as Mr. Wilson acknowledged, a mere Matter of Accomodation between the Northern and Southern States; that is, if you will permit us to import Africans as slaves, we will consent that you may export Americans, as Soldiers; for this new Constitution clearly admits, by the 2d Clause of the 6th Article, which says, "that this Constitution and the Laws of the United States which shall be made in pursuance thereof, and all treaties made, or which shall be made under the Authority of the United States, shall be the supreme Law of the Land, etc. any Thing in the Constitution or Laws of any State to the Contrary notwithstanding."

May not Treaties be immediately entered into with some of the Nations of Europe for assisting them with Troops, which, if they do not enlist voluntarily, may, by this Clause, be detached and transported to the West or East-indies, etc.?

Philadelphiensis II (Benjamin Workman)
Philadelphia *Freeman's Journal*, November 28, 1787[20]

From the proceedings of the convention, respecting *liberty of conscience*, foreign politicians might be led to draw a strange conclusion, viz. that the majority of that assembly were either men of *no* religion, or all of *one* religion; such a conclusion naturally follows their silence on that subject; they must either have been indifferent about religion, or determined to compel the whole continent to conform to their own. For my own part, I really think, that their conduct in this instance is inexplicable: it is impossible to divine what might have been their intentions.

{To illustrate this defect in the new constitution, by a familiar instance: we shall suppose that the negroes of Georgia, or some of the southern states, prompted by the love of *sacred liberty*, shall attempt to free themselves from cruel slavery, by a *noble appeal to arms*. In this case the Congress may order the militia of Pennsylvania to march off to quell the insurrection: now on such an occasion, what must the condition of that Pennsylvanian be, who, besides being conscientiously scrupulous against bearing arms, on any account whatever, has, over and above, made the *manumission of slavery*, a part of his religious creed? Miserable must be the state of such a man's mind indeed! More to be pitied is he, than the wretches against whom he is compelled to fight! The foregoing supposition is by no means an unnatural one; and truly, if the new constitution be adopted, I have little doubt, but the thing itself will some time or other be realized. I shall by way of digression add one sentiment, namely, that I should have no objection, that the slaves in the United States would free themselves to-morrow from their present *thraldom*, provided no lives be lost on this occasion; and with this proviso, I sincerely pray, that God may grant them success in their first attempt. Freedom is the birth-right of every man; and who is he that hath dared to rob his fellow men of this glorious privilege, with whom God will not enter into judgment?}

Before I dismiss this subject, I cannot help taking notice of the inconsistency of some Pennsylvanians, in respect to this new government. The very men, who should oppose it with all their influence, seem to be the most zealous for establishing it. Strange indeed! that the professed enemies of *negro* and every other species of *slavery*, should themselves join in the adoption of a constitution whose very basis is *despotism* and *slavery*, a constitution that militates so far against freedom, that even their own religious liberty may probably be destroyed by it. Alas! what frail, what inconsistent beings we are! To the catalogue of human weaknesses and mistakes, this is one to be added.

---

[20]Except for the text in braces, this item was also printed in the Philadelphia *Independent Gazetteer* on November 28. It was reprinted in the Boston *American Herald* on December 17. Workman had emigrated from Ireland in 1784 and served as a mathematics tutor at the University of Pennsylvania. Beginning in 1786 he edited a popular almanac.

A FEDERAL REPUBLICAN
A Review of the Constitution, Philadelphia, November 28, 1787

The next thing which we proceed to, is the importation of slaves, contained in the ninth section of the first article. It says, that "the migration or importation of such persons as any of the states now existing shall think proper to admit, shall not be prohibited by Congress prior to the year 1808, but a tax or duty may be imposed upon such importation, not exceeding ten dollars for each person." "The truth is, (says a citizen of America) Congress cannot prohibit the importation of slaves during that period; but the laws against the importation of them into any particular state stand unrepealed. An immediate abolition of slavery would be ruin upon the whites and misery upon the blacks in the southern states. The constitution therefore hath wisely left each state to pursue its own measures with respect to this article of legislation during the period of twenty-one years."[21] That the importation of slaves shall not be forbidden till that time may be very wise—but what hath that to do with the abolition of slavery? To prohibit the importation of slaves is not to abolish slavery. For all that is contained in this constitution, this country may remain degraded by this impious custom till the end of time.

PENNSYLVANIA RATIFYING CONVENTION DEBATES
November 28 to December 10, 1787

November 28

Thomas McKean:[22] I earnestly hope, sir, that the statutes of the federal government will last till they become the common law of the land, as excellent and as much valued as that which we have hitherto fondly denominated the birthright of an American. Such are the objects to which the powers of the proposed government extend. Nor is it entirely left to this evident principle, that nothing more is given than is expressed, to circumscribe the federal authority. For, in the ninth section of the first Article, we find the powers so qualified that not a doubt can remain. In the first clause of that section, there is a provision made for an event which must gratify the feelings of every friend to humanity. The abolition of slavery is put within the reach of the federal government; and when we consider the situation and circumstances of the Southern States, every man of candor will find more reason to rejoice that the power should be given at all, than to regret that its exercise should be postponed for twenty years.

---

[21]See A Citizen of America, October 17 (above).
[22]Taken from Alexander J. Dallas' notes of the debates in the *Pennsylvania Herald*, December 26, 1787.

December 3

William Findley:[23] The manner of numbering the inhabitants is dark—
"other Persons" (Article 1, section 2).

Article 1, section 9, 1st clause: Migration, etc. is unintelligible. It is unfortunate if this guarantees the importation of slaves or if it lays a duty on the
importation of other persons.

This is a reservation; and yet the power of preventing importation is
nowhere given.

Anthony Wayne:[24] What were the Southern States to gain by the Constitution? No restraint in the *Articles of Confederation.* In this [Constitution] the
restraint [is] 21 *years.* A duty amounting to a prohibition.

James Wilson:[25] Much fault has been found with the mode of expression,
used in the first clause of the ninth section of the first article. I believe I can
assign a reason, why that mode of expression was used, and why the term
slave was not directly admitted in this constitution;—and as to the manner of
laying taxes, this is not the first time that the subject has come into the view of
the United States, and of the legislatures of the several states. The gentleman
(Mr. Findley) will recollect, that in the present congress, the quota of the
foederal debt, and general expences, was to be in proportion to the value of
LAND, and other enumerated property, within the states. After trying this
for a number of years, it was found on all hands, to be a mode that could not
be carried into execution. Congress were satisfied of this, and in the year
1783, recommended, in conformity with the powers they possess'd under the
articles of confederation, that the quota should be according to the number
of free people, including those bound to servitude, and excluding Indians not
taxed.[26] {These were the very expressions used in 1783}, and the fate of this
recommendation was similar to all their other resolutions. It was not carried
into effect, but it was adopted by no fewer than eleven, out of thirteen states;
and it can not but be matter of surprise, to hear gentlemen, who agreed to
this very mode of expression at that time, come forward and state it as an
objection on the present occasion. It was natural, sir, for the late convention,
to adopt the mode after it had been agreed to by eleven states, and to use the
expression, which they found had been received as unexceptionable before.
With respect to the clause, restricting congress from prohibiting the migration or importation of such persons, as any of the states now existing, shall
think proper to admit, prior to the year 1808. The honorable gentleman says,
that this clause is not only dark, but intended to grant to congress, for that

---

[23]Taken from James Wilson's Notes, Historical Society of Pennsylvania.
[24]Anthony Wayne's Notes, Privately owned by H. Bartholomew Cox.
[25]Taken from Thomas Lloyd's *Debates of the Convention of the State of Pennsylvania, on the Constitution . . . .* (Philadelphia, 1788).

time, the power to admit the importation of slaves. No such thing was intended; but I will tell you what was done, and it gives me high pleasure, that so much was done. Under the present confederation, the states may admit the importation of slaves as long as they please; but by this article after the year 1808, the congress will have power to prohibit such importation, notwithstanding the disposition of any state to the contrary. I consider this as laying the foundation for banishing slavery out of this country, and though the period is more distant than I could wish, yet it will produce the same kind, gradual change, which was pursued in Pennsylvania. It is with much satisfaction I view this power in the general government, whereby they may lay an interdiction on this reproachful trade; but an immediate advantage is also obtained; for a tax or duty may be imposed on such importation, not exceeding ten dollars for each person; and, this sir, operates as a partial prohibition; it was all that could be obtained, I am sorry it was no more; but from this I think there is reason to hope, that yet a few years, and it will be prohibited altogether; and in the mean time, the new states which are to be formed, will be under the control of congress in this particular; and slaves will never be introduced amongst them. The gentleman says, that it is unfortunate in another point of view; it means to prohibit the introduction of white people from Europe, as this tax may deter them from coming amongst us; a little impartiality and attention will discover the care that the convention took in selecting their language. The words are the *migration or* importation of such persons, &c. shall not be prohibited by congress prior to the year 1808, but a tax or duty may be imposed on such importation; it is observable here, that the term migration is dropped, when a tax or duty is mentioned; so that congress have power to impose the tax, only on those imported.

Satirical Report of Robert Whitehill.[27] It has been said that Congress will have power, by the new constitution, to lay an impost on the *importation* of slaves, into these states; but that they will have no power to impose any tax upon the *migration* of Europeans. Do the gentlemen, sir, mean to insult our understandings, when they assert this? Or are they ignorant of the English language? If, because of their ignorance, they are at a loss, I can easily explain this clause for them—The words "*migration*" and "*importation*," sir, being *connected* by the *disjunctive* conjunction "*or*," certainly mean either migration, or importation; either the one, or the other; or both. Therefore, when we say " a tax may be laid upon such *importation*," we mean, either upon the *importation*, or *migration*; or upon both; for, because they are *joined together*, in the first instance, by the *disjunctive* conjunction *or*, they are both synonimous terms for the same thing—therefore, "*such importation*," because the *comparative* word, *such*, is used, means both importation, and migration.

---

[26]The speech as printed in Lloyd's *Debates* omits any reference to "three-fifths of other persons." The text immediately following in braces was marked for deletion in Lloyd's *Errata*.

December 4

James Wilson:[28] I recollect, on a former day, the honorable gentleman from Westmoreland (William Findley) and the honorable gentleman from Cumberland (Robert Whitehill) took exceptions against the first clause of the 9th section, Article I, arguing very unfairly, that because Congress might impose a tax or duty of ten dollars on the importation of slaves, within any of the United States, Congress might therefore permit slaves to be imported within this state, contrary to its laws. I confess I little thought that this part of the system would be excepted to.

I am sorry that it could be extended no further; but so far as it operates, it presents us with the pleasing prospect, that the rights of mankind will be acknowledged and established throughout the Union.

If there was no other lovely feature in the Constitution, but this one, it would diffuse a beauty over its whole countenance. Yet the lapse of a few years and Congress will have power to exterminate slavery from within our borders.

December 10

Thomas McKean:[29] Provision is made that Congress shall have power to prohibit the importation of slaves after the year 1808, but the gentlemen in opposition accuse this system of a crime, because it has not prohibited them at once. I suspect those gentlemen are not well acquainted with the business of the diplomatic body, or they would know that an agreement might be made, that did not perfectly accord with the will and pleasure of any one person. Instead of finding fault with what has been gained, I am happy to see a disposition in the United States to do so much.

EDMUND PRIOR TO MOSES BROWN
New York, December 1, 1787[30]

Thy favour of the 18th Ulto. I duly recd., and should have answered it Long since, but a Member of the Late Convention, who I had some acquaintance with, being Absent, I was unable to obtain that information I wished for, and altho he is yet away I shall nevertheless endeavour to reply to thine,

[27]This satirical piece is taken from "Puff" (probably Benjamin Rush) printed in the Philadelphia *Independent Gazetteer*, December 6, 1787. Several other satirical comments were printed about this speech.

[28]Taken from Thomas Lloyd's *Debates*.

[29]Taken from Thomas Lloyd's *Debates*.

[30]Moses Brown Papers, Rhode Island Historical Society. Prior was a New York City merchant and a Quaker.

The Great oversight of the Convention in respect to securing universal Liberty & Impartial Justice is generally attributed to the influence of the Southern Members, who had they duly adverted to the Publick declarations made in the days of their fear and distress, a very different determination in respect to Slavery would have taken place; With us it is however agreed that the State Legislatures will not be restrained from enacting such Laws for the prevention of the Odious traffick, as they may Judge expedient, for themselves, and I wish it may be the Case, hoping the advocates for the poor afflicted & oppress'd Africans will not be discouraged from pursuing their Laudable purpose—Its nevertheless allowed that should the Constitution be adopted the State of Massachusetts will no Longer be an Assylum to the Negroes, unless they Should, except that Article, in their adoption Notwithstanding our Testimony is so opposite to the sentiments of that body yet cannot see, how we shall move in the business, farther than a Patient gradual Perseverance, for the Work is evidently on its way, and I have no doubt will in time be effected, hope our Patience may keep Pace with the Success & we Steadily press forward—at times I have been possessed with a fear Least from the Cause being so good and the unrighteousness & Cruelty of Slavery, we should be induced to attempt to drive, & thereby be in danger of Shifting our ground, which would then become an uncertain foundation.

Thinc of the 19th with its inclosures[31] was very acceptable, I had no expectation of any State going so far yet, its an excellent example for the others and I hope they will, adopt or enact Similar Laws—It has been published here & in Jersey & have no Doubt but in Philada. also.

CATO VI (GEORGE CLINTON)
New York Journal, December 13, 1787[32]

The next objection that arises against this proffered constitution is, that the apportionment of representatives and direct taxes are unjust.—The words as expressed in this article are, "representatives and direct taxes shall be apportioned among the several states, which may be included in this union, according to their respective numbers, which shall be determined by adding to the whole number of free persons, including those bound to service for a term of years, and excluding Indians not taxed three fifths of all other persons." In order to elucidate this, it will be necessary to repeat the remark in my last number, that the mode of legislation in the infancy of free communities was by the collective body, and this consisted of free persons, or those

[31]Brown enclosed the Quaker petition to the Rhode Island assembly and the subsequent act prohibiting Rhode Island citizens from participating in the foreign slave trade. See Moses Brown to James Thornton, Sr., November 13, 1787 (Chapter 3).

[32]Reprinted in the New York *Daily Advertiser*, December 15.

whose age admitted them to the rights of mankind and citizenship—whose sex made them capable of protecting the state, and whose birth may be denominated Free Born, and no traces can be found that even women, children, and slaves, or those who were not sui juris, in the early days of legislation, meeting with the free members of the community to deliberate on public measures; hence is derived this maxim in free governments, that representation ought to bear a proportion to the number of free inhabitants in a community; this principle your own state constitution, and others, have observed in the establishment of a future census, in order to apportion the representatives, and to increase or diminish the representation to the ratio of the increase or diminution of electors. But, what aid can the community derive from the assistance of women, infants, and slaves, in their deliberation, or in their defence? and what motive therefore could the convention have in departing from the just and rational principle of representation, which is the governing principle of this state and of all America.

JAMES THORNTON, SR., TO MOSES BROWN
Byberry, Philadelphia County, December 17, 1787[33]

Thine dated the 13th. 11 mo. 1787, I received which was very acceptable. . . . your Christian Endeavour, with your Legislature for the Abolition of Slavery and trafick in the African trade being succesfull as the Law they made thereupon Evinces, is truely Salutary and wish might take place here, many friends here view the Transactions of the Convention respecting Leaveing the Trade open to Africa in the Same point of view as thee does, and have Occasionly mention'd to Leading men in State affairs, as one of the Grand reasons of our Objections to the proposed Constitution, but as a religious Society we can have Nothing to do with Seting up nor pulling down Governments but Live Peaceably under all Governments Set over us in Godliness and honesty: yet ought to Shew Publickly our disaprobation of Every Oppressive and unrighteous Act—done by men in power.

BENJAMIN RUSH TO ELIZABETH GRAEME FERGUSON
Philadelphia, December 25, 1787[34]

I rejoice to find that a spirit of humanity has at last reached the southern states upon the subject of the *slavery* of the Negroes. In one-and-twenty years the new government will probably put an end to the African trade forever in

---

[33]Moses Brown Papers, Rhode Island Historical Society. Thornton answers Brown's letter of November 13 (Chapter 3).
[34]William H. Welch Library, Johns Hopkins Medical School. Rush refers to "the persecution and slander" he was subjected to because of his long years of abolitionism.

America. O! Virtue, Virtue, who would not follow thee blindfold. The prospect of this glorious event more than repays me for all the persecution and slander to which my principles and publications exposed me about 16 or 17 years ago.

PUBLIUS (JAMES MADISON)
The Federalist 38, New York *Independent Journal*, January 12, 1788[35]

To the People of the State of New-York.
   . . . Is the importation of slaves permitted by the new Constitution for twenty years? By the old, it is permitted forever.

A COUNTRYMAN FROM DUTCHESS COUNTY V (HUGH HUGHES)
New York Journal, January 22, 1788

   . . . Should the new constitution be sufficiently corrected *by a substantial* bill of rights, an equitable representation, . . . and relinquishing every idea of drenching the bowels of Africa in gore, for the sake of enslaving its free-born innocent inhabitants, I imagine we might become a happy and respectable people. . . . I have no idea of marching 500 or 1000 miles to quell an insurrection of such emigrants as are proposed by the new constitution, to be introduced for one and twenty years.

PUBLIUS (JAMES MADISON)
The Federalist 42, *New York Packet*, January 22, 1788[36]

To the People of the State of New-York.
   It were doubtless to be wished that the power of prohibiting the importation of slaves, had not been postponed until the year 1808, or rather that it had been suffered to have immediate operation. But it is not difficult to account either for this restriction on the general government, or for the manner in which the whole clause is expressed. It ought to be considered as a great point gained in favor of humanity, that a period of twenty years may terminate for ever within these States, a traffic which has so long and so loudly upbraided the barbarism of modern policy; that within that period it will re-

---

   [35]Reprinted in the New York *Daily Advertiser* and *New York Packet*, January 15; *New York Journal*, January 26; Exeter, N.H., *Freeman's Oracle*, February 15; and in the second volume of the book edition of *The Federalist* published in New York City on May 28, 1788.
   [36]Reprinted in the New York *Independent Journal*, January 23; the New York *Daily Advertiser*, January 24; and the second volume of the book edition of *The Federalist* published in New York City on May 28, 1788.

ceive a considerable discouragement from the foederal Government, and may be totally abolished by a concurrence of the few States which continue the unnatural traffic, in the prohibitory example which has been given by so great a majority of the Union. Happy would it be for the unfortunate Africans, if an equal prospect lay before them, of being redeemed from the oppressions of their European brethren!

Attempts have been made to pervert this clause into an objection against the Constitution, by representing it on one side as a criminal toleration of an illicit practice, and on another, as calculated to prevent voluntary and beneficial emigrations from Europe to America. I mention these misconstructions, not with a view to give them an answer, for they deserve none; but as specimens of the manner and spirit in which some have thought fit to conduct their opposition to the proposed government.

PUBLIUS (JAMES MADISON)
The Federalist 54, *New York Packet*, February 12, 1788[37]

To the People of the State of New-York.
The next view which I shall take of the House of Representatives, relates to the apportionment of its members to the several States, which is to be determined by the same rule with that of direct taxes.

It is not contended that the number of people in each State ought not to be the standard for regulating the proportion of those who are to represent the people of each State. The establishment of the same rule for the apportionment of taxes, will probably be as little contested; though the rule itself in this case, is by no means founded on the same principle. In the former case, the rule is understood to refer to the personal rights of the people, with which it has a natural and universal connection. In the latter, it has reference to the proportion of wealth, of which it is in no case a precise measure, and in ordinary cases a very unfit one. But notwithstanding the imperfection of the rule as applied to the relative wealth and contributions of the States, it is evidently the least exceptionable among the practicable rules; and had too recently obtained the general sanction of America, not to have found a ready preference with the Convention.[38]

All this is admitted, it will perhaps be said: But does it follow from an admission of numbers for the measure of representation, or of slaves com-

[37]Reprinted in the New York *Independent Journal*, February 13, and in the second volume of the book edition of *The Federalist* published in New York City on May 28, 1788.
    [38]A reference to the proposed amendment to the Articles of Confederation that would have changed the method of apportioning federal expenses from a system based on land values to one based on population, including slaves counted as three-fifths of a free person. See Chapter 1.

bined with free citizens, as a ratio of taxation, that slaves ought to be included in the numerical rule of representation? Slaves are considered as property, not as persons. They ought therefore to be comprehended in estimates of taxation which are founded on property, and to be excluded from representation which is regulated by a census of persons. This is the objection, as I understand it, stated in its full force. I shall be equally candid in stating the reasoning which may be offered on the opposite side.

We subscribe to the doctrine, might one of our southern brethren observe, that representation relates more immediately to persons, and taxation more immediately to property, and we join in the application of this distinction to the case of our slaves. But we must deny the fact that slaves are considered merely as property, and in no respect whatever as persons. The true state of the case is, that they partake of both these qualities; being considered by our laws, in some respects, as persons, and in other respects, as property. In being compelled to labor not for himself, but for a master; in being vendible by one master to another master; and in being subject at all times to be restrained in his liberty, and chastised in his body, by the capricious will of another, the slave may appear to be degraded from the human rank, and classed with those irrational animals, which fall under the legal denomination of property. In being protected on the other hand in his life & in his limbs, against the violence of all others, even the master of his labor and his liberty; and in being punishable himself for all violence committed against others; the slave is no less evidently regarded by the law as a member of the society; not as a part of the irrational creation; as a moral person, not as a mere article of property. The Foederal Constitution therefore, decides with great propriety on the case of our slaves, when it views them in the mixt character of persons and of property. This is in fact their true character. It is the character bestowed on them by the laws under which they live; and it will not be denied that these are the proper criterion; because it is only under the pretext that the laws have transformed the negroes into subjects of property, that a place is disputed them in the computation of numbers; and it is admitted that if the laws were to restore the rights which have been taken away, the negroes could no longer be refused an equal share of representation with the other inhabitants.

This question may be placed in another light. It is agreed on all sides, that numbers are the best scale of wealth and taxation, as they are the only proper scale of representation. Would the Convention have been impartial or consistent, if they had rejected the slaves from the list of inhabitants when the shares of representation were to be calculated; and inserted them on the lists when the tariff of contributions was to be adjusted? Could it be reasonably expected that the southern States would concur in a system which considered their slaves in some degree as men, when burdens were to be imposed, but refused to consider them in the same light when advantages were to be conferred? Might not some surprize also be expressed that those who reproach the southern States with the barbarous policy of considering as property a part of their human brethren, should themselves contend that the government to which all the States are to be parties, ought to consider this unfortunate race more compleatly in the unnatural light of property, than the very laws of which they complain!

It may be replied perhaps that slaves are not included in the estimate of representatives in any of the States possessing them. They neither vote themselves, nor increase the votes of their masters. Upon what principle then ought they to be taken into the foederal estimate of representation? In rejecting them altogether, the Constitution would in this respect have followed the very laws which have been appealed to, as the proper guide.

This objection is repelled by a single observation. It is a fundamental principle of the proposed Constitution, that as the aggregate number of representatives allotted to the several States, is to be determined by a foederal rule founded on the aggregate number of inhabitants, so the right of choosing this allotted number in each State is to be exercised by such part of the inhabitants, as the State itself may designate. The qualifications on which the right of suffrage depend, are not perhaps the same in any two States. In some of the States the difference is very material. In every State, a certain proportion of inhabitants are deprived of this right by the Constitution of the State, who will be included in the census by which the Foederal Constitution apportions the representatives. In this point of view, the southern States might retort the complaint, by insisting, that the principle laid down by the Convention required that no regard should be had to the policy of particular States towards their own inhabitants; and consequently, that the slaves as inhabitants should have been admitted into the census according to their full number, in like manner with other inhabitants, who by the policy of other States, are not admitted to all the rights of citizens. A rigorous adherence however to this principle is waved by those who would be gainers by it. All that they ask is, that equal moderation be shewn on the other side. Let the case of the slaves be considered as it is in truth a peculiar one. Let the compromising expedient of the Constitution be mutually adopted, which regards them as inhabitants, but as debased by servitude below the equal level of free inhabitants, which regards the *slave* as divested of two fifths of the *man*.

After all may not another ground be taken on which this article of the Constitution, will admit of a still more ready defence. We have hitherto proceeded on the idea that representation related to persons only, and not at all to property. But is it a just idea? Government is instituted no less for protection of the property, than of the persons of individuals. The one as well as the other, therefore may be considered as represented by those who are charged with the government. Upon this principle it is, that in several of the States, and particularly in the State of New-York, one branch of the government is intended more especially to be the guardian of property, and is accordingly elected by that part of the society which is most interested in this object of government. In the Foederal Constitution, this policy does not prevail. The rights of property are committed into the same hands with the personal rights. Some attention ought therefore to be paid to property in the choice of those hands.

For another reason the votes allowed in the Foederal Legislature to the people of each State, ought to bear some proportion to the comparative wealth of the States. States have not like individuals, an influence over each other arising from superior advantages of fortune. If the law allows an opulent citizen but a single vote in the choice of his representative, the respect and consequence which he derives from his fortunate situation, very frequently guide the votes of others to the objects of his choice; and through this imperceptible channel the rights of property are conveyed into the public representation. A State possesses no such influence over other States. It is not probable that the richest State in the confederacy will ever influence the choice of a single representative in any other State. Nor will the representatives of the larger and richer States, possess any other advantage in the Foederal Legislature over the representatives of other States, than what may result from their superior number alone; as far therefore as their superior wealth and weight may justly entitle them to any advantage, it ought to be secured to them by a superior share of representation. The new Constitution is in this respect materially different from the existing confederation, as well as from that of the United Netherlands, and other similar confederacies. In each of the latter the efficacy of the foederal resolutions depends on the subsequent and voluntary resolutions of the States composing the Union. Hence the States, though possessing an equal vote in the public councils, have an unequal influence, corresponding with the unequal importance of these subsequent and voluntary resolutions. Under the proposed Constitution, the foederal acts will take effect without the necessary intervention of the individual States. They will depend merely on the majority of votes in the Foederal Legislature, and consequently each vote whether proceeding from a larger or a smaller State, or a State more or less wealthy or powerful, will have an equal weight and efficacy; in the same manner as the votes individually given in a State Legislature, by the representatives of unequal counties or other districts, have each a precise equality of value and effect; or if there be any difference in the case, it proceeds from the difference in the personal character of the individual representative, rather than from any regard to the extent of the district from which he comes.

Such is the reasoning which an advocate for the southern interests might employ on this subject: And although it may appear to be a little strained in some points, yet on the whole, I must confess, that it fully reconciles me to the scale of representation, which the Convention have established.

In one respect the establishment of a common measure for representation and taxation will have a very salutary effect. As the accuracy of the census to be obtained by the Congress, will necessarily depend in a considerable degree on the disposition, if not the co-operation of the States, it is of great importance that the States should feel as little bias as possible to swell or to reduce the amount of their numbers. Were their share of representation alone to be governed by this rule they would have an interest in exaggerating their inhabitants. Were the rule to decide their share of taxation alone, a contrary temptation would prevail. By extending the rule to both objects, the States will have opposite interests, which will controul and ballance each other; and produce the requisite impartiality.

HAMPDEN (WILLIAM FINDLEY)
Pittsburgh Gazette, February 16, 1788

To the character of being inconsistent, I shall add that of being mysterious and hard to be understood, or at least very liable of being misunderstood. What reader will say that the other persons, three-fifths of which are to be taken with a view to taxation and representation, or the clause respecting the raising of a revenue from, or prohibiting the importation of persons in the first and ninth sections, is expressed with candid clearness? If slaves, or emigrant servants only are designed, why are they not so expressed? Candor certainly required a manner of expression suitable to the people's uptakings.

DELIBERATOR
Philadelphia *Freeman's Journal*, February 20, 1788

. . . Congress may, under the sanction of that clause in the constitution which empowers them to regulate commerce, authorize the importation of slaves, even into those states where this iniquitous trade is, or may be prohibited by their laws or constitution.

BENJAMIN RUSH TO JEREMY BELKNAP
Philadelphia, February 28, 1788[39]

In answer to your question respecting the conduct & opinions of the quakers in Pennsylvania,[40] I am very happy in being able to inform you that

---

[39]Belknap Papers, Massachusetts Historical Society.
[40]On February 12 Belknap had asked Rush how Pennsylvania Quakers interpreted the slave trade clause of the Constitution (Chapter 3). Belknap became concerned when Antifederalist James Neal, a Quaker preacher from Kittery, Maine, criticized the slave-trade clause of the Constitution in the Massachusetts convention for prohibiting Congress from interfering with the importation of slaves for twenty years. Neal was answered by Theophilus Parsons "who construed that article into a dawn of hope for the final abolition of the horrid Traffick."

they are all (with an exception of three or four persons only) highly fœderal.— There was a respectable representation of that Society in our Convention, all of whom voted in favor of the New Constitution.[41] They consider very wisely that the Abolition of slavery in our country must be gradual in order to be effectual, and that the Section of the Constitution which will put it in the power of Congress twenty years hence to restrain it altogether, was a great point obtained from the Southern States. The appeals therefore that have been made to the humane & laudable prejudices of our quakers by our Antifoederal writers, upon the Subject of Negro Slavery, have been treated by that prudent Society with Silence and Contempt.

THE PENNSYLVANIA ABOLITION SOCIETY
Petition to the Constitutional Convention
Pennsylvania Gazette, March 5, 1788[42]

The following Memorial, drawn up by the Society for the gradual abolition of slavery in Philadelphia, was intended to be presented to the late Fœderal Convention, but was withheld, upon an assurance being given by a member of the convention that the great object of the memorial would be taken under consideration, and that the memorial, in the beginning of the deliberations of the convention, might alarm some of the southern states, and thereby defeat the wishes of the enemies of the African trade. While we rejoice in the step which has been taken by the convention to put a total stop to the commerce and slavery of the negroes one and twenty years hence, it is to be hoped the publication of the memorial may have some weight with individual states, to pass laws to prohibit that inhuman traffic, before the power of Congress over that part of the commerce of the states shall take place,

J. H.

To the Hon. the CONVENTION of the United States of America, now assembled in the City of Philadelphia.

The MEMORIAL of the Pennsylvania Society for promoting the Abolition of Slavery, and the Relief of free Negroes unlawfully held in Bondage.

The Pennsylvania Society for promoting the abolition of slavery, and the relief of free negroes unlawfully held in bondage, rejoice with their fellow citizens in beholding a Convention of the states assembled for the purpose of amending the fœderal constitution.

---

[41]"Undeniable Facts" stated that eight of the sixty-nine members of the Pennsylvania Convention were Quakers, all of whom voted to ratify the Constitution (Philadelphia *Independent Gazetteer*, January 15, 1788).

[42]Reprinted in the Philadelphia *Independent Gazetteer*, March 7.

They recollect with pleasure, that among the first acts of the Illustrious Congress of the year 1774, was a resolution for prohibiting the importation of African slaves.

It is with deep distress they are forced to observe, that the peace was scarcely concluded, before the African trade was revived, and American vessels employed in transporting the inhabitants of Africa to cultivate, as slaves, the soil of America, before it had drank in all the blood which had been shed in her struggle for liberty.

To the revival of this trade the Society ascribe part of the obloquy with which foreign nations have branded our infant states. In vain will be the pretensions of the United States to a love of liberty, or a regard for national character, while they share in the profits of a commerce, that can only be conducted upon rivers of human tears and blood.

By all the attributes therefore of the Deity, which are offended by this inhuman traffic—by the union of our whole species in a common ancester, and by all the obligations which result from it—by the apprehensions and terror of the righteous vengeance of God in national judgments—by the certainty of the great and awful day of retribution—by the efficacy of the prayers of good men, which would only insult the majesty of heaven if offered up in behalf of our country, while the iniquity we deplore continues among us—by the sanctity of the christian name—by the pleasures of domestic connections and the pangs which attend their dissolution—by the captivity and sufferings of our fellow citizens in Algiers, which seem to be intended by divine providence to awaken us to a sense of the injustice and cruelty of dooming our African brethren to perpetual slavery and misery—by a regard to the consistency of principles and conduct which should mark the citizens of republics—by the magnitude and intensity of our desires to promote the happiness of those millions of intelligent beings, who will probably cover this immense continent with rational life—and by every other consideration that religion, policy and humanity can suggest—the Society implore the present Convention to make the suppression of the African trade in the United States a part of their important deliberations.

June 2, 1787.

PHILADELPHIA *INDEPENDENT GAZETTEER*
March 7, 1788[43]

A correspondent says, it is an impudent falsehood to declare that the people called Quakers are generally attached to the new constitution. It is most certain that at their meetings of business they have determined not to

---

[43]Reprinted in the Baltmore *Maryland Gazette*, March 14; the *Massachusetts Centinel*, March 19; and the *New Hampshire Spy*, March 21.

support it. Every considerate person must know it be against their principles, as one must suppose it to be against the principles of all the sincere professors of christianity, to raise a man to a throne, in opposition to a lawful government,[44] who notoriously holds negroes in slavery without any design of liberating them, and who sells them, when his necessities urge him to it, as if they were beasts.

PENNSYLVANIA GAZETTE
March 19, 1788[45]

Though there is *very little* opposition to the proposed foederal constitution in South-Carolina, it appears that a principal ground of objection with its opponents *there* is, that it will finally invest the foederal legislature with a power *to regulate or prevent* the importation of slaves. The Minority of Pennsylvania, who were always friends to the abolition of negro slavery, and the states of Rhode-Island and Massachusetts, who consider slaves as *freed* by coming into their jurisdiction, can never expect to agree with the gentlemen in Carolina, who oppose on such principles.

THE FEDERAL COMMITTEE OF ALBANY
An Impartial Address, April 1788[46]

... *Objection.* That slaves are computed in apportioning Representatives.

Answer. Agreeable to the New System, taxation and representation must go together. These objectors should have been so candid as to add, that all direct taxes must be laid, on each state, in proportion to the number of its inhabitants; that by this computation five slaves will pay taxes equal to three free men; which will be a great advantage to New-York and the eastern states, who have very few slaves.

[44]A reference to the argument that the Constitution unconstitutional supplanted the Articles of Confederation and to the anticipated election of George Washington as America's first president. Antifederalists repeatedly charged that the president would be an elected monarch.

[45]Reprinted eight times by May 5 from New Jersey to South Carolina.

[46]The full title of this pamphlet is *An Impartial Address, to the Citizens of the City and County of Albany: or the 35 Anti-Federal Objections Refuted* written by the Federal Committee of the City of Albany (Albany, 1788).

New York Morning Post
April 11, 1788[47]

By the new system of government, proposed by the late American Convention, the poor Africans (as if the States, in their bustle about liberty, had discovered a right to enslave them) are doomed to endure a continuance of depredation, rapine, and murder, for 21 years to come. The Congress being, for that time, absolutely precluded from interference with that most flagrant of natural justice.

James Pemberton to John Pemberton
Philadelphia, April 20, 1788[48]

. . . as yet Six States only have adopted the new plan, one of which (the Massachusetts) has acceded to it with divers exceptions, or recommendations of amendment, and great uneasiness, and opposition appear among the Politicians in this and other places, numerous publications daily coming out on the subject, so that its establishment remains doubtful, and the first principles being in divers respects erroneous, and particularly so in regard to the flagrant evil of Slavery, and its infamous traffic, it can not be expected that a Government on such an unjust foundation can be durable; animosity, dissentions, and commotions will be most likely to attend it. . . .

we have obtained a good [antislave trade] law here, but much labour still remains necessary among the Slave-holders in these new States, as also to promote the moral & religious wellfare of the Negroes who have been restored to their just rights of freedom.

Rhode Island: An Example to Follow
Philadelphia *Independent Gazetteer*, May 1, 1788

A correspondent says, that the state of Rhode-Island deserves applause and imitation for her wisdom and virtue in three very important matters. In the first place by an act of Assembly the state of Rhode-Island declared that all the negroes born there after March, 1784, were absolutely and at once free. The assembly also lately passed an act which laid a heavy fine upon any citizen of the state who should carry negroes from Africa to any part of the world whatever, and which made the vessel engaged in the horrid traffic li-

---

[47]Printed under a London dateline, this paragraph was reprinted in Massachusetts, Pennsylvania, Maryland, and three times in Virginia.
[48]Pemberton Papers, Historical Society of Pennsylvania.

able to forfeiture. In this holy and glorious zeal for a persecuted part of the human species (in which she has been considerably followed by the state of Pennsylvania) she deserves the esteem of the whole Christian world, and will draw down upon herself the blessings of Heaven. Secondly, The state of Rhode-Island a long time ago manifested a just indignation against the dangerous society of the Cincinnati,[49] and declared that no members of that society should hold an office in the state. Thirdly, They have submitted the new constitution to all the freemen of the state, who have rejected it by a large majority, and the motion which was made in the assembly for calling a convention, was rejected by a majority of 27. The new constitution therefore is cast out of that state (to use the strong expression of the prophet) AS A MENSTRUOUS CLOTH. In this procedure the house of representatives in Massachusetts Bay seem inclined to support her, and perhaps the people of New-Hampshire. If therefore the friends of liberty and human nature will unite, they may baffle the dark and wicked conspiracy which has been formed to enslave this country, notwithstanding Maryland has adopted the new government, and South Carolina probably will adopt it.

According to the poet.

The wise and active conquer difficulties
By daring to attempt them, sloth and folly,
Shiver and shrink at sight of toil and hazard,
And make the impossibility they fear.

[49]At the end of the war, officers of the Continental Army formed themselves into a permanent society named after the victorious Roman general Cincinnatus, who surrendered his military commission and returned to his farm. Ostensibly the society was to raise funds for the support of widows and children of soldiers killed in the war. Additionally, the society promoted a stronger Union with more powers for Congress. Membership in the society was limited to officers, who passed their membership along to their first-born sons. Vehement opposition arose to the society because of its aristocratic overtones.

FEDERAL FARMER: LETTER XVIII (ELBRIDGE GERRY)
New York, May 2, 1788[50]

I might observe more particularly upon several other parts of the constitution proposed; but it has been uniformly my object in examining a subject so extensive, and difficult in many parts to be illustrated, to avoid unimportant things, and not to dwell upon points not very material. The rule for apportioning requisitions on the states, having some time since been agreed to by eleven states, I have viewed as settled. The stipulation that congress, after twenty one years may prohibit the importation of slaves, is a point gained, if not so favourable as could be wished for.

JAMES PEMBERTON TO JOHN PEMBERTON
Philadelphia, May 3, 1788[51]

It is generally agreed, that the conclusion of the Convention on the Subject [of the slave trade], will not restrain the Assemblys of the Separate States from passing any prohibitory laws which they may judge expedient to abolish that infamous traffic; but it would have been far more consistent with former declarations, and the principles of Justice in that body to have manifested their abhorrence of it, then would their proposed System have yielded some hope of Stability.

JAMES PEMBERTON TO JAMES PHILLIPS
Philadelphia, May 4, 1788[52]

The late interesting intelligence of the combined Efforts of benevolence raised in your kingdom, are circulated thro. these States by means of our news papers, which have been found to be the most ready and useful mode; I have not been inattentive to the distribution of the Treaties in such manner as I judged would be most effectual to promote the design of their publication; there is much labour still necessary here, particularly in the southern states to awaken them to a sense of their inequity and of their danger; The Convention have not only fallen greatly short of the wishes of the multi-

[50]Federal Farmer was perhaps the most successful Antifederalist essayist in the country. In early November he published a forty-page pamphlet that went through several printings. The second Federal Farmer pamphlet, first advertised for sale on May 2, 1788, was entitled *An Additional Number of Letters from the Federal Farmer to the Republican Leading to a fair Examination of the System of Government Proposed by the Late Convention . . . .* The authorship of Federal Farmer is uncertain. I believe Elbridge Gerry to be the author.
[51]Pemberton Papers, Historical Society of Pennsylvania.
[52]Pemberton Papers, Historical Society of Pennsylvania.

tudes, but erred against conviction, and their acknowledged duty; If the new system of Government takes place one of the first objects should be to alter and amend it, and on this head, it is expected there will be pressing Solicitations; Maryland has lately acceded to the plan making the seventh state, it seems yet doubtful whether Virginia, Carolina, New Hampshire, and New York will come in to it.

A CAUTION
Philadelphia *Independent Gazetteer,* May 6, 1788[53]

Whereas, in the year 1787, some vessels were fitted out at the port of Philadelphia, for the iniquitous purpose of stealing the inhabitants of Africa, from all the endearments of domestic life;[54] one of which vessels has succeeded in obtaining a number of poor blacks, and has taken them to a port in the West Indies, where they are under the iron hand of oppression. From this shameful traffic, this horrid source, the proprietors of the vessel have purchased some West India produce, which, after landing at Wilmington, they have brought up to this city, and offered for sale.

It is a grateful circumstance to the supporters of the common rights of mankind, that the virtuous inhabitants of the city, reprobate the horrid idea.—A correspondent hopes, that the citizens will further testify their disapprobation of the practice, by turning with indignation from the purchase of any property, thus basely procured by men so lost to the common feelings of humanity; notwithstanding the *patriotic* convention, at which a *Washington* presided, have declared that this abominable traffic shall be continued for TWENTY years by the people of America!

AN AMERICAN (TENCH COXE)
Pennsylvania Gazette, May 21, 1788[55]

To the Honorable the Members *of the* Convention *of* VIRGINIA.

By the special delegation of the people of your respectable commonwealth, you are shortly to determine on the fate of the proposed constitution of foederal government. First invited to that important measure by the resolutions of your legislature, from the wisest considerations, America, confiding in the steadiness of your patriotism, and feeling that new weight is daily given

[53]Reprinted in the Philadelphia *Freeman's Journal,* May 14; *New York Journal,* May 16; *Boston Gazette,* May 19; and Winchester *Virginia Gazette,* May 21.

[54]In March 1788 the Pennsylvania legislature passed an act prohibiting the fitting out of vessels in any of the state's ports for the purpose of engaging in the slave trade.

[55]Reprinted three times in Virginia and once each in Maryland, Massachusetts, and Rhode Island.

to your original inducements, doubts not it is now to receive your sanction. But before the awful determination which is to call *the American union* once more into political existence shall be finally taken, permit one of the most respectful of your countrymen to trespass a few minutes on your time and patience.

The qualities of the proposed government have been so fully explained, and it will receive such further exposition in your honorable body, that it is needless to attempt a regular discussion of the subject. This paper shall therefore be confined to *a few particular considerations* that have been already mentioned by others, or which may now be suggested for the first time.

It has been urged by some sensible and respectable men, that your extensive state will not be properly represented in the foederal senate. Permit me to remind you, that while you have but one vote of thirteen in the present union, you will have twelve in ninety one in the new confederacy.[56] Suffer me to observe too, that as the United States are *free governments*, it might not have been very unreasonable if the people of Virginia could have given only the same number of votes at *an election for foederal purposes*, as they can give at *a state election*. If the citizens of Virginia find it *wise and prudent*, that *free* persons *only* shall be taken into consideration in electing their *state* legislature, would it appear extraordinary that citizens of the United States should think *the same rule* proper in electing the *foederal* representatives. By the present arrangement, you may enjoy the weight and power of *five* votes and a half for 168,000 slaves, being three fifths of your whole number of blacks.[57] Were these to be deducted from the votes of Virginia in the foederal house of representatives, it would leave little more than one vote in thirteen in that house. In the present Congress, as before observed, and in the proposed senate, a thirteenth vote is allotted to Virginia. Taking the number of free citizens, which is the proper rule of representation in *free governments*, Virginia, in the foederal representation, would have about as many votes as New York, and *fewer* than Massachusetts or Pennsylvania. It will be proper to consider too the effect of the erection of Kentucke into a separate state, and of her becoming another member of the new confederacy. When that *certain event* shall take place, Virginia will fall *considerably short* of the proportion of one in fourteen of the free white inhabitants of the United States. Impartially considering this true state of things, the opinion that Virginia will hold a share of the powers of the new government, less than she is entitled to, will appear to be erroneous. If, on examination, these facts shall be found to be stated with

[56]Virginia had ten of the sixty-five representatives in the first U.S. House of Representatives and two of the first twenty-six senators.

[57]"An American" was using the Constitution's ratio of no more than one representative for every 30,000 inhabitants. By an estimate used in the Constitutional Convention, the slave population of Virginia in 1787 was 280,000, three-fifths of which was 168,000.

accuracy and candor, and the observations and reasonings upon them shall appear just and fair, we confidently trust your honorable house will not consider the proposed constitution as exceptionable in that particular. . . .

It has been objected to the proposed fœderal constitution, that it tends to render our country more vulnerable, by admitting the further importation of slaves. To persons not accurately acquainted with the whole of the American constitutions, this objection may appear of weight. But when it is canvassed before so enlightened an assembly as the Convention of Virginia, the mistake will be instantly discovered. It will be remembered that ten of the states, and Virginia among the number, have already prohibited the further importation of slaves, and that the powers of the legislature of *each state*, even after the adoption of the constitution, will not only remain *competent to prohibition of the slave trade*, but (if they find the measure wise and safe) to the emancipation of the slaves already among us. It may be added further, that the exercise of this power of the state governments can *in no wise* be controuled or restrained by the fœderal legislature. . . .

It will be urged, perhaps, that property should be represented, and that though Virginia has only 252,000 free inhabitants, your representation should still be greater than that of Massachusetts and Pennsylvania, because you are richer. But surely this argument will not be urged by the friends of *equal liberty among the people*. It will not be objected *openly* against the proposed constitution, that it secures *the equal liberties of the poor*. But suppose for a moment a claim for a representation of property were admissible before an assembly of *the free and equal citizens of America*, will not Virginia enjoy the advantage of two votes *more* in the fœderal government than either Massachusetts or Pennsylvania, though each of those states has 108,000 free citizens *more than yours*. If we were represented *by that only rule of republics*, for your *ten* representatives, Massachusetts would have *more than fourteen*, and Pennsylvania the same number, while both of them are limited to *eight*. Here then we see *the balance of property* said to be in favor of Virginia has procured her three fourths as much *extra* power, as *the lives, liberties and property of all the people of Massachusetts or Pennsylvania*. Power has been given to your state *with no sparing hand*. You (suffer me respectfully to say so) of all the members of the union, appear to have the least cause of complaint. Permit me to remind you of the objections made *on this ground* by Mr. Martin, of Maryland.[58] The opposition *there* asserted that the great states had too large a share of power, and you have the most of all. The same sentiments were urged in the Connecticut Convention. Is it probable then that an allotment of power *more favorable to*

---

[58]See Luther Martin, *Genuine Information* V, Baltimore *Maryland Gazette*, January 11 (Chapter 5). On April 9 George Nicholas wrote to David Stuart that Antifederalist Luther Martin's *Genuine Information* would benefit the cause of ratification in Virginia, "particularly those parts where he speaks of the slaves and the advantages which this government gives to the large states."

you would be made by a new Convention? I submit to your candor whether you ought to ask a greater share. A comparison, in point of wealth and re-sources, between your state and any other, is a matter I wish to touch with delicacy. I mean not to offend, but you would despise a freeman, that would decline *the decent expression of his thoughts* on so momentous an occasion. I would submit to you, whether the energy of 250,000 whites in a southern climate, surrounded by more than as many slaves, can be, *or rather whether it is*, equal to that of the same number in a northern climate? Whether two or three negroes in Virginia will be found equal to one yeoman or manufacturer of Pennsylvania or Massachusetts? Whether the ships, mercantile capitals, houses, and monied corporations of Philadelphia, with her growing manu-factures and connexions in foreign commerce, may not be placed in the scale against *the balance* of wealth you may be thought to possess, when Kentucke shall become an independent member of the American union.

5

The South Debates
Slavery and
the Constitution

T HE SOUTH WAS NOT UNITED ON THE
*issues of slavery. Virginia and Maryland, with a surplus of slaves, wanted the foreign
slave trade closed immediately. Georgia and South Carolina, and to a lesser extent
North Carolina, all in need of additional slaves, wanted the foreign trade to remain
open at least for a period of time. Because a prohibition of the foreign slave trade
would cause an immediate increase in the price of all slaves in America, those
slaveowners with a surplus of slaves ready for sale would reap profits from a prohibition of slave importations. The continuation of the African slave trade, in essence,
depressed the market for slaves already in America. Virginia Antifederalists, though
keenly aware of the law of supply and demand, never made this mercenary argument
for closing the foreign slave trade. They concentrated on the strategic perspective.
Slaves, they reminded their fellow Southerners, were dangerous during peacetime
and would assuredly join the enemy during war. Only occasionally did Southerners
publicly denounce the slave trade as morally abominable, while Federalists in the
Deep South disparaged the biased stance taken by Virginia Antifederalists to increase
the value of their surplus slave property.*

*Most Southerners who opposed the closing of the foreign slave trade argued
that slaves were needed to develop their economies. During the last two years of the
war, the British army had campaigned vigorously in the South. Charleston was captured in 1780 and Lord Cornwallis marched erratically through Georgia, South
Carolina, North Carolina, and Virginia. Wherever the army went, slaves escaped
from their owners and sought British protection. Many emigrated to Nova Scotia at
the end of the war. Some slaves ran away to Florida, while others joined the Ameri-*

can army in hopes of earning their freedom at war's end. With farmlands devastated, the supply of productive slaves diminished, and the ill effects of the postwar deflation, many Southerners viewed the importation of slaves as their economic panacea. Rawlins Lowndes of Charleston argued that without an infusion of new slaves into its economy, South Carolina "would degenerate into one of the most contemptible [states] in the union." Lowndes, in fact, boldly asserted that the foreign slave trade "could be justified on the principles of religion, humanity and justice." To transport "a set of human beings from a bad country to a better, was fulfilling every part of those principles." With few slaves of their own, Northerners, according to Lowndes, begrudged Southerners theirs. Lachlan McIntosh of Georgia suggested that the Southern states in general and Georgia in particular should ratify the Constitution for only the limited term of twenty years. At that time, with the constitutional barrier eliminated, the Northern-dominated Congress would certainly prohibit the continuation of the foreign slave trade, much to the detriment of the Deep South. The Southern states could then determine whether they would ratify the Constitution unconditionally, extend their limited-term adoption, or abandon the Union.

Southern Federalists found the Constitution's slavery provisions easy to defend. Northerners had conceded a great deal: they gave representation to a species of property that they themselves did not possess; they allowed the slave trade to remain open for twenty years with no guarantee that Congress would close the trade after 1808; they gave slaveowners the right to recover their runaway property in any part of the Union; and, because it had no specific authority, Congress could never emancipate slaves. In a political world of compromise, the Southern delegates to the Constitutional Convention could have accomplished little more. South Carolina Convention delegate Charles Cotesworth Pinckney suggested that "considering all circumstances, we have made the best terms for the security of this species of property it was in our power to make. We would have made better if we could, but on the whole I do not think them bad."

Southern delegates to the Constitutional Convention gratefully acknowledged the accommodating spirit of New England delegates. David Ramsay in Charleston told his Massachusetts correspondent that "Your delegates never did a more politic thing than in standing by those of South Carolina about negroes. Virginia deserted them & was for an immediate stoppage of further importation. The Dominion [i.e., Virginia] has lost much popularity by the conduct of her delegates on this head. The language is 'the Eastern [New England] states can soonest help us in case of invasion & it is more our interest to encourage them & their shipping than to join with or look up to Virginia." Ramsay told his Southern audience that avarice had led New Englanders to support a continuation of the slave trade for twenty years because they would derive commercial advantages. This same Yankee avarice could be counted on in 1808—New Englanders in Congress would never vote to cut off the importation of slaves as long as more slaves were needed to produce exportable staples that would be shipped by Yankee merchants in Northern-owned vessels. "Their interests," Ramsay assured his Southern readers, "will therefore coincide with our's." In the meantime, as many slaves could be imported as were desired and other sources were also available—"*the natural increase of those we already have, and the influx from our north-ern neighbours, who are desirous of getting rid of their slaves, will afford a sufficient number for cultivating*" all Southern lands.

Luther Martin, a Maryland delegate to the Constitutional Convention, was the only Southerner in the public debate over the ratification of the Constitution to condemn this New England–Deep South coalition, which he publicly denounced as an evil bargain. He warned his readers "*that national crimes can only be, and frequently are, punished in this world by national punishments, and that the continuance of the slave trade, and thus giving it a national sanction and encouragement, ought to be considered as justly exposing us to the displeasure and vengeance of Him, who is equal Lord of all, and who views with equal eye, the poor African slave and his American master!*" Martin suggested that Americans ought to prohibit the importation of slaves in their federal constitution "*and to authorize the general government from time to time, to make such regulations as should be thought most advantageous for the gradual abolition of slavery, and the emancipation of the slaves which are already in the States.*" Martin's arguments were so objectionable to most Southerners that Federalists sometimes alluded to them as a reason for adopting the Constitution.

Martin felt confident that the slave trade would never be ended if it were allowed to remain open for twenty years. When Americans fought Britain for their liberties, "*we warmly felt for the common rights of men.*" With freedom gained, however, "*we are daily growing more insensible to those rights.*" Those states that prohibited the importation of slaves would reconsider their decisions when they saw slave-importing states prosper. Soon, all of the Southern states would re-enter the slave trade.

On rare occasions, a Southern Federalist alluded to the happy prospect of the eventual end of slavery. In a Virginia newspaper article aimed at Quakers, a correspondent looked forward to the new federal government which "*would eagerly embrace the opportunity not only of putting an end to the importation of slaves, but of abolishing slavery forever.*" Southern Antifederalists encouraged slaveowners to believe that the Constitution would endanger slavery. George Mason and Patrick Henry both raised such a specter in the Virginia ratifying convention. Mason complained that "*there is no clause in the Constitution that will prevent the Northern and Eastern states from meddling with our whole property of that kind. There is a clause to prohibit the importation of slaves after twenty years, but there is no provision made for securing to the Southern states those they now possess. It is far from being a desirable property. But it will involve us in great difficulties and infelicity to be now deprived of them. There ought to be a clause in the Constitution to secure us that property, which we have acquired under our former laws, and the loss of which would bring ruin on a great many people.*" Mason also criticized the clause in the Constitution that forbade any amendment that would prohibit the importation of slaves before 1808. Therefore, the Constitution delayed the end of the foreign slave trade while endangering slavery in the South. The delegates to the Constitutional Convention "*have done what they ought not to have done, and have left undone what they ought to have done.*"

Patrick Henry suggested that Congress could and would use its "power to provide for the general defence and welfare" as the mechanism to abolish slavery altogether. Although he deplored slavery, Henry saw "that prudence forbids its abolition." But "the majority of Congress is to the North, and the slaves are to the South. In this situation, I see a great deal of the property of the people of Virginia in jeopardy, and their peace and tranquillity gone away." It was impossible, in Henry's judgment, to emancipate the slaves "without producing the most dreadful and ruinous consequences." The institution of slavery ought to be left untouched—"We ought to possess them in the manner we have inherited them from our ancestors, as their manumission is incompatible with the felicity of the country. But we ought to soften, as much as possible, the rigour of their unhappy fate."

James Madison and Governor Edmund Randolph defended the foreign slave clause of the Constitution and vehemently objected to Mason's and Henry's emotionally charged arguments about imminent emancipation. According to Madison, the slave trade was left open until 1808 only because Georgia and South Carolina "would not have entered into the Union of America, without the temporary permission of that trade. And if they were excluded from the Union, the consequences might be dreadful to them and to us. We are not in a worse situation than before. That traffic is prohibited by our [Virginia] laws, and we may continue the prohibition. The Union in general is not in a worse situation. Under the articles of Confederation, it might be continued forever: But by this clause an end may be put to it after twenty years. There is therefore an amelioration of our circumstances." Furthermore, Congress could tax all slave imports, but not to the extent that it "would amount to a prohibition." Madison strongly denied that the Constitution empowered the federal government to interfere with slavery. If ever Congress attempted an emancipation, it would be viewed as "an usurpation of power." "There is no power to warrant it. . . . Such an idea never entered into any American breast, nor do I believe it ever will," except for those people "who substitute suspicions to reason."

Governor Randolph took a strangely paradoxical approach to the emancipation issue. He felt morally outraged with Patrick Henry's defense of slavery, while he assured his fellow delegates that no clause in the Constitution authorized an abolition of slavery. "I hope that there is none here, who considering the subject in the calm light of philosophy, will advance an objection dishonorable to Virginia; that at the moment they are securing the rights of their citizens, an objection is started that there is a spark of hope, that those unfortunate men now held in bondage, may, by the operation of the General Government, be made free." Randolph, however, assured the convention delegates that no Virginia delegate to the Constitutional Convention "had the smallest suspicion of the abolition of slavery."

Zachariah Johnston, a farmer from western Virginia, stood alone among Federalists in the Virginia ratifying convention in predicting and defending the abolition of slavery regardless of the new Constitution. Antifederalists "tell us that they see a progressive danger of bringing about emancipation. The principle has begun since the revolution. Let us do what we will, it will come round. Slavery has been the foundation of that impiety and dissipation which have been so much disseminated among our countrymen. If it were totally abolished, it would do much good."

In the North Carolina ratifying convention, Federalist James Iredell gave lip-service to the plight of slaves: "When the entire abolition of slavery takes place, it will be an event which must be pleasing to every generous mind, and every friend of human nature; but we often wish for things which are not attainable." Antifederalist James Galloway felt more uneasy about emancipation. "The property of the Southern States consists principally of slaves. If they mean to do away [with] slavery altogether, this property will be destroyed." Galloway feared that Northerners through the Constitution intend "to bring forward manumission. If we must manumit our slaves, what country shall we send them to? It is impossible for us to be happy, if, after manumission, they are to stay among us."

CHARLESTON *COLUMBIAN HERALD*
July 26, 1787[1]

Extract of a letter from a Gentleman in Philadelphia, to his Friend in this City, dated July 4, 1787.

". . . Have just heard from undoubted authority, that a member of the Convention will propose this week, that no slave whatever be imported into any of the states for the term of twenty-five years."

GEORGE MASON
Objections to the Constitution, September 1787

George Mason, a wealthy Virginia planter, was the prime author of the Virginia Declaration of Rights and state constitution of 1776. He represented Virginia in the Constitutional Convention where he was one of three delegates who refused to sign the Constitution on the last day of the Convention, at least in part, because the Constitution did not include a bill of rights. He was one of the Antifederal leaders in the state ratifying convention.

Manuscript copies of Mason's objections were circulated throughout the country for two months before they were published independently in three newspapers: the Massachusetts Centinel, *November 21, 1787;* Virginia Journal, *November 22, and the Winchester* Virginia Gazette, *November 23. At least twenty-five other newspapers as well several pamphlet anthologies reprinted the objections. The excerpt printed here was taken from an enclosure to a letter sent by Mason to George Washington on October 7, 1787 (Washington Papers, Library of Congress).*

. . . The general Legislature is restrained from prohibiting the further Importations of Slaves for twenty odd Years; tho' such Importations render the United States weaker, more vulnerable, and less capable of Defence.

[1]Reprinted seventeen times from New Hampshire to Virginia by September 12.

REPORT OF THE NORTH CAROLINA DELEGATES
TO THE CONSTITUTIONAL CONVENTION
Philadelphia, September 18, 1787[2]

In the Course of four Months Severe and painful application and anxiety, the Convention have prepared a plan of Government for the United States of America which we hope will obviate the defects of the present Foederal Union and procure the enlarged purposes which it was intended to effect. Inclosed we have the honor to send you a Copy, and when you are pleased to lay this plan before the General Assembly we entreat that you will do us the justice to assure that honorable body that no exertions have been wanting on our part to guard & promote the particular Interest of North Carolina. You will Observe that the representation in the Second Branch of the National Legislature is to be According to Numbers, that is to say, According to the whole Number of white Inhabitants added to three fifths of the blacks; . . . we had many things to hope from a National Government and the Chief thing we had to fear from such a Government was the Risque of unequal or heavy Taxation but we hope You will believe as we do that the Southern States in General and North Carolina in Particular are well Secured on that head by the Proposed System. It is provided in the 9th. Section of Article the first that no Capitation or other direct Tax shall be laid except in Proportion to the Number of Inhabitants, in which Number five Blacks are only Counted as three.—If a land Tax is laid we are to Pay at the same Rate, for Example, fifty Citizens of North Carolina can be taxed no more for all their Lands than fifty Citizens in one of the eastern States. This must be greatly in our favour for as Most of their Farms are Small & many of them live in Towns, we certainly have, one with another, land of twice the Value that they Possess. When it is also considered that five Negroes are only to be charged the Same Poll Tax as three whites the advantage must be considerably increased Under the Proposed Form of Government. The Southern States have also a much better Security for the Return of Slaves who might endeavour to escape than they had under the original Confederation—

JAMES MADISON TO GEORGE WASHINGTON
New York, October 18, 1787[3]

[Commenting on George Mason's objections to the Constitution.]

My memory fails me also if [in the Constitutional Convention] he did not acquiesce in if not vote for, the term allowed for the further importation of slaves.

[2]To Governor Richard Caswell, Governors' Letterbooks, North Carolina Archives.
[3]Washington Papers, Library of Congress.

James Madison to Thomas Jefferson
New York, October 24, 1787[4]

. . . I return to the third object abovementioned, the adjustment of the different interests of different parts of the Continent. Some contended for an unlimited power over trade including exports as well as imports, and over slaves as well as other imports; some for such a power, provided the concurrence of two thirds of both Houses were required; Some for such a qualification of the power, with an exemption of exports and slaves, others for an exemption of exports only. The result is seen in the Constitution. S. Carolina & Georgia were inflexible on the point of the slaves.

George Lee Turberville to James Madison
Richmond, December 11, 1787[5]

Another objection (and that I profess appears very weighty with me) is the want of a Council of State to assist the President—to detail to you the various reasons that lead to this opinion is useless. You have seen them in all the publications almost that pretend to analyse this system—most particularly in Colo. Masons We have heard from *private persons* that a system of government was engrossed—which had an Executive council—and that the priviledge of importing slaves (another great evil) was not mention'd in it—but that a Coalition took place between the members of the small states—& those of the southern states—& they barter'd the Council for the Priviledge—and the present plan thus defective—owes its origin to this Junction—if this was the case it takes greatly off from the confidence that I ever conceived to be due to this Convention—such conduct wou'd appear rather like the attempt of a party to carry an interested measure in a state legislature than the production of the United Wisdom—Virtue—& Uprightness of America called together to deliberate upon a form of Government that will affect themselves & their latest Posterity. . . .

For what Reason—or to answer what republican Veiw is it, that the way is left open for the importation of Negro slaves for twenty one Yrs?

---

[4]Jefferson Papers, Library of Congress.
[5]Madison Collection, New York Public Library. Turberville, a planter, represented Richmond County in the Virginia House of Delegates.

LACHLAN MCINTOSH TO JOHN WEREAT
Skidoway Island, Ga., December 17, 1787[6]

Lachlan McIntosh was a wealthy and influential Camden County, Ga., planter. Starting with a small investment before the Revolution, he regularly purchased slaves and received increasingly large grants from Georgia's royal government. Evidence indicates he was a demanding master who worked his slaves hard. But in 1775, McIntosh had a change of heart. Swept up in the Revolutionary movement, McIntosh became a leader opposing British imperial policies. In January 1775 a number of like-minded Georgians met in the Altamaha River village of Darien and drew up a petition protesting Parliament's arbitrary rule designed to "subject and enslave" American colonists. After listing a series of complaints, the petition, drafted by McIntosh, addressed a problem closer to home. "To show the world that we are not influenced by any contracted or interested motives, but a general philanthropy for all mankind, of whatever climate, language, or complexion, we hereby declare our disapprobation and abhorrence of the unnatural practice of Slavery in America, (however the uncultivated state of our country, or other specious arguments may plead for it,) a practice founded in injustice and cruelty, and highly dangerous to our liberties, (as well as lives,) debasing part of our fellow-creatures below men, and corrupting the virtue and morals of the rest; and is laying the basis of that liberty we contended for (and which we pray the Almighty to continue to the latest posterity) upon a very wrong foundation. We therefore resolve, at all times to use our utmost endeavours for the manumission of our Slaves in this Colony, upon the most safe and equitable footing for the masters and themselves."

The war for independence was fought and won, but no freedom came for Georgia slaves. Lachlan McIntosh, like many other Georgia planters, faced new battles: first to overcome the destruction caused by the war and the confiscation of slaves by the British and then to survive the postwar depression that began in 1785. By the end of 1787 McIntosh's ideological fervor for universal liberty had abated. New lands were available for cultivation. Slaves would be needed. Promises of and support for emancipation receded. McIntosh even defended the foreign slave trade and criticized the Constitution because it allowed its prohibition in 1808. In the letter that follows, McIntosh suggests to John Wereat (soon to be elected president of the Georgia ratifying Convention) that Georgia adopt the Constitution for a twenty-year period. At that time—coinciding with the time when Congress would have the authority to prohibit the foreign-slave trade—Georgians could decide whether they wanted to stay in or abandon the Union. Philanthropy and humanity had given way to avarice and exploitation. McIntosh's vision of economic growth and prosperity totally overwhelmed his earlier view of the "injustice and cruelty" of slavery.

. . . as I had a wish to be in this [Georgia's ratifying] Convention, I drew up the enclosed compromise as a memorandum for myself, which I hold some hopes might meet with the wishes of all parties, either with or without the

[6]McIntosh Papers, Georgia Historical Society.

annexed conditions, and be adopted not only by our own but some other states, especially the Southern States, who are more particularly interested as they are, and ever will continue from their extent and other circumstances, the minority in Congress. Therefore it may be thought prudent, at least for them at this time, to avoid the rocks on both sides of the question instead of binding ourselves and posterity forever to adopt the Constitution only for a certain period of time during which they will have a fair trial of its effects, and at the expiration of that time be at liberty and have it in their own power to adopt it again if they please for another period, either without or with any amendments they may find necessary, which probably will hereafter be done by conventions, as the precedent is now set which is a new and far better method of settling public differences than the old way of cutting one another's throats. If we bind ourselves and our posterity now, by adopting this Constitution without any conditions or limitation of time, any efforts made thereafter for redress of grievances must be termed rebellion, as it will be impossible to obtain amendments in the mode proposed when the majority, which is observed will ever be against the Southern States, find it their interest to continue them, and men of influence are once fixed in their saddles.

It is known to have been the intention of the Eastern and Northern States to abolish slavery altogether when in their power, which, however just, may not be convenient for us so soon as for them, especially in a new country and hot climate such as Georgia. Let us therefore keep the proper time for it in our own power while we have it. This Constitution prolongs the time for 20 years more, which is one reason for fixing upon that period in the enclosed limits, as well as to pay off our national incumbrances, which it is conceived may be done in that time when we have given up all our purse strings for that purpose without regard to our own particular engagements.

LUTHER MARTIN
Genuine Information V, Baltimore *Maryland Gazette*
January 11, 1788

Luther Martin, a native of New Jersey, had served as Maryland attorney general since 1778. He represented Maryland in the Constitutional Convention and voted against ratification in the state ratifying convention in April 1788. On November 29, 1787, Martin and other Maryland delegates to the Constitutional Convention appeared upon request before the Maryland House of Delegates to explain what had happened in the Constitutional Convention. Martin's speech before the assembly was expanded and reorganized in his "Genuine Information," which was printed in twelve installments in the Baltimore Maryland Gazette *between December 28, 1787, and February 8, 1788. Eight newspapers throughout the country reprinted some or all of the installments. The entire "Genuine Information" was reprinted in Philadelphia as a pamphlet in April 1788. This particular installment was reprinted in the* Pennsylvania Packet, *January 18;* Pennsylvania Herald,

January 23; Philadelphia Independent Gazetteer, *January 24;* New York Journal, *January 25 and 26; and* State Gazette of South Carolina, *April 28.*

Mr. Martin's Information to the House of Assembly, continued.

With respect to *that part* of the *second* section of the *first* article, which relates to the *apportionment of representation* and *direct taxation*, there were considerable objections made to it, besides the great objection of *inequality*—It was urged, that no principle could justify taking *slaves* into computation in *apportioning* the number of *representatives* a State should have in the government—that it involved the absurdity of *increasing* the power of a State in making laws for *free men* in *proportion* as that State *violated* the *rights of freedom*—That it might be proper to take slaves into consideration, when *taxes* were to be apportioned, because it had a tendency to *discourage slavery*; but to take them into account in *giving representation* tended to *encourage* the *slave trade*, and to make it the *interest* of the States to *continue* that *infamous traffic*—That slaves could not be taken into account as *men*, or *citizens*, because they were not admitted to the *rights of citizens* in the States which adopted or continued slavery—If they were to be taken into account as *property*, it was asked, what peculiar circumstance should render this property (of *all others* the most *odious* in its nature) entitled to the *high privilege* of conferring *consequence* and *power* in the *government* to its possessors, rather than *any other* property—and why *slaves* should, as property, be taken into account rather than *horses, cattle, mules,* or any *other species*—and it was observed by an honorable member from Massachusetts [Elbridge Gerry], that he considered it as dishonorable and humiliating to enter into compact with the *slaves* of the *southern States*, as it would be with the *horses* and *mules* of the *eastern.* It was also objected, that the *numbers* of representatives appointed by this section to be sent by the particular States to compose the *first* legislature, were not precisely *agreeable* to the *rule* of representation adopted by this system, and that the numbers in this section are *artfully lessened* for the *large* States, while the *smaller* States have their *full proportion* in order to prevent the *undue influence* which the *large* States will have in the government from being *too aparent*; and I think, Mr. Speaker, that this objection is *well founded.*—I have taken some pains to obtain information of the numbers of free men and slaves in the different States, and I have reason to believe, that if the estimate was *now* taken, which is directed, and one delegate to be sent for every thirty thousand inhabitants, that Virginia would have at least *twelve* delegates, Massachusetts *eleven*, and Pennsylvania *ten*, instead of the numbers stated in *this section*; whereas the *other* States, I believe, would not have more than the numbers there allowed them, nor would Georgia, most probably at present, send more than *two*—If I am right, Mr. Speaker, upon the enumeration being made, and the representation being apportioned according to the rule prescribed, the *whole number* of delegates would be *seventy-one*, *thirty-six* of which would be a *quorum* to do business; the delegates of Virginia, Massachusetts, and Pennsylvania, would amount to *thirty-three* of that quorum—Those three States will, therefore, have *much more* than *equal* power and influence in *making* the laws and regulations, which are to affect this continent, and will have a *moral certainty* of *preventing* any laws or regulations which *they disapprove*, although they might be thought ever so *necessary* by a *great majority* of the States.[7]

SOUTH CAROLINA HOUSE OF REPRESENTATIVES
Debate over the Calling of a State Ratifying Convention
January 16–18, 1788[8]

All of the state legislatures called specially elected conventions to consider the new Constitution. Only in South Carolina, however, did the legislature seriously debate the merits of the Constitution before calling its convention. During these legislative debates over the Constitution, sixty-six-year old Rawlins Lowndes of Charleston severely criticized the concessions made by Southern delegates in the Constitutional Convention on slavery and commercial legislation. If the Constitution were adopted with these provisions, Lowndes predicted that "the sun of the Southern states would set, never to rise again." Lowndes offered the only defense of the foreign slave trade not merely as a necessity, but as a positive good. Only a handful of delegates supported Lowndes; General Charles Cotesworth Pinckney and his cousin Charles Pinckney, both of whom had been delegates to the Constitutional Convention, defended the new form of government. The legislature overwhelmingly voted to call a state convention, which met in May 1788 and ratified the Constitution.

January 16

Rawlins Lowndes . . . believed that they [the South Carolina delegates to the Constitutional Convention] had done every thing in their power to procure for us a proportionate share in this new government; but the very little which they had gained proved what we might expect in the future; and that the interest of the Northern states would so predominate, as to divest us of any pretensions to the title of a republic. In the first place, what cause was there for jealousy of our importing negroes? Why confine us to 20 years, or rather why limit us at all? For his part he thought this trade could be justified on the principles of religion, humanity and justice; for certainly to translate a set of human beings from a bad country to a better, was fulfilling every part of those principles. But they don't like our slaves, because they have none themselves, and therefore want to exclude us from this great advantage; why should

[7]On April 9, Federalist George Nicholas of Virginia wrote to David Stuart stating: "I have seen Luther Martin's publication, or at least part of it, and think it will be of great service if we could have it in Richmond; particularly those parts where he speaks of the slaves and the advantages which this government gives to the large states. Cannot you procure it." Madison Collection, New York Public Library.

[8]*Debates Which Arose in the House of Representatives of South Carolina, on the Constitution Framed for the United States* . . . (Charleston, 1788).

the southern states allow of this without the consent of nine states? (Judge Nathaniel Pendleton observed, that only three states, Georgia, South Carolina, and North Carolina, allowed the importation of negroes, Virginia had a clause in her constitution for this purpose,[9] and Maryland, he believed, even before the war, prohibited them.[10]) Mr. Lowndes observed, that we had a law prohibiting the importation of negroes for three years,[11] a law he greatly approved of, but there was no reason offered why the southern states might not find it necessary to alter their conduct, and open their ports—Without negroes this state would degenerate into one of the most contemptible in the union, and cited an expression that fell from General Pinckney, on a former debate, that whilst there remained one acre of swamp land in South Carolina, he should raise his voice against restricting the importation of negroes. Even in granting the importation for 20 years, care had been taken to make us pay for this indulgence, each negro being liable on importation to pay a duty not exceeding ten dollars, and in addition to this were liable to a capitation tax. Negroes were our wealth, our only natural resource, yet behold how our kind friends in the North were determined soon to tie up our hands, and drain us of what we had.—The Eastern [New England] states drew their means of subsistence in a great measure from their shipping, and on that head they had been particularly careful not to allow of any burthens—they were not to pay tonnage or duties, no not even the form of clearing out—all ports were free and open to them! Why then call this a reciprocal bargain, which took all from one party to bestow it on the other? (Major Butler observed, that they were to pay five percent impost.) This Mr. Lowndes proved must fall upon the consumer. They are to be the carriers, and we being the consumers, therefore all expences would fall upon us.

Edward Rutledge: The gentleman had complained of the inequality of the taxes between the northern and southern states—then ten dollars a head was imposed on the importation of negroes, and that those negroes were afterwards taxed. To this it was answered, that the ten dollars per head, was an

[9]See notes 26 and 31 below.

[10]Beginning in 1695 the Maryland colonial legislature passed various acts levying taxes on the importation of slaves. Not until 1783, however, did the legislature prohibit the importation of slaves, specifying that any slave imported into the state contrary to this act "shall thereupon immediately cease to be a slave, and shall be free."

[11]On March 28, 1787 the South Carolina legislature passed an act prohibiting the importation of slaves for three years. Another act passed the same day provided penalties for importing slaves during the allotted period—the imported slaves would be forfeited and sold, and the importer would "be liable to a penalty of one hundred pounds, to the use of the State, for every such negro or slave so imported and brought in." Similar acts were passed by South Carolina in 1788, 1792, 1794, 1796, 1800, and 1802. In each case, however, visitors and new settlers to the state could bring their slaves with them from their previous residence. In December 1803, South Carolina legalized the importation of slaves.

equivalent to the 5 per cent. on imported articles; and as to their being afterwards taxed, the advantage is on our side; or, at least not against us. In the
northern states the labor is performed by white people, in the southern by
black. All the free people, (and there are few others) in the northern states,
are to be taxed by the new constitution; whereas only the free people and
three-fifths of the slaves in the southern states are to be rated in the apportioning of taxes.

January 17

General Charles Cotesworth Pinckney: As we found it necessary to give
very extensive powers to the federal government both over the persons and
estates of the citizens, we thought it right to draw one branch of the legislature immediately from the people, and that both wealth and numbers should
be considered in the representation. We were at a loss for some time for a rule
to ascertain the proportionate wealth of the states; at last we thought that the
productive labour of the inhabitants was the best rule for ascertaining their
wealth; in conformity to this rule, joined to a spirit of concession, we determined that representatives should be apportioned among the several states,
by adding to the whole number of free persons three-fifths of the slaves.—
We thus obtained a representation for our property, and I confess I did not
expect that we should have been told on our return, that we had conceded too
much to the Eastern States, when they allowed us a representation for a specie
of property which they have not among them.

The General then said he would make a few observations on the objections which the gentleman had thrown out on the restrictions that might be
laid on the African trade after the year 1808. On this point your delegates had
to contend with the religious and political privileges of the eastern and middle
states, and with the interested and inconsistent opinion of Virginia, who was
warmly opposed to our importing more slaves. I am of the same opinion now
as I was two years ago, when I used the expressions the gentleman has quoted,
that while there remained one acre of swamp land uncleared of in South Carolina I would raise my voice against restricting the importation of negroes. I
am as thoroughly convinced as that gentleman is, that the nature of our climate, and the flat, swampy situation of our country obliges us to cultivate our
lands with negroes, and that without them S. Carolina would soon be a desert
waste. You have so frequently heard my sentiments on this subject that I need
not now repeat them. It was alledged by some of the members who opposed
an unlimited importation, that slaves increased the weakness of any state who
admitted them; that they were a dangerous species of property which an invading enemy could easily turn against ourselves and the neighbouring states,
and that as we were allowed a representation for them in the House of Representatives, our influence in government would be increased in proportion as
we were less able to defend ourselves. Shew some period, said the members
from the Eastern states when it may be in our power to put a stop, if we

please, to the importation of this weakness, and we will endeavor for your convenience, to restrain the religious and political prejudices of our people on this subject. The middle states and Virginia made us no such proposition; they were for an immediate and total prohibition. We endeavored to obviate the objections that were made in the best manner we could, and assigned reasons for our insisting on the importation, which there is no occasion to repeat, as they must occur to every gentleman in the house: A committee of the states was appointed in order to accommodate this matter, and after a great deal of difficulty, it was settled on the footing recited in the constitution.[12]

By this settlement we have secured an unlimited importation of negroes for twenty years; nor is it declared that the importation shall be then stopped; it may be continued—we have a security that the general government can never emancipate them, for no such authority is granted, and it is admitted on all hands, that the general government has no powers but what are expressly granted by the constitution; and that all rights not expressed were reserved by the several states. We have obtained a right to recover our slaves in whatever part of America they may take refuge, which is a right we had not before. In short, considering all circumstances, we have made the best terms for the security of this species of property it was in our power to make. We would have made better if we could, but on the whole I do not think them bad.

Robert Barnwell: I now come to the last point for consideration, I mean the clause relative to the negroes; and here I am particularly pleased with the constitution; it has not left this matter of so much importance to us open to immediate investigation; no, it has declared that the United States shall not at any rate consider this matter for 21 years, and yet gentlemen are displeased with it. Congress has guaranteed this right for that space of time, and at its expiration may continue it as long as they please. This question then arises, what their interest will lead them to do; the Eastern states, as the hon. gentleman says, will become the carriers of America, it will therefore certainly be their interest to encourage exportation to as great extent as possible; and if the quantum of our products will be diminished by the prohibition of negroes, I appeal to the belief of every man, whether he thinks those very carriers will themselves dam up the sources from whence their profit is derived. To think so is so contradictory to the general conduct of mankind, that I am of opinion, that without we ourselves put a stop to them that the traffic for negroes will continue for ever.

Mr. Barnwell concluded by declaring that this constitution was in his opinion, like the laws of Solon, not the best possible to be formed, but the best that our situation will admit of—He considered it as the Panacea of America, whose healing power will pervade the continent, and sincerely believed that its ratification is a consummation devoutly to be wished.

[12]See Chapter 2.

January 18

[On the reason the South Carolina delegates in the Constitutional Convention opposed adding a bill of rights.]

General Pinckney: Another reason weighed particularly with the members from this state against the insertion of a bill of rights, such bills generally begin with declaring, that all men are by nature born free, now we should make that declaration with a very bad grace, when a large part of our property consists in men who are actually born slaves.

PENU BOWEN TO JOSEPH WARD
Charleston, January 16, 1788[13]

To renew our correspondence—after enquiring after the health & happiness of yourself & family—as you are both a political & speculative man—I think of nothing by which to amuse or interest you so much, as some account of our parliamentary Debates on the subject of the new federal constitution. The House took it up in order to qualify themselves to act intelligibly upon the question, whether to recommend it to their Constituents to adopt the model of procedure relative thereto pointed out by Congress &c. I assure you twas very interesting to me, to take a view of the house of Assembly, & hear the great and principal speakers of the State. They are a more numerous representation than yours & make a much better appearance; and if there be not, particular instances of superior or equal abilities in public speaking—at the same time there are fewer by many, of indifferent, ordinary or low. I think their Orators are not so correct, & Ciceronian as yours in New England—yet they have more fire. They really want method & propriety; in ease however, & fluency with rapidity they exceed you. They are in favour, & fond of the constitution in question—except in one instance—as ostensible at least. There is but one speaker against it to 8 or 10 Capital members in favour of it. The opposer is old Mr. Lowndes—and he seems to be heartily and zealously engaged. However his difficulties & objections appear not very forceable or weighty—and indeed the old Gentleman does not seem possessed of any surpassing talents to heighten or set off the defects he is afraid of. He appears to me rather as a set, obstinate, almost superannuated character—and am told he always opposes new things, & raises up bugbears & scarecrows. He seemed most horridly afraid of the influence—a preponderance of it—from the Northern States—and particularly expatiated upon & banded about the matter of the prohibition of the Negroe Trade—after 20 years—and here he advanced a sentiment which you, I know, will tremendously reprobate—Viz. that he in

[13]Ward Papers, Chicago Historical Society.

his conscience believ'd slavery to be defensible upon all principles i.e. principles of policy, morality & religion—and his argument was that of bringing them into better situations than they are taken from—as to information—maintenance &c. Another great character, tho' an advocate for the frame of Government in general, yet with severest asperity reflected upon the principles, as well as understandings, yea, & honesty of the people of the Northward, for pretending to meddle or have any thing to do about this business of the Negroe Trade and said he would have it go out by way of protest accompanying the ratification of the constitution. No one indeed undertook pointedly to defend or justify the clause or oppose the old Gentleman in his remarks on that subject. But to do justice to the politicks & principles of the Assembly—the opposer was fairly, fully & abundantly answer'd, refuted, born down, & almost silenced—and good degrees of candour with sentiments truely federal & urbanical, were thrown out, yea & espoused. I was really pleased & almost charmed with the respectful & conciliating spirit that was in general manifested toward my native Country [New England]—from the principle characters of respectability & influence here—the question was not called for but I dare say twill go in favour—nearly nem con:[14]

LUTHER MARTIN
Genuine Information VII, Baltimore *Maryland Gazette*, January 18, 1788[15]

By the *ninth* section of this article, the importation of such persons as any of the States now existing, shall think proper to admit, shall not be prohibited prior to the year one thousand eight hundred and eight, but a duty may be imposed on such importation not exceeding ten dollars for each person.

The design of this clause is to prevent the general government from prohibiting the importation of slaves, but the same reasons which caused them to strike out the word "*national*," and not admit the word "*stamps*,"[16] influ-

[14]*Nemine contradicente* (no one contradicting—unanimously).

[15]Reprinted *Pennsylvania Packet*, January 25; *Pennsylvania Herald*, January 26; Philadelphia *Independent Gazetteer*, January 28; Philadelphia *Freeman's Journal*, January 30; *New York Journal*, February 27, March 1, 7; Boston *American Herald*, March 31, April 3; Charleston *City Gazette*, April 14 (excerpt); *State Gazette of South Carolina*, May 8, 15; and in a pamphlet edition.

[16]Early in the Convention the delegates agreed to abandon the Articles of Confederation and to create a national government. Sensitive to the effect that certain words might have on the people, the delegates soon voted to change all references to "the national government" to "the government of the United States." The New Jersey plan, submitted by delegates from the smaller states, called for the retention of the Articles of Confederation with additional powers given to Congress, including the power to levy stamp taxes. The new federal Congress was given extensive tax powers, but stamp taxes were never mentioned in the Constitution.

enced them here to guard against the word "*slaves*," they anxiously sought to avoid the admission of expressions which might be odious in the ears of Americans, although they were very willing to admit into their system those *things* which the *expressions* signified: And hence it is, that the clause is so worded, as really to authorise the general government to impose a duty of ten dollars on every foreigner who comes into a State to become a citizen, whether he comes *absolutely free*, or *qualifiedly* so as a servant—although this is contrary to the design of the framers, and the duty was only meant to extend to the importation of *slaves*.

This clause was the subject of a great diversity of sentiment in the convention;—as the system was reported by the committee of detail, the provision was general, that such importation should not be prohibited, without confining it to any particular period.—This was rejected by eight States—Georgia, South-Carolina, and I think North-Carolina voting for it.

We were then told by the delegates of the two first of those States, that their States would never agree to a system which put it in the power of the general government to prevent the importation of slaves, and that they, as delegates from those States, must withhold their assent from such a system.

A committee of one member from each State was chosen by ballot, to take this part of the system under their consideration, and to endeavour to agree upon some report which should reconcile those States;—to this committee also was referred the following proposition, which had been reported by the committee of detail, to wit, "No *navigation* act shall be passed without the assent of *two-thirds* of the members present in each house;" a proposition which the *staple* and *commercial* States were solicitous to *retain*, lest their *commerce* should be placed too much under the power of the *eastern* States, but which these last States were as anxious to *reject*.—This committee, of which also I had the honour to be a member, met and took under their consideration the subjects committed to them; I found the *eastern* States, notwithstanding their *aversion* to *slavery*, were very willing to indulge the southern States, at least with a temporary liberty to prosecute the *slave trade*, provided the southern States would in their turn gratify them, by laying no *restriction on navigation acts*; and after a very little time, the committee, by a great majority, agreed on a report, by which the general government was to be prohibited from preventing the importation of slaves for a limited time, and the restrictive clause relative to navigation acts was to be omitted.

This report was adopted by a majority of the convention, but not without considerable opposition.—It was said, that we had but just assumed a place among independent nations, in consequence of our opposition to the attempts of Great-Britain to *enslave us*—that this opposition was grounded upon the preservation of *those rights*, to which God and Nature had entitled *us*, not in *particular*, but in *common* with *all the rest of mankind*—That we had *appealed* to the *Supreme Being* for his *assistance*, as the *God of freedom*, who could not but *approve* our efforts to preserve the *rights* which he had thus *imparted to his creatures*—that now, when we scarcely had risen from our *knees*, from *supplicating* his *aid* and *protection*—in *forming our government* over a *free people*, a government formed pretendedly on the *principles* of *liberty* and for *its preservation*,—in *that* government to have a provision, not only putting it out of *its power* to *restrain* and *prevent* the *slave trade*, but *even encouraging that most infamous traffic*, by giving the *States power* and *influence* in the union, in *proportion* as they *cruelly and wantonly sport with the rights of their fellow creatures*, ought to be considered as a *solemn mockery of*, and *insult to, that God* whose protection we had then implored, and could not fail to hold us up in *detestation*, and render us *contemptible* to every *true friend* of liberty in the world.—It was said, it ought to be considered that *national* crimes can *only be*, and *frequently are, punished* in this world by *national punishments*, and that the *continuance* of the slave trade, and thus giving it a *national sanction* and *encouragement*, ought to be considered as *justly exposing* us to the *displeasure* and *vengeance* of *Him*, who is equal Lord of all, and who views with equal eye, the poor *African slave* and his *American master!*

LUTHER MARTIN
Genuine Information VIII, Baltimore *Maryland Gazette*, January 22, 1788[17]

Mr. MARTIN's Information to the House of Assembly, continued.

It was urged that by this system, we were giving the general government full and absolute power to regulate commerce, under which general power it would have a right to *restrain*, or *totally prohibit* the *slave trade*—it must appear to the world absurd and disgraceful to the last degree, that we should *except* from the exercise of that power, the *only branch* of *commerce*, which is *unjustifiable in its nature*, and *contrary* to the *rights of mankind*—That on the contrary, we ought *rather to prohibit expressly* in our *constitution*, the *further importation* of *slaves*; and to *authorize* the general government from time to time, to make such regulations as should be thought most advantageous for the *gradual abolition* of *slavery*, and the *emancipation* of the *slaves* which are already in the States.

That *slavery* is *inconsistent* with the *genius* of *republicanism*, and has a tendency to *destroy* those *principles* on which it is *supported*, as it *lessens the sense* of the *equal rights* of *mankind*, and habituates us to *tyranny* and *oppression*.—It was further urged, that by this system of government, every State is to be protected both from *foreign invasion* and from *domestic insurrections*; that from this consideration, it was of the *utmost importance* it should have a power to restrain the importation of slaves, since in *proportion* as the number of slaves were encreased in any State, in the *same* proportion the State is *weakened* and

[17]Reprinted Philadelphia *Independent Gazetteer*, February 11; *New York Journal*, March 7, 12, 14; Boston *American Herald*, April 3 (excerpt); *State Gazette of South Carolina*, May 15, 19 (excerpt); and in a pamphlet edition.

exposed to foreign invasion, or domestic insurrection, and *by so much the less* will it be able to protect itself against *either*; and therefore will by so much the more, want aid from, and be a burthen to, the union.—It was further said, that as in this system we were giving the general government a power under the idea of national character, or national interest, to regulate even our *weights* and *measures*, and have prohibited all possibility of *emitting paper money*, and *passing instalment laws, &c.*—It must appear still more extraordinary, that we should prohibit the government from interfering with the slave trade, than which *nothing* could so *materially affect* both our *national honour* and *interest*.— These reasons influenced me both on the committee and in convention, most decidedly to oppose and vote against the clause, as it now makes a part of the system.

You will perceive, Sir, not only that the general government is prohibited from interfering in the slave trade *before* the year eighteen hundred and eight, but that there is no provision in the constitution that it shall *afterwards* be prohibited, nor any security that such prohibition will ever take place— and I think there is great reason to believe that if the importation of slaves is permitted until the year eighteen hundred and eight, it will not be prohibited afterwards—At *this time* we do not generally hold this commerce in so *great* abhorrence as we have done.—When our *own* liberties were at stake, we *warmly* felt for the *common rights of men*—The danger being thought to be past, which threatened ourselves, we are daily growing *more insensible* to those rights—In those States who have restrained or prohibited the importation of slaves, it is only done by legislative acts which may be repealed—When those States find that they must in their *national character* and *connection* suffer in the *disgrace*, and share in the *inconveniences* attendant upon that detestable and iniquitous traffic, they may be desirous also to share in the *benefits* arising from it, and the odium attending it will be greatly effaced by the sanction which is given to it in the general government.

DAVID RAMSAY TO BENJAMIN LINCOLN
Charleston, January 29, 1788[18]

Our Assembly is now sitting & have unanimously agreed to hold a convention. By common consent the merits of the foederal constitution were freely discussed on that occasion for the sake of enlightening our citizens. Mr Lownds was the only man who made direct formal opposition to it. His objections were local & proceeded from an illiberal jealousy of New:England men. He urged that you would raise freights on us & in short that you were

---

[18]Lincoln Papers, Massachusetts Historical Society. Born in Pennsylvania, Ramsay graduated from the College of New Jersey (Princeton) in 1765. He studied medicine with Benjamin Rush and received a medical degree from the College of Philadelphia (the University of Pennsylvania) in 1772. After practicing medicine in Maryland for a year he moved to Charleston, S.C. From 1776 to 1790 he served in the South Caro-

too cunning for our honest people. That your end of the continent would rule the other. That the sun of our glory would set when the new constitution operated. He has not one foederal idea in his head nor one that looks beyond Pedee.[19] He is said to be honest & free of debt but he was an enemy to Independence & though our President in 1778 he was a British subject in 1780. His taking protection was rather the passive act of an old man than otherwise. He never aided nor abetted the British government directly but his example was mischievous. His opposition has poisoned the minds of some. I fear the numerous class of debtors more than any other. On the whole I have no doubt that it will be accepted by a very great majority of this State. The sentiments of our leading men are of late much more foederal than formerly. This honest sentiment was avowed by the first characters. "New England has lost & we have gained by the war her suffering citizens ought to be our carriers though a dearer freight should be the consequence." Your delegates never did a more political thing than in standing by those of South Carolina about negroes. Virginia deserted them & was for an immediate stoppage of further importation. The dominion has lost much popularity by the conduct of her delegates on this head. The language now is "the Eastern states can soonest help us in case of invasion & it is more our interest to encourage them & their shipping than to join with or look up to Virginia." In short sir a revolution highly favorable to union has taken place. Foederalism & liberality of sentiment has gained great ground. Mr Lownds still thinks you are a set of sharpers—does not wonder that you are for the new constitution as in his opinion you will have all the advantage. You begrudge us our negroes in his opinion. But he is almost alone.

CIVIS RUSTICUS
Virginia Independent Chronicle, January 30, 1788

The following "objections to the Constitution of Government formed by the Convention," are stated to be Col. Mason's.

I shall remark on them with that freedom which every person has a right to exercise on publications, but, with that deference, which is due to this respectable and worthy gentleman; to whose great and eminent talents, profound judgment, and strength of mind, no man gives a larger credit, than he, who presumes to criticise his objections—these, falling from so great a height,

---

lina assembly (except for a year in British captivity and three years in the Confederation Congress.) Lincoln, a Hingham, Mass., farmer, had been a major general during the Revolution. He commanded the forces that surrendered to the British at Charleston, S.C., in 1780. In 1787 he led the Massachusetts militia that suppressed Shays's Rebellion.

[19]The Pedee River flows south easterly into the Atlantic Ocean as it runs parallel to South Carolina's northern border.

from one of such authority, may be supposed, if not taken notice of, to contain arguments unanswerable—not obtruding themselves on my mind in that forcible manner, I submit to the decision of the public, whether, what is now offered, contain declamation or reason; cavil, or refutation. . . .

12th. Not restraining for twenty years the importation of Africans will not effect us—This gives South-Carolina and Georgia that privilege, if it be their pleasure to avail themselves of it—Is not this objection, the excess of criticism?

CIVIS (DAVID RAMSAY)
Charleston *Columbian Herald*, February 4, 1788

David Ramsay wrote this essay to counter the arguments of his fellow Charlestonian representative Rawlins Lowndes in the debate in the South Carolina assembly over the calling of a state ratifying convention. Drafted "in a few hours," Ramsay wrote "in a summary way & in a plain stile for the benefit of common people." The essay was also printed as a twelve-page pamphlet entitled An Address to the Freemen of South-Carolina, on the Subject of the Federal Constitution, *which was distributed to members of the South Carolina legislature. Ramsay also sent copies of the pamphlet to Benjamin Rush in Philadelphia and Benjamin Lincoln in Massachusetts. The essay was reprinted in the* Pennsylvania Mercury, *April 3; the Fredericksburg* Virginia Herald, *April 17;* Maryland Journal, *April 25;* Winchester *Virginia Centinel,* April 30, *and the May issue of the Philadelphia* American Museum.

To the Citizens of South Carolina.

The eastern states, by the revolution, have been deprived of a market for their fish, of their carrying-trade, their ship building, and almost of every thing but their liberties. As the war has turned out so much in our favor, and so much against them, ought we to begrudge them the carrying of our produce, especially when it is considered, that by encouraging their shipping, we increase the means of our own defence. Let us examine also the federal constitution, by the principle of reciprocal concession. We have laid a foundation for a navigation act.—This will be a general good; but particularly so to our northern brethren. On the other hand, they have agreed to change the federal rule of paying the continental debt, according to the value of land as laid down in the confederation, for a new principle of apportionment, to be founded on the numbers of inhabitants in the several states respectively. This is an immense concession in our favor. Their land is poor; our's rich; their numbers great; our's small; labour with them is done by white men, for whom they pay an equal share; while five of our negroes only count as equal to three of their whites. This will make a difference of many thousands of pounds in settling our continental accounts. It is farther objected, that they have stipulated for a right to prohibit the importation of negroes after 21 years. On this subject observe, as they are bound to protect us from domestic violence, they think we ought not to increase our exposure to that evil, by an unlimited importation of slaves. Though Congress may forbid the importation of negroes after 21 years, it does not follow that they will. On the other hand, it is probable that they will not. The more rice we make, the more business will be for their shipping: their interest will therefore coincide with our's. Besides, we have other sources of supply—the importations of the ensuing 20 years, added to the natural increase of those we already have, and the influx from our northern neighbours, who are desirous of getting rid of their slaves, will afford a sufficient number for cultivating all the lands in this state.

A Virginian
Virginia Independent Chronicle, February 13, 1788[20]

A hint to the people called Quakers *in Virginia*
Gentlemen, Have you considered the plan of the new Constitution? If you have, I think you certainly disapprove it, especially in two points. 1st. As it admits of the importation of slaves to America for a limited time; for admitting slavery to be justifiable, it would be very impolitic to allow of any more of the poor Africans to be brought amongst us; instead whereof, I think it would better become us all as men and Christians, to endeavor to release those already under our care from the grievous burthens they are labouring under, than to permit any more to be subjected to the like sufferings.

George Nicholas
Charlottesville, Va., February 16, 1788[21]

The next objection is that if this government is adopted the property that we have in slaves may be lost or injured. So far is this from being true that we can venture to say that the new government will be the best security that we can have for retaining that property. Congress could pass no act which would injure that property but in one of three ways either
    1st. by passing an act of emancipation: or
    2dly. by permitting the other states to harbour the fugitives or
    3dly. by imposing such taxes on them as would oblige the owners to discharge them.

---

[20]For an answer to this item, see "One of the People Called Quakers in the State of Virginia," *Virginia Independent Chronicle*, March 12 (below).
[21]Reuben T. Durrett Collection, George Nicholas, Department of Special Collections, University of Chicago Library. Part of the manuscript letter is missing and thus the addressee is not known. Nicholas, a lawyer, represented Albemarle County in the Virginia House of Delegates and ratifying convention.

They could not pass an act for this emancipation because both Congress and the different state legislatures are forbid to pass *ex post facto* laws and therefore if the new government should take place neither Congress or a state legislature could pass an act to deprive any man or set of men of property which they hold under the general laws of the land. And therefore if this government had taken place prior to the last session of our assembly they could not have passed a law for the emancipation of Robt. Mooreman's Negroes.[22] Neither could Congress secondly injure you by permitting them to be harboured and protected in the other states for by an express clause in the constitution all slaves escaping from one state into another shall be delivered up. Nor could they in the third instance injure you by the mode of imposing the taxes. A poll tax is the only tax they could impose which could affect our slaves and the constitution is so guarded in that respect that we can receive no injury by that means. It is expressly declared that no capitation or poll tax shall be imposed except in proportion to the enumeration therein directed which is that we should be charged with only three fifths of the number of our slaves. It never could be the interest therefore of the states which have no slaves to impose a poll tax. Because in case of a poll tax if Massachussets had one thousand white inhabitants, and Virginia also one thousand inhabitants but one half of them white and the other half black; Massachussets would be obliged to pay Congress for her whole number whereas Virginia would pay for only eight hundred. Thus by this constitution this part of our property is much better secured and the possessors of it less liable to oppressive taxes than even under our state government.

REPUBLICUS
Kentucke Gazette, March 1, 1788

This leads me to Art. 1 Sect. 9. "The migration or importation of such persons, as any of the States now existing, shall think proper to admit, shall not be prohibited by the Congress prior to the year 1808; (twenty years hence) but a tax, or duty may be imposed on such importation, not exceeding ten dollars for each person." An excellent clause this, in an Algerine constitution; but not so well calculated (I hope) for the latitude of America. It is not to be disguised that by "such persons," slaves are principally, if not wholly intended: and shall this be found among the principles of a free people, and making a radical part of the grand base, on which they would erect an edifice sacred to

---

[22]Quaker Charles Moorman, a Louisa County planter, died on October 12, 1778 leaving a will that provided for the emancipation of his slaves. Not until 1782, however, did the Virginia legislature enact a manumission law allowing slaveowners to free their slaves. (See Chapter 1 for this law.) Because a question arose as to the legality of Moorman's actions, the legislature adopted an act in December 1787 confirming "the freedom of certain negroes late the property of Charles Moorman, deceased."

liberty. "Tell it not in Gath!"[23] O that no envious surge might ever roll it to the eastern side of the atlantic! Unhappy africans! what have they done? Have they murdered our citizens or burnt our settlements? Have they butchered, scalped, and exhausted every device of torture, on our defenceless women, and innocent children; as the savage mescriants of our own country have done? No, no! Then, why deprive them of the greatest of all blessings, liberty, "without which," says Dr. Price "man is a beast, and life a curse";[24] while coward-like, we court, caress, and cringe to our murderers. Ignorant, and comparatively innocent, till we taught them the diabolical arts of destruction, captivity, and death; and provided them with the infernal means of carrying them into practice; and all this to furnish ourselves with slaves, at the guilty expence oftimes, of the blood of, ten times the number of those thus enslaved, who lost their lives in the gallant, the virtuous defence of themselves, and families. Has this guilt ever been attoned? and do we boast of being advocates for liberty? shocking absurdity! More absurd still than a licence for such an execrable trade, should be radically woven into, and become an essential part of our national constitution, a constitution formed by a chosen assembly of our most eminent and respectable citizens; and where a personage presided, second to no individual of the human family.

The boast of America.—The wonder of Europe.—

O liberty! O virtue! O my country.

Tell us, ye who can thus, coolly, reduce the impious principle of slavery, to a constitutional system: ye professed violators of liberties of mankind: where will ye stop? what security can you give, that, when there shall remain no more black people, ye will not enslave others, white as yourselves? when Africa is exhausted, will ye spare America? and is not twenty years (taking into the accompt the slain with the more unhappy captives, victims to perpetual slavery) sufficient to depopulate her inmost forests? Or is this only an ill boding prelude, sounded in the ears, and designedly introductory to the fate of these (yet unhappy) states, who gave you existence; and who even now, while you are thus ungratefully soaring toward the summit of *Aristocracy*, are honouring you with their confidence? I shudder at the catastrophe? awake my fellow citizens! and let this infamous clause, together with the principle which gave it birth, be not only expunged out of your constitution: but contemned, eradicated, torn from your heart forever.

---

[23]II Samuel 1:20.

[24]Richard Price, *Observations on the Nature of Civil Liberty, the Principles of Government, and the Justice and Policy of War with America* . . . (London, 1776), section I, "Of the Nature of Liberty in General," pp. 5–6. *Observations* first appeared in London in February 1776 and within two months over 60,000 copies (in fourteen editions) were sold. Later in 1776, it was reprinted twice in Philadelphia, and once each in Boston, New York, and Charleston.

ONE OF THE PEOPLE CALLED QUAKERS IN THE STATE OF VIRGINIA
Virginia Independent Chronicle, March 12, 1788[25]

"*A Virginian*" might have a right to expect, and would perhaps have received, the thanks of "*the people called Quakers in Virginia*," for the "*hint*" he hath given them, if they thought it was wholly dictated by an unfeigned regard for their interests and happiness: but its seeming want of candor, the criterion, by which a plain simple people, lovers of truth, are led to judge, inclines them to think that it springs from some other motive.

He tells the Quakers, that they should "*disapprove of the new constitution*"—"*because it admits of the importation of slaves to America for a limited time.*" Hence it would seem, as if he inferred, and would have them to believe that the new constitution would introduce slaves into Virginia contrary to the inclination of the people: which the Quakers apprehend is not the case. Virginia indeed, may import slaves, but she may, as she now does, also prohibit,[26] and which it is reasonable to expect she will continue to do; and therefore, the Quakers, or any other society opposed to the slave trade, have nothing to apprehend on that score; and more especially, when it is considered that the late convention, used every means in their power, to prevail upon the Carolina's and Georgia, the only states in the union, that at present import slaves, at once to put an end to this unjust traffic; but the representatives of these states being inflexible in their opposition thereto, occasioned the limited importation as the best compromise that could be made; hence it is but just to conclude, that the new foederal government, if established, would eagerly embrace the opportunity not only of putting an end to the importation of slaves, but of abolishing slavery forever.

MARCUS V (JAMES IREDELL)
Norfolk and Portsmouth Journal, March 19, 1788[27]

Answers to Mr. Mason's *Objections* to the New Constitution, Recommended by the late Convention at Philadelphia. . . .
Xth. Objection.

---

[25]This item answers "A Virginian," *Virginia Independent Chronicle*, February 13 (above).

[26]In 1778 the Virginia legislature prohibited the importation of slaves by sea or land; any slaves so imported could not be bought or sold. Heavy fines were specified for those who engaged in such actions (Hening, ix, 471–72).

[27]Iredell was a prominent Federalist lawyer in Edenton, N.C. He had previously served on the state superior court and as attorney general, and in 1788 was a member of the Council of State. He was later appointed to the U.S. Supreme Court. His five-part, unnumbered Marcus series first appeared in the *Norfolk and Portsmouth Journal* on February 20. The series was reprinted in a North Carolina newspaper and as a pamphlet.

"The general Legislature is restrained from prohibiting the further importation of slaves for twenty odd years, though such importations render the United States weaker, more vulnerable, and less capable of defence."
Answer.

If all the States had been willing to adopt this regulation, I should, as an individual, most heartily have approved of it, because, even if the importation of slaves in fact rendered us stronger, less vulnerable, and more capable of defence, I should rejoice in the prohibition of it, as putting a stop to a trade which has already continued too long for the honor and humanity of those concerned in it. But as it was well known that South-Carolina and Georgia thought a further continuance of such importations useful to them, and would not perhaps otherwise have agreed to the new Constitution, those States which had been importing till they were satisfied, could not with decency have insisted upon their relinquishing advantages [which they] themselves had already enjoyed. Our situation makes it necessary to bear the evil as it is. It will be left to the future Legislatures to allow such importations or not. If any, in violation of their clear conviction of the injustice of this trade, persist in pursuing it, this is a matter between God and their own consciences. The interests of humanity will however have gained something by a prohibition of this inhuman trade, though at the distance of twenty odd years.

WILLIAM PIERCE
Gazette of the State of Georgia, March 20, 1788

Extract of a letter from the Hon. William Pierce, Esq. to St. George Tucker, Esq. dated New York, Sept. 28, 1787.[28]
. . . Many objections have been already started to the Constitution because it was not founded on a Bill of Rights; but I ask how such a thing could have been effected; I believe it would have been difficult in the extreme to have brought the different states to agree in what probably would have been proposed as the very first principle, and that is, "that all men are born equally free and independent." Would a Virginian have accepted it in this form? Would he not have modified some of the expressions in such a manner as to have injured *the strong sense of them*, if not to have buried them altogether in *ambiguity and uncertainty?*

[28]Pierce, a Savannah merchant, had represented Georgia in the Constitutional Convention. He left the Convention early, however, to attend the Confederation Congress then meeting in New York City. Tucker was a judge of the General Court of Virginia.

A NATIVE OF VIRGINIA
Observations upon the Proposed Plan of Federal Government
Petersburg, Va., April 2, 1788

Representatives and direct taxes shall be apportioned among the several States which may be included within this Union, according to their respective numbers, which shall be determined by adding to the whole number of free persons, including those bound to service for a term of years, and excluding Indians not taxed, three-fifths of all other persons. . . .

Every free person counts one, every five slaves count three. By this regulation our consequence in the Union is increased, by an increase of numbers in the Congress. But some objectors argue that this arrangement is unjust; and that it bears hard upon the southern States, who have been accustomed to consider their slaves merely as property; as a subject for, not as agents to taxation; and therefore by adding three fifths of our slaves to the free persons, our numbers are increased; and consequently by how much is that increase, by so much is the increase of our federal burthen. It is true, that slaves are property,—but are they not persons too? Does not their labour produce wealth? And is it not by the produce of labour, that all taxes must be paid? The Convention justly considered them in the light of persons, rather than property: But at the same time conceiving their natural forces inferior to those of the whites; knowing that they require freemen to overlook them, and that they enfeeble the State which possesses them, they equitably considered five slaves only of equal consequence with three free persons. What rule of federal taxation so equal, and at the same time so little unfavourable to the southern States, could the Convention have established, as that of numbers so arranged? Suppose the value of the lands in the respective States had been adopted as the measure: Let us see what then would have been the consequence. The northern States are comparatively small to the southern, and are very populous; whilst to the southward, the inhabitants are scattered over a great extent of territory. Any given number of men in the latter States possess much greater quantities of land, than the like number in the former. It is true the lands to the northward sell for a greater price than those to the southward, but the difference in price is by no means adequate to the difference in quantity; consequently an equal number of men to the southward would have to pay a much greater federal tax than the like number to the northward.

By the 8th article of Confederation, the value of lands is made the measure of the federal quotas. Virginia in consequence is rated something above Massachusetts, whose number of white inhabitants is nearly double.

After all, this point is perhaps of no great consequence. The Congress probably will rarely, if ever, meddle with direct taxation, as the impost duties will in all likelihood answer all the purposes of government, or at any rate the post-office, which is daily increasing, and a tax upon instruments of writing, will supply any deficiency. . . .

Sect. 9. *The migration or importation of such persons as any of the States now existing shall think proper to admit, shall not be prohibited by the Congress prior to the year one thousand eight hundred and eight, but a tax or duty may be imposed on such importation, not exceeding ten dollars for each person.*

This clause is a proof of deference in the members of the Convention, to each other, and of concession of the northern to the southern States. There is no doubt but far the greater part of that Convention hold domestic slavery in abhorrence. But the members from South-Carolina and Georgia, thinking slaves absolutely necessary for the cultivation and melioration of their States, insisted upon this clause. But it affects not the law of Virginia which prohibits the importation of slaves. . . .

No person held to service or labour in one State, under the laws thereof, escaping into another, shall, in consequence of any law or regulation therein, be discharged from such service or labour, but shall be delivered up on claim of the party to whom such service or labour may be due.

The convenience, justice, and utility, of these sections, are obvious.

At present, slaves absconding and going into some of the northern States, may thereby effect their freedom: But under the Federal Constitution they will be delivered up to the lawful proprietor.

GEORGE MASON TO THOMAS JEFFERSON
Gunston Hall, May 26, 1788[29]

I make no Doubt that You have long ago received Copys of the new Constitution of Government, framed last Summer, by the Delegates of the several States, in general Convention at Philadelphia.—Upon the most mature Consideration I was capable of, and from Motives of sincere Patriotism, I was under the Necessity of refusing my Signature, as one of the Virginia Delegates; and drew up some general Objections; which I intended to offer, by Way of Protest; but was discouraged from doing so, by the precipitate, & intemperate, not to say indecent Manner, in which the Business was conducted, during the last week of the Convention, after the Patrons of this new plan found they had a decided Majority in their Favour; which was obtained by a Compromise between the Eastern, & the two Southern States, to permit the latter to continue the Importation of Slaves for twenty odd Years; a more favourite Object with them, than the Liberty and Happiness of the People.

[29]Jefferson Papers, Library of Congress.

VIRGINIA RATIFYING CONVENTION DEBATES
June 11–25, 1788[30]

The Virginia ratifying convention met on June 2, 1788, with the attention of the entire country focused upon it. Eight of the necessary nine states had already ratified the Constitution. The New York and New Hampshire conventions were scheduled to meet two weeks later, but no matter what they did, Virginia was critical. Without Virginia, the new Union was doomed.

The debates in the Virginia convention were expected to be spectacular—and they were. James Madison, Governor Edmund Randolph, Chancellor Edmund Pendleton, future Chief Justice John Marshall and George Nicholas led Federalists in defending the new Constitution against the aging statesman George Mason, former Congressman James Monroe, and the indefatigable Patrick Henry. The division between Federalists and Antifederalists was too close to predict the outcome as the debate focused primarily on the lack of a bill of rights and the necessity of amending the Constitution before it was ratified.

Henry and Mason did everything they could to raise doubts in the minds of convention delegates about the advisability of adopting the Constitution. They condemned the foreign slave trade clause that prohibited Congress from banning this odious traffic for twenty years, but they also raised the specter of a congressionally required national emancipation of slaves justified through a broad interpretation of the general welfare clause, the necessary and proper clause, and the interests of national defense. Madison and Randolph vehemently countered these charges. Taking contrary positions from those espoused in the Constitutional Convention, they praised the foreign slave trade clause for allowing the imminent prohibition of this nefarious business. They also assured the delegates that no one in the Constitutional Convention suggested the idea of emancipation.

On June 25, the Virginia convention ratified the Constitution by a majority of ten votes. Slavery would be secure under the new Constitution.

June 11

George Mason: If the objections be removed—If those parts which are clearly subversive of our rights be altered, no man will go further than I will to advance the Union. We are told in strong language, of dangers to which we will be exposed unless we adopt this Constitution. Among the rest, domestic safety is said to be in danger. This Government does not attend to our domestic safety. It authorises the importation of slaves for twenty odd years, and thus continues upon us that nefarious trade. Instead of securing and protecting us, the continuation of this detestable trade, adds daily to our weakness. Though this evil is increasing, there is no clause in the Constitution that will prevent the Northern and Eastern States from meddling with our whole prop-

---

[30]*Debates and Other Proceedings of the Convention of Virginia. . . .* (3 vols., Petersburg, Va., 1788–1789).

erty of that kind. There is a clause to prohibit the importation of slaves after twenty years, but there is no provision made for securing to the Southern States those they now possess. It is far from being a desirable property. But it will involve us in great difficulties and infelicity to be now deprived of them. There ought to be a clause in the Constitution to secure us that property, which we have acquired under our former laws, and the loss of which would bring ruin on a great many people.

Henry Lee: The Honorable Gentleman abominates it [the Constitution], because it does not prohibit the importation of slaves, and because it does not secure the continuance of the existing slavery! Is it not obviously inconsistent to criminate it for two contradictory reasons? I submit to the consideration of the Gentleman, whether, if it be reprehensible in the one case, it can be censurable in the other?

June 12

Patrick Henry: Sure I am, that the dangers of this system are real, when those who have no similar interests with the people of this country, are to legislate for us—when our dearest interests are left in the power of those whose advantage it may be to infringe them. How will the quotas of troops be furnished? *Hated* as requisitions are, your Federal officers cannot collect troops like dollars, and carry them in their pockets. You must make those *abominable* requisitions for them, and the scale will be in proportion to the number of your blacks, as well as your whites, unless they violate the constitutional rule of apportionment. This is not calculated to rouse the fears of the people. It is founded in truth. How oppressive and dangerous must this be to the Southern States who alone have slaves? This will render their proportion infinitely greater than that of the Northern States. It has been openly avowed that this shall be the rule. I will appeal to the judgments of the Committee, whether there be danger.

June 17

(*The first clause, of the ninth section, read.*)
George Mason: This is a fatal section, which has created more dangers than any other.—The first clause, allows the importation of slaves for twenty years. Under the royal Government, this evil was looked upon as a great oppression, and many attempts were made to prevent it; but the interest of the African merchants prevented its prohibition. No sooner did the revolution take place, than it was thought of. It was one of the great causes of our separation from Great-Britain. Its exclusion has been a principal object of this State, and most of the States in the Union.[31] The augmentation of slaves weak-

---

[31]During the eighteenth century, the Virginia legislature passed several laws placing high duties on the importation of slaves, but the Crown disallowed some of them.

ens the States; and such a trade is diabolical in itself, and disgraceful to mankind. Yet by this Constitution it is continued for twenty years. As much as I value an union of all the States, I would not admit the Southern States into the Union, unless they agreed to the discontinuance of this disgraceful trade, because it would bring weakness and not strength to the Union. And though this infamous traffic be continued, we have no security for the property of that kind which we have already. There is no clause in this Constitution to secure it; for they may lay such a tax as will amount to manumission. And should the Government be amended, still this detestable kind of commerce cannot be discontinued till after the expiration of twenty years.—For the fifth article, which provides for amendments, expressly excepts this clause. I have ever looked upon this as a most disgraceful thing to America. I cannot express my detestation of it. Yet they have not secured us the property of the slaves we have already. So that "They have done what they ought not to have done, and have left undone what they ought to have done."

James Madison: I should conceive this clause to be impolitic, if it were one of those things which could be excluded without encountering greater evils.—The Southern States would not have entered into the Union of America, without the temporary permission of that trade. And if they were excluded from the Union, the consequences might be dreadful to them and to us. We are not in a worse situation than before. That traffic is prohibited by our laws, and we may continue the prohibition. The Union in general is not in a worse situation. Under the articles of Confederation, it might be continued forever: But by this clause an end may be put to it after twenty years. There is therefore an amelioration of our circumstances. A tax may be laid in the mean time; but it is limited, otherwise Congress might lay such a tax as would amount to a prohibition. From the mode of representation and taxation, Congress cannot lay such a tax on slaves as will amount to manumission. Another clause secures us that property which we now possess. At present, if any slave elopes to any of those States where slaves are free, he becomes emancipated by their laws. For the laws of the States are uncharitable to one another in this respect. But in this Constitution, "No person held to service, or labor, in one State, under the laws thereof, escaping into another, shall in consequence of any law or regulation therein, be discharged from such service or labor; but shall be delivered up on claim of the party to

---

In 1772 the House of Burgesses petitioned George III, beseeching him to remove all restraints on the royal governor that prevented him from assenting to such laws. Since the imperial government did not change its policy, the new state constitution of 1776 accused George III of an "inhuman use of his negative" to prevent the exclusion of slaves. Thomas Jefferson included a similar clause in his draft of the Declaration of Independence, only to have Congress delete it. (See Chapter 1.) In 1778 and 1785 the Virginia legislature prohibited the importation of slaves, and by 1788 all of the states (except Georgia) had similar prohibitions or had imposed high duties on slave imports. Mason himself had attacked the slave trade in the Fairfax Resolves (1774), in the Constitutional Convention, and in his published objections to the Constitution.

whom such service or labour may be due."—This clause was expressly inserted to enable owners of slaves to reclaim them. This is a better security than any that now exists. No power is given to the General Government to interpose with respect to the property in slaves now held by the States. The taxation of this State being equal only to its representation, such a tax cannot be laid as he supposes. They cannot prevent the importation of slaves for twenty years; but after that period they can. The Gentlemen from South-Carolina and Georgia argued in this manner:—"We have now liberty to import this species of property, and much of the property now possessed, has been purchased, or otherwise acquired, in contemplation of improving it by the assistance of imported slaves. What would be the consequence of hindering us from it? The slaves of Virginia would rise in value, and we would be obliged to go to your markets." I need not expatiate on this subject. Great as the evil is, a dismemberment of the Union would be worse. If those States should disunite from the other States, for not indulging them in the temporary continuance of this traffic, they might solicit and obtain aid from foreign powers.

John Tyler warmly enlarged on the impolicy, iniquity, and disgracefulness of this wicked traffic. He thought the reasons urged by Gentlemen in defence of it, were inconclusive, and ill-founded. It was one cause of the complaints against British tyranny, that this trade was permitted. The revolution had put a period to it; but now it was to be revived. He thought nothing could justify it. This temporary *restriction* on Congress militated, in his opinion, against the arguments of Gentlemen on the other side, that what was not given up was retained by the States; for that if this restriction had not been inserted, Congress could have prohibited the African trade. The power of prohibiting it, was not expressly delegated to them; yet they would have had it by implication, if this restraint had not been provided. This seemed to him to demonstrate most clearly the necessity of restraining them by a Bill of Rights, from infringing our unalienable rights. . . .

Mr. *Henry* insisted, that the insertion of these restrictions on Congress, was a plain demonstration, that Congress could exercise powers by implication. The Gentleman had admitted that Congress could have interdicted the African trade, were it not for this restriction. If so, the power not having been expressly delegated, must be obtained by implication. He demanded, where then was their doctrine of reserved rights? He wished for negative clauses to prevent their assuming any powers but those expressly given.—He asked, why it was omitted to secure us that property in slaves, which we held now? He feared its omission was done with design. They might lay such heavy taxes on slaves, as would amount to emancipation, and then the Southern States would be the only sufferers. His opinion was confirmed by the mode of levying money. Congress, he observed, had power to lay and collect taxes, imposts and excises. Imposts (or duties) and excises were to be uniform. But this uniformity did not extend to taxes.—This might compel the Southern States to liberate their negroes. He wished this property therefore to be guarded. He considered the clause which had been adduced by the Gentleman as a security for this property, as no security at all. It was no more than this—That a run-away negro could be taken up in Maryland or New-York. This could not prevent Congress from interfering with that property by laying a grievous and enormous tax on it, so as to compel owners to emancipate their slaves rather than pay the tax. He apprehended it would be productive of much stock-jobbing, and that they would play into one another's hands in such a manner as that this property would be lost to this country.

George Nicholas wondered that Gentlemen who were against slavery, would be opposed to this clause, as after that period the slave trade would be done away. He asked, if Gentlemen did not see the inconsistency of their arguments? They object, says he, to the Constitution, because the slave trade is laid open for twenty odd years; and yet they tell you, that by some latent operation of it, the slaves who are so now, will be manumitted! At the same moment it is opposed for being promotive and destructive of slavery!—He contended that it was advantageous to Virginia, that it should be in the power of Congress to prevent the importation of slaves after twenty years, as it would then put a period to the evil complained of.

As the Southern States would not confederate without this clause, he asked, if Gentlemen would rather dissolve the Confederacy than to suffer this temporary inconvenience, admitting it to be such? Virginia might continue the prohibition of such importation during the intermediate period; and would be benefited by it, as a tax of ten dollars, on each slave, might be laid; of which she would receive a share. He endeavoured to obviate the objection of Gentlemen, that the restriction on Congress was a proof that they would have power not given them, by remarking, that they would only have had a general superintendency of trade, if the restriction had not been inserted. But the Southern States insisted on this exception to that general superintendency for twenty years. It could not therefore have been a power by implication, as the restriction was an exception from a delegated power. The taxes could not, as had been suggested, be laid so high on negroes as to amount to emancipation; because taxation and representation were fixed according to the census established in the Constitution. The exception of taxes, from the uniformity annexed to duties and excises, could not have the operation contended for by the Gentleman; because other clauses had clearly and positively fixed the census. Had taxes been uniform it would have been universally objected to, for no one object could be selected without involving great inconveniences and oppressions. But, says Mr. *Nicholas*, is it from the General Government we are to fear emancipation? Gentlemen will recollect what I said in another house, and what other Gentlemen have said that advocated emancipation. Give me leave to say, that that clause is a great security for our slave tax. I can tell the Committee, that the people of our country are reduced to beggary by the taxes on negroes.—Had this Constitution been adopted, it would not have been the case. The taxes were laid on all our negroes. By this system two-fifths are exempted. He then added, that he had imagined Gentlemen would not support here what they had opposed in another place.

Patrick Henry replied, that though the proportion of each was to be fixed by the census, and three-fifths of the slaves only were included in the enumeration, yet the proportion of Virginia being once fixed, might be laid on blacks and blacks only. For the mode of raising the proportion of each State being to be directed by Congress, they might make slaves the sole object to raise it of. Personalities he wished to take leave of: They had nothing to do with the question, which was solely whether that paper was wrong or not.

George Nicholas replied, that negroes must be considered as persons or property. If as property, the proportion of taxes to be laid on them was fixed in the Constitution: If he apprehended a poll tax on negroes, the Constitution had prevented it. For, by the census, where a white man paid ten shillings, a negro paid but six shillings. For the exemption of two fifths of them reduced it to that proportion.

(The 2d, 3d, and 4th clauses read.)

George Mason said, that Gentlemen might think themselves secured by the restriction in the fourth clause, that no capitation or other direct tax should be laid but in proportion to the census before directed to be taken. But that when maturely considered it would be found to be no security whatsoever. It was nothing but a direct assertion, or mere confirmation of the clause which fixed the ratio of taxes and representation. It only meant that the quantum to be raised of each State, should be in proportion to their numbers in the manner therein directed. But the General Government was not precluded from laying the proportion of any particular State on any one species of property they might think proper. For instance, if 500,000 dollars were to be raised, they might lay the whole of the proportion of the Southern States on the blacks, or any one species of property: So that by laying taxes too heavily on slaves, they might totally annihilate that kind of property. No real security could arise from the clause which provides, that persons held to labor in one State, escaping into another, shall be delivered up. This only meant, that runaway slaves should not be protected in other States. As to the exclusion of *ex post facto* laws, it could not be said to create any security in this case. For laying a tax on slaves would not be *ex post facto*.

James Madison replied, that even the Southern States, who were most affected, were perfectly satisfied with this provision, and dreaded no danger to the property they now hold. It appeared to him, that the General Government would not intermeddle with that property for twenty years, but to lay a tax, on every slave imported, not exceeding ten dollars; and that after the expiration of that period they might prohibit the traffic altogether. The census in the Constitution was intended to introduce equality in the burdens to be laid on the community. No Gentleman objected to laying duties, imposts, and excises, uniformly. But uniformity of taxes would be subversive of the principles of equality: For that it was not possible to select any article which would be easy for one State, but what would be heavy for another. That the proportion of each State being ascertained, it would be raised by the General Government in the most convenient manner for the people, and not by the selection of any one particular object.[32] That there must be some degree of confidence put in agents, or else we must reject a state of civil society altogether. Another great security to this property, which he mentioned, was, that five States were greatly interested in that species of property, and there were other States which had some slaves, and had made no attempt, or taken any step to take them from the people. There were a few slaves in New-York, New-Jersey and Connecticut:[33] These States would probably oppose any attempts to annihilate this species of property. He concluded, by observing, that he would be glad to leave the decision of this to the Committee [of the Whole of the Virginia convention].

Edmund Randolph: But the insertion of the negative restrictions has given cause of triumph it seems, to Gentlemen. They suppose, that it demonstrates that Congress are to have powers by implication. I will meet them on that ground. I persuade myself, that every exception here mentioned, is an exception not from general powers, but from the particular powers therein vested. To what power in the General Government is the exception made, respecting the importation of negroes? Not from a general power, but from a particular power expressly enumerated. This is an exception from the power given them of regulating commerce.

June 24

Patrick Henry: Among ten thousand implied powers which they may assume, they may, if we be engaged in war, liberate every one of your slaves if they please. And this must and will be done by men, a majority of whom have not a common interest with you. They will therefore have no feeling for your interests. It has been repeatedly said here, that the great object of a national Government, was national defence. That power which is said to be intended for security and safety, may be rendered detestable and oppressive. If you give power to the General Government to provide for the general defence, the means must be commensurate to the end. All the means in the possession of the people must be given to the Government which is intrusted with the public defence. In this State there are 236,000 blacks, and there are many in several other States. But there are few or none in the Northern States, and yet if the

[32]Antifederalists argued that Congress should have only the power to assess each state a quota of the total federal expenses, but that the states should be allowed to decide how best to raise their federal tax assessments. Only if the states refused to pay their quotas would the federal government be allowed to levy direct taxes within a state.

[33]According to the federal census of 1790, New York had 21,324 slaves, New Jersey 11,423, and Connecticut 2,764.

Northern States shall be of opinion, that our numbers are numberless, they may call forth every national resource. May Congress not say, that every black man must fight?—Did we not see a little of this last war?—We were not so hard pushed, as to make emancipation general. But acts of Assembly passed, that every slave who would go to the army should be free.[34] Another thing will contribute to bring this event about—slavery is detested—we feel its fatal effects—we deplore it with all the pity of humanity. Let all these consider-ations, at some future period, press with full force on the minds of Congress. Let that urbanity, which I trust will distinguish America, and the necessity of national defence:—Let all these things operate on their minds. They will search that paper, and see if they have power of manumission.—And have they not, Sir?—Have they not power to provide for the general defence and welfare?—May they not think that these call for the abolition of slavery?—May they not pronounce all slaves free, and will they not be warranted by that power? There is no ambiguous implication, or logical deduction—The paper speaks to the point. They have the power in clear unequivocal terms; and will clearly and certainly exercise it. As much as I deplore slavery, I see that prudence forbids its abolition. I deny that the General Government ought to set them free, because a decided majority of the States have not the ties of sympathy and fellow-feeling for those whose interest would be affected by their emancipa-tion. The majority of Congress is to the North, and the slaves are to the South. In this situation, I see a great deal of the property of the people of Virginia in jeopardy, and their peace and tranquillity gone away. I repeat it again, that it would rejoice my very soul, that every one of my fellow beings was emancipated. As we ought with gratitude to admire that decree of Heaven, which has numbered us among the free, we ought to lament and deplore the necessity of holding our fellow-men in bondage. But is it practicable by any human means, to liberate them, without producing the most dreadful and ruinous consequences? We ought to possess them in the manner we have inherited them from our ancestors, as their manumission is incompatible with the felicity of the country. But we ought to soften, as much as possible, the rigour of their unhappy fate. I know that in a variety of particular instances,

[34]In 1775 the legislature passed an act stipulating that, with certain exceptions, "all free male persons, hired servants, and apprentices" between the ages of sixteen and fifty were liable to serve in the militia. Some runaway slaves enlisted as soldiers. In 1777 the legislature, seeking to end this practice, required that any black or mu latto wishing to enlist should produce a certificate from a justice of the peace of his home county certifying that he was a freeman. During the course of the Revolution, many slaveowners "caused their slaves to enlist . . . as substitutes for free persons" by informing the recruiting officers that these slaves were freemen. After the term of enlistment, some slaveowners tried to force these enlistees back into slavery, "con-trary to the principles of justice, and to their own solemn promise." Consequently, in 1783, the legislature freed these enlistees if they had served faithfully.

the Legislature listening to complaints, have admitted their emancipation.[35] Let me not dwell on this subject. I will only add, that this, as well as every other property of the people of Virginia, is in jeopardy, and put in the hands of those who have no similarity of situation with us. This is a local matter, and I can see no propriety in subjecting it to Congress.

Edmund Randolph: That Honorable Gentleman [Patrick Henry], and some others, have insisted that the abolition of slavery will result from it, and at the same time have complained, that it encourages its continuation. The inconsistency proves in some degree, the futility of their arguments. But if it be not conclusive, to satisfy the Committee [of the whole of the ratifying convention] that there is no danger of enfranchisement taking place, I beg leave to refer them to the paper itself. I hope that there is none here, who considering the subject in the calm light of philosophy, will advance an objection dishonorable to Virginia; that at the moment they are securing the rights of their citizens, an objection is started that there is a spark of hope, that those unfortunate men now held in bondage, may, by the operation of the General Government, be made *free*. But if any Gentleman be terrified by this apprehension, let him read the system. I ask, and I will ask again and again, till I be answered (not by declamation) where is the part that has a tendency to the abolition of slavery? Is it the clause which says, that "the migration or importation of such persons as any of the States now existing, shall think proper to admit, shall not be prohibited by Congress prior to the year 1808?" This is an exception from the power of regulating commerce, and the restriction is only to continue till 1808. Then Congress can, by the exercise of that power, prevent future importations; but does it affect the existing state of slavery? Were it right here to mention what passed in Convention on the occasion, I might tell you that the Southern States, even South-Carolina herself, conceived this property to be secure by these words. I believe, whatever we may think here, that there was not a Member of the Virginia delegation who had the smallest suspicion of the abolition of slavery. Go to their meaning. Point out the clause where this formidable power of emancipation is inserted. But another clause of the Constitution proves the absurdity of the supposition. The words of the clause are, "No person held to service or labor in one State, under the laws thereof, escaping into another, shall in consequence of any law or regulation therein, be discharged from such service or labor; but shall be delivered up on claim of the party to whom such service or labor may be due." Every one knows that slaves are held to service and labor.

[35]In 1779 and 1780 the legislature, acting upon "applications," passed acts freeing individual slaves. In 1782 the legislature adopted an act which allowed owners to manumit their slaves under certain restrictions without having to petition the legislature for a special act. (See Chapter 1 for this act.) Despite this act, some slaves still had to petition the legislature to make certain that wills were properly executed. The 1782 act contributed to an increase in the number of free blacks in Virginia. In 1782 there were fewer than 3,000, while in 1790 there were 12,866.

And when authority is given to owners of slaves to vindicate their property, can it be supposed they can be deprived of it? If a citizen of this State, in consequence of this clause, can take his runaway slave in Maryland, can it be seriously thought, that after taking him and bringing him home, he could be made free?

George Mason: With respect to commerce and navigation, he [Edmund Randolph] has given it as his opinion, that their regulation, as it now stands, was a *sine qua non* of the Union, and that without it, the States in Convention would never concur. I differ from him. It never was, nor in my opinion ever will be, a *sine qua non* of the Union. I will give you, to the best of my recollection, the history of that affair. This business was discussed at Philadelphia for four months, during which time the subject of commerce and navigation was often under consideration; and I assert, that eight States out of twelve, for more than three months, voted for requiring two-thirds of the members present in each House to pass commercial and navigation laws. True it is, that afterwards it was carried by a majority, as it stands. If I am right, there was a great majority for requiring two-thirds of the States in this business, till a compromise took place between the Northern and Southern States; the Northern States agreeing to the temporary importation of slaves, and the Southern States conceding, in return, that navigation and commercial laws should be on the footing on which they now stand.[36] If I am mistaken, let me be put right. These are my reasons for saying that this was not a *sine qua non* of their concurrence. The Newfoundland fisheries will require that kind of security which we are now in want of: The Eastern States therefore agreed at length, that treaties should require the consent of two-thirds of the members present in the Senate.

James Madison: I was struck with surprise when I heard him [Patrick Henry] express himself alarmed with respect to the emancipation of slaves. Let me ask, if they should even attempt it, if it will not be an usurpation of power? There is no power to warrant it, in that paper. If there be, I know it not. But why should it be done? Says the Honorable Gentlemen for the general welfare—It will infuse strength into our system. Can any Member of this Committee suppose, that it will increase our strength? Can any one believe, that the American Councils will come into a measure which will strip them of their property, discourage, and alienate the affections of, five-thirteenths of

---

[36]On August 6, 1787, the Committee of Detail, of which Edmund Randolph was a member, reported the first draft of the Constitution. The report forbade Congress from prohibiting the importation of slaves and it required that navigation acts be passed by two-thirds of the members of each house. Northern delegates generally opposed these two provisions. Consequently, on August 24 a committee of eleven (one from each state) reported a compromise, which denied Congress the power to prohibit the foreign slave trade before 1800 and deleted the section requiring a two-thirds majority for enacting navigation acts. The Convention accepted the committee report but changed the date concerning the slave trade to 1808.

the Union. Why was nothing of this sort aimed at before? I believe such an idea never entered into any American breast, nor do I believe it ever will, unless it will enter into the heads of those Gentlemen who substitute unsupported suspicions to reasons.

Patrick Henry: The Honorable Gentleman who was up some time ago, exhorts us not to fall into a repetition of the defects of the Confederation. He said we ought not to declare that each State retains every power, jurisdiction and right, which is not expressly delegated, because experience has proved the insertion of such a restriction to be destructive, and mentioned an instance to prove it.[37] That case, Mr. Chairman, appears to me to militate against himself.—Passports[38] would not be given by Congress—and why? Because there was a clause in the Confederation which denied them implied powers. And says he, shall we repeat the error? He asked me where was the power of emancipating slaves. I say it will be implied, unless implication be prohibited. He admits that the power of granting passports will be in the new Congress without the insertion of this restriction—Yet he can shew me nothing like such a power granted in that Constitution. Notwithstanding he admits their right to this power by implication, he says that I am unfair and uncandid in my deduction, that they can emancipate our slaves, though the word emancipation be not mentioned in it. They can exercise power by implication in one instance, as well as in another. Thus by the Gentleman's own argument, they can exercise the power though it be not delegated.

June 25

James Innes: We are told that the New-Englanders mean to take our trade from us, and make us hewers of wood and carriers of water; and the next moment that they will emancipate our slaves! But how inconsistent is this? They tell you that the admission of the importation of slaves for twenty years, shews that their policy is to keep us weak, and yet the next moment they tell you, that they intend to set them free! If it be their object to corrupt and enervate us, will they emancipate our slaves? Thus they complain and argue

[37]Article 2 of the Articles of Confederation stated that "Each state retains its sovereignty, freedom and independence, and every Power, Jurisdiction and right, which is not by this confederation expressly delegated to the United States, in Congress assembled."

[38]Henry here refers to an incident in early 1782 when Congress granted passports for British merchants to export tobacco from Virginia. When the vessels arrived in Virginia, Governor Benjamin Harrison refused to recognize the passports. On May 20, 1782, the Virginia House of Delegates adopted five resolutions protesting the passports. Some of the delegates argued that, under the Articles of Confederation, only the states, not Congress, had the power to issue passports. Randolph, then serving as Virginia attorney general, disagreed with this interpretation. By mid-June the Virginia legislature relented and allowed Congress' view to go "into due effect."

against it on contradictory principles.—The Constitution is to turn the world
topsy turvy to make it answer their various purposes.

Zachariah Johnston: They tell us that they see a progressive danger of
bringing about emancipation. The principle has begun since the revolution.
Let us do what we will, it will come round. Slavery has been the foundation of
that impiety and dissipation which have been so much disseminated among
our countrymen. If it were totally abolished, it would do much good.

MANY (ARTHUR CAMPBELL)
Virginia Independent Chronicle, June 18, 1788[39]

In a serious hour, and in the presence of the Governor of the Universe,
what reasonable excuse can then be made, for permitting, and that constitutionally, depredations on a distant and inoffensive people, for the term of
twenty-one years.—Seemingly in the same spirit, and with the same narrow
policy, is the clause expressed ambiguously for the admission of NEW STATES
into the Union. Art. 11 of the confederation is expressed in a different style.

There are many good things, excellent regulations, set forth in the new
plan, that will long be respected by a grateful, and enlightened people. But
when we turn our eyes to the dark side of the picture, a sigh, a tear, a lamen-
tation, may be excited, for the imperfections that beset the best of men, and
that attends the wisest institutions, whilst we are destined to act on the present
theatre of human affairs.

Such are the sentiments of MANY.

NORTH CAROLINA RATIFYING CONVENTION DEBATES
July 24–29, 1788[40]

*The North Carolina convention met on July 21, 1788, about a month after
neighboring Virginia had adopted the Constitution. Contrary to the expectation that
North Carolina would follow Virginia's lead, the heavily Antifederalist convention
voted not to ratify the Constitution until amendments (including a bill of rights)
were accepted.*

*The Antifederalist convention delegates raised several objections to the slavery
provisions of the Constitution. The foreign slave trade should have been immediately
closed and they anticipated a general emancipation of slaves. With no viable plans for
colonization, the thought of manumitted blacks filled most delegates with horror.
Former Constitutional Convention delegates William R. Davie and Richard Dobbs
Spaight, the former who owned about twenty slaves and the latter about ninety,*

---

[39]Reprinted in the Philadelphia *Independent Gazetteer* on October 2. Campbell
was a planter in Washington County, Virginia.

[40]*Proceedings and Debates of the Convention of North-Carolina . . .* (Edenton, N.C.,
1789).

explained the difficulties faced in drafting the Constitution and argued that slavery would be even more secure under the new government.

July 24

The first three clauses of the 2nd section read.

William Goudy: This clause of taxation will give an advantage to some states over the others. It will be oppressive to the Southern States. Taxes are equal to our representation. To augment our taxes, and increase our burdens, our negroes are to be represented. If a state has fifty thousand *negroes*, she is to send one representative for them. I wish not to be represented with negroes, especially if it increases my burdens.

William R. Davie: I will endeavor to obviate what the gentleman last up said. I wonder to see gentlemen so precipitate and hasty on a subject of such awful importance. It ought to be considered, that some of us are slow of apprehension, or not having those quick conceptions, and luminous understanding, of which other gentlemen may be possessed. The gentleman "does not wish to be represented with negroes." This, sir, is an unhappy species of population; but we cannot at present alter their situation. The Eastern States had great jealousies on this subject. They insisted that their cows and horses were equally entitled to representation; that the one was property as well as the other. It became our duty, on the other hand, to acquire as much weight as possible in the legislation of the Union; and as the Northern States were more populous in whites, this only could be done by insisting that a certain proportion of our slaves should make a part of the computed population. It was attempted to form a rule of representation from a compound ratio of wealth and population; but, on consideration, it was found impracticable to determine the comparative value of lands, and other property, in so extensive a territory, with any degree of accuracy; and population alone was adopted as the only practicable rule or criterion of representation. It was urged by the deputies of the Eastern States, that a representation of two fifths would be of little utility, and that their entire representation would be unequal and burdensome—that, in a time of war, slaves rendered a country more vulnerable, while its defence devolved upon its free inhabitants. On the other hand, we insisted that, in time of peace, they contributed, by their labor, to the general wealth, as well as other members of the community—that, as rational beings, they had a right of representation, and, in some instances, might be highly useful in war. On these principles the Eastern States gave the matter up, and consented to the regulation as it has been read. I hope these reasons will appear satisfactory. It is the same rule or principle which was proposed some years ago by Congress, and assented to by twelve of the states.[41] It may wound

---

[41]Only eleven states had adopted this 1783 proposed amendment to the Articles of Confederation, which would have changed the method of apportioning federal

the delicacy of the gentleman from Guilford (Mr. Goudy), but I hope he will endeavor to accommodate his feelings to the interest and circumstances of his country.

James Galloway said, that he did not object to the representation of negroes, so much as he did to the fewness of the number of representatives. He was surprised how we came to have but five, including those intended to represent negroes. That, in his humble opinion, North Carolina was entitled to that number independent of the negroes.

Richard Dobbs Spaight endeavored to satisfy him, that the Convention had no rule to go by in this case—that they could not proceed upon the ratio mentioned in the Constitution till the enumeration of the people was made—that some states had made a return to Congress of their numbers, and others had not—that it was mentioned that we had had time, but made no return—that the present number was only temporary—that in three years the actual census would be taken, and our number of representatives regulated accordingly.

July 26

1st clause of the 9th section read.

Joseph M'Dowall wished to hear the reasons of this restriction.

Richard Dobbs Spaight answered, that there was a contest between the Northern and Southern States; that the Southern States, whose principal support depended on the labor of slaves, would not consent to the desire of the Northern States to exclude the importation of slaves absolutely; that South Carolina and Georgia insisted on this clause, as they were now in want of hands to cultivate their lands; that in the course of twenty years they would be fully supplied; that the trade would be abolished then, and that, in the mean time, some tax or duty might be laid on.

Joseph M'Dowall replied, that the explanation was just such as he expected, and by no means satisfactory to him, and that he looked upon it as a very objectionable part of the system.

James Iredell: I rise to express sentiments similar to those of the gentleman from Craven [Mr. Spaight]. For my part, were it practicable to put an end to the importation of slaves immediately, it would give me the greatest pleasure; for it certainly is a trade utterly inconsistent with the rights of humanity, and under which great cruelties have been exercised. When the entire abolition of slavery takes place, it will be an event which must be pleasing to every generous mind, and every friend of human nature; but we often wish

---

expenses among the states from the value of land to an apportionment based on population, counting three-fifths of the slaves. (See Chapter 1 for the debate in Congress over this amendment.)

for things which are not attainable. It was the wish of a great majority of the Convention to put an end to the trade immediately; but the states of South Carolina and Georgia would not agree to it. Consider, then, what would be the difference between our present situation in this respect, if we do not agree to the Constitution, and what it will be if we do agree to it. If we do not agree to it, do we remedy the evil? No, sir, we do not. For if the Constitution be not adopted, it will be in the power of every state to continue it forever. They may or may not abolish it, at their discretion. But if we adopt the Constitution, the trade must cease after twenty years, if Congress declares so, whether particular states please so or not; surely, then we can gain by it. This was the utmost that could be obtained. I heartily wish more could have been done. But as it is, this government is nobly distinguished above others by that very provision. Where is there another country in which such a restriction prevails? We, therefore, sir, set an example of humanity, by providing for the abolition of this inhuman traffic, though at a distant period. I hope, therefore, that this part of the Constitution will not be condemned because it has not stipulated for what was impracticable to obtain.

Richard Dobbs Spaight further explained the clause. That the limitation of this trade to the term of twenty years was a compromise between the Eastern States and the Southern States. South Carolina and Georgia wished to extend the term. The Eastern States insisted on the entire abolition of the trade. That the state of North Carolina had not thought proper to pass any law prohibiting the importation of slaves, and therefore its delegation in the Convention did not think themselves authorized to contend for an immediate prohibition of it.

James Iredell added to what he had said before, that the states of Georgia and South Carolina had lost a great many slaves during the war, and that they wished to supply the loss.

James Galloway: The explanation given to this clause does not satisfy my mind. I wish to see this abominable trade put an end to. But in case it be thought proper to continue this abominable traffic for twenty years, yet I do not wish to see the tax on the importation extended to all persons whatsoever. Our situation is different from the people to the north. We want citizens; they do not. Instead of laying a tax, we ought to give a bounty to encourage foreigners to come among us. With respect to the abolition of slavery, it requires the utmost consideration. The property of the Southern States consists principally of slaves. If they mean to do away slavery altogether, this property will be destroyed. I apprehend it means to bring forward manumission. If we must manumit our slaves, what country shall we send them to? It is impossible for us to be happy, if, after manumission, they are to stay among us.

James Iredell: The worthy gentleman, I believe, has misunderstood this clause, which runs in the following words: "The migration or importation of such persons as any of the states now existing shall think proper to admit, shall not be prohibited by the Congress prior to the year 1808; but a tax or duty may be imposed on such importation, not exceeding ten dollars for each persons." Now, sir, observe that the Eastern States, who long ago have abolished slaves, did not approve of the expression *slaves*; they therefore used another, that answered the same purpose. The committee will observe the distinction between the two words *migration* and *importation*. The first part of the clause will extend to persons who come into this country as free people, or are brought as slaves. But the last part extends to slaves only. The word *migration* refers to free persons; but the word *importation* refers to slaves, because free people cannot be said to be imported. The tax, therefore, is only to be laid on slaves who are imported, and not on free persons who migrate. I further beg leave to say that the gentleman is mistaken in another thing. He seems to say that this extends to the abolition of slavery. Is there any thing in this Constitution which says that Congress shall have it in their power to abolish the slavery of those slaves who are now in the country? Is it not the plain meaning of it, that after twenty years they may prevent the future importation of slaves? It does not extend to those now in the country. There is another circumstance to be observed. There is no authority vested in Congress to restrain the states, in the interval of twenty years, from doing what they please. If they wish to prohibit such importation, they may do so. Our next Assembly may put an entire end to the importation of slaves.

July 29

Article 4th. The last clause read.

James Iredell begged leave to explain the reason of this clause. In some of the Northern States they have emancipated all their *slaves*. If any of our slaves, said he, go there, and remain there a certain time, they would, by the present laws, be entitled to their freedom, so that their masters could not get them again. This would be extremely prejudicial to the inhabitants of the Southern States; and to prevent it, this clause is inserted in the Constitution. Though the word *slave* is not mentioned, this is the meaning of it. The northern delegates, owing to their particular scruples on the subject of slavery, did not choose the word *slave* to be mentioned.

6

Slavery and the New Nation

THROUGHOUT THE DEBATE OVER THE *ratification of the Constitution, Federalists insisted that a tariff on imported goods would provide the federal government with most of its needed revenue. Consequently, when the first federal Congress assembled, one of its first and most pressing concerns was to enact an impost bill that would lay a five-percent tariff on most imports, with certain exceptions that would have higher or lower rates. Toward the end of the month-long debate over these exceptions, Virginia Representative Josiah Parker, obviously under instructions from James Madison, proposed a ten-dollar tax on each imported slave, the maximum permitted under the Constitution.*

Parker argued that the slave trade—"this irrational and inhuman traffic"— should be prohibited because it violated the principles of the Revolution; but, because the Constitution forbade such a closure, the tariff was the next best alternative. James Madison supported the duty from "the dictates of humanity, the principles of the people, the national safety and happiness, and prudent policy." "It is hoped," he said, "that by expressing a national disapprobation of this trade, we may destroy it, and save ourselves from reproaches, and our posterity the imbecility ever attendant on a country filled with slaves." In a rare statement of support for a broad interpretation of the powers granted Congress by the Constitution, Madison suggested that "It is a necessary duty of the general government to protect every part of the empire against danger, as well internal as external; every thing therefore which tends to encrease this danger, though it may be a local affair, yet if it involves national expence or safety, becomes of concern to every part of the union, and is a proper subject for the consideration of those charged with the general administration of the government." This was

*perilously close to the interpretation of the Constitution that Patrick Henry had warned
would be used to abolish slavery.*

*Representatives from the Deep South and from New England united again
and argued that the ten-dollar-tax proposal came too late in the debate, that it was a
partial tax aimed against Georgia, that it emanated from Virginia, which would
benefit from a restriction of the slave trade, and that slaves should not be taxed as
property. If such a tariff were enacted, Representative James Jackson of Georgia
demanded that a similar duty should be laid on white indentured servants immigrat-
ing into the Northern states.*

*Representative Jackson condemned "the fashion of the day, to favor the liberty
of slaves." Blacks, he argued, were better off as slaves than free. In fact American
slaves were better off than if they remained in Africa, where they would have been
enslaved but without benevolent masters "bound by the ties of interest and law, to
provide for their support and comfort in old age, or infirmity."*

*Faced with a prolonged debate which threatened to delay the passage of the
impost bill, Madison relented and agreed to seek a separate bill later in the session. At
Madison's prompting, Parker withdrew his motion. A separate bill for a ten-dollar
tax was introduced by Parker on September 19, 1789. The bill was read and postponed "until the next session." No further consideration of a tax on imported slaves
occurred until 1804 when South Carolina reopened its foreign slave trade.*

*The debate over the slave clauses of the Constitution invigorated abolitionist
societies in Philadelphia, New York, and New England in their efforts to gain gradual
emancipation in their states and in others if possible, to ameliorate the condition of
slaves throughout America, to educate the children of slaves and freedmen, and to
protect freedmen from kidnappings and sale back into slavery. Quakers in Philadelphia and the New York and Pennsylvania abolition societies petitioned the first federal
Congress to take action against the African slave trade and in favor of emancipation.
The petitions were submitted to a select committee. On March 16, 1790, the House
of Representatives considered the committee's report, and, for the next week, a vitu-
perative debate threatened the foundations of the Union. Southern representatives
clearly stated that any attempt to interfere with the foreign slave trade would be
unconstitutional; any attempt to emancipate the slaves by law "would never be submitted to by the southern states without a civil war." Representative Thomas Scott of
Pennsylvania, however, suggested that if he were a federal judge and slaves came
before him, he would go as far as he could to free them. Representative James Jackson
warned that federal judges with such opinions would soon be killed in his home state
of Georgia. William Loughton Smith of South Carolina argued that the Southern
states entered the Union "from political, not from moral motives"; his constituents
did not want to learn morals from self-righteous Quakers. Smith's colleague from
South Carolina Thomas Tudor Tucker vehemently asserted that "No authority is
expressly given" in the Constitution to interfere with slavery. Consequently, he asked,
"why should we trouble ourselves to enquire whether, by construction or implication,
we can find some color for interfering in a business which we cannot effectually accomplish? We have gone much too far in explaining the constitution; and if we continue
on the same plan, there is danger that we shall at length persuade ourselves, that*

every power which is not expressly refused is given to us [i.e., Congress]. . . . Whether the practice of holding negroes in slavery be wrong or not, is not for us to decide, because we have no power to prevent it." If slavery was wrong, there was "little doubt but the states themselves will in due time put an end to it."

On March 22, the Committee of the Whole House completed its consideration of the select committee's report. The report of the Committee of the Whole House, in essence, emasculated the select committee's report. Northern representatives moved to consider this revised report, but a majority of the House felt that too much time had already been spent on a divisive issue that, in any case, Congress was unable to resolve. On motion of James Madison, it was agreed to put both reports on the journals of the House. The report of the Committee of the Whole reflected "the meaning of the constitution" and as such should "go on the journals for the information of the public." Furthermore, Madison felt that the report would "tend to quiet the apprehensions of the southern states, by recognizing that Congress had no power whatever to prohibit the importation of slaves prior to the year 1808, or to attempt to manumit them at any time."

The general attitude of Americans in the North and South toward slavery continued to drift apart and solidify. The anti-slave petitions debated in Congress in March 1790 caused a great "uproar" in the Southern states. In Charleston, it was said to "be more safe for a man to proclaim through this city that there was no God, than that slave-holding was inconsistent with his holy law; for from the clergy down to the peasant, I have heard them defend this inhuman diabolical practice with indignation against every friend to their freedom."[1] By the end of the first federal Congress it was apparent that despite the strong opposition to the African slave trade, Congress would not act to ameliorate the terrible conditions endured by Africans in the slave trade, much less to abolish the trade. Southerners found support from Northerners to postpone action indefinitely.

The Debate Over a Tax on Imported Slaves
House of Representatives, May 13, 1789[2]

The house resolved itself into a committee of the whole on the impost bill.

Fisher Ames (Mass.) moved to insert china, crockery-ware and gun-powder; he thought them articles of luxury.

Thomas Fitzsimons (Pa.) desired the gentleman to change the expression from crockery into earthen and stone-ware, which being done, the committee agreed to insert china, earthen, and stone-ware at seven one-half per cent.

---

[1] Extract of a letter from a gentlemen in Charleston, S.C., to his friend in New-Jersey, dated March 31, Philadelphia *Freeman's Journal*, August 11, 1790.
[2] The debates are taken from Charlene Bangs Bickford, Kenneth R. Bowling, and Helen E. Veit, eds., *Debates in the House of Representatives, First Session: April–May 1789*, Volume X of *The Documentary History of the First Federal Congress* (Baltimore, 1992), 643–51.

ad valorem, but negatived gun-powder. The committee afterwards added look-ing-glasses and brushes.

Josiah Parker (Va.) moved to insert a clause in the bill, imposing a duty on the importation of slaves of ten dollars each person. He was sorry that the constitution prevented Congress from prohibiting the importation altogether; he thought it a defect in that instrument that it allowed of such actions, it was contrary to the revolution principles, and ought not to be permitted; but as he could not do all the good he desired, he was willing to do what lay in his power. He hoped such a duty as he moved for would prevent, in some degree, this irrational and inhuman traffic; if so, he should feel happy from the success of his motion.

William L. Smith (S.C.) hoped that such an important and serious proposition as this would not be hastily adopted; it was a very late moment for the introduction of new subjects. He expected the committee had got through the business, and would rise without discussing any thing further; at least if gentlemen were determined on considering the present, he hoped they would delay for a few days, in order to give time for an examination of the subject. It was certainly a matter big with the most serious consequences to the state he represented; he did not think any one thing that had been discussed was so important to them, and the welfare of the union, as the question now brought forward, but he was not prepared to enter on any argument, and therefore requested the motion might either be withdrawn or laid on the table.

Roger Sherman (Conn.) approved of the object of the motion, but he did not think this bill was proper to embrace the subject. He could not reconcile himself to the insertion of human beings, as an article of duty, among goods, wares and merchandize. He hoped it would be withdrawn for the present, and taken up hereafter as an independent subject.

James Jackson (Ga.) observing the quarter from which this motion came, said it did not surprize him, though it might have that effect on others. He recollected that Virginia was an old settled state, and had her complement of slaves, so she was careless of recruiting her numbers by this means, the natural increase of her imported blacks were sufficient for their purposes; but he thought gentlemen ought to let their neighbours get supplied before they imposed such a burthen upon the importation. He knew this business was viewed in an odious light to the eastward, because the people were capable of doing their own work, and had no occasion for slaves; but gentlemen will have some feeling for others; they will not try to throw all the weight upon others who have assisted in lightening their burthens; they do not wish to charge us for every comfort and enjoyment of life, and at the same time take away the means of procuring them; they do not wish to break us down at once.

He was convinced from the inaptitude of the motion, and the want of time to consider it, that the candor of the gentleman would induce him to withdraw it for the present; and if ever it came forward again, he hoped it would comprehend the white slaves [i.e., indentured servants] as well as black, who were imported from all the gaols [i.e., jails] of Europe; wretches, convicted of the most flagrant crimes, were brought in and sold without any duty whatever. He thought that they ought to be taxed equal to the Africans, and had no doubt but the constitutionality and propriety of such a measure was equally apparent as the one proposed.

Thomas Tudor Tucker (S.C.) thought it unfair to bring in such an important subject at a time when debate was almost precluded. The committee had gone through the impost bill, and the whole union were impatiently expecting the result of their deliberations, the public must be disappointed, and much revenue lost, or this question cannot undergo that full discussion which it deserves.

We have no right, said he, to consider whether the importation of slaves is proper or not, the Constitution gives us no power in that point, it is left to the states to judge of that matter as they see fit. But if it was a business the gentleman was determined to discourage he ought to have brought his motion forward sooner, and even then not have introduced it without previous notice. He hoped the committee would reject the motion, if it was not withdrawn; he was not speaking so much for the state he represented as for Georgia, because the state of South Carolina had a prohibitory law, which could be renewed when its limitation expired.[3]

Josiah Parker had ventured to introduce the subject after full deliberation, and did not like to withdraw it. Although the gentleman from Connecticut (Mr. Sherman) had said, that they ought not to be enumerated with goods, wares, and merchandize, he believed they were looked upon by the African traders in this light; he knew it was degrading the human species to annex that character to them; but he would rather do this than continue the actual evil of importing slaves a moment longer. He hoped Congress would do all that lay in their power to restore human nature its inherent privileges, and if possible wipe off the stigma which America labored under. The inconsistency in our principles, with which we are justly charged, should be done away; that we may shew by our actions the pure beneficence of the doctrine we held out to the world in our declaration of independence.

Roger Sherman thought the principles of the motion and the principles of the bill were inconsistent; the principle of the bill was to raise revenue, the

[3]On March 28, 1787, the South Carolina legislature passed an act prohibiting the importation of slaves for three years. Another act that passed the same day provided penalties for importing slaves during the allotted period—the imported slaves would be forfeited and sold, and the importer would "be liable to a penalty of one hundred pounds, to the use of the State, for every such negro or slave so imported and brought in." Similar acts were passed by South Carolina in 1788, 1792, 1794, 1796, 1800, and 1802. In each case, however, visitors and new settlers to the state could bring their slaves with them from their previous residence. In December 1803, South Carolina legalized the importation of slaves.

principle of the motion to correct a moral evil. Now, considering it as an object of revenue, it would be unjust, because two or three states would bear the whole burthen, while he believed they bore their full proportion of all the rest. He was against receiving the motion into this bill, though he had no objection to taking it up by itself, on the principles of humanity and policy; and therefore would vote against it if it was not withdrawn.

Fisher Ames joined the gentleman last up, no one could suppose him favorable to slavery, he detested it from his soul, but he had some doubts whether imposing a duty on the importation, would not have the appearance of countenancing the practice; it was certainly a subject of some delicacy, and no one appeared to be prepared for the discussion, he therefore hoped the motion would be withdrawn.

Samuel Livermore (N.H.) was not against the principle of the motion, but in the present case he conceived it improper, if negroes were goods, wares or merchandize they came within the title of the bill, if they were not, the bill would be inconsistent, but if they are goods, wares, or merchandize the 5 per cent ad valorem, will embrace the importation and the duty of 5 per cent is nearly equal to 10 dollars per head, so there is no occasion to add it even on the score of revenue.

James Jackson said it was the fashion of the day, to favor the liberty of slaves; he would not go into a discussion of the subject, but he believed it was capable of demonstration that they were better off in their present situation, than they would be if they were manumitted; what are they to do if they are discharged? Work for a living? Experience has shewn us they will not. Examine what is become of those in Maryland, many of them have been set free in that state; did they turn themselves to industry and useful pursuits? No, they turn out common pick pockets, petty larceny villains, and is this mercy forsooth to turn them into a way in which they must lose their lives, for when they are thrown upon the world, void of property and connections, they cannot get their living but by pilfering. What is to be done for compensation? Will Virginia set all her negroes free? Will they give up the money they cost them, and to whom? When this practice comes to be tried there, the sound of liberty will lose those charms which make it grateful to the ravished ear. But our slaves are not in a worse situation than they were on the coast of Africa? It is not uncommon there for the parents to sell their children in peace; and in war the whole are taken and made slaves together. In these cases it is only a change of one slavery for another; and are they not better here, where they have a master bound by the ties of interest and law, to provide for their support and comfort in old age, or infirmity, in which, if they were free, they would sink under the pressure of woe for want of assistance.

He would say nothing on the partiality of such a tax, it was admitted by the avowed friends of the measure; Georgia, in particular would be oppressed. On this account it would be the most odious tax Congress could impose.

James Schureman (N.J.) hoped the gentleman would withdraw his motion, because the present was not the time or place for introducing the business;

he thought it had better be brought forward in the house, as a distinct proposition. If the gentleman persisted in having the question determined, he would move the previous question if he was supported.

James Madison (Va.): I cannot concur with gentlemen who think the present an improper time or place to enter into a discussion of the proposed motion; if it is taken up in a separate view, we shall do the same thing at a greater expence of time. But the gentlemen say that it is improper to connect the two objects, because they do not come within the title of the bill, but this objection may be obviated by accommodating the title to the contents; there may be some inconsistency in combining the ideas which gentlemen have expressed, that is, considering the human race as a species of property; but the evil does not arise from adopting the clause now proposed, it is from the importation to which it relates. Our object in enumerating persons on paper with merchandize, is to prevent the practice of actually treating them as such, by having them in future, forming part of the cargoes of goods, wares, and merchandize to be imported into the United States, the motion is calculated to avoid the very evil intimated by the gentleman.

It has been said that this tax will be partial and oppressive; but suppose a fair view is taken of this subject, I think we may form a different conclusion. But if it be partial or oppressive, are there not many instances in which we have laid taxes of this nature? Yet are they not thought to be justified by national policy? If any article is warranted on this account, how much more are we authorized to proceed on this occasion? The dictates of humanity, the principles of the people, the national safety and happiness, and prudent policy requires it of us; the constitution has particularly called our attention to it—and of all the articles contained in the bill before us, this is one of the last I should be willing to make a concession upon so far as I was at liberty to go, according to the terms of the constitution or principles of justice—I would not have it understood that my zeal would carry me to disobey the inviolable commands of either.

I understood it had been intimated, that the motion was inconsistent or unconstitutional, I believe, sir, my worthy colleague [Josiah Parker] has formed the words with a particular reference to the constitution, any how, so far as the duty is expressed, it perfectly accords with that instrument; if there are any inconsistencies in it, they may be rectified; I believe the intention is well understood, but I am far from supposing the diction improper. If the description of the persons does not accord with the ideas of the gentleman from Georgia (Mr. Jackson) and his idea is a proper one for the committee to adopt, I see no difficulty in changing the phraseology.

I conceive the constitution in this particular, was formed in order that the government, whilst it was restrained from laying a total prohibition, might be able to give some testimony of the sense of America, with respect to the African trade. We have liberty to impose a tax or duty upon the importation of such persons as any of the states now existing shall think proper to admit; and this liberty was granted, I presume, upon two considerations—the first was, that until the time arrived when they might abolish the importation of slaves, they might have an opportunity of evidencing their sentiments, on the policy and humanity of such a trade; the other was that they might be taxed in due proportion with other articles imported; for if the possessor will consider them as property, of course they are of value, and ought to be paid for. If gentlemen are apprehensive of oppression from the weight of the tax, let them make an estimate of its proportion, and they will find that it very little exceeds five per cent. ad valorem, so that they will gain very little by having them thrown into that mass of articles, whilst by selecting them in the manner proposed, we shall fulfill the prevailing expectation of our fellow citizens, and perform our duty in executing the purposes of the constitution. It is to be hoped, that by expressing a national disapprobation of this trade, we may destroy it, and save ourselves from reproaches, and our posterity the imbecility ever attendant on a country filled with slaves.

I do not wish to say any thing harsh, to the hearing of gentlemen who entertain different sentiments from me, or different sentiments from those I represent; but if there is any one point in which it is clearly the policy of this nation, so far as we constitutionally can, to vary the practice obtaining under some of the state governments it is this; but it is certain a majority of the states are opposed to this practice, therefore, upon principle, we ought to discountenance it as far as is in our power.

If I was not afraid of being told that the representatives of the several states, are the best able to judge of what is proper and conducive to their particular prosperity, I should venture to say that it is as much the interest of Georgia and South Carolina, as of any in the union. Every addition they received to their number of slaves, tends to weaken them and renders them less capable of self defence; in case of hostilities with foreign nations, they will be the means of inviting attack instead of repelling invasion. It is a necessary duty of the general government to protect every part of the empire against danger, as well internal as external; every thing therefore which tends to encrease this danger, though it may be a local affair, yet if it involves national expence or safety, becomes of concern to every part of the union, and is a proper subject for the consideration of those charged with the general administration of the government. I hope in making these observations, I shall not be understood to mean that a proper attention ought not to be paid to the local opinions and circumstances of any part of the United States, or that the particular representatives are not best able to judge of the sense of their immediate constituents.

If we examine the proposed measure, by the agreement there is between it, and the existing state laws, it will shew us that it is patronized by a very respectable part of the union. I am informed that South-Carolina has prohibited the importation of slaves, for several years yet to come; we have the satisfaction then of reflecting that we do nothing more than their own laws do at this moment. This is not the case with one state. I am sorry that her situation is such as to seem to require a population of this nature, but it is impossible in the nature of things, to consult the national good without doing what we do not wish to do, to some particular part.

Perhaps gentlemen contend against the introduction of the clause, on too slight grounds, if it does not comport with the title of the bill, alter the latter, if it does not conform to the precise terms of the constitution amend it. But if it will tend to delay the whole bill, that perhaps will be the best reason for making it the object of a separate one. If this is the sense of the committee I shall submit.

Elbridge Gerry (Mass.) thought all duties ought to be laid as equal as possible. He had endeavoured to inforce this principle yesterday, but without the success he wished for, he was bound by the principle of justice therefore to vote for the proposition; but if the committee were desirous of considering the subject fully by itself he had no objection, but he thought when gentlemen laid down a principle, they ought to support it generally.

Aedanus Burke (S.C.) said gentlemen were contending for nothing, that the value of a slave averaged about £80 and the duty on that sum at five per cent. would be ten dollars, as Congress could go no farther than that sum, he conceived it made no difference whether they were enumerated or left in the common mass.

James Madison: If we contend for nothing, the gentlemen who are opposed to us, do not contend for a great deal; but the question is, whether the five per cent. ad valorem, on all articles imported, will have any operation at all upon the introduction of slaves, unless we make a particular enumeration on this account; the collector may mistake, for he would not presume to apply the term goods, wares, and merchandize to any person whatsoever. But if that general definition of goods, wares, and merchandize are supposed to include African slaves, why may we not particularly enumerate them, and lay the duty pointed out by the Constitution, which as gentlemen tell us, is no more than five per cent. upon their value; this will not encrease the burthen upon any, but it will be that manifestation of our sense, expected by our constituents, and demanded by justice and humanity.

Theodorick Bland (Va.) had no doubt of the propriety or good policy of this measure. He had made up his mind upon it, he wished slaves had never been introduced into America; but if it was impossible at this time to cure the evil, he was very willing to join in any measures that would prevent its extending farther. He had some doubts whether the prohibitory laws of the states were not in part repealed. Those who had endeavoured to discountenance this trade, by laying a duty on the importation, were prevented by the Constitution from continuing such regulation, which declares, that no state shall lay any impost or duties on imports. If this was the case, and he suspected pretty strongly that it was, the necessity of adopting the proposition of his colleague was now apparent.

Roger Sherman said the Constitution does not consider these persons as a species of property; it speaks of them as persons, and says, that a tax or duty may be imposed on the importation of them into any state which shall permit the same, but they have no power to prohibit such importation for twenty years. But Congress have power to declare upon what terms persons coming into the United States shall be entitled to citizenship, the rule of naturalization must, however, be uniform. He was convinced there were others ought to be regulated in this particular, the importation of whom was of an evil tendency, he meant convicts particularly. He thought that some regulation respecting them was also proper; but it being a different subject, it ought to be taken up in a different manner.

James Madison was led to believe, from the observation that had fell from the gentlemen, that it would be best to make this the subject of a distinct bill: he therefore wished his colleague would withdraw his motion, and move in the house for leave to bring in a bill on the same principles.

Josiah Parker consented to withdraw his motion, under a conviction, that the house was fully satisfied of its propriety. He knew very well that these persons were neither goods nor wares, but they were treated as articles of merchandize. Altho' he wished to get rid of this part of his property, yet he should not consent to deprive other people of theirs by any act of his without their consent.

Anti-Slave Petitions to the First Federal Congress[4]

PETITION FROM THE YEARLY MEETING OF QUAKERS
from Pennsylvania, New Jersey, Delaware, and the Western Parts
of Maryland and Virginia, October 3, 1789

To the Senate and House of Representatives of the United States.

The Address of the people called Quakers, in their annual assembly convened:

Firmly believing that unfeigned righteousness in public as well as private stations is the only sure ground of hope for the Divine blessing, whence alone rulers can derive true honor, establish sincere confidence in the hearts of the people, and feeling their minds animated with the ennobling principle of universal goodwill to men, find a conscious dignity and felicity in the harmony and success attending the exercise of a solid, uniform virtue, short of

---

[4]The text of the debates is taken from the New York *Daily Advertiser*, March 18, 20, 22 and 24, 1790, and the *New-York Daily Gazette*, March 26 and 27, 1790. For the entire debate, see Helen E. Veit, Charlene Bangs Bickford, Kenneth R. Bowling, and William Charles diGiacomantonio, eds., *Debates in the House of Representatives, Second Session: January–March 1790*, Volume XII of *The Documentary History of the First Federal Congress* (Baltimore, 1994).

which the warmest pretentions to public spirit, zeal for our country, and the rights of men, are fallacious and illusive.

Under this persuasion, as professors of faith in that ever blessed, all-perfect Lawgiver, whose injunctions remain of undiminished obligation on all who profess to believe in him, "whatsoever ye would that men should do unto you, do you even so unto them;" we apprehend ourselves religiously bound to request your serious Christian attention to the deeply interesting subject whereon our religious society, in their annual assembly, on the tenth month, 1783, addressed the then Congress,[5] who, though the Christian rectitude of the concern was by the Delegates generally acknowledged, yet not being vested with the powers of legislation, they declined promoting any public remedy against the gross national iniquity of trafficking in the persons of fellowmen; but divers of the Legislative bodies of the different States, on this Continent, have since manifested their sense of the public detestation due to the licentious wickedness of the African trade for slaves, and the inhuman tyranny and blood guiltiness inseparable from it; the debasing influence whereof most certainly tends to lay waste the virtue, and, of course, the happiness of the people.

Many are the enormities, abhorrent to common humanity and common honesty, which, under the Federal countenance given to this abominable commerce, are practised in some of the United States, which we judge it not needful to particularize to a body of men, chosen as eminently distinguished for wisdom as extensive information. But we find it indispensably incumbent on us as a religious body, assuredly believing that both the true temporal interest of nations, and eternal well-being of individuals, depend on doing justly, loving mercy, and walking humbly before God, the creator, preservor, and benefactor of men, thus to attempt to excite your attention to the affecting subject; earnestly desiring that the Infinite Father of Spirits may so enrich your minds with his love and truth, and so influence your understandings, by that pure wisdom which is full of mercy and good fruits, as that a sincere and impartial inquiry may take place, whether it be not an essential part of the duty of your exalted station to exert upright endeavors, to the full extent of your power, to remove every obstruction to public righteousness, which the influence of artifice of particular persons, governed by narrow, mistaken views of self-interest, has occasioned; and whether, notwithstanding such seeming impediments, it be not in reality within your power to exercise justice and mercy, which, if adhered to, we cannot doubt, must produce the abolition of the slave trade.

We consider this subject so essentially and extensively important, as to warrant a hope that the liberty we now take will be understood, as it really is, a compliance with a sense of religious duty; and that your Christian endeavors to remove reproach from the land may be efficacious to sweeten the labor, and lessen the difficulties incident to the discharge of your important trust.

[5]See Chapter 1 for this petition.

PETITION FROM THE PENNSYLVANIA ABOLITION SOCIETY
Philadelphia, February 3, 1790

That from a regard for the happiness of Mankind, an Association was formed several years since in this State by a number of her Citizens, of various religious denominations for promoting the *Abolition of Slavery* & for the relief of those unlawfully held in bondage. A just and accurate Conception of the true Principles of liberty, as it spread through the land, produced accessions to their numbers, many friends to their Cause, & a legislative Co-operation with their views, which, by the blessing of Divine Providence, have been successfully directed to the *relieving from bondage a large number of their fellow Creatures of the African Race.* They have also the Satisfaction to observe, that in consequence of that Spirit of Philanthropy & genuine liberty which is generally diffusing its beneficial Influence, similar Institutions are gradually forming at home & abroad.

That mankind are all formed by the same Almighty being, alike objects of his Care & equally designed for the Enjoyment of Happiness the Christian Religion teaches us to believe, & the Political Creed of Americans fully coincides with the Position. Your Memorialists, particularly engaged in attending to the Distresses arising from Slavery, believe it their indispensable Duty to present this Subject to your notice—They have observed with great Satisfaction, that many important & salutary Powers are vested in you for "promoting the Welfare & *securing the blessings of liberty to the People of the United States.*" And as they conceive, that these blessings ought rightfully to be administered, *without distinction of Colour,* to all descriptions of People, so they indulge themselves in the pleasing expectation, that nothing, which can be done for the relief of the unhappy objects of their care will be either omitted or delayed—

From a persuasion that equal liberty was originally the Portion, & is still the Birthright of all Men, & influenced by the strong ties of Humanity & the Principles of their Institution, your Memorialists conceive themselves *bound to use all justifiable endeavours to loosen the bands of Slavery* and promote a general Enjoyment of the blessings of Freedom. Under these Impressions they earnestly intreat your serious attention to the Subject of Slavery, that you will be pleased to countenance the *Restoration of liberty* to those unhappy Men, who alone, in this land of Freedom, are degraded into perpetual Bondage, and who, amidst the general Joy of surrounding Freemen, are groaning in Servile Subjection, that you will devise means for removing this *Inconsistency from the Character of the American People*; that you will promote Mercy and Justice towards this distressed Race, & that you will Step to the very verge of the Powers vested in you for discouraging every Species of Traffick in the Persons of our fellow Men.

REPORT OF THE SELECT COMMITTEE OF
THE HOUSE OF REPRESENTATIVES
February 12, 1790

The Committee to whom were referred sundry memorials from the people called Quakers, and also, a memorial from the Pennsylvania Society for promoting the Abolition of Slavery, submit the following report:

That, from the nature of the matters contained in these memorials, they were induced to examine the powers vested in Congress, under the present Constitution, relating to the Abolition of Slavery, and are clearly of opinion,

First. That the General Government is expressly restrained from prohibiting the importation of such persons "as any of the States now existing shall think proper to admit, until the year one thousand eight hundred and eight."

Secondly. That Congress, by a fair construction of the Constitution, are equally restrained from interfering in the emancipation of slaves, who already are, or who may, within the period mentioned, be imported into, or born within, any of the said States.

Thirdly. That Congress have no authority to interfere in the internal regulations of particular States, relative to the instructions of slaves in the principles of morality and religion; to their comfortable clothing, accommodations, and subsistence; to the regulation of their marriages, and the prevention of the violation of the rights thereof, or to the separation of children from their parents; to a comfortable provision in cases of sickness, age, or infirmity; or to the seizure of, transportation, or sale of free negroes; but have the fullest confidence in the wisdom and humanity of the Legislatures of the several States, that they will revise their laws from time to time, when necessary, and promote the objects mentioned in the memorials, and every other measure that may tend to the happiness of slaves.

Fourthly. That, nevertheless, Congress have authority, if they shall think it necessary, to lay at any time a tax or duty, not exceeding ten dollars for each person of any description, the importation of whom shall be by any of the States admitted as aforesaid.

Fifthly. That Congress have authority to interdict, or (so far as it is or may be carried on by citizens of the United States, for supplying foreigners) to regulate the African trade, and to make provision for the humane treatment of slaves, in all cases while on their passage to the United States, or to foreign ports, so far as respects the citizens of the United States.

Sixthly. That Congress have also authority to prohibit foreigners from fitting out vessels in any port of the United States, for transporting persons from Africa to any foreign port.

Seventhly. That the memorialists be informed, that in all cases to which the authority of Congress extends, they will exercise it for the humane objects of the memorialists, so far as they can be promoted on the principles of justice, humanity, and good policy.

DEBATES IN THE HOUSE OF REPRESENTATIVES
March 16–23, 1790

March 16

Elias Boudinot (N.J.) moved to take up the report of the committee on the memorial of the people called Quakers; after some opposition his motion was agreed to. The report was then read.

James Jackson (Ga.) rose and observed, that he had been silent on the subject of the Report's coming before the committee [of the whole house], because he wished the principles of the resolutions to be examined fairly and to be decided on their true grounds. He was against the propositions generally, and would examine the policy, the justice and the use of them; and he hoped if he could make them appear in the same light to others as they did to him by fair argument; that the gentlemen in opposition were not so determined in their opinions, as not to give up their present sentiments.

On the principle of emancipation this question arises: What is to be done with the slaves when freed? Two propositions present themselves. Either by incorporating them with the class of citizens or by colonizing them—one or the other of those alternatives must be carried into execution. Mr. Jefferson, our secretary of state (speaking in his notes on Virginia)[6] on the first head, declares it to be impolitic. I know not, Sir, whether I accurately deliver his words; but as well as my memory serves me they are, that "deep rooted prejudices entertained by the whites—ten thousand recollections by the blacks of the injuries they have sustained—new provocations—the real distinctions which nature has made, and many other circumstances would divide us into parties, and produce convulsions which would never end but with the extermination of the one or the other race." To these he adds, physical and moral objections, as the difference of colour, and so forth. Sir, the observations of this learned gentleman are not merely theoretical—We are taught the truth and justice of them by experience. . . .

Still however, will some gentlemen insist that emancipation, and the prevention of importation is policy, because some of the states have given into the measure. If Sir, the northern states to Virginia, have found it their interest, ought not the southern states to be allowed the same privilege of finding their interest. Will Congress and those states not concerned in the event, undertake to decide for those states which are! Is it not an interference with their local politics, which Congress are not warranted in, nay, are prevented by the constitution? Is it not an infraction of that sacred compact which brought us together, that compact which brought us mutually to relinquish a share of our interests to preserve the remainder? Did not the southern states

[6]For Jefferson's *Notes on the State of Virginia*, see Chapter 7.

for this very principle give into what might be termed the navigation law of the eastern and northern states,[7] and having accomplished that, will they break the compact, the principle on which it was obtained? . . . The custom, the habit of slavery is established, and Congress cannot interfere without endangering the whole system of government: that excellent constitution which we have so happily effected. Cannot the southern states be left to themselves on this subject? Is it supposed that they are not capable of discerning their interest and of receiving improvement? Is their humanity less than that of their more northern neighbours, that they must volunteer this business and exercise humanity for them? . . .

Does the justice of the interference stand on better grounds? I think not. For instance—I hold one thousand acres of tide rice land, on the Altamaha. On the expectation of importations these one thousand acres are worth three guineas per acre—take away this expectation of importation, and you take away that value altogether—restrict that importation, and you diminish that value one half. In the exact proportion as you injure the free importation in that ratio, sir, do you injure this property. Numbers in South-Carolina and Georgia are in this predicament. How, sir, are they to be compensated? Have those Friends [the Quakers] a purse sufficient, and are they willing to carry their justice and humanity so far as to give it? Have Congress a treasury sufficient to indemnify those holders? I do not believe they have, and how, sir, is justice to be done without that compensation? The same objection arises to emancipation; the same compensation justice requires. . . .

March 17

William L. Smith (S.C.) said he lamented much that this subject had been brought before the house—that he had deprecated it from the beginning because he foresaw that it would produce a very unpleasant discussion—that it was a subject of such nature as to excite the alarms of the southern members who could not view, without anxiety, any interference in it on the part of Congress. He remarked, that as they were resolved into a committee of the whole on the powers of Congress respecting slavery and the slave-trade, in consequence of certain memorials from the people called Quakers and the Pennsylvania society for the abolition of slavery, the whole subject, as well as the contents of those memorials, was under consideration: he should

[7]A reference to the compromise in the Constitutional Convention in which New England delegates agreed to a provision that would prohibit Congress from closing the slave trade immediately in exchange for Southern support for a provision allowing commercial legislation to be passed by a simple majority vote (instead of a two-thirds majority vote) in each house of Congress. See Chapter 2 for this compromise.

therefore enter into the business at large and offer some comments on the contents of the memorials.

The memorial from the Quakers contained, in his opinion, a very indecent attack on the character of those states which possessed slaves; it reprobated slavery as bringing down reproach on the southern states, and expatiated on the detestation due to the licentious wickedness of the African trade, and the inhuman tyranny, and blood-guiltiness inseparable from it. He could not but consider it as calculated to fix a stigma of the blackest nature on the character of the state he had the honor to represent, and to hold its citizens up to public view as men divested of every principle of honor and humanity. Considering it in that light, he felt it incumbent on him not only to refute those atrocious calumnies, but to resent the improper language made use of by the memorialists. Before he entered into the discussion he begged to observe that when any class of men deviated from their own religious principles, and officiously came forward in a business with which they had no concern and attempted to dictate to Congress, he could not ascribe their conduct to any other cause, but to an intolerant spirit of persecution: this application came with the worst grace possible from the Quakers, who professed never to intermeddle in politics, but to submit quietly to the laws of the country. . . .

The memorial from the Pennsylvania society applied in express terms for an emancipation of slaves, and the report of the committee appeared to hold out the idea that Congress might exercise the power of emancipation after the year 1808; for it said that Congress could not emancipate slaves prior to that period. He remarked that either the power of manumission still remained with the several states, or it was exclusively vested in congress; for no one would contend that such a power could be concurrent in the several states and the United States. He then shewed that the state governments clearly retained all the rights of sovereignty which they had before the establishment of the constitution, unless they were exclusively delegated to the United States; and this could only exist, where the constitution, granted in express terms an exclusive authority to the Union, or where it granted in one instance an authority to the Union, and in another prohibited the states from exercising the like authority, or where it granted an authority to the Union, to which a similar authority in the states would be repugnant. He applied these principles to the case in question, and asked, whether the constitution had, in express terms, vested the Congress with the power of manumission; or whether it restrained the states from exercising that power; or whether there was any authority given to the Union, with which the exercise of this right by any state would be inconsistent? If these questions were answered in the negative, it followed that Congress had not an exclusive right to the power of manumission. Had it a concurrent right with the states? no gentlemen would assert it, because the absurdity was obvious; for a state regulation on the subject might differ from a federal regulation, in which case one or the other must give way: as the laws of the United States were paramount to those of the individual states, the federal regulations would abrogate that of the states, consequently the states would thus be divested of a power which it was evident they now had and might exercise whenever they thought proper. But admitting that Congress had authority to manumit the slaves in America, and were disposed to exercise it, would the southern states acquiesce in such a measure without a struggle? Would the citizens of that country tamely suffer their property to be torn from them? Would even the citizens of the other states which did not possess this property, desire to have all the slaves let loose upon them? Would not such a step be injurious even to the slaves themselves? It was well known that they were an indolent people, improvident, averse to labor; when emancipated, they would either starve or plunder. Nothing was a stronger proof of the absurdity of emancipation than the fanciful schemes which the friends to the measure had suggested: one was to ship them off the country, and colonize them in some foreign region; this plan admitted that it would be dangerous to retain them within the United States after they were manumitted; but surely it would be inconsistent with humanity to banish these people to a remote country, and to expel them from their native soil, and from places to which they had a local attachment; it would be no less repugnant to the principles of freedom, not to allow them to remain here, if they desired it: how could they be called freemen, if they were against their consent to be expelled from the country? Thus did the advocates for emancipation acknowledge that the blacks when liberated, ought not to remain here to stain the blood of the whites by a mixture of the races.

Another plan was to liberate all those who should be born after a certain limited period: such a scheme would produce this very extraordinary phenomenon, that the mother would be a slave, and her child would be free—These young emancipated negroes, by associating with their enslaved parents, would participate in all the debasement which slavery was said to occasion. But allowing that a practicable scheme of general emancipation could be devised, there can be no doubt that the two races would still remain distinct. It was known from experience that the whites had such an idea of their superiority over the blacks that they never even associated with them; even the warmest friends to the blacks kept them at a distance, and rejected all intercourse with them. Could any instance be quoted of their intermarrying? the Quakers asserted that nature had made all men equal, and that the difference of color could not place negroes on a worse footing in society than the whites; but had any of them ever married a negro, or would any of them suffer their children to mix their blood with that of a black—they would view with abhorrence such an alliance.

Mr. Smith then read some extracts from Mr. Jefferson's notes on Virginia, proving that negroes were by nature an inferior race of beings; and that the whites would always feel a repugnance at mixing their blood with that of the blacks. Thus, he proceeded, that respectable author, who was desirous of countenancing emancipation, was on a consideration of the subject induced candidly to avow that the difficulties appeared insurmountable. The friends to manumission had said that by prohibiting the further importation of slaves and by liberating those born after a certain period, a gradual emancipation might take place, and that in process of time the very color would be extinct and there would be none but whites. He was at a loss to learn how that consequence would result. If the blacks did not intermarry with the whites, they would remain black to the end of time; for it was not contended that liberating them would whitewash them; if they did intermarry with the whites, then the white race would be extinct, and the American people would be all of the mulatto breed. In whatever light therefore the subject was viewed, the folly of emancipation was manifest. He trusted these considerations would prevent any further application to Congress on this point, and would so far have weight with the committee as to reject the clause altogether, or at least to declare in plain terms that Congress have no right whatever to manumit the slaves of this country.

Various objections, said he, had at different times been alledged against the abominable practice, as it had been called, of one man exercising dominion over another; but slavery was no new thing in the world—the Romans, the Greeks, and other nations of antiquity, held slaves at the time christianity first dawned on society, and the professors of its mild doctrines never preached against it. Here Mr. Smith read a quotation from the Roman and Grecian history, and from some accounts of the government and manners of the people of Africa, before they had any knowledge of the African traders, from which he said it appeared that slavery was not disapproved of by the apostles when they went about diffusing the principles of christianity; and that it was not owing to the African trade, as had been alledged, that the people of Africa made war on each other.

Another objection against slavery was, that the number of slaves in the southern states weakened that part of the Union, and in case of invasion would require a greater force to protect it. Negroes, it was said, would not fight: but he would ask whether it was owing to their being black or to their being slaves; if to their being black, then unquestionably emancipating them would not remedy the evil, for they would still remain black; if it was owing to their being slaves, he denied the position; for it was an undeniable truth, that in many countries slaves made excellent soldiers. In Russia, Hungary, Poland, the peasants were slaves, and yet were brave troops. In Scotland, not many years ago, the Highland peasants were absolute slaves to their lairds, and they were renowned for their bravery. The Turks were as much enslaved as the negroes—their property and lives were at the absolute disposal of the Sultan, yet they fought with undaunted courage. Many other instances might be quoted, but those would suffice to refute the fact. Had experience proved that the negroes would not make good soldiers? He did not assert that they would, but they had never been tried; discipline was every thing; white militia made but indifferent soldiers before they were disciplined. It was well known that according to the present art of war, a soldier was a mere machine, and he did not see why a black machine was not as good as a white one; in one respect the black troops would have the advantage of appearing more horrible in the eyes of the enemy. But admitting that they would not fight, to what would the argument lead? Undoubtedly to shew that the Quakers, Moravians, and all the non resisting and non fighting sects, constituted the weakness of a country. Did they not contribute to strengthen the country against invasion by staying at home and joining the invader as soon as he was successful? But they furnished money, he should be told and paid substitutes—and did not the slaves by encreasing the agriculture of the country add to its wealth, and thereby encrease its strength? did they not moreover perform many laborious services in the camp and in the field, assist in transporting baggage, conveying artillery, throwing up fortifications, and thus encrease the numbers in the ranks by supplying their places in these services? Nor was it necessary that every part of an empire should furnish fighting men, one part supplied men, another money—one part was strong in population, another in valuable exports, which added to the opulence of the whole. Great Britain obtained no soldiers from her East and West India settlements, were they therefore useless? She was obliged to send troops to protect them, but their valuable trade furnished her with means of paying those troops.

Another objection was, that the public opinion was against slavery: how did that appear? Were there any petitions on the subject excepting that from the Pennsylvania society and a few Quakers? And were they to judge for the whole continent? Were the citizens of the northern or southern states to dictate to Congress on a measure in which the southern states were so deeply interested? There were no petitions against slavery from the southern states, and they were the only proper judges of what was for their interest. The toleration of slavery in the several states was a matter of internal regulation and policy, in which each state had a right to do as she pleased, and no other state had any right to intermeddle with her policy or laws. If the citizens of the northern states were displeased with the toleration of slavery in the southern states, the latter were equally disgusted with some things tolerated in the former. He had mentioned on a former occasion the dangerous tenets and pernicious practices of the sect of Shaking Quakers, who preached against matrimony, and whose doctrine and example, if they prevailed, would either depopulate the United States, or people it with a spurious race. However the people of South Carolina reprobated the gross and immoral conduct of these Shakers, they had not petitioned Congress to expel them from the continent, though they thought such a measure would be serviceable to the United States. The legislature of South Carolina had prohibited theatrical representations, deeming them improper, but they did not trouble Congress with an application to abolish them in New-York and Philadelphia—The southern citizens might also consider the toleration of Quakers as an injury to the community, because in time of war they would not defend their country from the enemy, and in time of peace they were interfering in the concerns of others, and doing every thing in their power to excite the slaves in the southern states to insurrection; notwithstanding which the people of those states had not required the assistance of Congress to exterminate the Quakers.

But he could not help observing that this squeamishness was very extraordinary at this time. The northern states knew that the southern states had slaves before they confederated with them. If they had such an abhorrence for slavery, why said Mr. Smith, did they not cast us off and reject our alliance? The truth was, that the most informed part of the citizens of the northern states knew, that slavery was so ingrafted into the policy of the southern states, that it could not be eradicated without tearing up by the roots their happiness, tranquility and prosperity—that if it were an evil, it was one for which there was no remedy, and therefore, like wise men, they acquiesced in it: we, on the other hand, knew that the Quaker doctrines had taken such deep root in some of the States that all resistance to them must be useless: we therefore made a compromise on both sides, we took each other with our mutual bad habits and respective evils, for better for worse; the northern States adopted us with our slaves, we adopted them with their Quakers. There was then an implied compact between the Northern and Southern people that no step should be taken to injure the property of the latter, or to disturb their tranquility. It was therefore with great pain he had viewed the anxiety of some of the members to pay such uncommon respect to the memorialists as even to set aside the common rules of proceeding, and attempt to commit the memorials the very day they were presented, though the southern members had solicited one day's delay. Such proceedings had justly raised an alarm in the minds of his southern colleagues; and feeling that alarm, they would have acted a dishonorable part to their constituents had they not expressed themselves with that warmth and solicitude which some gentlemen had disapproved.

A proper consideration of this business, must convince every candid mind, that emancipation would be attended with one or other of these consequences; either that a mixture of the races would degenerate the whites, without improving the blacks, or that it would create two separate classes of people in the community involved in inveterate hostility, which would terminate in the massacre and extirpation of one or the other, as Moors were expelled from Spain and the Danes from England. The negroes would not be benefited by it; free negroes never improve in talents, never grow rich, and continue to associate with the people of their own colour. This is owing either to the natural aversion the whites entertain towards them, and an opinion of the superiority of their race, or to the natural attachment the blacks have to those of their own colour; in either case it proves that they will after manumission continue as a distinct people, and have separate interests. The author [Thomas Jefferson] already quoted, has proved that they are an inferior race even to the Indians.

After the last war a number of negroes which had been stolen from the southern states, and carried to England, either quitted the persons who carried them there, or were abandoned by them. Unable to provide for themselves, and rejected from the society of the common people of England, they were begging about the streets of London in great numbers; they supplicated captains of vessels to carry them back to their owners in America, preferring slavery there, to freedom in England. Many of them were shipped to Africa by the humanity of the English, and were either butchered or made slaves by their savage countrymen, or reshipped for sale to the plantations.

But some persons have been of opinion that if the further importation of slaves could be prohibited, there would be a gradual extinction of the species. Having shewn the absurdity of liberating the postnati without extending it to all the slaves old and young, and the greater absurdity and even impracticability of extending it to all, I shall say a few words with regard to the extinction; that would be impossible, because they increase—to occasion an extinction, Congress must prohibit all intercourse between the sexes; this would be an act of humanity they would not thank us for, nor would they be persuaded that it was for their own good, or Congress must, like Herod, order all the children to be put to death as soon as born. If then nothing but evil would result from emancipation, under the existing circumstances of the country, why should Congress stir at all in the business, or give any countenance to such dangerous applications. We have been told that the government ought to manifest a disposition inimical to this practise which the people reprobate. If some citizens, from misinformation and ignorance have imbibed prejudices against the southern states, if ill-intentioned authors have related false facts, and gross misrepresentations tending to traduce the character of the whole state, and to mislead the citizens in other states, is that a sufficient reason why a large territory is to be depopulated, merely to gratify the wish of some misinformed individuals? But what have the citizens of the other states to do with our slaves? Have they any right to interfere with our internal policy?

This is not an object of general concern, for I have already proved that it does not weaken the Union; but admit that it did, will the abolition of slavery strengthen South-Carolina? It can only be cultivated by slaves; the climate, the nature of the soil, ancient habits, forbid it by the whites; experience convinces us of the truth of this. Great-Britain made every attempt to settle Georgia by whites alone and failed, and was compelled at length to introduce slaves; after which, that state increased very rapidly in opulence and importance. If the slaves are emancipated, they will not remain in that country—remove the cultivators of the soil, and the whole of the low country, all the fertile rice and indigo swamps will be deserted, and become a wilderness. What then becomes of its strength? Will such a scheme increase it? Instead of increasing the population of the whites, there will be no whites at all; if the low country is deserted, where will be the commerce, the valuable exports of that country, the large revenue raised from its imports and from the consumption of the rich planters? In a short time the northern and eastern states will supply us with their manufacturers; if you depopulate the rich low country of South-Carolina and Georgia, you will give us a blow which will immediately recoil on yourselves. Suppose 140,000 slaves in those states, which require annually five yards of cloth each, making 700,000 yards at half a dollar a yard, this makes 350,000 dollars, besides the articles of linen, flannel, oznaburgh, blankets, molasses, sugar and rum for the use of the negroes; now, either the eastern and middle states will supply us with all these articles or they will receive the benefit of the impost on them if they are imported from foreign countries. Without the rice swamps of Carolina, Charleston would decay, so would the commerce of that city: this would injure the back country. If you injure the southern states, the injury would reach our northern and eastern brethren; for the states are links of one chain: if we break one the whole must fall to pieces. Thus it is manifest that in proportion to the increase of our agriculture will our wealth be increased; the increase of which will augment that of our sister states, which will either supply us with their commodities, or raise a large revenue upon us, or be the carriers of our produce to foreign markets.

It has been said that the toleration of slavery brings down reproach on America. It only brings reproach on those who tolerate it, and we are ready to bear our share. We know that none but prejudiced and uncandid persons, who have hastily considered the subject and are ignorant of the real situation of the southern states, throw out these insinuations. We found slavery ingrafted in the very policy of the country when we were born, and we are persuaded of the impolicy of removing it; if it be a moral evil, it is like many others which exist in all civilized countries and which the world quietly submits to. Humanity has been a topic of declamation on this subject: that sentiment has different operations on different individuals, and he had it in his power to shew, that humanity first gave origin to the transportation of slaves from Africa into America. Bartholomew de las Casas, bishop of Chiapa, a Spaniard renowned for his humanity and virtues, in order to save the Indians in South America from slavery, prevailed on his monarch to substitute Africans, which were accordingly purchased on the coast of Africa and shipped to the Spanish colonies to work in the mines: this appears in Robertson's history of America, which Mr. Smith quoted. At this day the Spaniards give considerable encouragement to the transportation of slaves into their islands. Mr. Smith read the edict for that purpose.

Another objection is, that slavery vitiates and debases the mind of the owner of this sort of property. Where, said he, is the proof of this allegation? Do the citizens of the southern states exhibit more ferociousness in their manners, more barbarity in their dispositions than those of the other states? Are crimes more frequently committed there? A proof of the absurdity of this charge may be found in the writings of those who wish to disseminate this mischievous idea, and yet, in their relation of facts, contradict it themselves. They lay down general principles which they take upon credit from others, or which they publish with sinister views, and when they enter into a detail of the history of those states, they overset their own doctrines. Thus, one writer tells us, that the southern citizen who is educated in principles of superiority to the slaves which surround him, has no idea of government, obedience and good order, till he mingles with the hardy and free spirited yeomanry of the north, and that after mixing with them, he will return home with his mind more enlarged, his views more liberalized, and his affections rectified, and become a more generous friend to the rights of human nature: but hear what the eastern traveller is to learn by visiting the enslaved regions of the south: He will see, says the same writer immediately after, industry crowned with affluence, independence, hospitality, liberality of manners; and notwithstanding the prevalence of domestic slavery, he will find the noblest sentiments of freedom and independence to predominate; he will extol their enterprize, art and ingenuity; and will reflect that nature is wise, and that Providence in the distribution of its favors, is not capricious. . . .

It was well known that when the African slaves were brought to the coast for sale, it was customary to put to death all those who were not sold; the abolition of the slave trade would therefore cause the massacre of the people. The cruel mode of transportation was another motive to this abolition; but it was to be presumed that the merchants would so far attend to their own interests as to preserve the lives and the health of the slaves on the passage: all voyages must be attended with inconveniencies, and those from Africa to America not more than others. As to their confinement on board, it was no more than was necessary; as to the smallness of space allotted them, it was more than was allowed to soldiers in a camp; for the measurement of cubical air breathed by the Africans compared with that of soldiers in a camp, was in favor of the former as thirty to seventeen; it was full as much as was allotted in ships of war to seamen who by the laws of England were frequently on their return to their families after a long and dangerous voyage, seized by violence, hurried away by a press-gang, and forced on another voyage more tedious and perilous than the first to a hot and sticky climate, where several hundreds of them were stowed away in the hold of a vessel. In cases of disobedience the Captain had a right, for slight offences, to inflict on them corporal punishment without the intervention of a court martial, and in other cases they were punishable by very severe laws, executed by martial courts established for that purpose. The same may be observed of the soldiers, who were frequently flogged severely for trivial offences, instances have been known of their being put under the care of a surgeon, after receiving a small part of the intended flagellation, to refit them for the residue.

Having thus removed the force of the observations which have been advanced against the toleration of slavery by a misguided and misinformed humanity, I shall only add, that I disapprove of the whole of the report; because it either states some power sufficiently expressed in the constitution which is unnecessary, or it sets forth some power which I am clear Congress don't possess. The concluding paragraph is an extraordinary one. In what mode are the memorialists to be informed of our humane dispositions? Are we to send a special committee to inform them? Or is the Speaker to write them a letter, or the Sergeant at Arms with the mace to wait on them? In short, Mr. Chairman, the whole of this business has been wrong from the beginning to end, and as one false step generally leads to others, so has the hasty commitment of these memorials involved us in all this confusion and embarrassment. I hope therefore if any kind of report is agreed to, it will be something like that proposed by my colleague.

March 22

Mr. Boudinot said, altho' he most heartily approved of many of the arguments and doctrines of his hon. friend from Pennsylvania [Thomas Scott], yet he could not go all lengths with him. He thought with him, that our time had been taken up, and great labour had been used in arguments that no wise related to the merits of the question before the committee, but he could not agree that the clause in the constitution relating to the want of power in Congress to prohibit the importation of such persons, as any of the states now existing shall think proper to admit, prior to the year 1808, and authorising a tax or duty on such importation not exceeding 10 dollars for each person, did not extend to negro slaves. Candor required that he should acknowledge, that this was the express design of the constitution, and therefore Congress could not interfere in prohibiting the importation or promoting the emancipation of them prior to that period. Mr. Boudinot observed, that he was well informed that the tax or duty of 10 dollars, was provided instead of the 5 per cent. ad valorem, and was so expressly understood by all parties in the convention. That therefore it was the interest and duty of Congress to impose this tax, or it would not be doing justice to the states, or equalizing the duties throughout the Union. If this was not done, Merchants might bring their whole capitals into this branch of trade and save paying any duties whatever. Mr. Boudinot had hoped, that the great lengths which the hon. gentleman from Pennsylvania had carried the argument, would have convinced gentlemen in the opposition, of the propriety, if not the necessity of the resolutions on the table. Is it not prudent now, while the design of the framers of the constitution, is well known and while the best information can be obtained, for Congress to declare their sense of it, on points which the gentlemen say involve their great and essential interests, especially when the gentleman from Pennsylvania, gives so different a construction to it, from what the gentlemen from the southward think right. Is it not advantageous to the southern states to have an explicit declaration calming their fears, and preventing unnecessary jealousies on this subject—can there be any foundation for alarm, when Congress expressly declares, that they have no power of interference prior to the year 1808. But gentlemen say that they have been charged with impropriety of conduct, in discovering so much warmth and earnestness, on a subject with which their dearest interests are so intimately connected. That all men are led by interest and they are justified in pursuing the same line of conduct. . . .

But when gentlemen attempt to justify this unnatural traffic, or to prove the lawfulness of slavery, they should advert to the genius of our government and the principles of the revolution. By the declaration of Congress in 1775, setting forth the causes and necessity of taking up arms, they say, "If it was possible for men who exercise their reason to believe that the divine author of our existence, intended a part of the human race to hold an absolute property in, and an unbounded power over others, marked out by his infinite goodness and wisdom, as the objects of a legal domination never rightfully resistible however severe and oppressive, the inhabitants of these Colonies, might at least require from the parliament of Great-Britain, some evidence that this dreadful authority over them, had been granted to that body." And by the declaration of Independence in 1776, Congress declare, "We hold these truths to be self-evident: that all men are created equal; that they are endowed by their creator, with certain unalienable rights; that among these are life, liberty, and the pursuit of happiness."

This then is the language of America in the day of distress. Mr. Chairman, I would not be understood, to contend the right of Congress at this time to prohibit the importation of slaves, whatever might have been the principles of the revolution or the genius of the government, by the present constitution we are clearly and positively restrained till the year 1808, and I am sure that no gentleman in this committee would have the most distant wish to wound this instrument of our connection.

But there is a wide difference between justifying this ungenerous traffick and supporting a claim to property, vested at the time of the constitution, and guaranteed thereby. Besides it would be inhumanity itself to turn these unhappy people loose to murder each other or to perish for want of the necessaries of life. I never was an advocate for so extravagant a conduct.

Many arguments were pointed against the danger of our emancipating these slaves, or even holding up an idea that we had a power so to do—and much time was taken up to disprove this right in Congress. As no claim of this kind is contended for, but the resolutions already passed expressly contradict it, I shall make no farther observations on them. . . .

The gentleman last up (Mr. Smith) said that it was now acknowledged, that one of the memorials had asked something contrary to the Constitution. I have never acknowledged this. The language is, that Congress would go to "the very verge of the Constitution" to accomplish the business; but there is no request to exceed it.

Thomas Scott (Pa.) observed that the subject before the committee had agitated the minds of many civilized nations, for a number of years: therefore what is said, and more particularly what is finally done in Congress at this time, will in some degree form the political character of America on the subject of slavery. What is said will form the characters of the speakers, and what is done will in a degree form the character of the American people on this subject.

Sir, said Mr. Scott, I perceive that most of the arguments advanced on this occasion, have gone against the emancipation of such as are already slaves in America: but this question is not before the committee; the report under consideration involves no such idea; it is granted on all hands that Congress have no authority to intermeddle in that business; and I believe that the several states, with whom this authority really rests, will from time to time make such advances in the premises, as justice to the master and slave, the dictates of humanity, sound policy, and the state of society, will require or admit: and here I rest contented. The arguments, therefore, which have been urged on this point, merit no answer.

An advocate for slavery, in its fullest latitude, at this age of the world, and on the floor of the American Congress too, is, with me, a phenomenon in politics; yet such advocates have appeared, and many arguments have been advanced on that head, to all which I will answer only by calling upon this committee, and upon every person who has heard them, to believe them if they can! With me they defy, yea, mock all belief.

Sir, the question before the committee is not, what will Congress do with respect to the African slave trade? but, what have Congress authority to do? If this is a question at all to be examined, and on which we are to come to any resolutions expressive of the powers of Congress, I think we ought to express those powers fully; and inasmuch as we have already agreed upon certain things that Congress cannot do, and some that may constitutionally be done, I think it may be well to proceed to a full declaration of all the powers of Congress relative to the subject. I think this declaration might contribute considerably to prevent the necessity of exercising those powers; and I believe those powers, when examined, will not only extend to a retention of the proposition immediately under consideration, but much farther.

Sir, it is said Congress can in no wise interfere with the African slave trade, because there is a clause in the constitution that says the migration or importation of persons, &c. shall not be prohibited by Congress prior to, &c. but a tax or duty may be laid on each, &c. and it is argued that person, here, means African slave: if it does, I believe it is the first place it ever did, and therefore can by no means admit it; nor can I think it satisfactory to be told that there was an understanding between the northern and southern members, in the national convention, on this subject. When we are considering our constitutional powers, we must judge of them by the face of the instrument under which we sit, and not by the certain understandings that the framers of that instrument may be supposed to have had of each other, and which never transpired. In a word, sir, I think it a very poor compliment to the convention, to suppose they couched the idea of an African slave under the term person; and at any rate, the constitution was in no degree obligatory until ratified by a certain number of state conventions, who I presume cannot be supposed to be acquainted with this understanding in the national convention, and consequently must have ratified it upon its own merits, as apparent on its face. I had the honor of a seat in one of those conventions, and gave my assent to its ratification on those principles: I did then, I do now, and ever shall consider of the powers of Congress by the expression of the constitution, in the same latitude I should do was the constitution a thousand years old, and every man in the convention that formed it in his grave; which I can easily conceive might, in the course of providence, have been the case before ever a Congress met under the constitution, and in that case I believe we should not now have found ourselves at a loss for want of contemporary witnesses or expositors.

Sir, I acknowledge that, by this clause of the constitution, Congress is denied the power of prohibiting the migration or importation of persons, but may impose a tax or duty; and, I say, as well on the white as black person. But, sir, some certain inadmissible qualities may be adherent to persons which, from the necessity of things, must and will amount to an exclusion even of the persons themselves, such as plague or pestilential disease; suppose the inhabitants of a neighbouring island were infected with such a disease, and some one state in the union, from motives of enmity to this country, and with design to depopulate it, should pass a law authorising and encouraging the migration or importation of those persons with this infectious quality adherent to them, and that Congress had timely notice of this nefarious business, would an interference be unconstitutional? I think not, sir.

Suppose a state, from enmity as before expressed, and with a design to destroy the union, should set about introducing some hundred thousand alien enemies, with their arms and accoutrements, and Congress had timely notice, would not the enmity, the arms and accoutrements of those persons, be readily granted inadmissible qualities, and amount to an exclusion of the persons? I think it would be granted: and if the importation was not prevented, I should be more inclined to impute it to want of physical than constitutional power. In consistency with this mode of reasoning, I believe if Congress should at any time be of opinion that a state of slavery, being attached to a person, was a quality altogether inadmissible into America, they would not be barred (by this clause of the constitution) of prohibiting that baneful quality: I believe as in the first case the plague, and in the second the enmity and arms, so in the third the state of slavery, may (notwithstanding any thing in this clause) be declared by Congress, qualities, conditions or adherents, or what you please to call them, which, whilst attached to any person, the person himself cannot be admitted. So much for this clause of the constitution.

But, sir, by another clause, Congress have power to regulate trade, &c. Under this head, not only the proposition now under consideration, but any other, or further regulations, which to Congress may seem expedient, are fully and clearly in their power: Nay, sir, if those wretched Africans are to be considered as property, as some gentlemen would have it, and consequently as subjects of trade and commerce, they and their masters so far lose the benefits of their personality, that Congress may at pleasure declare them contraband goods, and so prohibit the trade altogether.

Again, sir, Congress have power to establish a rule of naturalization. Under this head, it is clear that this rule is at the arbitrary will of Congress, and that they may, whenever they please, declare (by law) that every person, whether black, white, blue or red, who from foreign parts can only get his or her foot on the American shore, within the territory of the United States, shall to all intents and purposes be not only free persons, but free citizens. This doctrine is clearly and fully proved by the bill on the subject of naturalization, which was read this morning; in which bill it is provided, as a part of the conditions of naturalization, that he shall be a free white person, plainly implying that if this provision had not been made, the black man slave, as well as the white man free, could have availed himself of that law by fulfilling the conditions thereof.

Moreover, sir, Congress have power to define and punish piracies and felonies on the high seas. Under this head no doubt Congress may, when they please, declare by law, that an American going to the coast of Africa and there receiving on board of any vessel any person in chains and fetters of iron, or bound with cords, or in any manner under confinement, or without his or her own free will and consent, certified as Congress may direct, and carrying such person to any other part of the world (whether sold as a slave or not) is guilty of piracy and felony on the high seas, and on conviction shall suffer death accordingly without benefit of clergy; and Congress may perhaps go as far with respect to foreigners who may land persons within the territory of the United States on any other terms than what they may please to proscribe.

So much as to the powers of Congress I consent that the world should know; I consent those people in the gallery (about whom so much has been said) should know that there is at least one member on this floor who believes that Congress have ample powers to do all they have asked respecting the African slave trade; nor do I doubt but Congress will (whenever necessity and polity dictates the measure) exercise those powers, and I believe that the importation of one cargo of slaves would go far towards inducing such laws as I have mentioned: but I believe also, sir, that this necessity is not like to happen, the states will severally, I think, do what is right in the premises. If the question was, what will Congress now do on this subject? there is not a gentleman from the southern states more ready to say they will do nothing than I am; I think there is no necessity of acting; but as it is what is their powers? I would wish those powers, if expressed at all, to be fully expressed.

As much of the time of the committee has been spent on this subject, and perhaps we are past the proper place to bring forward any new propositions, I shall content myself with voting for the retention of the proposition before you until we come into the house, when something more expressive of the full powers of Congress will doubtless be brought forward.

REPORT OF THE COMMITTEE OF THE WHOLE
March 22, 1790

That the migration or importation of such persons as any of the States now existing shall think proper to admit, cannot be prohibited by Congress, prior to the year one thousand eight hundred and eight.

That Congress have no authority to interfere in the emancipation of slaves, or in the treatment of them within any of the States; it remaining with the several States alone to provide any regulations therein, which humanity and true policy may require.

That Congress have authority to restrain the citizens of the United States from carrying on the African trade, for the purpose of supplying foreigners with slaves, and of providing, by proper regulations, for the humane treatment, during their passage, of slaves imported by the said citizens into the States admitting such importations.

March 23

It was then moved that the house should take up the report of the committee of the whole on the memorials of the people called Quakers.

James Madison (Va.) said, that although he thought the proposition expressed important constitutional truths, yet it was improper for the house to pass them in the form of abstract declarations. Three or four memorials, from respectable bodies of citizens, had been presented; the respect due to them required that they should be committed; the committee had made a report, reciting the powers and restrictions in the constitution as to the African trade, &c. and concluding with a proposed resolution in answer to the memorialists. This report of the select committee had been considered by a committee of the whole, and the recital had been so amended as to speak the meaning of the constitution; but as the resolution, which the recital was intended to introduce, had been rejected, he was of opinion the recital ought to fall along with it. He wished, at the same time, however, that the proceedings of the committee of the whole might go on the journals for the information of the public. This would sufficiently answer the two purposes of shewing the sense of the body on the subject of the memorials, and of carrying an implication that Congress will exercise the powers vested by the constitution whenever it should be requisite. He wished the entry to be made for another reason: it would be the best answer to the arguments used by the gentlemen who had so warmly opposed the proceeding, and tend to quiet the apprehensions of the southern states, by recognizing that Congress had no power whatever to prohibit the importation of slaves prior to the year 1808, or to attempt to manumit them at any time. He accordingly moved,

"That the proceedings of the committee of the whole should be entered on the journals, as the best method of putting an end to the business. . . ."

The question being put, 25 appeared against, and 24 for taking up the report; it being previously understood, that negativing this motion was preparing the way for the one proposed by Mr. Madison.

The question on Mr. Madison's motion was now put. . . .

The reports were accordingly ordered to be entered on the journals, viz. The report of the select committee, together with the report of the committee of the whole.

THOMAS COLE, PETER BASSNETT MATHEWS, AND MATTHEW WEBB
Petition to the South Carolina Senate, January 1, 1791[8]

Over 100,000 blacks lived in South Carolina in 1790—more than 43 percent of the state's total population. Only 1,801 were free (586 of them lived in Charleston). Slaves and free blacks had no access to the state's courts, but were subject to special magistrates' and freeholders' courts. The three petitioners here did not base their claim for additional rights on a federal bill of rights (which had not yet been ratified) but on Article I, section 2 of the United States Constitution, which counted them as free persons for the purpose of representation. The state of South Carolina received the benefit of the petitioners' free status in its representation to the U.S. House of Representatives and the petitioners paid taxes to the state. The petitioners asked for the right to testify in court and to enter into contracts. They were willing to take any oath and perform any duty to defend the state. They did not expect to "be put on an equal footing with the Free White Citizens of this State in general." But, because free blacks were unprotected by state constitutional guarantees, they were not accorded the right to petition the legislature, and the South Carolina senate refused to accept their petition.

To the Honorable David Ramsay Esquire President and to the rest of the Honorable New Members of the Senate of the State of South Carolina.

The Memorial of Thomas Cole Bricklayer, P. B. Mathews and Matthew Webb Butchers on behalf of themselves & other Free Men of Colour.

Humbly sheweth That in the Enumeration of Free Citizens by the Constitution of the United States for the purpose of Representation of the Southern States in Congress, Your Memorialists have been considered under that description as part of the Citizens of this State. Although by the Fourteenth and Twenty Ninth Clauses in an Act of Assembly made in the Year 1740 and intitled an Act for the better Ordering and Governing Negroes and other Slaves in this Province commonly called The Negroe Act now in force, Your Memorialists are deprived of the Rights and Privileges of Citizens by not having it in their power to give Testimony on Oath in prosecutions on behalf of the State from which cause many Culprits have escaped the Punishment due to their

---

[8]Petition Series, South Carolina Department of Archives and History.

Atrocious Crimes (nor can they give their Testimony in recovering Debts due to them, or in establishing Agreements made by them within the meaning of the Statutes of Frauds and Perjuries in force in this State except in cases where Persons of Colour are concerned, whereby they are subject to great Losses and repeated Injuries without any means of redress.

That by the said Clauses in the said Act, They are debarred of the Rights of Free Citizens by being subject to a Trial without the benefit of a Jury and subject to Prosecution by testimony of Slaves without Oath by which they are placed on the same footing.

Your Memorialists shew that they have at all Times since the Independence of the United States contributed and do now contribute to the support of government by chearfully paying their Taxes proportionable to their Property with others who have been during such period, and now are in full enjoyment of the Rights and Immunities of Citizens Inhabitants of a Free Independent State.

That as your Memorialists have been and are considered as Free-Citizens of this State they hope to be treated as such. They are ready and willing to take and subscribe to such Oath of Allegiance to the States as shall be prescribed by this Honorable House, and are also willing to take upon them any duty for the preservation of the Peace in the City or any other occasion if called on.

Your Memorialists do not presume to hope that they shall be put on an equal footing with the Free White Citizens of this State in general. They only humbly solicit such Indulgence as the Wisdom and Humanity of this Honorable House shall dictate in their favor by Repealing the Clauses [in] the Act beforementioned and substituting such a Clause as will effectually Redress the grievances which your Memorialists humbly submit in this their Memorial but under such restrictions as to your Honorable House shall seem proper.

May it therefore please your Honors to take your Memorialists Case into tender consideration, and make such Acts or insert such Clauses for the purpose of relieving your Memorialists from the unmeritted grievance they now Labour under as in your Wisdom shall seem meet. And as in duty bound your Memorialists will ever pray—

Racism in the North

THE NEW YORK GUBERNATORIAL ELECTION OF 1792

In the 1792 New York gubernatorial election, the abolitionist views of Chief Justice of the United States John Jay proved to be a substantial detriment in his attempt to defeat incumbent George Clinton. John C. Wynkoop, an ardent supporter of the Revolution and former Clintonian lieutenant in Kingston, believed that the chief justice would "not have many Votes from the Dutch Inhabitants" of Ulster County, because "a great majority" of them "possess many Slaves, and as Mr. Jay is President of the Society for the abolition of Slavery, they will probably vote against him for that reason alone." Wynkoop regretted the prejudice of his fellow Ulstermen and believed that Jay's sentiments were "founded on the eternal Law of Nature and Nature's God." John C. Dongan, a Staten Island lawyer, warned Jay that his opponents "descend to the lowest subterfuges of craft and chicane, to mislead the ignorant and unwary. . . . It is said that it is your desire to rob every Dutchman of the property he possesses most dear to his heart, his slaves." [9]

Jay decided to meet this opposition straightforwardly by declaring that "every man of every color and description has a natural right to freedom." He believed that emancipation should be accomplished "in such a way as may be consistent with the justice due to them, with the justice due to their master, and with the regard due to the actual state of society. These considerations unite in convincing me that the abolition of slavery must necessarily be gradual." [10] Jay defended the New York Abolition Society and his association with it, which he had severed when he became chief justice. He explained that the Society tried "to promote by virtuous means the extension of the blessings of liberty, to protect a poor and friendless race of men, their wives and children from the snares and violence of men-stealers, to provide instruction for children who were destitute of the means of education, and who, instead of pernicious, will now become useful members of society." [11]

Although opposition to Jay's abolitionist sentiments also thrived in Columbia County, Peter Van Schaack, a former Loyalist during the Revolution, held out some hope that a change was imminent. Federalist writer "A.B." suggested that Jay did not advocate emancipation "without the consent of the proprietors, or an adequate compensation. Slavery," Jay believed, "is indeed odious, and the practice of it in a free country, much to be lamented; but as the laws of society have tolerated the practice, it is but reasonable, that the abolition should be effected in such a way, as not to interfere with the regard that is due to private property." [12] A reported dialogue between two farmers in Wallkill, Ulster County, brought out the worst kind of bigotry. "You know," farmer Abraham said, "he is for making the negroes free, and let them stay amongst us, that they may mix their blood with white people's blood, and so to make the whole country bastards and out-laws." Farmer David responded. "As for Mr. Jay's making the negroes free, they said it should not cost our treasury a penny, and that the owners of them would have 'a full compensation,' for that as negroes were generally thieves, idlers and squanderers of their masters' property, it would be proper to set them free and discharge them; and for farmers to do their own work, and take care that every thing was safe and in its place; for that farmers generally smoaked too

[9]Dongan to Jay, February 27, 1792, Henry P. Johnston, ed., *The Correspondence and Public Papers of John Jay* (4 vols., New York, 1891), III, 413n.
[10]John Jay to John C. Dongan, *ibid.*, 414.
[11]*Ibid.*, 415.
[12]"Tammany," The Guardian, No. V, *New York Journal*, March 28, 1792.

much, took too much leisure, and trusted too much to their negroes, and this was a loss to them and the state, in a few years, which amounted to more than the negroes were worth. Hence, sir, you see how a full compensation may be had."[13]

Van Schaack ironically suggested that one method of gaining support for Jay would be to publicize Governor Clinton's stance on slavery. If Clinton opposed emancipation he could be charged accordingly and perhaps would lose support from those people with abolitionist sentiments. If Clinton supported emancipation, that would tend to neutralize the opposition to Jay. The governor, in fact, did support emancipation and was the original vice president of the New York Abolition Society, but because of the strong opposition to emancipation in the Hudson River Valley—Clinton's base of support among yeoman farmers—the governor did not publicize his abolitionist sentiments. When Clinton's position on slavery was disclosed, Jay narrowed the gap in the election and eventually received a majority of the votes cast. But because of irregularities in voting and reporting in four counties, Clinton was declared the victor.

John C. Wynkoop to Peter Van Schaack
Kingston, February 23, 1792[14]

My head is so crammed with political, moral and electioneering stuff, that I can neither write grammatically nor intelligibly, without much Repetition and Tautology. Your and, let me add, my Friend Mr. Jay, who is proposed as the illustrious opponent to Gov. Clinton at the next Election, will, I fear, not have many Votes from the *Dutch Inhabitants* in *this* County. A great majority of these possess *many slaves*, and as Mr. Jay, is President of the society for the abolition of slavery, they will probably vote against him for that reason alone. The Justice and Propriety of his sentiments on this important subject, which is founded on the eternal Law of Nature and Nature's God, and which affects the Freedom and of Course the Happiness of thousands of our fellow Creatures, differing from ourselves only in *complexion*, have no *Weight* in the minds of men, many of whom are rich in property of this Kind (if it deserves the Name of property) and which being darkened and jaundiced by the delusive tho' too powerful and condensed Claim of self-interest, cannot, as yet, admit the *Rays* of *Truth* This is a long sentence, but in my mind, a serious one.

You will therefore in your calculation of electioneering success, set down this County as *against* you. . . .

---

[13]*New York Journal*, April 21, 1792.
[14]Van Schaack Papers, Library of Congress.

JOHN JAY TO JOHN C. DONGAN
New York, February 27, 1792[15]

As to my sentiments and conduct relative to the abolition of slavery, the fact is this:—In my opinion, every man of every color and description has a natural right to freedom, and I shall ever acknowledge myself to be an advocate for the manumission of slaves in such way as may be consistent with the justice due to them, with the justice due to their master, and with the regard due to the actual state of society. These considerations unite in convincing me that the abolition of slavery must necessarily be gradual.

On being honored with the commission I now hold [chief justice of the United States], I retired from the [New York Abolition] Society to which you allude, and of which I was President, it appearing to me improper for a judge to be a member of such associations. That Society I fear has been misrepresented, for instead of censure they merit applause. To promote by virtuous means the extension of the blessings of liberty, to protect a poor and friendless race of men, their wives and children from the snares and violence of men-stealers, to provide instruction for children who were destitute of the means of education, and who, instead of pernicious, will now become useful members of society—are certainly objects and cares of which no man ought to be censured; and these are the objects and the cares of that benevolent society.

PETER VAN SCHAACK TO JOHN C. WYNKOOP
Kinderhook, March 13, 1792[16]

I have received your favor and leaving the matters of Business to take their fate I proceed to the more important subject of Politics, trusting that however prudence may induce you to passivity of conduct, your mind has not so far lost its natural vivacity as to be under the Influence of a Stoical apathy at the present momentous crisis. I own to you that I have not felt myself so warmly Interested in the Event of any former Election as I now do in that of the approaching one as it respects the first Magistrate of the State. My Zeal is the more excited by the detestable Calumnies which with the Venerable Character of Mr. Jay is assail'd, and whilst sycophants and Dependants are imposing upon Credulity and folly to obstruct the Honor & Interest of the Community, it is difficult to remain an unconcerned Spectator. The manumission business is here as with you made an Engine of to tarnish the Illustrious Character of Mr. Jay; but I trust its pernicious effect will in a great measure be

---

[15]Henry P. Johnston, ed., *The Correspondence and Public Papers of John Jay* (4 vols., New York, 1891), III, 414.
[16]Van Schaack Papers, Library of Congress.

defeated. People are already coming to their senses and feel the Impropriety of opposing a man so unexceptionable upon grounds so Questionable at least, if not wicked. They consider that if it *should really* be true that they have not an absolute right to convert a part of God's rational Creatures into Brutes, It would add to the Iniquity of the Practice if they were to oppose a man who holds a doctrine so full of Philanthropy even if it should in this Instance be misapplied. Such as opposed Great Britain lest the Consequence however remote of a present submission should be the loss of Liberty to their Posterity, begin to doubt how far it is consistent in them to maintain in Hereditary Bondage a part of their fellow creatures and they feel the force of the Divine command whatever "ye would that men should do unto you even so do ye unto them." Some indeed there are whose avarice induces them to deny "that these Creatures so black & with such flat Noses are a Part of the Human Species and these I suppose will by and by trace up their Origin to a different source from that of our first Parents in Paradise or prove that Noah would not admit such Wretches into the Ark. Minds like these will easily be brought to conclude that the poor black wretches are not contemplated by the Supreme Being in the Decalogue and particularly that it would be no transgression of the Sixth Commandment to destroy one of them any more than to slaughter a Bullock. I really do not see how these consequences can be avoided by those who maintain the legitimacy of Slavery. One would imagine that the present tragical scenes in the West Indies would bear an awfull admonition to people of this cast, scenes which by the by have been predicted long since by those who have opposed this horrid practice. Ardit Electagee. I wish this may not prove a presage to our southern neighbours—It is to be imputed to the laudable exertions of the Quakers that something like this does not also threaten *us*.

To me it appears evident that the ground of the present Opposition to Mr. Jay will defeat the purpose of those who make it by bringing about more speedily perhaps than it would otherwise have happened the very Event they deprecate. Reformation is generally the consequence of a bigotted maintenance of false tenets whether in Religion or Morality. I do not precisely know what part Mr. Jay has acted in this business. As to the Injustice of Slavery, I presume that none among you is so hardened as to deny it; as to the Inexpediency of a general Manumission some worthy Characters among us maintain it with some force of argument; but I do not find any who do not Contemplate a gradual Manumission as a desirable object. I hope the County of Ulster and especially the Ancient town of Kingston is not less enlightened or less liberal than the little Village of Kinderhook; and that the Whigs will not have to receive lessons of Liberty from The Tories. But this would not be the only paradox arising out of this subject for while Slavery can have no Existence in the *Venal and corrupt* Island of Great Britain, it is cherished and advocated in the free country of America: thus while the Americans invite the slavish Europeans to quit their native Country and repose in the sunshine of Liberty here; the unfortunate Africans have to fly from this free country to regain their Liberty in England where a free born American cannot so much as breathe. Is there not something monstrously selfish in this kind of Liberty and is it what was pretended as the spirit of resistance and which being believed to Exist as it was pretended procured so many Friends in every Part of the Globe to the American Cause? By the by has Mr. Jay's competitor ever disclosed his sentiments upon this Question? surely it is of Importance enough to merit his animadversion if wrong; and if in his opinion right it would be more heroick tho' to his own disadvantage to assent to the truth, than to rest under the Veil of concealment & mistery seeking Preferment from the possessors of slaves. If Governor C. has ever declared himself it is unknown to me and therefore if I injure him it is unintentional. How is Mr. Addison affected? I should hope that the Sons of Science and the Friends of the liberal Arts will be pretty much United and will be a more honorable tho' perhaps not more numerous Band than the Masters of Negroes, Mulattos and Mestees.

I do indeed hear that it is attempted with these Masters to Coalesce the Reverend Clergy on pretence that Mr. Jay had Occasion'd a certain clause in the CONSTITUTION; but I trust that this pious Body of men will consider the seeming disqualification as a Real and substantial privilege and not forget that they are the servants of him whose "Kingdom was not of this World. . . ."[17]

I was much pleas'd to hear of a Letter from Major [Peter Van] Gaasbeck to some Gentlemen at Albany wherein he declared himself personally for Mr. Jay altho' he apprehended the Manumission Business would be an Insuperable bar. I trust this dark Cloud will vanish and that the rays of truth will shine in spite of it. But if it should not I shall at least rejoice that I have acted in Concert with so many Respectable Characters as support Mr. Jay.

JOHN C. WYNKOOP TO PETER VAN SCHAACK
Kingston, April 17, 1792[18]

I did not receive your letter of the 13th. of last Month before *Yesterday*. . . .

Your reflections on the Justice of a gradual Manumission, or rather Abolition of Slavery, are elegant and profound. They are founded on the eternal Law of Nature and of Nature's God. I say with the firm and enlightened Congress in their Declaration of Independence that "We hold these Truths to be self-evident; that all men are created equal; that they are endowed by their creator with certain unalienable Rights, that among these are life, liberty and the pursuit of Happiness." The difference of Complexion, of Manners,

---

[17]Jay was the primary draftsman of the state constitution of 1777 which prohibited ministers from political office.
[18]Van Schaack Papers, Library of Congress.

Customs and Climates, makes no Difference in the impartial Eye of Philosophy, between the Rights of human Nature, and the jetty Affrican, the copper coloured American, the tawny Spaniard and the fair European, in regard to Freedom all are *equal*. Mr. Jay's being president of the Manumission Society, as I informed you in my last, prejudiced many People against him in this County. They now begin to think it *an amicable Error* (if any) but what has made *Numbers* change their sentiments in Mr. Jay's Favor, is a Circumstance not generally known but which merits your particular Attention, I mean, that *Clinton* is *at least* as much for *manumitting* the Slaves as his Competitor. This Fact appears in the Journal of Assembly of *1785*, when a bill had passed both Houses for Abolishing Slavery but made some odious Distinctions between the Blacks when freed and the Whites, such as disqualifying the former from being *Electors* &c. The Council of Revision objected to the Bill because the Blacks *would not* (if the Bill passed) be entitled *to all the* Priviledges of *Freemen*. The Bill was accordingly lost. This Argument has an amazing Weight here, and if properly enforced will have the same with your Dutch Friends.

"A.B."
New York Journal, **March 28, 1792**

I do not mean to exculpate Mr. Jay from the charge of being a member of the manumission society, which has afforded a copious field for declamation and abuse; but I am persuaded, that the principles and actions of that society have been greatly misrepresented; and it has been for some time expected, that in justice, both to themselves and Mr. Jay, they would have laid them fairly before the public. I am not a member of the society, and perhaps, ill qualified to undertake its defence, but I am certain, that an "absolute manumission of slaves" is not contemplated by them, without the consent of the proprietors, or an adequate compensation. Slavery, is indeed odious, and the practice of it in a free country, much to be lamented; but as the laws of society have tolerated the practice, it is but reasonable, that the abolition should be effected in such a way, as not to interfere with the regard that is due to private property.

The wisdom, humanity and justice of the legislature, will, without doubt sooner or later, place this business upon a proper footing: In the mean time, the manumission society, by the gentle methods of example and reason, have endeavoured to incite their fellow men, to bestow freedom where it was in their power; and have exerted themselves to guard those upon whom it had been voluntarily conferred, from violence and fraud; they have endeavoured to enforce those acts of the legislature, that prohibit the inhuman slave trade, which in too many instances, has been the means of tearing parents from their children, husbands from their wives; and is destructive of every feeling that ought to characterize a human being. In addition to these objects of the society, let me mention the institution and endowment of schools, by which the children of negroes may be rescued from ignorance and idleness, and become not only useful to the public in this life, but acquainted with those which appertain to their welfare in a future state. I trust, that such principles and practices cannot fail, when understood, of meeting with approbation. If Mr. Jay should be rendered unpopular on this account, the situation of the public mind is much to be regretted. We may boast if we please, of our free and excellent constitution, and call ourselves the most enlightened people to be found upon the globe; but in spite of all our boasting, we shall furnish the most irrefragable proof, that we are under the dominion of selfish passion, and involved in more than Cimmerian darkness.

THE FUGITIVE SLAVE LAW
February 12, 1793

Late in the Constitutional Convention when a provision was considered for the extradition of criminals from one state to another, Pierce Butler of South Carolina was able to insert a clause calling for the return of runaway slaves. This fugitive slave clause was similar to a provision in the Northwest Ordinance adopted two months earlier in Congress. (See Chapter 1 for the Northwest Ordinance.)

The Constitution's fugitive slave clause put the burden for retrieving a runaway on the slaveowner, rather than on the state into which the slave had escaped. But no specific guidelines explained the procedures to be followed in seizing and returning runaways, and no penalties were prescribed for individuals assisting runaways or protections afforded free blacks who might be kidnapped and transported away into enslavement.

Enacting a fugitive slave law was not a high priority for Congress. Not until 1793 did Congress attempt to explain the procedures that should be followed in returning runaway slaves. According to the act, a slaveowner or his authorized agent could seize a runaway anywhere in the country and take him before a federal or state judge. If convinced that the captured individual was a fugitive, the judge would issue a certificate allowing the slave to be returned. Those obstructing the recovery of runaways were subject to a $500 fine.

Because of the financial burden on slaveowners, the cumbersome administrative procedures, and the dependence on the public opinion of the local area in which the runaways were seized, few fugitive slaves were recovered from the North under its provisions. The law, however, was used in recovering escaped slaves in the South, and it had a stifling effect on those who might assist runaways. Because of the difficulties enforcing the act and because the act ignored the rights of free blacks, agents of slaveowners too often kidnapped free blacks and sent them into slavery either in the South or the West Indies. Several attempts to amend the fugitive slave law failed in the early national period.

An Act respecting fugitives from justice, and persons escaping from the service of their masters.

. . . Section 3. *And be it also enacted,* That when a person held to labor in any of the United States, or in either of the Territories on the Northwest or South of the river Ohio, under the laws thereof, shall escape into any other of the said States or Territory, the person to whom such labor or service may be due, his agent or attorney, is hereby empowered to seize or arrest such fugitive from labor, and to take him or her before any Judge of the Circuit or District Courts of the United States, residing or being within the State, or before any magistrate of a county, city, or town corporate, wherein such seizure or arrest shall be made, and upon proof to the satisfaction of such Judge or magistrate, either by oral testimony or affidavit taken before and certified by a magistrate of any such State or Territory, that the person so seized or arrested, doth, under the laws of the State or Territory from which he or she fled, owe service or labor to the person claiming him or her, it shall be the duty of such Judge or magistrate to give a certificate thereof to such claimant, his agent, or attorney, which shall be sufficient warrant for removing the said fugitive from labor to the State or Territory from which he or she fled.

Section 4. *And be further enacted,* That any person who shall knowingly and willingly obstruct or hinder such claimant, his agent or attorney, in so seizing or arresting such fugitive from labor, or shall rescue such fugitive from such claimant, his agent or attorney, when so arrested pursuant to the authority herein given or declared; or shall harbor or conceal such person after notice that he or she was a fugitive from labor, as aforesaid, shall for either of the said offences, forfeit and pay the sum of five hundred dollars. Which penalty may be recovered by and for the benefit of such claimant, by action of debt, in any Court proper to try the same, saving moreover to the person claiming such labor or service his right of action for or on account of the said injuries, or either of them.

ABOLITION OF THE FOREIGN SLAVE TRADE
March 2, 1807

The Constitution provided that the foreign slave trade could not be prohibited by Congress before 1808. In his annual message on December 2, 1806, President Jefferson encouraged Congress to enact a law prohibiting the slave trade as of January 1, 1808, and that slave traders be warned that no vessel should enter into that trade that could not finish its business by the end of 1807.

A congressional committee considered Jefferson's recommendation and reported a bill that was intensely debated. Three issues demanded the most attention: what should be done with smuggled slaves that were seized in the illegal trade, what penalty should be allotted to smugglers, and should the federal government be allowed to regulate the domestic coastal trade of slaves.

The draft bill called for seized slaves to be sold into slavery by federal officials. Northern representatives did not want the federal government to be an active participant in such an immoral act. Other suggestions were proposed. Southerners vehemently opposed letting free blacks loose into the South and suggested sending confiscated slaves back to Africa. Northerners objected to exposing innocent blacks to another dangerous transatlantic voyage that would probably end in either death or enslavement in Africa. Northerners recommended sending the slaves to the Northern states where they would be sold as indentured servants and, after being trained appropriately, would receive their freedom. This proposal received little support from anyone—Northerners did not want the burden of more freedmen among them; Jeffersonian Republicans did not want an expanded role for the federal government. A compromise agreed that forfeited slaves would be delivered to state authorities to dispose of as they saw fit. Everyone knew that this meant that illegally-transported free men and women would be sold into slavery. Northerners satisfied themselves that the elimination of the federal government from any direct sale of slaves was palatable.

The draft bill provided the death penalty for convicted smugglers. Peter Early of Georgia, chairman of the committee that drafted the slave trade bill, spoke against the severity of the punishment arguing that no Southerner would inform or testify in court against smugglers under those conditions. On December 31, 1806, Early stated that "A large majority of people in the Southern states do not consider slavery as even an evil." Only a small reflecting minority pictured slavery as a political evil. The existence of freedmen amidst slavery was the evil most feared by Southerners. Abolitionism and emancipation were the greatest evils because they would lead to interracial warfare and carnage. Only the continuation of slavery prevented such a disaster. A compromise substituted a five to ten-year sentence and a fine of up to $10,000 for convicted smugglers.

After heated debates, the representatives agreed that federal authorities could regulate the coastal slave trade in ships above a certain size. Southerners feared that this "entering-wedge" would lead eventually to the complete regulation of the domestic slave trade and would provide the "pretext of universal emancipation." John Randolph of Roanoke argued that "He had rather lose the bill, he had rather lose all the bills of the session, he had rather lose every bill passed since the establishment of the government [in 1789], than agree to the provision contained in this slave bill." Despite this tirade, and a threat to lobby President Jefferson for a veto of the bill, the provision for federal regulation of the coastal slave trade stood.

Congress passed the amended bill on March 2, 1807, by a vote of 63 to 49. President Jefferson signed the act the following day. Much like the fugitive slave law of 1793, however, the law prohibiting the importation of slaves from Africa was unenforceable wherever public opinion sympathized with smugglers. By the end of the War of 1812, more than 10,000 slaves were being imported annually, and the sectional divisiveness of the country intensified.

An Act to prohibit the importation of slaves into any port or place within the jurisdiction of the United States, from and after the first day of January, in the year of our Lord one thousand eight hundred and eight.

Be it enacted, etc., That, from and after the first day of January, one thousand eight hundred and eight, it shall not be lawful to import or bring into the United States or the territories thereof, from any foreign kingdom, place, or country, any negro, mulatto, or person of color, with intent to hold, sell, or dispose of such negro, mulatto, or person of color, as a slave, or to be held to service or labor. . . .

7

———————◆———————

Slavery and the Founders: Three Perspectives

T

HE AMERICAN REVOLUTIONARY ERA
*was a time when slavery might have been abolished peacefully without dismembering
the Union. The rhetoric and the reality of fighting for liberty spurred in the American consciousness a devotion to freedom and a concomitant sense of guilt in the continued
enslavement of a race of people. Tragically, by 1787 this unique chance to rid America
of slavery vanished.*

*No one can explain with certainty why emancipation failed to occur. One criti-
cal factor, however, was enormously important in the continuation of slavery. Certainly,
Northern leaders were not willing to endanger the Union or forgo the Constitution
in order to attack slavery. Another critical factor was the failure of the powerful and
influential Southern leadership to participate in the emancipation movement. Active
support of gradual emancipation by George Washington, Thomas Jefferson, and James
Madison might have been sufficient to mount a serious attack on slavery. Each of
these three men acknowledged the injustice of slavery; all advocated its abolition; and
all personally held slaves. Yet none in his lifetime did anything of substance to free his
own slaves or support the cause of emancipation.*

*Why would highly principled men such as Washington, Jefferson, and Madi-
son refuse to act on their convictions? Their motivations were complex. Each man had
self-interested and public-spirited concerns that conflicted with a philosophical antipathy for slavery. Each man had a plantation that supported him and his family
financially while he devoted time and energy to public service. Southern society valued economic independence. To reject financial security and turn one's back on an
established social system would take a great deal of dedication to the cause, not to*

mention personal sacrifice. Each man was aware of his standing in his community, state, and section of the country. Support of emancipation might be detrimental politically, and their political careers were immensely important for Virginia and for the country at this critical time. Each man worried that the slavery issue, if pushed too hard, could lead to a backlash within Virginia and might threaten the Union itself. Each man, born and reared in a slave society, was uncertain that emancipation would benefit the white population, the South, the country, or even the slaves. All of these issues concerned them. Sometimes they were legitimate concerns; at other times they merely helped rationalize the status quo.

After the Revolution, advocates of emancipation regularly sought Washington's endorsement for their proposals. In 1783, the Marquis de Lafayette asked Washington to join him in purchasing a plantation and freeing its slaves to show that freedmen could succeed. A year later, historian William Gordon took the occasion of Lafayette's visit to Mount Vernon to write Washington that "You wished to get rid of all your Negroes, & the Marquis wisht that an end might be put to the slavery of all of them. I should rejoice beyond measure could your joint counsels & influence produce it, & thereby give the finishing stroke & the last polish to your political characters." In 1785 Methodist ministers asked Washington to sign their petition to the Virginia legislature seeking emancipation. On all of these occasions, Washington awkwardly backed away from a commitment. Although he had come to oppose slavery and wanted to divorce himself from the peculiar institution, he refused to be a champion of emancipation. He waited for the development of a strong emancipation movement in Virginia that he could endorse, not realizing that such a movement was unlikely without his early and active support. Thus, Washington continued to struggle with the inconsistency of his theoretical opposition to slavery and his real world situation.

Instead of freeing his slaves or endorsing emancipation, Washington tried to treat his slaves as humanely as the system allowed. He encouraged marriages and family life among slaves, he did not buy or sell them, he made some slaves overseers, and he prepared young slaves for their eventual freedom. In an undated memorandum, Washington stated his goal: "To make the Adults among them as easy & as comfortable in their circumstances as their actual state of ignorance & improvidence would admit; & to lay a foundation to prepare the rising generation for a destiny different from that in which they were born; afforded some satisfaction to my mind, & could not I hoped be displeasing to the justice of the Creator."

Washington once told a visiting Englishman "that nothing but the rooting out of slavery can perpetuate the existence of our union by consolidating it in a common bond of principle." If, however, slavery divided America, Washington revealed to Edmund Randolph, "he had made up his mind to move and be of the northern."

In December 1793, Washington wrote Arthur Young, an English agricultural reformer, that he wanted to bring "good" English farmers to Mount Vernon as tenants. Washington would retain only the mansion house farm itself; the other four Mount Vernon farms would be leased to the English farmers. "Many of the Negroes, male and female, might be hired by the year as labourers" to work the land. In this way, Washington could "liberate a certain species of property which I possess, very repugnantly to my own feelings." In essence, rather than freeing his slaves directly and hiring them himself, Washington searched for a buffer that would insulate him from the criticism of his fellow Virginians.

As President of the United States, Washington occasionally felt that freeing his slaves might serve as a beacon leading other Southerners to free their slaves, but he also felt that such an act by him might drive the North and the South further apart. With Washington as an example, the North might redouble its abolition efforts, while a beleaguered South might become increasingly more defensive. Rather than risk such divisiveness, Washington again avoided any public stance on slavery. Only in his will would he make a statement freeing his slaves.

Jefferson and Madison, as well as many other slaveholders in the North and in the South, believed that blacks were inherently inferior to whites. This racism was unshakeable and deeply affected their attitude toward emancipation. Although profoundly troubled by slavery, they felt that emancipation would succeed only if three conditions were met: (1) a gradual plan of compensated emancipation had to be accompanied by (2) a systematic colonization of the freedmen apart from white America and (3) the freed and colonized slaves would have to be replaced by white immigrant labor from Europe. Both Jefferson and Madison consistently supported such a three-point program, although they did little to implement it. When asked to lead, they declined; and as the years passed, their enthusiasm waned; their hope turned to despair. They realized that the country they had been so instrumental in founding was saddled with an insoluble moral dilemma. The father of the Declaration of Independence and the father of the Constitution and the Bill of Rights realized that their great experiment in seeking equality and justice for all had failed in this regard. Their personal tragedy (as well as the country's) was that the Revolutionary fervor for liberty might have led to the elimination of slavery if they had been as courageous in their opposition to slavery as they and the other founding fathers had been against Great Britain. They were willing to risk all for their own liberty. They were unwilling to take similar risks for the liberty of enslaved blacks.

Jefferson epitomized the changing sentiment of Southern leaders sympathetic to emancipation. Throughout the Revolution, he noticed a new attitude toward slavery in America. "I think a change already perceptible since the origin of the present revolution," Jefferson wrote in his Notes on the State of Virginia. *"The spirit of the master is abating, that of the slave rising from the dust, his condition mollifying, the way I hope preparing, under the auspices of heaven, for a total emancipation, and that this is disposed, in the order of events, to be with the consent of the masters, rather than by their extirpation."*

By 1785, however, Jefferson put his hope in the next generation of leaders— the "young men grown & growing up. These have sucked in the principles of liberty as it were with their mother's milk, and it is to them I look with anxiety to turn the fate of this question." A year later he suggested that "we must await with patience the workings of an overruling providence, and hope that that is preparing the deliverance of these our suffering brethren." Twenty years later, in 1807, he continued to look to the distant unforseeable future when "all the members of the human family

*may, in the time prescribed by the Father of us all, find themselves securely estab-
lished in the enjoyments of life, liberty, & happiness." In 1814 he felt benevolently
that the status quo must continue, and so "until more can be done for them, we should
endeavor, with those whom fortune has thrown on our hands, to feed & clothe them
well, protect them from ill usage, require such reasonable labor only as is performed
voluntarily by freemen, and be led by no repugnancies to abdicate them, and our
duties to them"—but not to free them. He acknowledged that most Southerners thought
of slaves only "as legitimate subjects of property as their horses and cattle," and his
faith in the future generation vanished as he saw "an apathy unfavorable to every
hope" of emancipation permeate the South. Despite appeals to him to become a "rallying point," he refused to take any public action or even write a "testament" endorsing
emancipation. The cause, he was told, "rests with you," but he was unwilling to
accept the challenge. "This enterprise is for the young; for those who can follow it up,
and bear it through to its consummation. It shall have all my prayers, and these are
the only weapons of an old man."*

*In 1820 he captured the South's dilemma in a frightening metaphor: "we
have the wolf by the ear, and we can neither hold him, nor safely let him go. Justice is
in one scale, and self-preservation in the other." Finally, just six weeks before his
death, Jefferson cautioned against any precipitous actions. "A good cause is often
injured more by ill-timed efforts of its friends than by the arguments of its enemies.
Persuasion, perseverance, and patience are the best advocates on questions depending
on the will of others. The revolution in public opinion which this cause requires, is not
expected in a day, or perhaps in an age; but time, which outlives all things, will
outlive this evil also."*

Thomas Jefferson and Slavery

THOMAS JEFFERSON
Notes on the State of Virginia[1]

. . . To emancipate all slaves born after passing the act. The bill reported
by the revisors[2] does not itself contain this proposition; but an amendment
containing it was prepared, to be offered to the legislature whenever the bill
should be taken up, and further directing, that they should continue with
their parents to a certain age, then be brought up, at the public expence, to
tillage, arts or sciences, according to their geniusses, till the females should
be eighteen, and the males twenty-one years of age, when they should be
colonized to such place as the circumstances of the time should render most

[1]Merrill D. Peterson, ed., *Thomas Jefferson: Writings* (New York, 1984), 264–67, 269–70,
288–89.
[2]In October 1776 the Virginia legislature appointed a committee of five chaired by Jefferson to draft a bill for the general revision of laws. The committee brought in its report in June
1779, and it served as draft legislation for the next decade.

proper, sending them out with arms, implements of household and of the handicraft arts, feeds, pairs of the useful domestic animals, &c. to declare them a free and independant people, and extend to them our alliance and protection, till they shall have acquired strength; and to send vessels at the same time to other parts of the world for an equal number of white inhabitants; to induce whom to migrate hither, proper encouragements were to be proposed. It will probably be asked, Why not retain and incorporate the blacks into the state, and thus save the expence of supplying, by importation of white settlers, the vacancies they will leave? Deep rooted prejudices entertained by the whites; ten thousand recollections, by the blacks, of the injuries they have sustained; new provocations; the real distinctions which nature has made; and many other circumstances, will divide us into parties, and produce convulsions which will probably never end but in the extermination of the one or the other race.—

To these objections, which are political, may be added others, which are physical and moral. The first difference which strikes us is that of colour. Whether the black of the negro resides in the reticular membrane between the skin and scarf-skin, or in the scarf-skin itself; whether it proceeds from the colour of the blood, the colour of the bile, or from that of some other secretion, the difference is fixed in nature, and is as real as if its seat and cause were better known to us. And is this difference of no importance? Is it not the foundation of a greater or less share of beauty in the two races? Are not the fine mixtures of red and white, the expressions of every passion by greater or less suffusions of colour in the one, preferable to that eternal monotony, which reigns in the countenances, that immoveable veil of black which covers all the emotions of the other race? Add to these, flowing hair, a more elegant symmetry of form, their own judgment in favour of the whites, declared by their preference of them, as uniformly as is the preference of the Oranootan for the black women over those of his own species. The circumstance of superior beauty, is thought worthy attention in the propagation of our horses, dogs, and other domestic animals; why not in that of man? Besides those of colour, figure, and hair, there are other physical distinctions proving a difference of race. They have less hair on the face and body. They secrete less by the kidnies, and more by the glands of the skin, which gives them a very strong and disagreeable odour. This greater degree of transpiration renders them more tolerant of heat, and less so of cold, than the whites. Perhaps too a difference of structure in the pulmonary apparatus, which a late ingenious experimentalist (Crawford) has discovered to be the principal regulator of animal heat, may have disabled them from extricating, in the act of inspiration, so much of that fluid from the outer air, or obliged them in expiration, to part with more of it.[3]

---

[3]Adair Crawford (1748–1795), an English physician, published his *Experiments and Observations on Animal Heat, and the Inflammation of Combustible Bodies; being an Attempt to resolve these Phenomena into a General Law of Nature* in London in 1779.

They seem to require less sleep. A black, after hard labour through the day, will be induced by the slightest amusements to sit up till midnight, or, later, though knowing he must be out with the first dawn of the morning. They are at least as brave, and more adventuresome. But this may perhaps proceed from a want of forethought, which prevents their seeing a danger till it be present. When present, they do not go through it with more coolness or steadiness than the whites. They are more ardent after their female: but love seems with them to be more an eager desire, than a tender delicate mixture of sentiment and sensation. Their griefs are transient. Those numberless afflictions, which render it doubtful whether heaven has given life to us in mercy or in wrath, are less felt, and sooner forgotten with them. In general, their existence appears to participate more of sensation than reflection. To this must be ascribed their disposition to sleep when abstracted from their diversions, and unemployed in labour. An animal whose body is at rest, and who does not reflect, must be disposed to sleep of course.

Comparing them by their faculties of memory, reason, and imagination, it appears to me, that in memory they are equal to the whites; in reason much inferior, as I think one could scarcely be found capable of tracing and comprehending the investigations of Euclid; and that in imagination they are dull, tasteless, and anomalous. It would be unfair to follow them to Africa for this investigation. We will consider them here, on the same stage with the whites, and where the facts are not apocryphal on which a judgment is to be formed. It will be right to make great allowances for the difference of condition, of education, of conversation, of the sphere in which they move. Many millions of them have been brought to, and born in America. Most of them indeed have been confined to tillage, to their own homes, and their own society: yet many have been so situated, that they might have availed themselves of the conversation of their masters; many have been brought up to the handicraft arts, and from that circumstance have always been associated with the whites. Some have been liberally educated, and all have lived in countries where the arts and sciences are cultivated to a considerable degree, and have had before their eyes samples of the best works from abroad. The Indians, with no advantages of this kind, will often carve figures on their pipes not destitute of design and merit. They will crayon out an animal, a plant, or a country, so as to prove the existence of a germ in their minds which only wants cultivation. They astonish you with strokes of the most sublime oratory; such as prove their reason and sentiment strong, their imagination glowing and elevated. But never yet could I find that a black had uttered a thought above the level of plain narration; never see even an elementary trait of painting or sculpture. In music they are more generally gifted than the whites with accurate ears for tune and time, and they have been found capable of imagining a small catch. (*The instrument proper to them is the Banjar, which they brought hither from Africa, and which is the original of the guitar, its chords being precisely the four lower chords of the guitar.*) Whether they will be equal to the composition of a more extensive run of melody, or of complicated harmony, is yet to be proved. Misery is often the parent of the most affecting touches in poetry.—Among the blacks is misery enough, God knows, but no poetry. Love is the peculiar *oestrum* of the poet. Their love is ardent, but it kindles the senses only, not the imagination. Religion indeed has produced a Phyllis Whately;[4] but it could not produce a poet. The compositions published under her name are below the dignity of criticism. The heroes of the Dunciad are to her, as Hercules to the author of that poem. Ignatius Sancho has approached nearer to merit in composition; yet his letters do more honour to the heart than the head.[5] They breathe the purest effusions of friendship and general philanthropy, and shew how great a degree of the latter may be compounded with strong religious zeal. He is often happy in the turn of his compliments, and his stile is easy and familiar, except when he affects a Shandean fabrication of words. But his imagination is wild and extravagant, escapes incessantly from every restraint of reason and taste, and, in the course of its vagaries, leaves a tract of thought as incoherent and eccentric, as is the course of a meteor through the sky. His subjects should often have led him to a process of sober reasoning: yet we find him always substituting sentiment for demonstration. Upon the whole, though we admit him to the first place among those of his own colour who have presented themselves to the public judgment, yet when we compare him with the writers of the race among whom he lived, and particularly with the epistolary class, in which he has taken his own stand, we are compelled to enroll him at the bottom of the column. This criticism supposes the letters published under his name to be genuine, and to have received amendment from no other hand; points which would not be of easy investigation. The improvement of the blacks in body and mind, in the first instance of their mixture with the whites, has been observed by every one, and proves that their inferiority is not the effect merely of their condition of life. . . .

The opinion, that they are inferior in the faculties of reason and imagination, must be hazarded with great diffidence. To justify a general conclusion, requires many observations, even where the subject may be submitted to the Anatomical knife, to Optical glasses, to analysis by fire or by solvents. How much more then where it is a faculty, not a substance, we are examining; where it eludes the research of all the senses; where the conditions of its existence are various and variously combined; where the effects of those which are present or absent bid defiance to calculation; let me add too, as a circumstance of great tenderness, where our conclusion would degrade a whole race

[4]African-born, Phillis Wheatley (1753?–1784) was kidnapped and taken as a slave to Boston around 1761. She began writing poetry at the age of thirteen and was recognized as a prodigy. She traveled to England in 1773 where she achieved prominence. She published in 1773 *Poems on Various Subjects, Religious and Moral.*

[5]Born on board a slave trading vessel in the Caribbean, Sancho (1729–1780) was brought to England at the age of two. After escaping from his owners, he was befriended by the Duchess of Montagu whom he served as a butler. Late in life Sancho wrote letters to prominent artists and literary figures. His letters were published in book form in England in 1782.

of men from the rank in the scale of beings which their Creator may perhaps have given them. To our reproach it must be said, that though for a century and a half we have had under our eyes the races of black and of red men, they have never yet been viewed by us as subjects of natural history. I advance it therefore as a suspicion only, that the blacks, whether originally a distinct race, or made distinct by time and circumstances, are inferior to the whites in the endowments both of body and mind. It is not against experience to suppose, that different species of the same genus, or varieties of the same species, may possess different qualifications. Will not a lover of natural history then, one who views the gradations in all the races of animals with the eye of philosophy, excuse an effort to keep those in the department of man as distinct as nature has formed them? This unfortunate difference of colour, and perhaps of faculty, is a powerful obstacle to the emancipation of these people. Many of their advocates, while they wish to vindicate the liberty of human nature, are anxious also to preserve its dignity and beauty. Some of these, embarrassed by the question "What further is to be done with them?" join themselves in opposition with those who are actuated by sordid avarice only. Among the Romans emancipation required but one effort. The slave, when made free, might mix with, without staining the blood of his master. But with us a second is necessary, unknown to history. When freed, he is to be removed beyond the reach of mixture. . . .

It is difficult to determine on the standard by which the manners of a nation may be tried, whether *catholic*, or *particular*. It is more difficult for a native to bring to that standard the manners of his own nation, familiarized to him by habit. There must doubtless be an unhappy influence on the manners of our people produced by the existence of slavery among us. The whole commerce between master and slave is a perpetual exercise of the most boisterous passions, the most unremitting despotism on the one part, and degrading submissions on the other. Our children see this, and learn to imitate it; for man is an imitative animal. This quality is the germ of all education in him. From his cradle to his grave he is learning to do what he sees others do. If a parent could find no motive either in his philanthropy or his self-love, for restraining the intemperance of passion towards his slave, it should always be a sufficient one that his child is present. But generally it is not sufficient. The parent storms, the child looks on, catches the lineaments of wrath, puts on the same airs in the circle of smaller slaves, gives a loose to his worst of passions, and thus nursed, educated, and daily exercised in tyranny, cannot but be stamped by it with odious peculiarities. The man must be a prodigy who can retain his manners and morals undepraved by such circumstances. And with what execration should the statesman be loaded, who permitting one half the citizens thus to trample on the rights of the other, transforms those into despots, and these into enemies, destroys the morals of the one part, and the amor patriae of the other. For if a slave can have a country in this world, it must be any other in preference to that in which he is born to live and labour for another: in which he must lock up the faculties of his nature, contribute as far as depends on his individual endeavours to the evanishment of the human race, or entail his own miserable condition on the endless generations proceeding from him. With the morals of the people, their industry also is destroyed. For in a warm climate, no man will labour for himself who can make another labour for him. This is so true, that of the proprietors of slaves a very small proportion indeed are ever seen to labour. And can the liberties of a nation be thought secure when we have removed their only firm basis, a conviction in the minds of the people that these liberties are of the gift of God? That they are not to be violated but with his wrath? Indeed I tremble for my country when I reflect that God is just: that his justice cannot sleep for ever: that considering numbers, nature and natural means only, a revolution of the wheel of fortune, an exchange of situation, is among possible events: that it may become probable by supernatural interference! The Almighty has no attribute which can take side with us in such a contest.—But it is impossible to be temperate and to pursue this subject through the various considerations of policy, of morals, of history natural and civil. We must be contented to hope they will force their way into every one's mind. I think a change already perceptible, since the origin of the present revolution. The spirit of the master is abating, that of the slave rising from the dust, his condition mollifying, the way I hope preparing, under the auspices of heaven, for a total emancipation, and that this is disposed, in the order of events, to be with the consent of the masters, rather than by their extirpation.

THOMAS JEFFERSON TO RICHARD PRICE
Paris, August 7, 1785[6]

Virginia. This is the next state to which we may turn our eyes for the interesting spectacle of justice in conflict with avarice & oppression: a conflict wherein the sacred side is gaining daily recruits from the influx into office of young men grown & growing up. These have sucked in the principles of liberty as it were with their mother's milk, and it is to them I look with anxiety to turn the fate of this question.

THOMAS JEFFERSON TO JEAN NICOLAS DÉMEUNIER
Paris, June 26, 1786[7]

What a stupendous, what an incomprehensible machine is man! Who can endure toil, famine, stripes, imprisonment & death itself in vindication of his own liberty, and the next moment be deaf to all those motives whose power supported him thro' his trial, and inflict on his fellow men a bondage, one

---

[6]Julian P. Boyd, ed., *The Papers of Thomas Jefferson* (Princeton, 1953), VIII, 357.
[7]Peterson, ed., *Thomas Jefferson*, 592.

hour of which is fraught with more misery than ages of that which he rose in rebellion to oppose. But we must await with patience the workings of an overruling providence, & hope that that is preparing the deliverance of these our suffering brethren. When the measure of their tears shall be full, when their groans shall have involved heaven itself in darkness, doubtless a god of justice will awaken to their distress, and by diffusing light & liberality among their oppressors, or at length by his exterminating thunder, manifest his attention to the things of this world, and that they are not left to the guidance of a blind fatality.

Eradicating "Absurd and False Ideas" Thomas Jefferson and Benjamin Bannaker

Benjamin Banneker, about sixty years old at the time of the writing of this letter, was the eldest child of a free black couple who owned a tobacco farm in Maryland. Banneker was a mathematician, astronomer, publisher, and surveyor. He was one of three men selected to survey the site for the new federal capital on the banks of the Potomac River. The Georgetown Weekly Ledger *had reported on the arrival of Banneker that he was "an Ethiopian whose abilities as surveyor and astronomer already prove that Mr. Jefferson's concluding that that race of men were void of mental endowment was without foundation."*

Banneker, as author and publisher, sent a manuscript copy of his second almanac to Jefferson to convince him that the failure of blacks to achieve intellectual success was caused by their enslavement and not by any natural inferiority. Jefferson responded favorably two weeks later. Banneker and his Quaker supporters had the correspondence published to aid the anti-slavery cause. On several occasions during his career, Jefferson's political opponents used this correspondence to imply that Jefferson supported emancipation. By 1809, however, Jefferson had come to believe that Banneker did not prepare his almanac independently and that his 1791 letter showed "him to have had a mind of very common stature indeed."

BENJAMIN BANNEKER TO THOMAS JEFFERSON
Baltimore County, Maryland, August 19, 1791[8]

Sir, I am fully sencible of the greatness of that freedom which I take with you on the present occasion; a liberty which seemed to me scarcely allowable, when I reflected on that distinguished, and dignifyed station in which you

[8]Several versions of Banneker's letter exist. The copy received by Jefferson is in the Massachusetts Historical Society and is printed in the Society's *Collections*, Seventh Series, Vol. 1 (Boston, 1900), 38–43.

stand; and the almost general prejudice and prepossession, which is so prevailent in the world against those of my complexion.

I suppose it is a truth too well attested to you to need a proof here, that we are a race of beings who have long laboured under the abuse and censure of the world, that we have long been looked upon with an eye of contempt, and that we have long been considered rather as brutish than human, and scarcely capable of mental endowments.

Sir, I hope I may safely admit, in consequence of that report which hath reached me, that you are a man far less inflexible in sentiments of this nature than many others, that you are measurably friendly and well disposed toward us, and that you are willing and ready to lend your aid and assistance to our relief from those many distresses and numerous calamities to which we are reduced.

Now, Sir, if this is founded in truth, I apprehend you will readily embrace every oppertunity to eradicate that train of absurd and false ideas and oppinions which so generally prevails with respect to us, and that your sentiments are concurrent with mine, which are that one universal Father hath given being to us all, and that he hath not only made us all of one flesh, but that he hath also without partiality afforded us all the same sensations, and endued us all with the same faculties, and that, however variable we may be in society or religion, however diversifyed in situation or colour, we are all of the same family, and stand in the same relation to him.

Sir, if these are sentiments of which you are fully persuaded, I hope you cannot but acknowledge that it is the indispensible duty of those who maintain for themselves the rights of human nature, and who profess the obligations of Christianity, to extend their power and influence to the relief of every part of the human race, from whatever burthen or oppression they may unjustly labour under, and this I apprehend, a full conviction of the truth and obligation of these principles should lead all to.

Sir, I have long been convinced that if your love for your selves and for those inesteemable laws which preserve to you the rights of human nature was founded on sincerity, you could not but be solicitous that every individual, of whatsoever rank or distinction, might with you equally enjoy the blessings thereof; neither could you rest satisfyed, short of the most active diffusion of your exertions, in order to their promotion from any state of degradation to which the unjustifyable cruelty and barbarism of men may have reduced them.

Sir, I freely and chearfully acknowledge that I am of the African race, and in that colour which is natural to them of the deepest dye;* and it is under a sense of the most profound gratitude to the Supreme Ruler of the universe that I now confess to you that I am not under that state of tyrannical thraldom and inhuman captivity to which too many of my brethren are doomed; but

*My father was brought here a slave from Africa.

that I have abundantly tasted of the fruition of those blessings which proceed from that free and unequalled liberty with which you are favoured, and which I hope you will willingly allow you have received from the immediate hand of that Being from whom proceedeth every good and perfect gift.

Sir, suffer me to recall to your mind that time in which the arms and tyranny of the British crown were exerted with every powerful effort in order to reduce you to a state of servitude; look back, I intreat you, on the variety of dangers to which you were exposed; reflect on that time in which every human aid appeared unavailable, and in which even hope and fortitude wore the aspect of inability to the conflict, and you cannot but be led to a serious and grateful sense of your miraculous and providential preservation; you cannot but acknowledge that the present freedom and tranquility which you enjoy you have mercifully received, and that it is the peculiar blessing of Heaven.

This, Sir, was a time in which you clearly saw into the injustice of a state of slavery, and in which you had just apprehensions of the horrors of its condition; it was now, Sir, that your abhorrence thereof was so excited that you publickly held forth this true and invaluable doctrine, which is worthy to be recorded and remember'd in all succeeding ages: "We hold these truths to be self-evident, that all men are created equal, and that they are endowed by their Creator with certain unalienable rights; that among these are life, liberty, and the pursuit of happyness."

Here, Sir, was a time in which your tender feelings for yourselves had engaged you thus to declare; you were then impressed with proper ideas of the great valuation of liberty, and the free possession of those blessings to which you were entitled by nature; but, Sir, how pitiable is it to reflect that, altho you were so fully convinced of the benevolence of the Father of mankind, and of his equal and impartial distribution of those rights and privileges which he had conferred upon them, that you should at the same time counteract his mercies in detaining by fraud and violence so numerous a part of my brethren under groaning captivity and cruel oppression; that you should at the same time be found guilty of that most criminal act, which you professedly detested in others, with respect to yourselves.

Sir, I suppose that your knowledge of the situation of my brethren is too extensive to need a recital here; neither shall I presume to prescribe methods by which they may be relieved, otherwise than by recommending to you and all others to wean yourselves from these narrow prejudices which you have imbibed with respect to them, and as Job proposed to his friends, "put your souls in their souls stead";[9] thus shall your hearts be enlarged with kindness and benevolence toward them, and thus shall you need neither the direction of myself or others in what manner to proceed herein.

And now, Sir, altho my sympathy and affection for my brethren hath caused my enlargement thus far, I ardently hope that your candour and generosity will plead with you in my behalf, when I make known to you that it

[9]Paraphrase of Job 16:4.

was not originally my design; but that having taken up my pen in order to direct to you, as a present, a copy of an almanack which I have calculated for the succeeding year, I was unexpectedly and unavoidably led thereto.

This calculation, Sir, is the production of my arduous study in this my advanced stage of life; for having long had unbounded desires to become acquainted with the secrets of nature, I have had to gratify my curiosity herein thro my own assiduous application to astronomical study, in which I need not recount to you the many difficulties and disadvantages which I have had to encounter.

And altho I had almost declined to make my calculation for the ensuing year, in consequence of that time which I had allotted therefor being taking up at the Federal Territory by the request of Mr. Andrew Ellicott, yet finding myself under several engagements to printers of this State, to whom I had communicated my design, on my return to my place of residence, I industriously apply'd myself thereto, which I hope I have accomplished with correctness and accuracy; a copy of which I have taken the liberty to direct to you, and which I humbly request you will favourably receive, and altho you may have the opportunity of perusing it after its publication, yet I chose to send it to you in manuscript previous thereto, that thereby you might not only have an earlier inspection, but that you might also view it in my own handwriting.

And now, Sir, I shall conclude, and subscribe myself, with the most profound respect.

THOMAS JEFFERSON TO BENJAMIN BANNEKER
Philadelphia, August 30, 1791 [10]

I thank you sincerely for your letter of the 19th instant and for the Almanac it contained. No body wishes more than I do to see such proofs as you exhibit, that nature has given to our black brethren, talents equal to those of the other colors of men, and that the appearance of a want of them is owing merely to the degraded condition of their existence, both in Africa & America. I can add with truth, that no body wishes more ardently to see a good system commenced for raising the condition both of their body & mind to what it ought to be, as fast as the imbecility of their present existence, and other circumstances which cannot be neglected, will admit. I have taken the liberty of sending your Almanac to Monsieur de Condorcet, Secretary of the Academy of Sciences at Paris, and member of the Philanthropic society, because I considered it as a document to which your whole colour had a right for their justification against the doubts which have been entertained of them.

---

[10]Peterson, ed., *Thomas Jefferson*, 982–83.

THOMAS JEFFERSON
Sixth Annual Presidential Message to Congress
Washington, D.C., December 2, 1806[11]

I congratulate you, fellow-citizens, on the approach of the period at which you may interpose your authority constitutionally, to withdraw the citizens of the United States from all further participation in those violations of human rights which have been so long continued on the unoffending inhabitants of Africa, and which the morality, the reputation, and the best interests of our country, have long been eager to proscribe. Although no law you may pass can take prohibitory effect till the first day of the year one thousand eight hundred and eight, yet the intervening period is not too long to prevent, by timely notice, expeditions which cannot be completed before that day.

THOMAS JEFFERSON TO MESSRS. THOMAS, ELLICOT, AND OTHERS
November 13, 1807[12]

Whatever may have been the circumstances which influenced our forefathers to permit the introduction of personal bondage into any part of these states, & to participate in the wrongs committed on an unoffending quarter of the globe, we may rejoice that such circumstances, & such a sense of them, exist no longer. It is honorable to the nation at large that their legislature availed themselves of the first practicable moment for arresting the progress of this great moral & political error: and I sincerely pray with you, my friends, that all the members of the human family may, in the time prescribed by the Father of us all, find themselves securely established in the enjoyments of life, liberty, & happiness.

"A Most Solemn Obligation"
Thomas Jefferson and Edward Coles

Twenty-eight-year-old Edward Coles, Jefferson's neighbor in Albemarle County, Va., served as secretary to President James Madison from 1809 to 1815. (Coles's elder brother Isaac had been Jefferson's secretary while he was president.) In this series of three letters, Coles requested that Jefferson champion the cause of emancipation. Jefferson, pleading old age, told Coles that the younger generation ought to lead the movement to eradicate slavery from Virginia.

---

[11]Peterson, ed., *Thomas Jefferson*, 528. In response to Jefferson's appeal, Congress on March 2, 1807, enacted a law prohibiting the African slave trade as of January 1, 1808. For this act, see Chapter 6.

[12]Jefferson Papers, Library of Congress.

Coles inherited a plantation and slaves when his father died in 1808. But the idea of owning human beings sickened him. Coles told Jefferson that he planned to abandon Virginia and move to the Northwest Territory where slavery was forbidden. Jefferson pleaded with Coles to remain in Virginia and become the "Missionary" of the doctrine of emancipation. After two exploratory trips, however, Coles left Virginia with seventeen slaves in 1819 and settled at Edwardsville, Illinois. While travelling down the Ohio River, Coles told his slaves for the first time that they were not only leaving Virginia behind them, but also slavery. Upon their arrival in Illinois, Coles executed formal deeds of emancipation and assisted his former slaves in the purchase of 160 acre farms.

Though nominally a free territory, many residents of Illinois participated in a de facto system of slavery in which blacks were held as indentured servants for extremely long terms. When their terms neared completion, blacks were often kidnapped and sold into slavery in the South. When Illinois became a state in 1818, Congress required the inhabitants to adopt a constitution that prohibited slavery, despite strong local support to the contrary.

In 1822, only three years after his arrival, Coles was elected governor of Illinois. Two years later he led the anti-slave forces in opposing the call of a state constitutional convention that would have legalized slavery. He urged the legislature to prohibit all semblances of slavery in Illinois.

EDWARD COLES TO THOMAS JEFFERSON
Washington, July 31, 1814[13]

I never took up my pen with more hesitation or felt more embarrassment than I now do in addressing you on the subject of this letter. The fear of appearing presumptuous distresses me, and would deter me from venturing thus to call your attention to a subject of such magnitude, and so beset with difficulties, as that of a general emancipation of the slaves of Virginia, had I not the highest opinion of your goodness and liberality, in not only excusing me for the liberty I take, but in justly appreciating my motives in doing so.

I will not enter on the right which man has to enslave his brother man, nor upon the moral and political effects of slavery on individuals or on society; because these things are better understood by you than by me. My object is to entreat and beseech you to exert your knowledge and influence in devising, and getting into operation, some plan for the gradual emancipation of slavery. This difficult task could be less exceptionably, and more successfully performed by the revered fathers of all our political and social blessings than by any succeeding statesmen; and would seem to come with peculiar propriety and force from those whose valor, wisdom, and virtue have done so much

[13]"The Jefferson Papers," *Collections* of the Massachusetts Historical Society, Seventh Series, Vol. I (Boston, 1900), 200–202.

in meliorating the condition of mankind. And it is a duty, as I conceive, that devolves particularly on you, from your known philosophical, and enlarged view of subjects, and from the principles you have professed and practiced through a long and useful life, pre-eminently distinguished, as well by being foremost in establishing on the broadest basis the rights of man, and the liberty and independence of your country, as in being throughout honored with the most important trusts by your fellow-citizens, whose confidence and love you have carried with you into the shades of old age and retirement. In the calm of this retirement you might, most beneficially to society, and with much addition to your own fame, avail yourself of that love and confidence to put into complete practice those hallowed principles contained in that renowned Declaration, of which you were the immortal author, and on which we bottomed our right to resist oppression and establish our freedom and independence.

I hope that the fear of failing, at this time, will have no influence in preventing you from employing your pen to eradicate this most degrading feature of British colonial policy, which is still permitted to exist, notwithstanding its repugnance as well to the principles of our revolution as to our free institutions. For however highly prized and influential your opinions may now be, they will be still much more so when you shall have been snatched from us by the course of nature. If, therefore, your attempt should now fail to rectify this unfortunate evil—an evil most injurious both to the oppressed and to the oppressor—at some future day when your memory will be consecrated by a grateful posterity, what influence, irresistible influence, will the opinions and writings of Thomas Jefferson have on all questions connected with the rights of man, and of that policy which will be the creed of your disciples. Permit me, then, my dear Sir, again to intreat you to exert your great powers of mind and influence, and to employ some of your present leisure in devising a mode to liberate one half of our fellowbeings from an ignominious bondage to the other, either by making an immediate attempt to put in train a plan to commence this goodly work, or to leave human nature the invaluable testament, which you are so capable of doing, how best to establish its rights; so that the weight of your opinion may be on the side of emancipation when that question shall be agitated,—and that it will be, sooner or later, is most certain. That it may be soon is my most ardent prayer; that it will be rests with you.

I will only add, as an excuse for the liberty I take in addressing you on this subject, which is so particularly interesting to me, that from the time I was capable of reflecting on the nature of political society, and of the rights appertaining to man, I have not only been principled against slavery, but have had feelings so repugnant to it as to decide me not to hold them; which decision has forced me to leave my native State, and with it all my relations and friends. This, I hope, will be deemed by you some excuse for the liberty of this intrusion, of which I gladly avail myself to assure you of the very great respect and esteem with which I am, my dear Sir, your very sincere and devoted friend.

Thomas Jefferson to Edward Coles
Monticello, August 25, 1814[14]

Your favor of July 31, was duly received, and was read with peculiar pleasure. The sentiments breathed thro the whole do honor to both the head and heart of the writer. Mine on the subject of the slavery of negroes have long since been in possession of the public, and time has only served to give them stronger root. The love of justice & the love of country plead equally the cause of these people, and it is a moral reproach to us that they should have pleaded it so long in vain, and should have produced not a single effort, nay I fear not much serious willingness to relieve them & ourselves from our present condition of moral and political reprobation. From those of the former generation who were in the fulness of age when I came into public life, which was while our controversy with England was on paper only, I soon saw that nothing was to be hoped. Nursed and educated in the daily habit of seeing the degraded condition, both bodily & mental, of those unfortunate beings, not reflecting that that degradation was very much the work of themselves & their fathers, few minds have yet doubted but that they were as legitimate subjects of property as their horses and cattle. The quiet & monotonous course of colonial life has been disturbed by no alarm, & little reflection on the value of liberty. And when alarm was taken at an enterprize on their own, it was not easy to carry them to the whole length of the principle which they invoked for themselves. In the first or second session of the Legislature after I became a member, I drew to this subject the attention of Colo. [Richard] Bland, one of the oldest, ablest, and most respected members, and he undertook to move for certain moderate extensions of the protection of the laws to these people. I seconded his motion, and, as a younger member, was more spared in the debate: but he was denounced as an enemy to his country, & was treated with the grossest indecorum. From an early stage of our revolution other and more distant duties were assigned to me, so that from that time till my return from Europe in 1789, and I may say till I returned to reside at home in 1809, I had little opportunity of knowing the progress of public sentiment here on this subject. I had always hoped that the younger generation receiving their early impressions after the flame of liberty had been kindled in every breast, and had become as it were the vital spirit of every American, that the generous temperament of youth, analogous to the motion of their blood, and above the suggestions of avarice, would have sympathized with oppression wherever found, and proved their love of liberty beyond their own share of it. But my

[14]Jefferson Papers, Library of Congress.

intercourse with them, since my return, has not been sufficient to ascertain that they had made towards this point the progress I had hoped. Your solitary but welcome voice is the first which has brought this sound to my ear; and I have considered the general silence which prevails on this subject as indicating an apathy unfavorable to every hope. Yet the hour of emancipation is advancing, in the march of time. It will come; and whether brought on by the generous energy of our own minds, or by the bloody process of St. Domingo, excited and conducted by the power of our present enemy [Great Britain], if once stationed permanently within our country, & offering asylum & arms to the oppressed, is a leaf of our history not yet turned over.

As to the method by which this difficult work is to be effected, if permitted to be done by ourselves, I have seen no proposition so expedient on the whole, as that of emancipation of those born after a given day, and of their education and expatriation after a proper age. This would give time for a gradual extinction of that species of labor and substitution of another, and lessen the severity of the shock which an operation so fundamental cannot fail to produce. The idea of emancipating the whole at once, the old as well as the young, and retaining them here; is of those only who have not the guide of either knowledge or experience of the subject. For, men, probably of any colour, but of this color we know, brought from their infancy without necessity for thought or forecast, are by their habits rendered as incapable as children of taking care of themselves, and are extinguished promptly wherever industry is necessary for raising young. In the mean time they are pests in society by their idleness, and the depredations to which this leads them. Their amalgamation with the other colour produces a degradation to which no lover of his country, no lover of excellence in the human character can innocently consent.

I am sensible of the partialities with which you have looked towards me as the person who should undertake this salutary but arduous work. But this, my dear Sir, is like bidding old Priam to buckle the armour of Hector "trementibus aevo humeris et inutile ferrumcingi." No, I have overlived the generation with which mutual labors and perils begat mutual confidence and influence. This enterprise is for the young; for those who can follow it up, and bear it through to its consummation. It shall have all my prayers, and these are the only weapons of an old man. But in the meantime are you right in abandoning this property, and your country with it? I think not. My opinion has ever been that, until more can be done for them, we should endeavor, with those whom fortune has thrown on our hands, to feed & clothe them well, protect them from ill usage, require such reasonable labor only as is performed voluntarily by freemen, and be led by no repugnancies to abdicate them, and our duties to them. The laws do not permit us to turn them loose, if that were for their good; and to commute them for other property is to commit them to those whose usage of them we cannot controul. I hope then, my dear Sir, you will reconcile yourself to your country and its unfortunate condition; that you will not lessen its stock of sound disposition by withdrawing your portion from the mass. That, on the contrary you will come forward in the public councils, become the Missionary of this doctrine truly Christian, insinuate & inculcate it softly but steadily thro' the medium of writing & conversation, associate others in your labors, and when the phalanx is formed, bring on & press the proposition perseveringly until its accomplishment. It is an encouraging observation that no good measure was ever proposed, which, if duly pursued, failed to prevail in the end. We have proof of this in the history of the endeavors in the British parliament to suppress that very trade which brought this evil on us. And you will be supported by the religious precept "be not weary in well-doing."[15] That your success may be as speedy and compleat, as it will be of honorable & immortal consolation to yourself, I shall as fervently & sincerely pray as I assure you of my great friendship and respect.

EDWARD COLES TO THOMAS JEFFERSON
Washington, September 26, 1814[16]

I must be permitted again to trouble you my dear Sir, to return my grateful thanks for the respectful and friendly attention shown to my letter in your answer on the 25th ulto. Your favorable reception of sentiments not generally avowed if felt by our Countrymen, but which have ever been so inseparably interwoven with my opinion and feelings as to become as it were the rudder that shapes my course even against a strong tide of interest and of local partialities, could not but be in the highest degree gratifying to me. And your interesting and highly prized letter, conveying them to me in such flattering terms, would have called forth my acknowledgements before this but for its having been forwarded to me to the Springs, and from thence it was again returned here before I received it, which was only a few days since.

Your indulgent treatment encourages me to add that I feel very sensibly the force of your remarks on the impropriety of yielding to my repugnancies in abandoning my property in Slaves and my native State. I certainly should never have been inclined to yield to them if I had supposed myself capable of being instrumental in bringing about a liberation, or that I could by my example meliorate the condition of these oppressed people. If I could be convinced of being in the slightest degree useful in doing either, it would afford me very great happiness, and the more so as it would enable me to gratify many partialities by remaining in Virginia. But never having flattered myself with the hope of being able to contribute to either, I have long since determined, and should, but for my bad health ere this, have removed, carry-

[15]II Thessalonians 3:13.
[16]Jefferson Papers, Library of Congress.

ing along with me those who had been my Slaves, to the country North West of the river Ohio.

Your prayers I trust will not only be heard with indulgence in Heaven, but with influence on earth. But I cannot agree with you that they are the only weapons of one at your age, nor that the difficult work of cleansing the escutchion of Virginia of the foul stain of slavery can best be done by the young. To effect so great and difficult an object great and extensive powers both of mind and influence are required, which can never be possessed in so great a degree by the young as by the old. And among the few of the former who might unite the disposition with the requisite capacity, they are too often led by ambitious views to go with the current of popular feeling, rather than to mark out a course for themselves, where they might be buffetted by the waves of opposition; and indeed it is feared these waves would in this case be too strong to be effectually resisted, by any but those who had gained by a previous course of useful employment the firmest footing in the confidence and attachment of their Country. It is with them, therefore, I am persuaded, that the subject of emancipation must originate; for they are the only persons who have it in their power effectually to arouse and enlighten the public sentiment, which in matters of this kind ought not to be expected to lead but to be led; nor ought it to be wondered at that there should prevail a degree of apathy with the general mass of mankind, where a mere passive principle of right has to contend against the weighty influence of habit and interest. On such a question there will always exist in society a kind of vis inertia to arouse and overcome, which require a strong impulse, which can only be given by those who have acquired a great weight of character, and on whom there devolves in this case a most solemn obligation. It was under these impressions that I looked to you, my dear Sir, as the first of our aged worthies, to awaken our fellow Citizens from their infatuation to a proper sense of Justice, and to the true interest of their Country, and by proposing a system for the gradual emancipation of our Slaves, at once to form a rallying point for its friends, who enlightened by your wisdom and experience, and supported and encouraged by your sanction and patronage, might look forward to a propitious and happy result. Your time of life I had not considered as an obstacle to the undertaking. Doctor Franklin, to whom, by the way, Pennsylvania owes her early riddance of the evils of Slavery, was as actively and as usefully employed on as arduous duties after he had passed your age as he had ever been at any period of his life.

With apologizing for having given you so much trouble on this subject, and again repeating my thanks for the respectful and flattering attention you have been pleased to pay to it, I renew the assurances of the great respect and regard which makes me most sincerely yours.

THOMAS JEFFERSON TO THOMAS COOPER
Monticello, September 10, 1814[17]

*Thomas Cooper, an English historian, had written Jefferson on August 17
suggesting that the two compare "the conditions of Great Britain and the
United States." Jefferson thought the exercise would be interesting, and in this letter
he compares the populations of the two countries.*

. . . Nor in the class of laborers do I mean to withhold from the comparison that portion whose color has condemned them, in certain parts of our
Union, to a subjection to the will of others. Even these are better fed in these
states, warmer clothed, & labor less than the journeymen or day laborers of
England. They have the comfort too of numerous families, in the midst of
whom they live, without want, or fear of it; a solace which few of the laborers
of England possess. They are subject, it is true, to bodily coercion; but are not
the hundreds of thousands of British soldiers & seamen subject to the same,
without seeing, at the end of their career, & when age & accident shall have
rendered them unequal to labor, the certainty, which the other has, that he
will never want? And has not the British seaman, as much as the African, been
reduced to this bondage by force, in flagrant violation of his own consent, and
of his natural right in his own person? and with the laborers of England generally, does not the moral coercion of want subject their will despotically to
that of their employer, as the physical constraint does the soldier, the seaman,
or the slave? But do not mistake me. I am not advocating slavery. I am not
justifying the wrongs we have committed on a foreign people, by the example
of another nation committing equal wrongs on their own subjects. On the
contrary there is nothing I would not sacrifice to a practicable plan of abolishing every vestige of this moral and political depravity. But I am at present
comparing the condition & degree of suffering to which oppression has reduced the man of one color, with the condition and degree of suffering to
which oppression has reduced the man of another color; equally condemning
both.

THOMAS JEFFERSON TO JOHN HOLMES
Monticello, April 22, 1820[18]

. . . But this momentous question [the Missouri Compromise], like a fire
bell in the night, awakened and filled me with terror. I considered it at once as
the knell of the Union. It is hushed, indeed, for the moment. But this is a
reprieve only, not a final sentence. A geographical line, coinciding with a

---

[17]Jefferson Papers, Library of Congress.
[18]Peterson, ed., *Thomas Jefferson*, 1434–35.

marked principle, moral and political, once conceived and held up to the angry passions of men, will never be obliterated; and every new irritation will mark it deeper and deeper. I can say, with conscious truth, that there is not a man on earth who would sacrifice more than I would to relieve us from this heavy reproach, in any *practicable* way. The cession of that kind of property, for so it is misnamed, is a bagatelle which would not cost me a second thought, if, in that way, a general emancipation and *expatriation* could be effected; and gradually, and with due sacrifices, I think it might be. But as it is, we have the wolf by the ear,[19] and we can neither hold him, nor safely let him go. Justice is in one scale, and self-preservation in the other. Of one thing I am certain, that as the passage of slaves from one State to another, would not make a slave of a single human being who would not be so without it, so their diffusion over a greater surface would make them individually happier, and proportionally facilitate the accomplishment of their emancipation, by dividing the burthen on a greater number of coadjutors. An abstinence too, from this act of power, would remove the jealousy excited by the undertaking of Congress to regulate the condition of the different descriptions of men composing a State. This certainly is the exclusive right of every State, which nothing in the constitution has taken from them and given to the General Government. Could Congress, for example, say, that the non-freemen of Connecticut shall be freemen, or that they shall not emigrate into any other State?

I regret that I am now to die in the belief, that the useless sacrifice of themselves by the generation of 1776, to acquire self-government and happiness to their country, is to be thrown away by the unwise and unworthy passions of their sons, and that my only consolation is to be, that I live not to weep over it. If they would but dispassionately weigh the blessings they will throw away, against an abstract principle more likely to be effected by union than by scisson, they would pause before they would perpetrate this act of suicide on themselves, and of treason against the hopes of the world.

THOMAS JEFFERSON TO JARED SPARKS
Monticello, February 4, 1824[20]

I duly received your favor of the 13th, and with it, the last number of the *North American Review*.[21] This has anticipated the one I should receive in course, but have not yet received, under my subscription to the new series. The article on the African colonization of the people of color, to which you invite my attention, I have read with great consideration. It is, indeed, a fine

---

[19]This metaphor has often been incorrectly cited as "we have the wolf by the ears." The manuscript, however, is clear that the word is singular.

[20]Peterson, ed., *Thomas Jefferson*, 1484–87.

[21]Sparks had sent Jefferson the January 1824 issue of the Boston *North American Review*, which contained the sixth annual report of the American Colonization Society.

one, and will do much good. I learn from it more, too, than I had before known, of the degree of success and promise of that colony.

In the disposition of these unfortunate people, there are two rational objects to be distinctly kept in view. First. The establishment of a colony on the coast of Africa, which may introduce among the aborigines the arts of cultivated life, and the blessings of civilization and science. By doing this, we may make to them some retribution for the long course of injuries we have been committing on their population. And considering that these blessings will descend to the "*nati natorum, et qui nascentur ab illis,*" we shall in the long run have rendered them perhaps more good than evil. To fulfil this object, the colony of Sierra Leone promises well, and that of Mesurado adds to our prospect of success. Under this view, the colonization society is to be considered as a missionary society, having in view, however, objects more humane, more justifiable, and less aggressive on the peace of other nations, than the others of that appellation.

The second object, and the most interesting to us, as coming home to our physical and moral characters, to our happiness and safety, is to provide an asylum to which we can, by degrees, send the whole of that population from among us, and establish them under our patronage and protection, as a separate, free and independent people, in some country and climate friendly to human life and happiness. That any place on the coast of Africa should answer the latter purpose, I have ever deemed entirely impossible. And without repeating the other arguments which have been urged by others, I will appeal to figures only, which admit no controversy. I shall speak in round numbers, not absolutely accurate, yet not so wide from truth as to vary the result materially. There are in the United States a million and a half of people of color in slavery. To send off the whole of these at once, nobody conceives to be practicable for us, or expedient for them. Let us take twenty-five years for its accomplishment, within which time they will be doubled. Their estimated value as property, in the first place, (for actual property has been lawfully vested in that form, and who can lawfully take it from the possessors?) at an average of two hundred dollars each, young and old, would amount to six hundred millions of dollars, which must be paid or lost by somebody. To this, add the cost of their transportation by land and sea to Mesurado, a year's provision of food and clothing, implements of husbandry and of their trades, which will amount to three hundred millions more, making thirty-six millions of dollars a year for twenty-five years, with insurance of peace all that time, and it is impossible to look at the question a second time. I am aware that at the end of about sixteen years, a gradual detraction from this sum will commence, from the gradual diminution of breeders, and go on during the remaining nine years. Calculate this deduction, and it is still impossible to look at the enterprise a second time. I do not say this to induce an inference that the getting rid of them is forever impossible. For that is neither my opinion nor my hope. But only that it cannot be done in this way. There is, I think, a way in which it can be done; that is, by emancipating the after-born, leaving them, on due compensation, with their mothers, until their services are worth their maintenance, and then putting them to industrious occupations, until a proper age for deportation. This was the result of my reflections on the subject five and forty years ago, and I have never yet been able to conceive any other practicable plan. It was sketched in the Notes on Virginia, under the fourteenth query. The estimated value of the new-born infant is so low, (say twelve dollars and fifty cents,) that it would probably be yielded by the owner gratis, and would thus reduce the six hundred millions of dollars, the first head of expense, to thirty-seven millions and a half; leaving only the expense of nourishment while with the mother, and of transportation. And from what fund are these expenses to be furnished? Why not from that of the lands which have been ceded by the very States now needing this relief? And ceded on no consideration, for the most part, but that of the general good of the whole. These cessions already constitute one fourth of the States of the Union. It may be said that these lands have been sold; are now the property of the citizens composing those States; and the money long ago received and expended. But an equivalent of lands in the territories since acquired, may be appropriated to that object, or so much, at least, as may be sufficient; and the object, although more important to the slave States, is highly so to the others also, if they were serious in their arguments on the Missouri question. The slave States, too, if more interested, would also contribute more by their gratuitous liberation, thus taking on themselves alone the first and heaviest item of expense.

In the plan sketched in the Notes on Virginia, no particular place of asylum was specified; because it was thought possible, that in the revolutionary state of America, then commenced, events might open to us some one within practicable distance. This has now happened. St. Domingo has become independent, and with a population of that color only; and if the public papers are to be credited, their Chief offers to pay their passage, to receive them as free citizens, and to provide them employment. This leaves, then, for the general confederacy, no expense but of nurture with the mother a few years, and would call, of course, for a very moderate appropriation of the vacant lands. Suppose the whole annual increase to be of sixty thousand effective births, fifty vessels, of four hundred tons burthen each, constantly employed in that short run, would carry off the increase of every year, and the old stock would die off in the ordinary course of nature, lessening from the commencement until its final disappearance. In this way no violation of private right is proposed. Voluntary surrenders would probably come in as fast as the means to be provided for their care would be competent to it. Looking at my own State only, and I presume not to speak for the others, I verily believe that this surrender of property would not amount to more, annually, than half our present direct taxes, to be continued fully about twenty or twenty-five years, and then gradually diminishing for as many more until their final extinction; and even this half tax would not be paid in cash, but by the delivery of an object which they have never yet known or counted as part of their property; and those not possessing the object will be called on for nothing. I do not go into all the details of the burthens and benefits of this operation. And who could estimate its blessed effects? I leave this to those who will live to see their accomplishment, and to enjoy a beatitude forbidden to my age. But I leave it with this admonition, to rise and be doing. A million and a half are within their control; but six millions, (which a majority of those now living will see them attain,) and one million of these fighting men, will say, "we will not go."

I am aware that this subject involves some constitutional scruples. But a liberal construction, justified by the object, may go far, and an amendment of the constitution, the whole length necessary. The separation of infants from their mothers, too, would produce some scruples of humanity. But this would be straining at a gnat, and swallowing a camel.

THOMAS JEFFERSON TO JAMES HEATON
Monticello, May 20, 1826[22]

The subject of your letter of April 20, is one on which I do not permit myself to express an opinion, but when time, place, and occasion may give it some favorable effect. A good cause is often injured more by ill-timed efforts of its friends than by the arguments of its enemies. Persuasion, perseverance, and patience are the best advocates on questions depending on the will of others. The revolution in public opinion which this cause requires, is not to be expected in a day, or perhaps in an age; but time, which outlives all things, will outlive this evil also. My sentiments have been forty years before the public. Had I repeated them forty times, they would only have become the more stale and threadbare. Although I shall not live to see them consummated, they will not die with me; but living or dying, they will ever be in my most fervent prayer. This is written for yourself and not for the public, in compliance with your request of two lines of sentiment on the subject.

[22]Peterson, ed., *Thomas Jefferson*, 1516.

James Madison and Slavery

James Madison to His Father
Philadelphia, September 8, 1783[23]

Honored Sir: Mr. Jones & myself being here transacting some private business which brought us from Princeton the end of last week, I here received your letter of the 22d. ulto. The favorable turn of my Mother's state of health is a source of great satisfaction to me, and will render any delay in my setting out for Virginia the less irksome to me. I shall return to Princeton tomorrow; my final leaving of which will depend on events, but can not now be at any very great distance. On a view of all circumstances I have judged it most prudent not to force Billey[24] back to Virginia even if [it] could be done; and have accordingly taken measures for his final separation from me. I am persuaded his mind is too thoroughly tainted to be a fit companion for fellow slaves in Virginia. The laws here do not admit of his being sold for more than 7 years. I do not expect to get near the worth of him; but cannot think of punishing him by transportation merely for coveting that liberty for which we have paid the price of so much blood, and have proclaimed so often to be the right, & worthy the pursuit, of every human being.

James Madison to Edmund Randolph
Orange, July 26, 1785[25]

I keep up my attention as far as I can command my time, to the course of reading which I have of late pursued & shall continue to do so. I am however far from being determined ever to make a professional use of it. My wish is if possible to provide a decent & independent subsistence, without encountering the difficulties which I foresee in that line. Another of my wishes is to depend as little as possible on the labour of slaves.

---

[23]William T. Hutchinson and William M. E. Rachal, eds., *The Papers of James Madison* (Chicago, 1971), VII, 304. At this time, Madison was serving as a Virginia delegate to Congress, which was then meeting in Princeton, N.J.

[24]At his birth in 1759, Billey was deeded in trust to James Madison, then a minor, by his grandmother, Rebecca Catlett Conway Moore. It is uncertain what happened to Billey, but it is thought that he was sold as an indentured servant in Philadelphia. Several years later, Madison felt that a runaway slave from his plantation might try to reach Billey in Philadelphia.

[25]Robert A. Rutland and William M. E. Rachal, eds., *The Papers of James Madison* (Chicago, 1973), VIII, 328.

JAMES MADISON
Memorandum on an African Colony for Freed Slaves
ca. October 20, 1789[26]

Without enquiring into the practicability or the most proper means of establishing a Settlement of freed blacks on the Coast of Africa, it may be remarked as one motive to the benevolent experiment that if such an asylum was provided, it might prove a great encouragement to manumission in the Southern parts of the U.S. and even afford the best hope yet presented of putting an end to the slavery in which not less than 600,000 unhappy negroes are now involved.

In all the Southern States of N. America, the laws permit masters, under certain precautions to manumit their slaves. But the continuance of such a permission in some of the States is rendered precarious by the ill effects suffered from freedmen who retain the vices and habits of slaves. The same consideration becomes an objection with many humane masters against an exertion of their legal right of freeing their slaves. It is found in fact that neither the good of the Society, nor the happiness of the individuals restored to freedom is promoted by such a change in their condition.

In order to render this change eligible as well to the Society as to the Slaves, it would be necessary that a compleat incorporation of the latter into the former should result from the act of manumission. This is rendered impossible by the prejudices of the Whites, prejudices which proceeding principally from the difference of colour must be considered as permanent and insuperable.

It only remains then that some proper external receptacle be provided for the slaves who obtain their liberty. The interior wilderness of America, and the Coast of Africa seem to present the most obvious alternative. The former is liable to great if not invincible objections. If the settlement were attempted at a considerable distance from the White frontier, it would be destroyed by the Savages who have a peculiar antipathy to the blacks: If the attempt were made in the neighbourhood of the White Settlements, peace would not long be expected to remain between Societies, distinguished by such characteristic marks, and retaining the feelings inspired by their former relation of oppressors & oppressed. The result then is that an experiment for providing such an external establishment for the blacks as might induce the humanity of Masters, and by degrees both the humanity & policy of the Governments, to forward the abolition of slavery in America, ought to be pursued on the Coast of Africa or in some other foreign situation.

[26]Charles F. Hobson and Robert A. Rutland, eds., *The Papers of James Madison* (Charlottesville, 1979), XII, 437–38.

ROBERT PLEASANTS TO JAMES MADISON
Virginia, June 6, 1791[27]

Being appointed by a late Meeting of the Humane, or Abolition society in this State, to procure a presentment of their Memorial to Congress, on the Subject of the Slave trade; and believing thou art a friend to general liberty, I conclude it will not be disagreeable to thee to endeavour to prevent, as far as may be in thy power, the Cruelties, and inhuman treatment, which, to the disgrace of professed Christians, hath been long practised towards the Inhabitants of Africa; I request therefore thou wilt inform, whether or not, thou would be willing to present the said Memorial to the Next Session of Congress; which hath for its object nothing more than what Congress have Resolved they have power to do.[28]

And Whilst I am mentioning the subject of the Slave-trade, perhaps it may not be improper to intimate, a strong desire I have of seeing some plan for a gradual abolition of Slavery promoted in this State, which appears to me, to be both a Moral and political Evil, that loudly Calls for redress. I am not insensible of the difficulties of abolishing old habits, especially when supported by emaginary Interest, yet knowing the Sentiments of divers Slave-holders, who are favourable to the design, I wish to have thy judgment on the propriety of a Petition to our Assembly, for a law declaring the Children of Slaves to be born after the passing such Act, to be free at the usual Ages of Eighteen and twenty one Years; and to enjoy such priviledges as may be consistent with justice and sound policy. This I conceive would be less liable to objections from interested Motives, than any other mode I have heard mentioned, or that occurred to me; and could it be effected, would gradually abolish an Evil of great Magnitude. And seeing we live in an enlightened age, when liberty is allowed to be the unalienable right of all mankind, it surely behooves us of the present generation, and more especially the Legislature, to endeavour to restore one of the most valuable blessings of life, to an injured and unhappy race of people, who we are taught to believe are of the same Original with our Selves, and equally the care of a Merciful Creator, that requires, we should do to others, as we would they should do to us. Thou art not insensible how hardly many of these poor Creatures are treated, and in general left without mental improvement; would it not then be Noble in thee, and others, who fill the first stations, and are favored with abilities & influence, to espouse the cause of the injured, the ignorant, and the helpless;

---

[27]Robert A. Rutland and Thomas A. Mason, eds., *The Papers of James Madison* (Charlottesville, 1983), XIV, 30–31. Pleasants was a prominent Henrico County Quaker merchant. He manumitted his slaves when they became adults.

[28]After a long, heated debate over petitions to abolish the foreign slave trade, the House of Representatives passed several resolutions one of which provided that Congress had authority to regulate the foreign slave trade to provide slaves with "humane treatment, during their passage." (For the debate in the House of Representatives and its resolutions, see Chapter 6.)

and become instrumental in promoting the Glorious time spoken of by the Prophet, when, "Righteousness shall cover the Earth, as the Water cover the Sea." Whereby the great end of Creation may be accomplished by each individual, and finally receive the Answer of, "Well done good and faithful servant &c."[29] Which is the sincere desire of one, who remains with Respect Thy Assured Friend.

JAMES MADISON TO ROBERT PLEASANTS
Philadelphia, October 30, 1791[30]

The petition relating to the Militia bill contains nothing that makes it improper for me to present it. I shall, therefore, readily comply with your desire on that subject.[31] I am not satisfied that I am equally at liberty with respect to the other petition. Animadversions, such as it contains and which the authorized object of the petitioners did not require, on the slavery existing in our country, are supposed by the holders of that species of property to lessen the value by weakening the tenure of it. Those from whom I derive my public station are known by me to be greatly interested in that species of property, and to view the matter in that light. It would seem that I might be chargeable at least with want of candour, if not of fidelity, were I to make use of a situation in which their confidence has placed me, to become a volunteer in giving a public wound, as they would deem it, to an interest on which they set so great a value. I am the less inclined to disregard this scruple, as I am not sensible that the event of the petition would in the last depend on the circumstance of its being laid before the House by this or that person.

Such an application as that to our own Assembly, on which you ask my opinion, is a subject in various respects, of great delicacy and importance. The consequences of every sort ought to be well weighed by those who would hazard it. From the view under which they present themselves to me, I can not but consider the application as likely to do harm rather than good. It may be worth your own consideration whether it might not produce successful attempts to withdraw[32] the privilege now allowed to individuals, of giving freedom to slaves. It would at least be likely to clog it with a condition that the persons freed should be removed from the Country [i.e., Virginia]; there

---

[29]Matthew 25:23.

[30]Robert A. Rutland and Thomas A. Mason, eds., *The Papers of James Madison*, XIV, 91–92.

[31]In a postscript to his June 6th letter, Pleasants asked Madison to submit a Quaker petition opposing Congress' proposed militia bill that had been printed for the public's consideration. The bill provided for full enlistment of "each and every able-bodied white male citizen" in each state's militia. No exemptions were made for conscientious objectors on religious grounds.

[32]At a later date, Madison put an asterisk after the words "withdraw" and "condition" (in the next sentence) and wrote in the margin: "it so happened." The increasing opposition to manumissions culminated in 1806 when the Virginia legislature passed an act requiring that freed slaves leave the state within twelve months.

being arguments of great force for such a regulation, and some would concur in it who in general, disapprove of the institution of slavery.

James Madison to Robert J. Evans
Montpellier, Va., June 15, 1819[33]

I have received your letter of the 3d instant, requesting such hints as may have occurred to me on the subject of an eventual extinguishment of slavery in the U.S.

Not doubting the purity of your views, and relying on the discretion by which they will be regulated, I cannot refuse such a compliance as will at least manifest my respect for the object of your undertaking.

A general emancipation of slaves ought to be 1. gradual. 2. equitable & satisfactory to the individuals immediately concerned. 3. consistent with the existing & durable prejudices of the nation.

That it ought, like remedies for other deep-rooted and widespread evils, to be gradual, is so obvious that there seems to be no difference of opinion on that point.

To be equitable & satisfactory, the consent of both the Master & the slave should be obtained. That of the Master will require a provision in the plan for compensating a loss of what he has held as property, guaranteed by the laws, and recognized by the constitution. That of the slave, requires that his condition in a state of freedom be preferable in his own estimation, to his actual one in a state of bondage.

To be consistent with existing and probably unalterable prejudices in the U.S., the freed blacks ought to be permanently removed beyond the region occupied by or allotted to a white population. The objections to a thorough incorporation of the two people, are with most of the Whites insuperable; and are admitted by all of them to be very powerful. If the blacks, strongly marked as they are by physical & lasting peculiarities, be retained amid the Whites, under the degrading privation of equal rights, political or social, they must be always dissatisfied with their condition as a change only from one to another species of oppression; always secretly confederated against the ruling & privileged class; and always uncontroulled by some of the most cogent motives to moral and respectable conduct. The character of the freed blacks, even where their legal condition is least affected by their colour, seems to put these truths beyond question. It is material also that the removal of the blacks to a distance precluding the jealousies & hostilities to be apprehended from a neighboring people stimulated by the contempt known to be entertained for their peculiar features; to say nothing of their vindictive recollections, or the predatory propensities which their state of society might foster. Nor is

[33]Madison Papers, Library of Congress.

it fair, in estimating the danger of collisions with the Whites, to charge it wholly on the side of the Blacks. There would be reciprocal antipathies doubling the danger.

The colonizing plan on foot, has as far as it extends, a due regard to these requisites; with the additional object of bestowing new blessings civil & religious on the quarter of the Globe most in need of them. The [American Colonization] Society proposes to transport to the African Coast, all free & freed blacks who may be willing to remove thither; to provide by fair means, & it is understood with a prospect of success, a suitable territory for their reception; and to initiate them into such an establishment as may gradually and indefinitely expand itself.

The experiment under this view of it, merits encouragement from all who regard slavery as an evil, who wish to see it diminished and abolished by peaceable & just means; and who have themselves no better mode to propose. Those who have most doubted the success of the experiment must at least have wished to find themselves in an error.

But the views of the Society are limited to the case of blacks already free, or who may be *gratuitously* emancipated. To provide a commensurate remedy for the evil, the plan must be extended to the great mass of blacks, and must embrace a fund sufficient to induce the Master as well as the slave to concur in it. Without the concurrence of the Master, the benefit will be very limited as it relates to the Negroes; and essentially defective as it relates to the U. States; and the concurrence of Masters, must for the most part, be obtained by purchase.

Can it be hoped that voluntary contributions, however adequate to an auspicious commencement, will supply the sums necessary to such an enlargement of the remedy? May not another question be asked? Would it be reasonable to throw so great a burden on the individuals distinguished by their philanthropy and patriotism?

The object to be obtained, as an object of humanity, appeals alike to all; as a national object, it claims the interposition of the nation. It is the nation which is to reap the benefit. The nation therefore ought to bear the burden.

Must then the enormous sums required to pay for, to transport, and to establish in a foreign land all the slaves in the U.S. as their masters may be willing to part with them, be taxed on the good people of the U.S. or be obtained by loans swelling the public debt to a size pregnant with evils next in degree to those of slavery itself?

Happily it is not necessary to answer this question by remarking that if slavery as a national evil is to be abolished, and it be just that it be done at the national expence, the amount of the expence is not a paramount consideration. It is the peculiar fortune, or rather a providential blessing of the U.S. to possess a resource commensurate to this great object, without taxes on the people, or even an increase of the Public debt.

I allude to the vacant territory the extent of which is so vast, and the vendible value of which is so well ascertained.

Supposing the number of slaves to be 1,500,000, and their price to average 400 dollars, the cost of the whole would be 600 millions of dollars. These estimates are probably beyond the fact; and from the number of slaves should be deducted 1. Those whom their Masters would not part with. 2. Those who may be gratuitously set free by their Masters. 3. Those acquiring freedom under emancipating regulations of the States. 4. Those preferring slavery where they are to freedom in an African settlement. On the other hand, it is to be noted that the expence of removal & settlement is not included in the estimated sum; and that an increase of the slaves will be going on during the period required for the execution of the plan.

On the whole, the aggregate sum needed may be stated at about 600 millions of dollars.

This will require 200 millions of acres, at 3 dollars per acre; or 300 millions at 2 dollars per acre; a quantity which, though great in itself, is perhaps not a third part of the disposable territory belonging to the U.S. And to what object so good, so great & so glorious, could that peculiar fund of wealth be appropriated? Whilst the sale of territory would, on one hand, be planting one desert with a free & civilized people, it would, on the other, be giving freedom to another people, and filling with them another desert. And, if in any instances wrong has been done by our forefathers to people of one colour in dispossessing them of their soil, what better atonement is now in our power than that of making what is rightfully acquired a source of justice & of blessings to a people of another colour?

As the revolution to be produced in the condition of the negroes must be gradual, it will suffice if the sale of territory keep pace with its progress. For a time at least the proceeds would be in advance. In this case it might be best, after deducting the expence incident to the surveys & sales, to place the surplus in a situation where its increase might correspond with the natural increase of the unpurchased slaves. Should the proceeds at any time fall short of the calls for their application, anticipations might be made by temporary loans to be discharged as the lands should find a market.

But it is probable that for a considerable period the sales would exceed the calls. Masters would not be willing to strip their plantations & farms of their laborers too rapidly. The slaves themselves connected, as they generally are by tender ties with others under other Masters, would be kept from the list of emigrants by the want of the multiplied consents to be obtained. It is probable indeed that for a long time a certain portion of the proceeds might safely continue applicable to the discharge of the debts or to other purposes of the nation, or it might be most convenient, in the outset, to appropriate a certain proportion only of the income from sales, to the object in view, leaving the residue otherwise applicable.

Should any plan similar to that here sketched be deemed eligible in itself, no particular difficulty is foreseen from that portion of the nation which with a common interest in the vacant territory has no interest in slave property. They are too just to wish that a partial sacrifice should be made for the general good; and too well aware that whatever may be the intrinsic character of that description of property, it is one known to the constitution, and as such could not be constitutionally taken away without just compensation. That part of the nation has indeed shewn a meritorious alacrity in promoting by pecuniary contributions the limited scheme for colonizing the Blacks, & freeing the nation from the unfortunate stain on it, which justifies the belief that any enlargement of the scheme, if founded on just principles, would find among them its earliest & warmest patrons.

It is evident however that in effectuating a general emancipation of slaves in the mode which has been hinted, difficulties of other sorts would be encountered. The provision for ascertaining the joint consent of the Masters & slaves; for guarding against unreasonable valuations of the latter; and for the discrimination of those not proper to be conveyed to a foreign residence, or who ought to remain a charge on masters in whose service they had been disabled or worn out, and for the annual transportation of such numbers, would require the mature deliberations of the national Councils. The measure implies also the practicability of procuring in Africa an enlargement of the district or districts, for receiving the exiles sufficient for so great an augmentation of their numbers.

Perhaps the Legislative provision best adapted to the case would be an incorporation of the Colonizing Society, or the establishment of a similar one, with proper powers, under the appointment & superintendence of the National Executive.

In estimating the difficulties however, incident to any plan of general emancipation, they ought to be brought into comparison with those inseparable from other plans, and be yielded to or not accordingly to the result of the comparison.

One difficulty presents itself which will probably attend every plan which is to go into effect under the Legislative provisions of the National Government. But whatever may be the defect of the existing powers of Congress, the Constitution has pointed out the way in which it can be supplied.[34] And it can hardly be doubted that the requisite power might readily be procured for attaining the great object in question, in any mode whatever approved by the nation.

If these thoughts can be of any aid in your search of a remedy for the great evil under which the nation labors, you are very welcome to them. You will allow me however to add that it will be more agreeable to me not to be publickly referred to in any use you may make of them.

---

[34]Madison is referring to an amendment to the Constitution.

George Washington and Slavery

GEORGE WASHINGTON TO ROBERT MORRIS
Mount Vernon, April 12, 1786[35]

I give you the trouble of this letter at the instance of Mr. Dalby of Alexandria; who is called to Philadelphia to attend what he conceives to be a vexatious lawsuit respecting a slave of his, which a Society of Quakers in the city (formed for such purposes) have attempted to liberate; The merits of this case will no doubt appear upon trial, but from Mr. Dalby's state of the matter, it should seem that this Society is not only acting repugnant to justice so far as its conduct concerns strangers, but, in my opinion extremely impolitickly with respect to the State, the City in particular; and without being able, (but by acts of tyranny and oppression) to accomplish their own ends. He says the conduct of this society is not sanctioned by Law: had the case been otherwise, whatever my opinion of the Law might have been, my respect for the policy of the State would on this occasion have appeared in my silence; because against the penalties of promulgated Laws one may guard; but there is no avoiding the snares of individuals, or of private societies. And if the practice of this Society of which Mr. Dalby speaks, is not discountenanced, none of those whose *misfortune* it is to have slaves as attendants, will visit the City if they can possibly avoid it; because by so doing they hazard their property; or they must be at the expence (and this will not always succeed) of providing servants of another description for the trip.

I hope it will not be conceived from these observations, that it is my wish to hold the unhappy people, who are the subject of this letter, in slavery. I can only say that there is not a man living who wishes more sincerely than I do, to see a plan adopted for the abolition of it; but there is only one proper and effectual mode by which it can be accomplished, and that is by Legislative authority; and this, as far as my suffrage will go, shall never be wanting. But when slaves who are happy and contented with their present masters, are tampered with and seduced to leave them; when masters are taken unawares by these practices; when a conduct of this sort begets discontent on one side and resentment on the other, and when it happens to fall on a man, whose purse will not measure with that of the Society, and he loses his property for want of means to defend it; it is oppression in the latter case, and not humanity in any; because it introduces more evils than it can cure.

I will make no apology for writing to you on this subject; for if Mr. Dalby has not misconceived the matter, an evil exists which requires a remedy; if he has, my intentions have been good, though I may have been too precipitate in this address.

---

[35]W. B. Allen, ed., *George Washington: A Collection* (Indianapolis, 1988), 318–19. Morris, perhaps the wealthiest and most powerful merchant in America, had been Superintendent of Finance during the Revolution. At this time he was a leader of the Pennsylvania Republican party, the forerunner of the Federalists.

GEORGE WASHINGTON TO JOHN FRANCIS MERCER
Mount Vernon, September 9, 1786[36]

I never mean (unless some particular circumstance should compel me to it) to possess another slave by purchase; it being among my first wishes to see some plan adopted, by which slavery in this country may be abolished by slow, sure, and imperceptible degrees.

GEORGE WASHINGTON
Undated Memorandum[37]

The unfortunate condition of the persons, whose labour in part I employed, has been the only unavoidable subject of regret. To make the Adults among them as easy & as comfortable in their circumstances as their actual state of ignorance & improvidence would admit; & to lay a foundation to prepare the rising generation for a destiny different from that in which they were born;[38] afforded some satisfaction to my mind, & could not I hoped be displeasing to the justice of the Creator.

GEORGE WASHINGTON
Last Will and Testament, July 9, 1799[39]

. . . Upon the decease of my wife, it is my Will and desire that all the Slaves which I hold in my own right, shall receive their freedom To emancipate them during her life, would, tho' earnestly wished by me, be attended with such insuperable difficulties on account of their Intermixture by Marriages with the Dower Negroes,[40] as to excite the most painful sensations, if not disagreeable consequences from the latter, while both descriptions are in the occupancy of the same Proprietor; it not being in my power, under the tenure by which the Dower Negroes are held, to manumit them. And whereas among those who will receive freedom according to this devise, there may be some, who from old age or bodily infirmities, and others who on account of their infancy, that will be unable to support themselves; it is my Will and desire that all who come under the first and second description shall be comfortably cloathed and fed by my heirs while they live; and that such of the

[36]John C. Fitzpatrick, ed., *The Writings of George Washington* . . . (39 vols., Washington, D.C., 1931–1944), XXIX, 5.
[37]Washington Papers, Rosenbach Library.
[38]According to his will (printed below), Washington tried to prepare slave children for their free status by teaching them to read and write and instructing them in "some useful occupation, agreeable to the Laws of the Commonwealth of Virginia."
[39]W. B. Allen, ed., *George Washington*, 667–68.
[40]A reference to the slaves which Martha Washington received from the estate of her first husband.

later description as have no parents living, or if living are unable, or unwilling to provide for them, shall be bound by the Court until they shall arrive at the age of twenty-five years; and in cases where no record can be produced, whereby their ages can be ascertained, the judgment of the Court upon its own view of the subject, shall be adequate and final. The Negroes thus bound, are (by their Masters or Mistresses) to be taught to read and write; and to be brought up to some useful occupation, agreeable to the Laws of the Commonwealth of Virginia, providing for the support of Orphan and other poor Children. And I do hereby expressly forbid the Sale, or transportation out of said Commonwealth, of any Slave I may die possessed of, under any pretence whatsoever. And I do moreover most pointedly, and most solemnly enjoin it upon my Executors hereafter named, or the Survivors of them, to see that this clause respecting Slaves, and every part thereof be religiously fulfilled at the Epoch at which it is directed to take place; without evation, neglect or delay, after the Crops which may then be on the ground are harvested, particularly as it respects the aged and infirm; Seeing that a regular and permanent fund be established for their Support so long as there are subjects requiring it; not trusting to the uncertain provision to be made by individuals. And to my Mulatto man William (calling himself William Lee) I give immediate freedom; or if he should prefer it (on account of the accidents which have befallen him, and which have rendered him incapable of walking or of any active employment) to remain in the situation he now is, it shall be optional in him to do so: In either case however, I allow him an annuity of thirty dollars during his natural life, which shall be independent of the victuals and cloaths he has been accustomed to receive, if he chuses the last alternative; but in full, with his freedom, if he prefers the first; and this I give him as a testimony of my sense of his attachment to me, and for his faithful services during the Revolutionary War.

Selected Bibliography

Aptheker, Herbert, ed. *A Documentary History of the Negro People in the United States*. New York, 1951.

Bailyn, Bernard. *The Ideological Origins of the American Revolution*. Cambridge, Mass., 1967.

Bergman, Peter M. and Jean McCarroll, comps. *The Negro in the Continental Congress*. New York, 1969.

Berlin, Ira. "The Revolution in Black Life," in Alfred F. Young, ed., *The American Revolution: Explorations in the History of American Radicalism*. DeKalb, Ill., 1976, pp. 349–82.

Berlin, Ira and Ronald Hoffman, eds. *Slavery and Freedom in the Age of the American Revolution*. Charlottesville, Va., 1981.

Brackett, Jeffery R. "The Status of the Slave, 1775–1789," in John Franklin Jameson, ed., *Essays in the Constitutional History of the United States in the Formative Period, 1775–1789*. Boston, 1889, pp. 263–311.

Bruns, Roger, ed. *Am I Not a Man and a Brother: The Antislavery Crusade of Revolutionary America, 1688–1788*. New York, 1977.

Butterfield, L. H., ed. *Letters of Benjamin Rush*. 2 vols. Princeton, 1951.

Cohen, William, "Thomas Jefferson and the Problem of Slavery," *Journal of American History*, 56 (1969), 503–26.

Coughtry, Jay. *The Notorious Triangle: Rhode Island and the African Slave Trade, 1700–1807*. Philadelphia, 1981.

Cushing, John D. "The Cushing Court and the Abolition of Slavery in Massachusetts: More Notes on the 'Quock Walker Case,'" *American Journal of Legal History*, V (1961), 118–44.

Davis, David Brion. *Was Thomas Jefferson an Authentic Enemy of Slavery?* Oxford, 1970.

———. *The Problem of Slavery in the Age of Revolution, 1770–1783.* Ithaca, N.Y., 1975.

diGiacomantonio, William. "'For the Gratification of a volunteering society': Antislavery and Pressure Group Politics in the First Federal Congress," *Journal of the Early Republic*, forthcoming.

Donnan, Elizabeth, ed. *Documents Illustrative of the History of the Slave Trade to America.* 4 vols. Washington, D.C., 1930–35.

———. "The New England Slave Trade After the Revolution," *New England Quarterly*, 3 (1930), 251–78.

Drake, Thomas E. *Quakers and Slavery in America.* New Haven, Conn., 1950.

DuBois, W. E. B. *The Suppression of the African Slave-Trade to the United States of America, 1638–1870.* New York, 1896.

Dumond, Dwight L. *Antislavery: The Crusade for Freedom in America.* Ann Arbor, Mich., 1961.

Farrand, Max, ed. *Records of the Federal Convention of 1787.* 4 vols. New Haven, Conn. 1937.

Finkelman, Paul. "Slavery and the Constitutional Convention: Making a Covenant with Death," in Richard Beeman, Stephen Botein, and Edward C. Carter II, eds., *Beyond Confederation: Origins of the Constitution and American National Identity.* Chapel Hill, N.C., 1987, pp. 188–225.

Franklin, John Hope. *From Slavery to Freedom.* New York, 1947.

Freehling, William W. "The Founding Fathers and Slavery," *American Historical Review*, 77 (1972), 81–93.

———. "Conditional Termination in the Early Republic," in Freehling's *The Road to Disunion: Secessionists at Bay, 1776–1854.* New York, 1990, pp. 121–43.

Frey, Sylvia R. *Water from the Rock: Black Resistance in a Revolutionary Age.* Princeton, 1991.

Greene, Jack P., "'Slavery or Independence': Some Reflections on the Relationship among Liberty, Black Bondage, and Equality in Revolutionary South Carolina," *South Carolina Historical Magazine*, 80 (1979), 193–214.

Greene, John C. "American Debate on the Negro's Place in Nature, 1780–1815," *Journal of the History of Ideas*, 15 (1954), 384–96.

Griswold, Charles L., Jr., "Rights and Wrongs: Jefferson, Slavery, and Philosophical Quandaries," in Michael J. Lacey and Knud Haakonssen, ed., *A Culture of Rights: The Bill of Rights in Philosophy, Politics, and Law, 1791 and 1991.* New York, 1991, pp. 144–214.

Hollander, Barnett. *Slavery in America: Its Legal History.* New York, 1963.

Hunt, Gaillard. "William Thornton and Negro Colonization," American Antiquarian Society *Proceedings*, new ser., 30 (1920), 32–61.

Jackson, Harvey H. "'American Slavery, American Freedom' and the Revolution in the Lower South: The Case of Lachlan McIntosh," *Southern Studies* (1980), 81–93.

Jordan, Winthrop. *White Over Black: American Attitudes Toward the Negro, 1550–1812*. Chapel Hill, N.C., 1968.

Kaplan, Sidney. *The Black Presence in the Era of the American Revolution, 1770–1800*. Washington, D.C., 1973.

Kolchin, Peter. *American Slavery, 1619–1877*. New York, 1993.

Livermore, George. *An Historical Research Respecting the Opinions of the Founders of the Republic on Negroes as Slaves, as Citizens, and as Soldiers*. 4th ed. Boston, 1863.

Lynd, Staughton. *Class Conflict, Slavery, and the United States Constitution: Ten Essays*. Indianapolis, Ind., 1967.

McColley, Robert. *Slavery and Jeffersonian Virginia*. Urbana, Ill., 1964.

MacEachren, Elaine. "Emancipation of Slavery in Massachusetts: A Reexamination, 1770–1790," *Journal of Negro History*, 55 (1970), 289–306.

MacLeod, Duncan J. *Slavery, Race, and the American Revolution*. London, 1974.

McManus, Edgar J. *A History of Negro Slavery in New York*. Syracuse, 1966.

McMaster, Richard K. "Liberty or Property? The Methodists Petition for Emancipation in Virginia, 1785," *Methodist History*, 10 (1971), 44–55.

Mathews, Donald G. *Slavery and Methodism: A Chapter in American Morality, 1780–1845*. Princeton, 1965.

Mellon, Matthew T. *Early American Views on Negro Slavery, from the Letters and Papers of the Founders of the Republic*. Boston, 1934.

Miller, John C., "Slavery," in Merrill D. Peterson, *Thomas Jefferson: A Reference Biography*. New York, 1986.

———. *The Wolf by the Ears: Thomas Jefferson and Slavery*. Charlottesville, Va., 1991 (originally published in New York, 1977).

Monaghan, Frank. "Anti-Slavery Papers of John Jay," *Journal of Negro History*, 17 (1932), 481–96.

Morgan, Edmund S. *American Slavery, American Freedom*. New York, 1975.

———. "Slavery and Freedom: The American Paradox," *Journal of American History*, 59 (1972), 5–29.

Mullen, Gerald W. *Flight and Rebellion: Slave Resistance in Eighteenth-century Virginia*. New York, 1972.

Nash, Gary B. *Race and Revolution*. Madison, Wis., 1990.

O'Brien, William. "Did the Jennison Case Outlaw Slavery in Massachusetts?" *William and Mary Quarterly*, 3rd ser., 17 (1960), 219–41.

Ohline, Howard A., "Slavery, Economic, and Congressional Politics, 1790." *Journal of Southern History*, 46 (1980), 335–60.

Quarles, Benjamin. *The Negro in the American Revolution*. Chapel Hill, N.C., 1961.

Robinson, Donald L. *Slavery in the Structure of American Politics, 1765–1820*. New York, 1970.

Ruchames, Louis, comp. *Racial Thought in America: A Documentary History*. Vol. 1, *From the Puritans to Abraham Lincoln*. Amherst, Mass., 1969.

Schmidt, Fredrika Teute, and Barbara Ripel Wilhelm, "Early Proslavery Petitions in Virginia," *William and Mary Quarterly*, 3rd ser., 30 (1973), 133–46.

Soderlund, Jean R. *Quakers & Slavery: A Divided Spirit.* Princeton, 1985.

Spector, Robert M. "The Quock Walker Cases (1781–1783): Slavery, Its Abolition, and Negro Citizenship in Early Massachusetts," *Journal of Negro History*, 53 (1968), 12–32.

Tise, Larry. *Proslavery: A History of the Defense of Slavery in America, 1701–1840.* Athens, Ga., 1988.

White, Shane. *Somewhat More Independent: The End of Slavery in New York City, 1770–1810.* Athens, Ga., 1991.

Wiecek, William M., "The Statutory Law of Slavery and Race in the Thirteen Mainland Colonies of British America," *William and Mary Quarterly*, 3rd ser., 34 (1977), 258–80.

———. "The Witch at the Christening: Slavery and the Constitution's Origins," in Leonard W. Levy and Dennis J. Mahoney, eds., *The Framing and Ratification of the Constitution.* New York, 1987, pp. 167–84.

Wilson, Douglas L. "Thomas Jefferson and the Character Issue," *Atlantic*, 270 (Nov. 1992), 57–74.

Wood, Peter H. "'Liberty is Sweet': African-American Freedom Struggles in the Years before White Independence," in Alfred F. Young, ed. *Beyond the American Revolution: Explorations in the History of American Radicalism.* DeKalb, Ill., 1993, pp. 149–84.

Zilversmit, Arthur. "Quock Walker, Mumbet, and the Abolition of Slavery in Massachusetts," *William and Mary Quarterly*, 3rd ser., 25 (1968), 614–24.

———. *The First Emancipation: The Abolition of Slavery in the North.* Chicago, 1967.

Index

"A Friend to the Rights of the People,"
97–98
Fugitive slave clause, 39–40, 65, 115;
criticism of, 67, 70–71, 74, 76;
praise of, 158, 162, 179, 184, 200;
and fugitive slave law, 238–39
Fuller, Abraham, 85

Galloway, James, 161, 198, 199
Gerry, Elbridge, 209; in Constitutional
Convention, 46, 50, 55, 60, 166.
See also "Federal Farmer"
Gordon, William, 244
Gorham, Nathaniel, 21; in Constitu-
tional Convention, 47, 50, 51, 62,
64; in Mass. convention, 85
Goudy, William, 197

Hall, Prince: petition of, 11–12
Hamilton, Alexander, 23, 44
"Hampden" (William Findley), 146
Harrison, Benjamin, 33
Hart, Levi, 95
Heath, William, 89–90
Heaton, James, 267
Henry, Patrick: in Va. convention, 159,
160, 186, 188–89, 190, 191–93,
195
Higginson, Stephen, 22
Hobart, John Sloss, 31
Holmes, John, 263–64
Holten, Samuel, 22
Hopkins, Samuel, 28; letters from, 5–6,
73, 95; as "Crito," 72, 72n
Hopkins, Stephen, 28
House of Representatives, U.S.: and
debate over tax on slave imports,
201–2, 203–10; and debate over
anti-slave petitions, 202, 210–30
Hughes, Hugh. *See* "A Countryman
from Dutchess County"

Imputed guilt, doctrine of, 69, 89–90,
96–97, 99, 108–9, 110–11
Ingersoll, Jared, 119, 121
Innes, James, 195–96
Iredell, James, 161, 198–99, 199, 199–
200, 200. *See also* "Marcus"

Jackson, James, 202, 204, 206, 214–15
Jay, John, 30, 231–38
Jefferson, Thomas, 24, 39; letters to,
163, 184, 252–55, 256–59, 261–62;
*A Summary View of the Rights of
British America*, 2–3; *Notes on the
State of Virginia*, 214, 217, 220,
245, 246–51, 266; and act to
prohibit foreign slave trade, 239,
240, 256; attitude of toward
slavery, 245–46, 246–67; letters
from, 251, 251–52, 255, 259–61,
263, 263–64, 264–67, 267
Jennison, Nathaniel, 17–18
Johnson, William Samuel, 52
Johnston, Zachariah, 160, 196
Jones, William, 89

Kidnappings of free blacks, 202
King, Rufus: in Constitutional
Convention, 44, 46–49, 51, 54, 55,
61, 63; in Mass. convention, 84, 85
Kosciusko, Thaddeus, 24

Lafayette, Marquis de: emancipation
plan of, 24–26, 244
"Landholder" (Oliver Ellsworth), 77,
78, 97
Langdon, John, 61, 63, 99, 99n
Lee, Arthur, 22
Lee, Henry, 186
Lee, Richard Henry, 79
Lee, William (slave), 278
Lettsom, John Coakley, 117
Lincoln, Benjamin, 175–76
Livermore, Samuel, 206
Livingston, Robert R., 31
Livingston, William, 62
Locke, John, 93
Lowndes, Rawlins: in S.C. assembly,
158, 167–68, 171–72, 175–76
Lusk, Thomas, 89, 91

M'Dowall, Joseph, 198
McIntosh, Lachlan, 158, 164–65
McKean, Thomas, 116, 135, 138
Madison, James: letters from, 36, 162,
163, 268, 271–72, 272–75; and